Photon and Electron Collisions with Atoms and Molecules

PHYSICS OF ATOMS AND MOLECULES

Recent volumes in this series:

ATOMS AND LIGHT: INTERACTIONS
John N. Dodd

COINCIDENCE STUDIES OF ELECTRON AND PHOTON IMPACT IONIZATION
Edited by Colm T. Whelan and H. R. J. Walters

DENSITY MATRIX THEORY AND APPLICATIONS, Second Edition
Karl Blum

ELECTRON COLLISIONS WITH MOLECULES, CLUSTERS, AND SURFACES
Edited by H. Ehrhardt and L. A. Morgan

INTRODUCTION TO THE THEORY OF LASER–ATOM INTERACTIONS, Second Edition
Marvin H. Mittleman

INTRODUCTION TO THE THEORY OF X-RAY AND ELECTRONIC SPECTRA OF FREE ATOMS
Romas Karazija

PHOTON AND ELECTRON COLLISIONS WITH ATOMS AND MOLECULES
Edited by Philip G. Burke and Charles J. Joachain

POLARIZATION BREMSSTRAHLUNG
Edited by V. N. Tsytovich and I. M. Ojringel

POLARIZED ELECTRON/POLARIZED PHOTON PHYSICS
Edited by H. Kleinpoppen and W. R. Newell

SELECTED TOPICS ON ELECTRON PHYSICS
Edited by D. Murray Campbell and Hans Kleinpoppen

THEORY OF ELECTRON–ATOM COLLISIONS, Part 1: Potential Scattering
Philip G. Burke and Charles J. Joachain

VUV AND SOFT X-RAY PHOTOIONIZATION
Edited by Uwe Becker and David A. Shirley

A Chronological Listing of Volumes in this series appears at the back of this volume.

Photon and Electron Collisions with Atoms and Molecules

Edited by

Philip G. Burke

The Queen's University of Belfast
Belfast, Northern Ireland

and

Charles J. Joachain

Free University of Brussels
Brussels, Belgium

Plenum Press • New York and London

Library of Congress Cataloging-in-Publication Data

Photon and electron collisions with atoms and molecules / edited by Philip G. Burke and Charles J. Joachain.
p. cm. -- (Physics of atoms and molecules)
Includes bibliographical references and index.
ISBN 0-306-45692-3
1. Electron-molecular collisions--Congresses. 2. Electron-atom collisions--Congresses. 3. Photoionization--Congresses. 4. Multiphoton processes--Congresses. I. Burke, P. G. II. Joachain, C. J. (Charles Jean) III. Series.
QC793.5.E628P46 1997
539.7'57--dc21 97-29100
CIP

Proceedings of the Second European Study Conference on Photon and Electron Collisions with Atoms and Molecules, held July 23 – 26, 1996, in Belfast, Northern Ireland

ISBN 0-306-45692-3

A Division of Plenum Publishing Corporation
233 Spring Street, New York, N. Y. 10013

http://www.plenum.com

10 9 8 7 6 5 4 3 2 1

Printed in the United States of America

PREFACE

Research on photon and electron collisions with atomic and molecular targets and their ions has seen a rapid increase in interest, both experimentally and theoretically, in recent years. This is partly because these processes provide an ideal means of investigating the dynamics of many particle systems at a fundamental level and partly because their detailed understanding is required in many other fields, particularly astrophysics, plasma physics and controlled thermonuclear fusion, laser physics, atmospheric processes, isotope separation, radiation physics and chemistry and surface science.

In recent years a number of important advances have been made, both on the experimental side and on the theoretical side. On the experimental side these include absolute measurements of cross sections, experiments using coincidence techniques, the use of polarised beams and targets, the development of very high energy resolution electron beams, the use of synchrotron radiation sources and ion storage rings, the study of laser assisted atomic collisions, the interaction of super-intense lasers with atoms and molecules and the increasing number of studies using positron beams. On the theoretical side these advances include the development of methods which allow accurate excitation and ionisation cross sections to be calculated at intermediate energies, the increasing ability to determine low energy collision cross sections for complex atomic and molecular targets, the development of non-perturbative approaches for studying multiphoton processes and ultra-fast femtosecond phenomena for many electron targets, the ability to accurately treat photon and electron collisions with molecules adsorbed on surfaces and the development of a fundamental understanding of chaotic phenomena and of collisions involving oriented and chiral molecules. Many of these theoretical advances have been made possible by the development and increasing applicability of high performance computers.

These important advances are discussed in detail in the present volume, which brings together 23 invited talks given by distinguished scientists at the PECAM II Conference on Photon and Electron Collisions with Atoms and Molecules held at the Queen's University of Belfast from 23-26 July 1996. This Conference followed the very successful PECAM I Conference held in Rome in July 1992. It also marked the end of a fruitful European Community Network in this area which brought together scientists from nine EC Laboratories and three Central and Eastern European Laboratories to carry out collaborative research in this field.

Finally, we are very grateful to the authors of the invited talks for their stimulating presentations at the Conference and for their efforts in preparing camera-ready manuscripts of their talks which are published in this volume. Through their efforts we believe that this volume provides a comprehensive overview of the most important recent advances in the field of electron and photon collisions with atoms and molecules together with their main applications.

P G Burke
C J Joachain

CONTENTS

EXCITATION OF ATOMS BY ELECTRON IMPACT: CURRENT STATUS AND FUTURE PROSPECTS

Klaus Bartschat

Department of Physics and Astronomy
Drake University
Des Moines, IA 50311, U.S.A.

INTRODUCTION

In recent years, much progress has been made in the experimental and theoretical studies of atomic collision processes. The preparation of spin-polarized collision partners and the measurement of orientation, alignment, and spin polarization parameters after the collision has enabled experimentalists to perform very detailed tests of theoretical models. In some cases, even "complete" or "perfect" experiments were performed, where all independent scattering amplitudes for a particular atomic collision process were determined experimentally. Since these amplitudes contain the maximum, quantum mechanically available information about the scattering event, the results of such measurements have presented important benchmark data for the assessment and further development of computer simulation programs. A general framework and a unified terminology for the description of spin-dependent electron–atom scattering has now been developed,[1] and the available experimental and theoretical data have been reviewed.[2]

At the same time, the rapid growth in the performance of computer hardware has allowed for the development of new numerical models to describe electron collision processes. Among the many methods that attempt to solve the Schrödinger equation subject to scattering boundary conditions, standard close-coupling-type approaches such as, among others, the "Convergent Close-Coupling" (CCC),[3] the "Intermediate Energy R-Matrix" (IERM),[4] or the "R-Matrix with Pseudo-States" (RMPS)[5] methods, have been extremely successful. The key advantage of those methods, when compared to perturbative approaches such as first- and higher-order plane- or distorted-wave approximations, lies in the fact that they are able to account for the effect of the target continuum states in the close-coupling expansion but still keep the full coupling between all the physical discrete as well as the bound and continuum pseudo-channels. In addition, the standard problem of pseudo-resonances can be diminished tremendously by increasing the number of states in the calculation.

The presentation in this chapter will focus on two topics. After a summary of the principal ideas behind the CCC, RMPS, and IERM methods, we first show some key examples of recent calculations for total cross sections obtained with these methods and compare the results with those from previous calculations and also from experiment. The second part deals with the present status of complete or nearly complete experiments. The importance for consistency checks within different sets of experimental data is demonstrated by comparing data from "over-complete" experiments, where the same physical observable is determined in different ways. This allows for a critical assessment not only of the reliability of theoretical predictions, but also of the quality of the experimental results. Finally, suggestions for future developments, both experimentally and theoretically, are made.

NUMERICAL METHODS: CCC, RMPS AND IERM

We now summarize the basic features of three recently developed close-coupling methods that account for the effect of the target continuum states in the expansion and keep the full coupling between all the physical discrete as well as the bound and continuum pseudo-channels.

The convergent close-coupling method (CCC)

The details of the CCC theory have been given by Bray and Stelbovics.[6] The method may be thought of as a standard close-coupling calculation where, in addition to the discrete target states, the target continuum is treated with the aid of positive energy pseudo-states. All of these states are obtained by diagonalising the target Hamiltonian in an orthogonal Laguerre basis. The usage of such a basis ensures that "completeness" is approached by simply increasing the basis size.

A key feature of the formalism lies in the fact that the coupled equations are formulated in momentum space, where they take the form of coupled Lippmann-Schwinger equations. These are solved separately, upon partial-wave expansion, for each total energy of interest. As such the method is not ideally suited for the study of detailed scattering behavior as a function of incident energy. However, modern computational resources involving large clusters of workstations allow for the parallel execution of the CCC program at many energies.

The R-matrix with pseudo-states method (RMPS)

The low-energy R-matrix method (for details, see Burke and Robb[7]) is another method to solve the close-coupling equations, this time in coordinate space. The important difference to the standard formulation is the division of configuration space into two regions, $r \leq a$ and $r > a$, where the R-matrix radius a is chosen in such a way that exchange effects between the projectile and the target electrons can be neglected in the external region. Here the coupled equations (without exchange) are solved for each collision energy and matched, at the boundary $r = a$, to the solution in the inner region. However, instead of solving a set of coupled integro-differential equations in the internal region for each collision energy, the $(N+1)$-electron wavefunction at energy E is expanded in terms of an energy independent basis set, ψ_k, as

$$\Psi_E = \sum_k A_{Ek}\psi_k \; . \tag{1}$$

The basis states ψ_k are constructed as

$$\psi_k = \mathcal{A} \sum_{ij} \Phi_i(1, \ldots, N)\, u_j(N+1)\, a_{ijk} + \sum_i \chi_i(1, \ldots, N+1)\, b_{ik}\,, \tag{2}$$

where the Φ_i are N-electron target states (either physical or pseudo-states), the u_j are members of a complete set of numerical continuum orbitals used to describe the motion of the scattered electron inside the box, and the χ_i are $(N+1)$-electron configurations. The latter are formed from the one-electron orbital basis used to describe the N-electron target. They are included to allow for electron correlation effects when the scattered electron is close to the nucleus and also to ensure completeness of the total trial wavefunction.

Each of the N-electron target states can be expanded in terms of a sum of orthonormal configurations

$$\Phi_i = \sum_{j=1}^{m} c_{ij} \phi_j\,. \tag{3}$$

The configurations ϕ_j are built up from one-electron orbitals, coupled together to give a function which is completely antisymmetric with respect to the interchange of the space and spin coordinates of any two electrons.

The one-electron orbitals used in constructing the ϕ_j configuration may be of two types: first, physical orbitals, which are usually the Hartree Fock orbitals, and second, suitably chosen non-physical pseudo-orbitals. As described in detail by Bartschat *et al.*,[8] the pseudo-orbitals can be used (i) to improve the description of the physical target states and (ii) to represent the effects of the infinite number of discrete and continuum states of the target atom or ion that cannot be included explicitly in the calculation. This idea is identical to that used in the CCC method. Although, due to the finite range of the pseudo-orbitals, these effects are only represented accurately in a certain region of configuration space (usually the R-matrix box), this is expected to be sufficient if one is interested in transitions between discrete states whose range is also restricted to within this box. In fact, the method also works for ionization if the box is made sufficiently large and some averaging over effective box sizes, determined by different ranges of the pseudo-orbitals, is included. For details, see Bartschat and Bray.[9,10]

The intermediate energy R-matrix method (IERM)

The intermediate energy R-matrix method (IERM) differs from the RMPS approach in that the target continuum orbitals and the motion of the projectile inside the box are described in terms of the same one-electron numerical continuum functions whose radial terms satisfy the differential equation

$$\left(\frac{d^2}{dr^2} - \frac{l(l+1)}{r^2} - 2V(r) + k_{nl}^2 \right) u_{nl}(r) = 0 \tag{4}$$

for each angular momentum, l, subject to the R-matrix boundary conditions

$$u_{nl}(0) = 0\,; \tag{5}$$

$$\frac{a}{u_{nl}(a)} \cdot \left.\frac{du_{nl}}{dr}\right|_{r=a} = b\,. \tag{6}$$

The potential $V(r)$ is chosen in such a way that, provided a suitable R-matrix boundary radius, a, is chosen, the first few $u_{nl}(r)$ will correspond to the bound target orbitals.

(For example, $V(r) = -Z/r$ for e–H scattering.) The reason for this choice of basis stems from the earlier work of Burke *et al.*[4] where the problem of including two continuum electrons within the R-matrix internal region and allowing for the escape of two electrons from this region was discussed. The two-electron IERM approach, for which results are shown in the next section, grew out of this work, the two-electron case being the simplest example to study.

One advantage of the IERM approach over the RMPS method lies in the fact that the atomic pseudo-orbitals are automatically orthogonal to the physical bound orbitals and there is no need to use any numerical orthogonalisation procedure. Therefore, it is possible to generate a large number of pseudo-orbitals without encountering linear dependence problems. On the other hand, the range of the pseudo-orbitals generated in the IERM approach is determined by the value of the R-matrix boundary being used. For example, if one chooses a boundary radius to envelope all the excited target states of interest, convergence of the cross section with respect to the number of pseudo-states will be most rapid for transitions involving the outermost orbitals, but usually slow for transitions involving the ground state, particularly the elastic scattering cross section.

Comparison of CCC and R-matrix approaches

There are two important differences between the CCC and the two R-matrix based approaches that should be pointed out. For the CCC method, as a straightforward way to solve the close-coupling equations, the computational effort required is roughly the same for each collision energy; some variation is possible depending on the number of states and the number of partial waves required for convergence of the observables of interest. On the other hand, the computational effort in the R-matrix approach consists of two parts: (i) a one-time effort to solve the internal region problem by diagonalizing the $(N+1)$-electron Hamiltonian and thereby determining, for example, the coefficients a_{ijk} and b_{ik} in Eq. (2) for the RMPS case, and (ii) the solution of the (no-exchange) coupled equations outside the box for each collision energy. Since the latter can be done very efficiently using newly developed packages such as FARM,[11] the R-matrix method is most suitable if results for a large number of collision energies are required, as for example in the study of resonances. On the other hand, the size of the Hamiltonian matrix to be diagonalized grows quickly with the size of the box and the maximum collision energy that needs to be treated. Consequently, the standard R-matrix method works best for transitions between low-lying discrete states and collision energies below the ionization threshold. However, intermediate energies up to several times the ionization energy can now be reached by both the RMPS method and the IERM approach, and transitions involving higher-lying Rydberg states can be calculated by subdividing the R-matrix internal region into a number of smaller sub-regions, as done in the two-dimensional R-matrix propagator approach.[4,12]

TOTAL CROSS SECTIONS

This section shows some key examples which allow for an assessment of the reliability of calculations for total cross sections in electron impact excitation.

Electron Scattering from Atomic Hydrogen

For many years, this has been the standard problem of atomic collision theory. From a theoretical point of view, the advantage of this problem lies in the fact that the non-relativistic target wavefunctions are known exactly. Hence, a detailed test of theoretical predictions by comparison with experimental data is essentially free from uncertainties in the description of the target structure. On the other hand, it is well known that the $(2p)^2P^o$ state in a close-coupling expansion only accounts for 65.8% of the static dipole polarizability of the $(1s)^2S$ ground state, and even the inclusion of all discrete P states still omits 18.6% of this long-range coupling effect.[13] Also, experiments on atomic hydrogen are very difficult, due to the necessary dissociation of molecular hydrogen to form an atomic beam and the short wavelengths in the Lyman series if light emitted by collisionally excited atoms is to be observed.

Recently, a benchmark calculation for elastic scattering and excitation of the $n = 2$ states from the ground state has been performed in collaboration between the Belfast and Flinders groups.[14] Results for the contribution from the $^1P^o$ partial wave symmetry to the total elastic and excitation cross sections as well as the converged 1s → 2s and 1s → 2p cross sections are shown in figure 1, where the latter are also compared with the experimental data of Williams.[15] The agreement between the completely independent theoretical calculations is nearly perfect, and so is the agreement with experiment. This indicates that theory has, indeed, solved this problem to an accuracy of better than 1%.

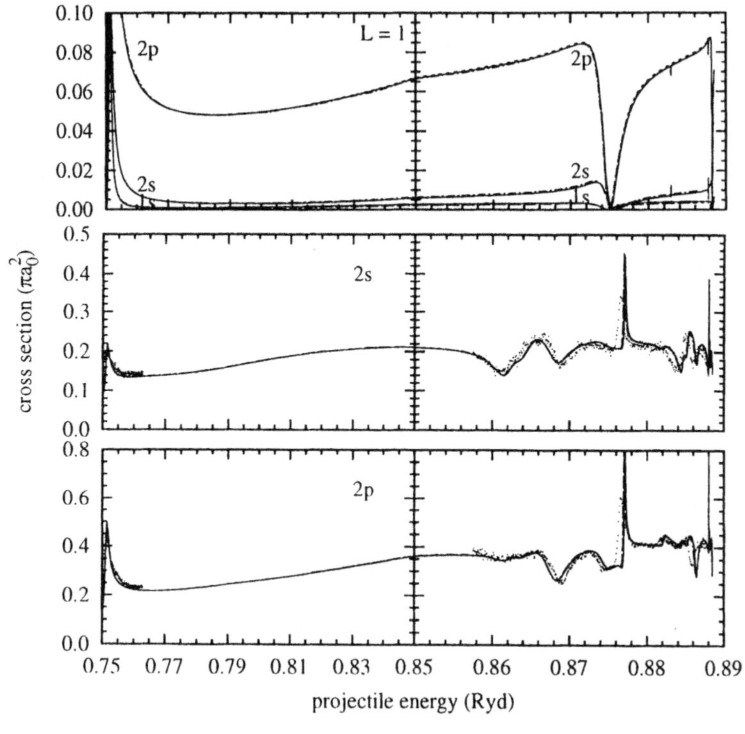

Figure 1: Partial wave contribution to the elastic and excitation cross sections for $e^- -$ H scattering from the $^1P^o$ symmetry (top) and total excitation cross sections for the 2s (center) and 2p (bottom) states for collision energies between the $n = 2$ and $n = 3$ thresholds. Solid line, RMPS; dashed line, IERM; dash-dotted line, CCC. The small dots represent the experimental data of Williams.[15] (From Bartschat *et al.*[14])

The potential of the CCC and RMPS models when applied to ionization is demonstrated in figure 2. Not only is the total ionization cross section (calculated as excitation to the positive-energy pseudo-states) in very good agreement with the experimental data over a wide range of incident energies, but so is the very sensitive ratio of cross section over excess energy for energies more than about 2 eV above threshold. This result becomes even more interesting, if one keeps in mind that neither of these methods was actually designed to treat the near-threshold region particularly well. Recent work using the IERM method indicates that it may even be possible to investigate the Wannier threshold law within this formalism.[16]

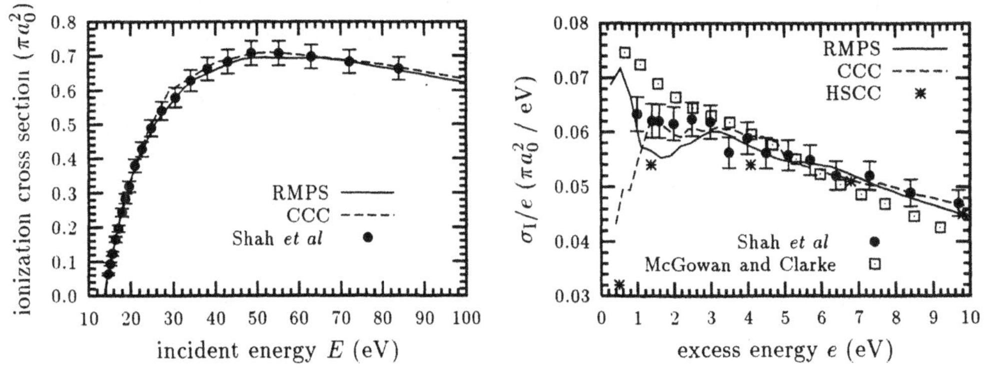

Figure 2: Total cross section (left) and ratio of cross section over excess energy (right) for electron impact ionization of H(1s). The RMPS and CCC calculations of Bartschat and Bray[10] are compared with hyperspherical close-coupling results of Kato and Watanabe[17] and experimental data of Shah *et al*[18] and of McGowan and Clarke.[19]

Electron Scattering from Helium

Helium is not only the second most abundant element in the universe, but it also has many academic and industrial applications. For these reasons obtaining reliable data for processes involving the helium atom is of immense importance, and electron scattering cross sections can provide valuable information. Since it is one of the simpler complex atoms for which a large amount of experimental data are available, helium also provides an ideal testing ground for developing a general method for electron scattering from complex atomic and ionic targets.

After a series of R-matrix calculations including an increasing number of physical target states, as well as numerous perturbation-type calculations based on varieties of the Born series, have been performed over the past two decades for e–He collisions, Fursa and Bray[20] were very successful in applying the CCC method to e–He scattering. Their calculation, which obtained excellent agreement with a variety of experimental data for elastic scattering, excitation and ionization, closed the "intermediate energy gap" between low energies, where close-coupling approaches that include only discrete states in the expansion are reasonably accurate, and high energies above several times the ionization potential, where simpler Born-type methods may be used with some confidence.

The RMPS method has also been applied to this collision problem.[8,21,22] That work represents the first application of this kind of close-coupling method to a truly complex target. (Although the theory outlined by Fursa and Bray is completely general, the actual calculation was performed in a frozen-core approximation which was found to be sufficiently accurate for the observables of interest in their paper.)

Figure 3 shows a comparison of RMPS and CCC results for the total cross sections of the $1^1S \rightarrow 2^3S$ and $1^1S \rightarrow 2^1P^o$ transitions with those from previous R-matrix calculations[23-25] and experiment. Once again, the agreement with experiment is very good. The agreement with the previous R-matrix calculations, however, is only good below the 2^1P^o threshold. For higher energies, particularly above the ionization threshold, the 5-state and even the 29-state results for the excitation cross sections often lie substantially above the RMPS and CCC results. This is to be expected since the calculations with discrete states only in the close-coupling expansion made no account for higher lying discrete states or the continuum ionization channels which become open at these energies. In addition, the earlier results for the $1^1S \rightarrow 2^3S$ transition seem to be strongly affected by a pseudo-resonance. On the other hand, the 29-state results are expected to be very accurate for incident energies below the $n = 5$ thresholds.

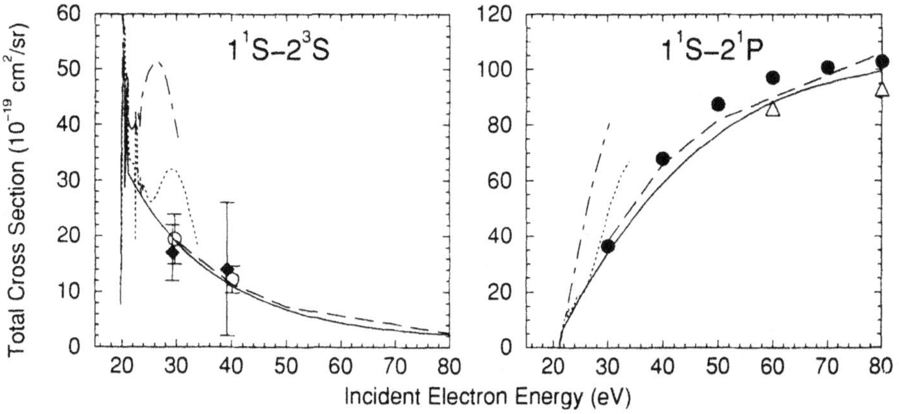

Figure 3: Integrated e–He cross sections for electron impact excitation of the 2^3S and 2^1P^o states from the ground state 1^1S. ———, RMPS;[22] — — —, CCC;[20] — · — ·, 5-state R-matrix calculation;[23] · · · · ·, 29-state R-matrix calculation.[24,25] Experimental data: open circles, Trajmar;[26] triangles, Chutjian and Srivastava;[27] diamonds, Hall *et al*;[28] solid circles, Donaldson *et al.*[29]

Electron Scattering from Beryllium

Accurate cross section results for electron collisions with neutral beryllium atoms are of great importance for applications in plasma physics.[30] The first comprehensive study was undertaken by Fon *et al*[31] who performed R-matrix calculations with 12 and 20 states in the close-coupling expansion. While the results were expected to be fairly accurate at low energies, some doubt about the reliability of the predictions remained for energies above the ionization threshold.

A recent RMPS calculation[32] verified, indeed, that the inclusion of the positive-energy pseudo-states to represent the target continuum is crucial in this energy range. This is demonstrated in figure 3 for the forbidden $(1s^22s^2)^1S \rightarrow (1s^22s2p)^3P^o$ transition. Inclusion of an increased number of pseudo-states not only makes the T-matrix averaging procedure,[33] used by Fon *et al* to eliminate the effect of pseudo-resonances, essentially unnecessary, but also reduces the cross sections by up to a factor of 2 ! A similar, though not quite as drastic effect was found for the optically allowed $(1s^22s^2)^1S \rightarrow (1s^22s2p)^1P^o$ transition. Further work on this system, urgently required for the applications mentioned above, is currently in progress.

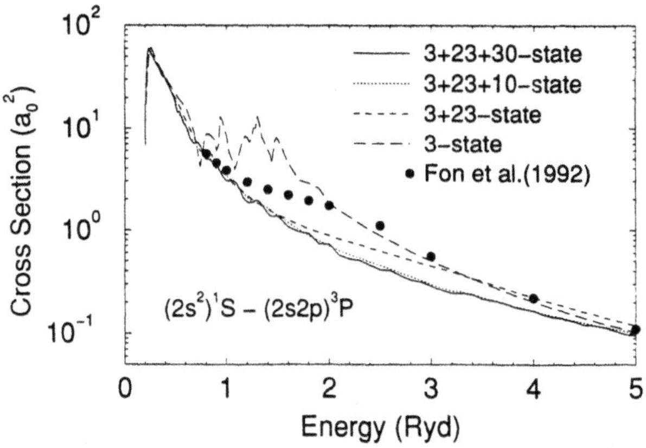

Figure 4: Integrated cross section for electron impact excitation of the $2^1S \rightarrow 2^3P^o$ transition in beryllium. The RMPS results[32] for various numbers of states included in the close-coupling expansion are compared with the results of Fon *et al.*[31]

Electron Scattering from Cesium

Given the success of the new close-coupling methods for light targets, the importance of further development work to include relativistic effects seems obvious. An open problem, for example, is demonstrated in figure 5 which shows the total cross section for elastic scattering and excitation of cesium atoms by electron impact. It is interesting to note that all calculations, from a non-relativistic 2-state model[34] (not shown) to 5-state and 8-state semi-relativistic (Breit-Pauli) calculations[35,36] and even a 5-state Dirac R-matrix model[37] seem to systematically overestimate this cross section when compared to experimental data.[38,39] Some preliminary work using the non-relativistic CCC approach[40] did not resolve this discrepancy either, but for more definite conclusions to be drawn a relativistic treatment, including a relativistic structure calculation, is required.

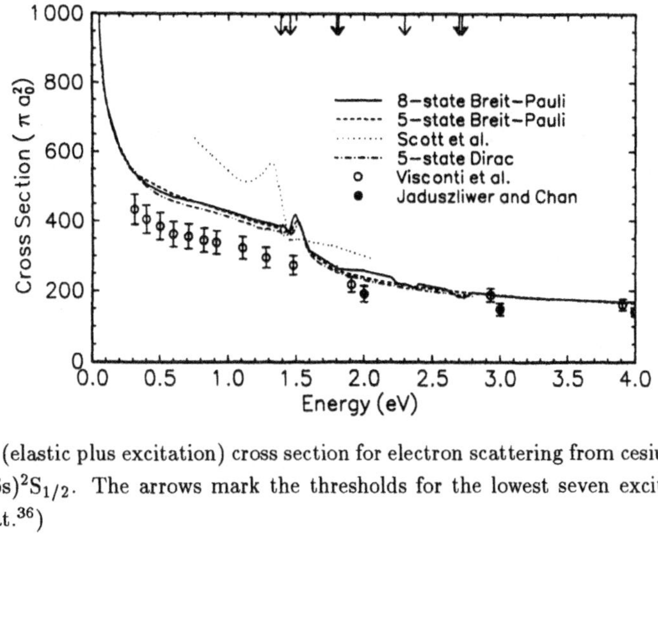

Figure 5: Total (elastic plus excitation) cross section for electron scattering from cesium atoms in their ground state $(6s)^2S_{1/2}$. The arrows mark the thresholds for the lowest seven excited target states. (From Bartschat.[36])

COMPLETE EXPERIMENTS

This section deals with some of the most detailed investigations of electron–atom collisions to date. These are the so-called "complete" or "perfect" scattering experiments that were called for by Bederson[41,42] more than 25 years ago. His papers stimulated the development of experiments, or rather combinations of experimental approaches, which resulted in a complete determination of the quantum mechanical scattering amplitudes. If all these amplitudes are known, a *unique* relationship between the vectors $|\psi\rangle_{in}$ and $|\psi\rangle_{out}$, which determine the initial and final states of the collision system, can be established. Benchmark data from such experiments allow for very detailed tests of theoretical models, since cancellation of errors due to averaging over key parameters, such as incident directions, scattering angle, velocity or temperature, cannot obscure potential problems in the theoretical description.

The current status of complete experiments has been reviewed by Andersen and Bartschat[1] who established a general framework to describe the corresponding collision processes and their experimental investigation in a common language. Furthermore, a compilation of available experimental and theoretical results for S → P excitation of (spin-polarized) target beams by (spin-polarized) electron beams was produced by Andersen *et al*[2], following the systematic approach developed earlier for unpolarized beams.[43] Below, we will demonstrate the power of the formalism by using some of the key examples discussed by Andersen and Bartschat.[1]

Electron Impact Excitation of He 1^1S → 2^1P^o

This process has served as the prime example for a complete experiment in electron–atom excitation. In the "natural coordinate system", where the z axis is chosen perpendicular to the scattering plane and the x axis is defined along the incident beam direction, the process is described by the two scattering amplitudes $f_{\pm 1}$, where the subscript indicates the magnetic quantum number M_L of the excited P state. Transitions to $M_L = 0$ are forbidden due to conservation of reflection symmetry in the scattering plane.[43] Hence, this process is determined by three parameters: the differential cross section, a parameter describing the relative size of the two amplitudes, and an angle that fixes their relative phase.

The two independent dimensionless parameters may be chosen as the angular momentum transfer $\langle\vec{\mathbf{L}}\rangle = (0, 0, L_\perp)$ with

$$L_\perp = \frac{|f_{+1}|^2 - |f_{-1}|^2}{|f_{+1}|^2 + |f_{-1}|^2} = \frac{|f_{+1}|^2 - |f_{-1}|^2}{\sigma} = w_+ - w_- \tag{7}$$

and the alignment angle γ of the major symmetry axis of the charge cloud in the scattering plane. This alignment angle, defined mod π, is related to the relative phase angle δ between the amplitudes through[43]

$$\delta = -2\,\gamma \pm \pi. \tag{8}$$

The parameters $(L_\perp, \gamma)$ are related to the Stokes vector $\vec{\mathbf{P}} = (P_1, P_2, P_3)$ for the light emitted in the direction perpendicular to the scattering plane in a subsequent P → S decay through[43]

$$P_1 + i\,P_2 = P_\ell\, e^{2i\gamma} = -\frac{2}{\sigma}\,|f_{+1}|\,|f_{-1}|\,e^{-i\delta}; \tag{9}$$

$$P_3 = -L_\perp. \tag{10}$$

A scheme of the experimental setup and the corresponding charge cloud are shown in figure 6.

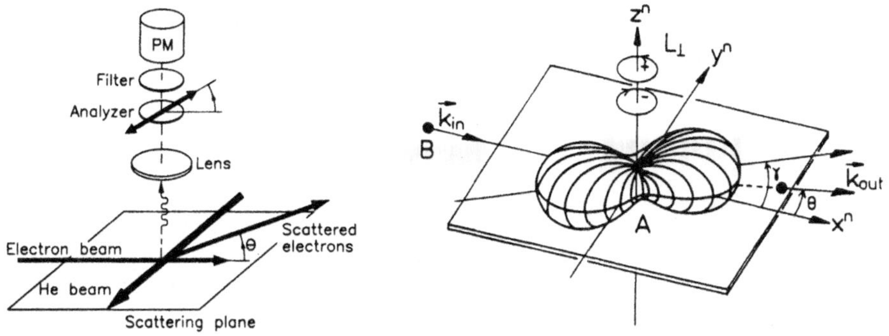

Figure 6: Basic setup of the electron–photon coincidence experiment with light polarization analysis developed at Stirling[44] (left) and the charge cloud of an excited P state[43] that can be determined in such experiments (right).

Determination of the parameter set

$$(\sigma; L_\perp; \gamma) \tag{11}$$

thus constitutes a perfect experiment, since the wavefunction of the He 2^1P^o state can be written as[1]

$$|\psi\rangle = \sqrt{\frac{\sigma}{2}(1+L_\perp)}\,|+1\rangle - e^{2i\gamma}\sqrt{\frac{\sigma}{2}(1-L_\perp)}\,|-1\rangle\,. \tag{12}$$

This wavefunction is normalized as

$$\langle\psi|\psi\rangle = \sigma = |f_{+1}|^2 + |f_{-1}|^2\,. \tag{13}$$

In this case, the degree of linear polarization, $P_\ell \equiv \sqrt{P_1^2 + P_2^2}$ is a shape parameter that describes the form of the electron charge cloud, with a length/width ratio in the scattering plane given by $\ell/w = (1+P_\ell)/(1-P_\ell)$. The total degree of polarization of the emitted light, $P = |\vec{\mathbf{P}}|$, is unity, i.e.,

$$P^2 = {L_\perp}^2 + {P_\ell}^2 = 1\,. \tag{14}$$

Figure 7 presents results for the differential cross section σ, the angular momentum transfer $L_\perp$, the alignment angle γ, and the degree of linear polarization P_ℓ for electron impact excitation of the $1^1S \rightarrow 2^1P^o$ transition at an incident electron energy of 50 eV. Due to experimental difficulties associated with the circular polarization measurement of the $2^1P^o \rightarrow 1^1S$ line at 58.4Å, the magnitude of P_3 was, in most experiments, obtained via $|P_3| = \sqrt{1-P_\ell^2}$, i.e., *assuming* completely polarized light according to Eq. (14). Then the sign was guessed by using older theoretical results and the *direct* measurement of P_3 by Khakoo *et al*[45] as a guideline.

As can be seen from figure 7, there is excellent agreement between the experimental data and the theoretical CCC[20] and RMPS[21] predictions. For the differential cross section σ of this optically allowed transition, the predictions from perturbation approaches, such as the "first-order many-body theory" (FOMBT)[46] and the "(first-order) distorted wave Born approximation" (DWBA)[47] are in reasonable agreement with those from the close-coupling theories and also with the experimental data. On the more detailed level of sublevel-resolved amplitudes and relative phases, however, there are major discrepancies between the FOMBT/DWBA results and experiment. As one might expect, these discrepancies diminish with increasing energy, but it was found[21] that the perturbative approaches are not sufficiently accurate even at an energy as high as 80 eV.

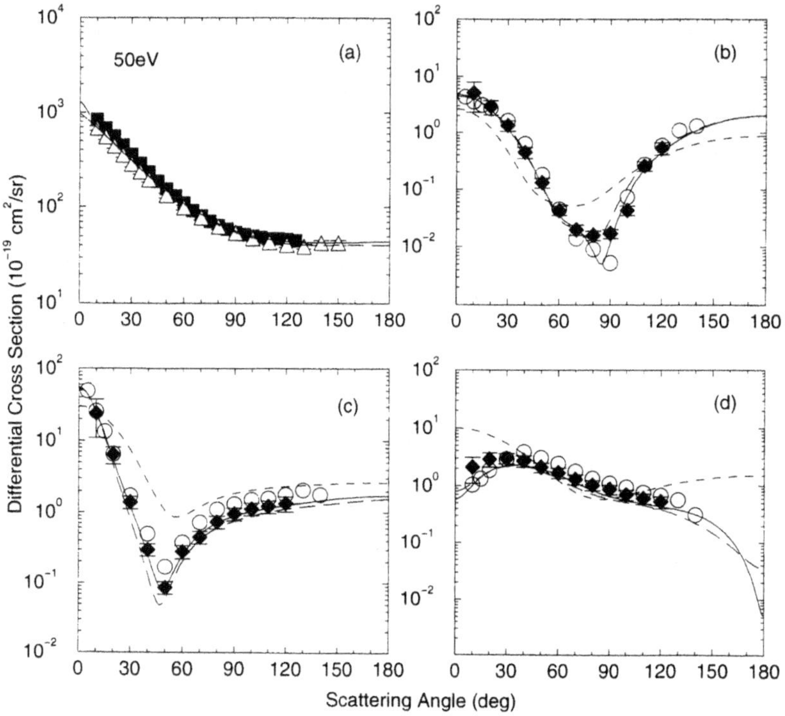

Figure 7: Differential cross section (a) and electron impact coherence parameters $L_\perp$ (b), γ (c), and P_ℓ (d) for electron impact excitation of the 2^1P^o state from the ground state 1^1S at an incident electron energy of 50 eV. The theoretical curves correspond to: ———, RMPS;[21] — — —, CCC;[20] -----, DWBA.[47,48] The sources for the experimental data are: diamonds;[28] circles (a);[49] circles (b–d);[50] squares;[48] solid triangles;[51] open triangles.[45]

Electron Impact Excitation of Na $2^2S \rightarrow 2^2P^o$

In this case, the projectile and target spins of $s = \frac{1}{2}$ double the number of independent scattering amplitudes from two to four, since there is the possibility of triplet (t) and singlet (s) scattering. The experimental determination of the independent amplitudes, $f^t_{\pm 1}$ and $f^s_{\pm 1}$, requires setups with spin-polarized target and projectile beams, as well as a spin analysis of at least one of the beams after the collision. The problem has been discussed in detail by Andersen and Bartschat.[52] They showed that a combination of data from experiments performed in the NIST,[53] Münster,[54] and Adelaide[55] groups would, in principle, allow for an extraction of these amplitudes, except for a four-fold ambiguity in the set of relative phases. Although these ambiguities could be resolved on a purely experimental basis, this is also possible with a high degree of confidence by using predictions from a sophisticated theoretical model such as a CCC calculation.[52,2]

Without going into further details, the extent to which agreement between experiment and theory has been achieved for this collision system is demonstrated in figure 8. This shows the spin-averaged angular momentum transfer $L_\perp$ as well as the spin-resolved contributions $L_\perp^t$ and $L_\perp^s$ in the triplet and singlet total spin channels, respectively, for electron impact excitation of the Na $2^2S \rightarrow 2^2P^o$ transition at an incident electron energy of 10.0 eV. The experimental data of the NIST group,[53] obtained in the time-reversed superelastic experiment where spin-polarized electrons de-excite laser-prepared spin-polarized sodium atoms, are compared with results from a "second-order distorted wave Born approximation" (DWB2),[56] a "close-coupling with optical potential" (CCO) model,[57] and a CCC calculation.[58] Note that all three theories predict the angular momentum transfer $L_\perp^t$ and, to a large extent, the spin-averaged $L_\perp$ in very good agreement with experiment. Only the CCC calculation, however, achieves nearly perfect agreement with the experimental data for the parameter $L_\perp^s$.

This example clearly shows the importance of performing such detailed experiments whose benchmark data may represent a real challenge for the theorists. The success of the CCC approach on this particular problem provided the confidence in this method to justify further development. This work then provided the computer programs needed for application of the CCC approach to many other collision problems of academic and practical importance.

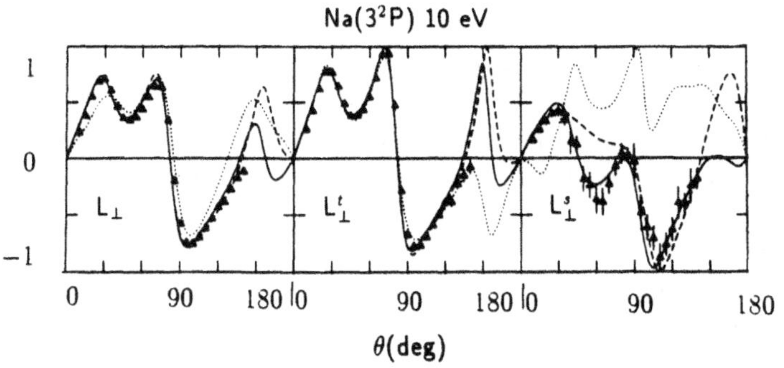

Figure 8: Angular momentum transfers $L_\perp$, $L_\perp^t$ and $L_\perp^s$ for electron impact excitation of the $3^2S \rightarrow 3^2P^o$ transition in sodium for an incident electron energy of 10 eV. The experimental data of the NIST group[53] are compared with DWB2[56] (······), CCO[57] (— — —) and CCC[58] (———) predictions.

Electron impact excitation of Hg $6^1S \rightarrow 6^3P_1^o$

This process is determined by six independent scattering amplitudes,[59] thereby requiring the determination of oneabsolute differential cross section, five relative magnitudes, and five relative phases in a complete experiment. The large number of independent parameters compared, for example, to He $1^1S \rightarrow 3^3P^o$ excitation is due to the fact that explicitly spin-dependent forces, such as the spin-orbit interaction or other relativistic effects, must be taken into account — in the $\mathcal{T}$ operator for the collision and/or in the description of the target states (e.g., by an intermediate coupling scheme). The additional degrees of freedom in this problem are illustrated in figure 9 which shows the charge cloud of the excited P state after impact excitation by spin-polarized electrons with polarization vector in the scattering plane. Note that the charge cloud has obtained a height (due to excitation of magnetic sublevels with $M_J = 0$ after spin-flips perpendicular to the scattering plane), and the symmetry axis is no longer restricted to this plane; instead, it may tilt away from the plane and even twist.[60]

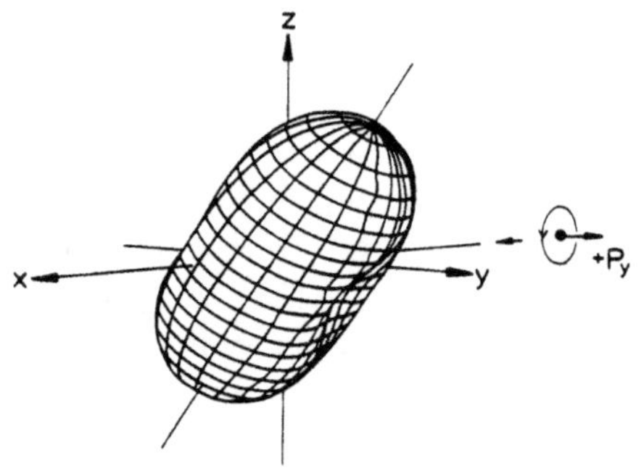

Figure 9: Example of a tilted and twisted charge cloud due to an in-plane spin polarized incident electron beam.[60] The figure is for Hg 6^3P_1 excitation with impact energy 8 eV, initial spin polarization $P_y = 1$ and scattering angle $\theta = 30°$.

This problem was also analyzed in detail by Andersen and Bartschat,[61] who simplified the previous analysis[59] (formulated in the collision frame with the quantization axis along the incident beam direction) significantly by performing the algebra once more in the natural coordinate system. They defined the complete set of parameters

$$(\sigma_u; w^{\uparrow},\ L_{\perp}^{+\uparrow},\ L_{\perp}^{+\downarrow},\ h^{\uparrow},\ h^{\downarrow};\ \gamma^{\uparrow},\ \gamma^{\downarrow},\ \Delta^{+},\ \Delta^{0},\ \delta^{\uparrow\downarrow}) \tag{15}$$

consisting of the differential cross section σ_u for unpolarized beams, angular momentum transfers $L_{\perp}^{+\uparrow,\downarrow}$, height parameters $h^{\uparrow,\downarrow}$, and alignment angles $\gamma^{\uparrow,\downarrow}$ for initial electron spin component "up" or "down" with respect to the scattering plane, respectively, a weight parameter $w^{\uparrow} = 1 - w^{\downarrow}$ to describe the relative cross sections for spin-up and spin-down electrons, and three additional angles $(\Delta^{+}, \Delta^{0}, \delta^{\uparrow\downarrow})$ which, like $\gamma^{\uparrow,\downarrow}$, are related to relative phases between complex scattering amplitudes.

Andersen and Bartschat[61] also introduced the "generalized Stokes parameters" (see figure 10) to analyze the emitted radiation in a systematic way. These parameters are defined in such a way that all four independent combinations of photon polarization analyzer and initial electron polarizations enter on an equal footing. Briefly, the quantity $I^{\hat{n}}_{\pm P_{x,y,z}}(\beta)$ is the light intensity transmitted by a linear polarization analyzer oriented at an angle β for incident electron polarizations in the x, y or z direction, with the light being observed in the direction denoted by $\hat{n}$, and similar definitions are made for the intensities transmitted by circular polarization analyzers. For example, in this notation the total intensity emitted in the y direction for unpolarized incident electrons can be constructed as

$$I_u^y = \tfrac{1}{4}\left[I^y_{+P_z}(45°) + I^y_{-P_z}(45°) + I^y_{+P_z}(135°) + I^y_{-P_z}(135°)\right] . \tag{16}$$

The generalized Stokes parameters Q_{ij}^{nm} measured with a photon detector in the direction $\hat{n}$ and electron spin polarization along the direction $\hat{m}$ are then defined by taking the three other independent linear combinations of the four intensities. The second subscript j =1,2,3 refers to the sign combinations $++--$, $+--+$ and $+-+-$, while the first subscript i =1,2,3 refers to the photon polarizer settings $(0°, 90°)$, $(45°, 135°)$ and (σ^-, σ^+), as for the standard Stokes vector. For example,

$$Q_{12}^{yz} \equiv \frac{I^y_{+P_z}(0°) - I^y_{-P_z}(0°) - I^y_{+P_z}(90°) + I^y_{-P_z}(90°)}{I^y_{+P_z}(0°) + I^y_{-P_z}(0°) + I^y_{+P_z}(90°) + I^y_{-P_z}(90°)} . \tag{17}$$

The standard three-component Stokes *vector* is thus replaced by a 3 × 3 *matrix*. As indicated in figure 10, some elements of this matrix correspond to the usual Stokes vector (P_1, P_2, P_3) measured for photon emission perpendicular to the scattering plane, plus a linear polarization P_4 measured in-plane for unpolarized beams. Most importantly, however, only very few elements of the full Stokes matrix contain new information which, in addition, can be obtained in several different ways. Furthermore, some of the elements should be identical, and even more should be zero. Consequently, measurements of the set of generalized Stokes parameters not only yield the new information, but also offer many built-in opportunities for consistency checks within the individual experimental data sets.

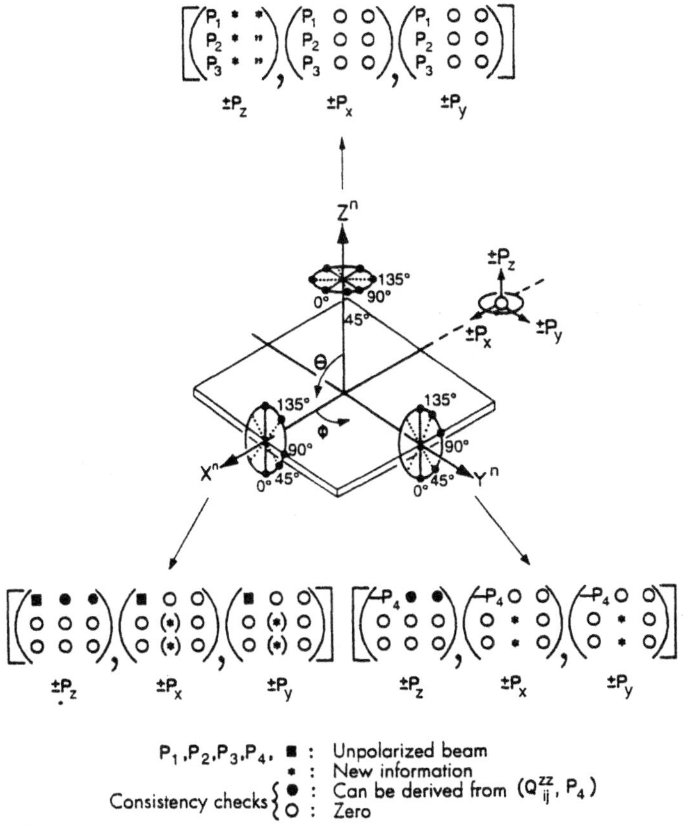

Figure 10: Definition of the generalized Stokes parameters and summary of the information that can be obtained from their measurement.[1]

This formalism was recently applied[62] to experimental data for electron impact excitation of the $(6s^2)^1S_0 \rightarrow (6s6p)^3P_1$ transition in Hg at an incident electron energy of 8 eV. The non-zero generalized Stokes parameter matrix elements for the z and y directions are shown in the left part of figure 11, together with theoretical results based on a 5-state Breit-Pauli R-matrix calculation[63] which is still the benchmark theory for this collision system. The agreement between experiment and theory is satisfactory, keeping in mind the complexity of the collision problem and the level of detail in the comparison. We also note that the experimental data for Q_{13}^{zz}, Q_{23}^{zz} and Q_{33}^{zz}, which should be identical within the error bars, are not completely consistent, although a tendency towards the same result is recognized.

Whereas a complete experiment also requires the measurement of some "generalized STU parameters"[64] that describe the change of an initial electron polarization vector in the collision, the measurement of the generalized Stokes parameters allows for the extraction of the subset ($w^{\uparrow}$, $L_{\perp}^{+\uparrow}$, $L_{\perp}^{+\downarrow}$, $h^{\uparrow}$, $h^{\downarrow}$; $\gamma^{\uparrow}$, $\gamma^{\downarrow}$) of parameters introduced in (15). The corresponding spin-averaged and spin-resolved coherence parameters are presented in the right part of figure 11. The vertical lines in the figure represent the scatter due to alternative ways of extracting the above subset, which is overdetermined by the twelve generalized Stokes parameters shown on the left.[62]

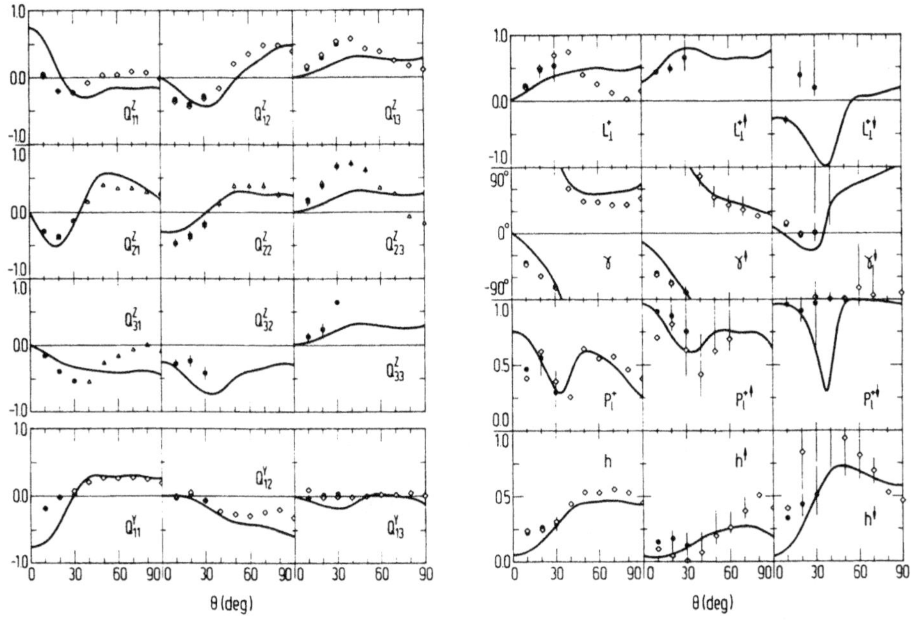

Figure 11: Generalized Stokes parameters (left) as well as spin-averaged and spin-resolved coherence parameters (right) for electron impact excitation of Hg(6s6p)^{3}P$_1$ at 8 eV. All data correspond to unpolarized electrons or transversal spin polarization along the z direction (perpendicular to the scattering plane). The upper index on the generalized Stokes parameters therefore refers to the direction of photon observation. The error bars on the experimental data ($\bullet$[62], $\diamond$[65,66,67]) reflect statistics and consistency tests (see text).

The coherence parameters are very different for the two spin directions. Note that at $\theta = 10°$, $L_{\perp}^{+\uparrow} > 0$ while $L_{\perp}^{+\downarrow} < 0$. The positive value of $L_{\perp}^{+\uparrow}$ agrees with well-established propensity rules,[43] while a *negative* value of $L_{\perp}^{+\downarrow}$ may seem surprising. However, both $L_{\perp}^{+\uparrow}$ and $L_{\perp}^{+\downarrow}$ may be non-zero for forward scattering, but $L_{\perp}^{+\uparrow}(0°) = -L_{\perp}^{+\downarrow}(0°)$ by symmetry requirements. The spin-resolved alignment angles $\gamma^{\uparrow}$ and $\gamma^{\downarrow}$ show no similarities, with the directions of the two major axes often perpendicular to each other (note that $\gamma^{\uparrow}(0°) = -\gamma^{\downarrow}(0°) \neq 0$). Finally, there is a large difference between the height parameters $h^{\uparrow}$ and $h^{\downarrow}$, with the theory predicting a maximum value of $h^{\downarrow} \approx 75\%$ near a scattering angle of 40° while $h^{\uparrow} \approx 25\%$. Spin-flips are thus very likely for spin "down" electrons, but those spin "down" electrons whose spin is not flipped tend to transfer a *negative* angular momentum to the atom.[62]

Electron impact excitation of He $1^1S \rightarrow 3^1D$

This process is determined by three independent scattering amplitudes, denoted by f_2, f_0 and f_{-2} in the natural frame.[1] Hence, the determination of one ***absolute*** differential cross section, two *relative* magnitudes, and two *relative* phases is required for a complete experiment. Although the four Stokes parameters (P_1, P_2, P_3, P_4) contain all information obtainable from the radiation pattern of the subsequent D → P decay, it is well known that they are not sufficient to define a complete experiment; instead, two possible solutions are obtained.[68]

The traditional *incomplete* parameterization of this case is the set

$$(\sigma_u; L_\perp, \rho_{00} = \tfrac{3}{2}h; \gamma, P_\ell) \tag{18}$$

with the latter four evaluated from the four Stokes parameters (P_1, P_2, P_3, P_4). A *complete* set of parameters,

$$(\sigma_u; L_\perp^+, L_\perp^-; \gamma^+, \gamma^-)\,. \tag{19}$$

was again introduced by Andersen and Bartschat[1] who performed an analysis involving the circular light polarization of the P → S cascade photon. Without going into the details of their analysis, the most important outcome was another possible consistency check involving *all four* Stokes parameters (P_1, P_2, P_3, P_4) *simultaneously.* The usual consistency check, $P = P_1^2 + P_2^2 + P_3^2 \leq 1$, is a much weaker condition and, therefore, of little practical importance for this case.

Figure 12 shows data obtained in electron–photon coincidence experiments with polarization analysis of the D → P photon,[69,70] compared with results from the CCC calculation of Fursa and Bray.[20] The left two columns of the figure show the data evaluation in terms of the standard set (18), while the new parameter set (19) is presented in the right column. Note that there is a *unique* relationship between the two subsets $(L_\perp^+, L_\perp^-)$ and $(L_\perp, \rho_{00})$ which both describe relative excitation cross sections between the magnetic sublevels with $M = 0, \pm 2$ of the excited D state. (Excitation of the $M = \pm 1$ sublevels is forbidden in the natural frame.) The inversion between the incomplete subset (P_ℓ, γ) and the complete subset (γ^+, γ^-) leaves the ambiguity mentioned above, but we see how a reliable theory can be used to distinguish between the "true" and the "ghost" solution with some degree of confidence.

Interestingly, there remained some problems in the data inversion for scattering angles of 60°, 80° and 100° degrees where the procedure actually failed and the most sensible solution was obtained by setting $\gamma^+ = \gamma^- = \gamma$.[1] A detailed analysis traced the problem back to the relatively large magnitude of P_ℓ. Based on the good agreement between theory and experiment for the parameter set $(L_\perp^+, L_\perp^-, \gamma)$ and the fact that the theoretical data are internally consistent (even if they do not describe Nature perfectly), one might suspect that the experimental data for P_ℓ are slightly too large. If this were true indeed, the agreement between experiment and theory might further improve in figure 12.

Finally, the ambiguity can be resolved experimentally in an electron–two-photon coincidence experiment. This has been demonstrated theoretically in several ways,[1,71] and first experimental data have just been obtained.[72,73]

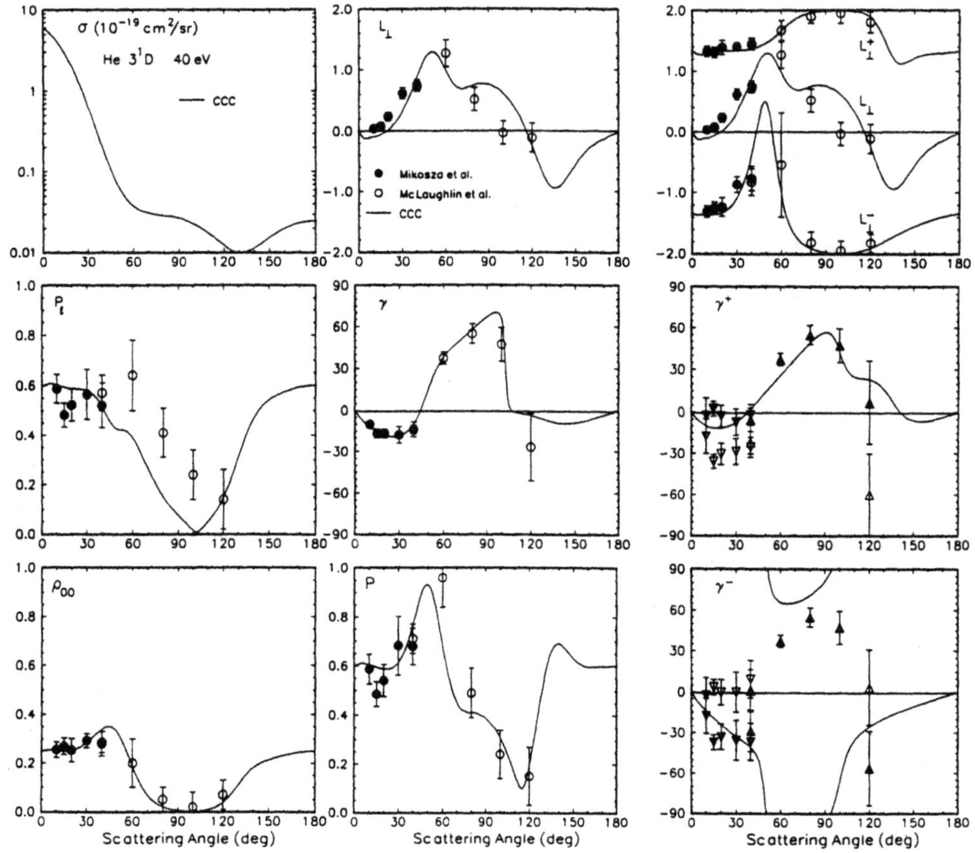

Figure 12: Differential cross section σ and sets of coherence parameters $(L_\perp, P_\ell, \gamma, \rho_{00}, P)$ and $(L_\perp^+, \gamma^+)$ and $(L_\perp^-, \gamma^-)$ for electron impact excitation of the He 1^1S $\rightarrow$ 3^1D transition at an incident electron energy of 40 eV. The experimental points have been calculated from the data of Mikosza[69] ("down" triangles) and McLaughlin *et al*[70] ("up" triangles). They are compared with results obtained from CCC scattering amplitudes of Fursa and Bray.[20] In the right column, the "true" (γ^+, γ^-) pair (full symbols) was guessed by using the theoretical results as a guide.[1]

RECOMMENDATIONS FOR FUTURE WORK

We have presented examples for total cross sections as well as angle-differential observables for electron impact excitation of atomic targets. The selection was determined by the aim to show not only the sometimes very close agreement between theoretical predictions and experimental data, but also to point out some cases where more work is clearly required. In fact, there are many more problem cases than Be, Cs or Hg. Besides targets of interest for plasma and atmospheric physics, such as C, N, and O, for which the amount of experimental data is very limited, important other examples include K, Cu, In, Tl and Pb, where there are large discrepancies between the available experimental data and theoretical predictions.[2] Furthermore, accurate intermediate-energy data are required for many electron–ion collision processes.

In light of the success of the CCC, RMPS, and IERM methods for relatively simple light targets, these methods should be tested on such more complex systems, including those where relativistic effects must be taken into account. While some initial steps have recently been taken in this direction, it is not clear at the present time when a full-relativistic calculation of this kind for e–Cs or e–Hg scattering will become possible. Given the rapid development of computer hardware, however, attempts to develop the necessary computer codes should be started immediately.

For many of the collision systems of interest, a close coordination between experimental and theoretical efforts would certainly increase the chances of success tremendously. If data from different measurements can be re-evaluated to come as closely as possible to the complete experiment, such data will serve as excellent benchmarks against which theoretical predictions can be checked. The most impressive example of this kind seems to be the e–Hg collision system where the determination of six complex amplitudes is a realistic possibility with presently available technology. Also, benchmark work using highly sophisticated theoretical methods will be very useful in the testing of newly developed approaches.

In addition to testing experimental setups and computer codes for discrete transitions, these techniques seem ready for the study of ionization and even ionization-excitation processes. Electron–photon coincidence experiments for the latter process, for example, allow for the extraction of phase information between scattering amplitudes that cannot be obtained in standard (e,2e) cross section setups.[74] First measurements of this kind have now been performed[75] and are expected to represent further challenges for theorists working in this field.

Acknowledgments

I would like to thank N. Andersen, I. Bray, J.T. Broad, P.G. Burke, V.M. Burke, G.F. Hanne, E.T. Hudson, M.P. Scott, and M. Uhrig for their contributions to this work. Financial support from the National Science Foundation under grant # PHY-9318377 and by NATO under grant CRG-930056 is gratefully acknowledged. Much of the work reported in this chapter was made possible through visiting fellowships granted to the author by the Queen's University of Belfast (Northern Ireland), the University of Münster (Germany) in connection with the SFB 216, and the Flinders University of South Australia.

REFERENCES

1. N. Andersen and K. Bartschat (1996), Complete experiments in electron–atom collisions, *Adv. At. Mol. Phys.* 36: in press
2. N. Andersen, K. Bartschat, J.T. Broad and I.V. Hertel (1996), Collisional alignment and orientation of atomic outer shells. III. Spin-resolved excitation, *Phys. Rep.*: in press
3. I. Bray and A.T. Stelbovics (1995), *Adv. Atom. Mol. Phys.* 35:209
4. P.G. Burke, C.J. Noble and M.P. Scott (1987), *Proc. Roy. Soc. A* 410:289
5. K. Bartschat, E.T. Hudson, M.P. Scott, P.G. Burke and V.M. Burke (1996), *J. Phys. B* 29:115
6. I. Bray and A.T. Stelbovics (1992), *Phys. Rev. A* 46:6995
7. P.G. Burke and W.D. Robb (1975), *Adv. At. Mol. Phys.* 11:143
8. K. Bartschat, E.T. Hudson, M.P. Scott, P.G. Burke and V.M. Burke (1996), Convergent R-matrix with pseudo-states calculation for e–H collisions, *Phys. Rev. A* 54: in press

9. K. Bartschat and I. Bray (1996), S-wave model for e–H scattering, *Phys. Rev. A* 54: in press
10. K. Bartschat and I. Bray (1996), Electron impact ionization of atomic hydrogen from the 1S and 2S states, *J. Phys. B* 29: in press
11. V.M. Burke and C.J. Noble (1995), *Comp. Phys. Commun.* 85:471
12. M. Le Dourneuf, J.M. Launay, and P.G. Burke (1990), *J. Phys. B* 23:L559
13. L. Castillejo, I.C. Percival and M.J. Seaton (1960), *Proc. Roy. Soc. London A* 254:259
14. K. Bartschat, I. Bray, P.G. Burke and M.P. Scott (1996), Benchmark calculations for e–H scattering between the n=2 and n=3 thresholds, *J. Phys. B* 29: in press
15. J.F. Williams (1988), *J. Phys. B* 21:2107
16. M.P. Scott and P.G. Burke (1996), Near-threshold e^-–H $^1S^e$ ionization, *PECAM II: Book of Abstracts*, 81
17. D. Kato and S. Watanabe (1995), *Phys. Rev. Lett.* 74:2443
18. M.B. Shah, D.S. Elliott and B. Gilbody (1987), *J. Phys. B* 20:3501
19. J.W. McGowan and E.M. Clarke (1968), *Phys. Rev.* 167:43
20. D.V. Fursa and I. Bray (1995), *Phys. Rev. A* 52:1279
21. K. Bartschat, E.T. Hudson, M.P. Scott, P.G. Burke and V.M. Burke (1996), Differential cross sections and electron impact coherence parameters for electron scattering from helium atoms, *J. Phys. B* 29: in press
22. E.T. Hudson, K. Bartschat, M.P. Scott, P.G. Burke and V.M. Burke (1996), Electron scattering from helium atoms. Phase shifts, resonance parameters and total cross sections, *J. Phys. B* 29: in press
23. W.C. Fon, K.A. Berrington, P.G. Burke and A.E. Kingston (1981), *J. Phys. B* 14:2921
24. P.M.J. Sawey, K.A. Berrington, P.G. Burke and A.E. Kingston (1990), *J. Phys. B* 23:4321
25. P.M.J. Sawey and K.A. Berrington (1993), *At. Data Nucl. Data Tables* 55:81
26. S. Trajmar (1973), *Phys. Rev. A* 8:191
27. A. Chutjian and S.K. Srivastava (1975), *J. Phys. B* 8:2360
28. R.I. Hall, G. Joyez, Y. Mazeau, J. Reinhard and C. Schermann (1973), *J. Physique (Orsay, France)* 34:827
29. F.G. Donaldson, M.A. Hender and J.W. McConkey (1972), *J. Phys. B* 5:1192
30. H.P. Summers, W.J. Dickson, A. Boileau, P.G. Burke, B. Denne-Hinnov, W. Fritsch, R. Giannella, N.C. Hawkes, M. von Hellermann, W. Mandl, N.J. Peacock, R.H.G. Reid, M.F. Stamp and P.R. Thomas (1991), *Plasma Physics and Controlled Fusion* 34:325
31. W.C. Fon, K.A. Berrington, P.G. Burke, V.M. Burke and A. Hibbert (1992), *J. Phys. B* 25:507
32. K. Bartschat, M.P. Scott and P.G. Burke (1996), Electron scattering from beryllium, *J. Phys. B* 29: in press
33. P.G. Burke, C.V. Sukumar and K.A. Berrington (1981), *J. Phys. B* 14:289
34. E. Karule (1972), *J. Phys. B* 5:2051
35. N.S. Scott, K. Bartschat, P.G. Burke, O. Nagy and W.B. Eissner (1984), *J. Phys. B* 17:3775
36. K. Bartschat (1993), *J. Phys. B* 26:3595
37. U. Thumm and D.W. Norcross (1993), *Phys. Rev. A* 47:305
38. P.J. Visconti, J.A. Slevin and K. Rubin (1971), *Phys. Rev. A* 3:1310
39. B. Jaduszliwer and Y.C. Chan (1992), *Phys. Rev. A* 47:197
40. K. Bartschat and I. Bray (1996), Calculation of electron-Cs scattering at intermediate energies, *Phys. Rev. A* 54: in press
41. B. Bederson (1969), *Comments At. Mol. Phys.* 1:41
42. B. Bederson (1969), *Comments At. Mol. Phys.* 1:65

43. N. Andersen, J.W. Gallagher, and I.V. Hertel (1988), *Phys. Rep.* 165:1
44. M.C. Standage and H. Kleinpoppen (1976), *Phys. Rev. Lett.* 36:577
45. M.A. Khakoo, K. Becker, J.L. Forand and J.W. McConkey (1986), *J. Phys. B* 19:L209.
46. D.C. Cartwright and G. Csanak (1988), *Phys. Rev. A* 38:2740
47. K. Bartschat and D.H. Madison (1988), *J. Phys. B* 21:2621
48. J.P. Beijers, D.H. Madison, J. van Eck and H.G.M. Heideman (1987), *J. Phys. B* 20:167
49. D.C. Cartwright, G. Csanak, S. Trajmar and D.F. Register (1992), *Phys. Rev. A* 45:1602
50. R. McAdams, M.T. Hollywood, A. Crowe and J.F. Williams (1980), *J. Phys. B* 13:3691
51. M. Eminyan, K.B. McAdam, J. Slevin and H. Kleinpoppen (1974), *J. Phys. B* 7:1519
52. N. Andersen and K. Bartschat (1993), *Comments At. Mol. Phys.* 29:157
53. J.J. McClelland, M.H. Kelley and R.J. Celotta (1989), *Phys. Rev. A* 40:2321
54. T. Hegemann, S. Schroll and G.F. Hanne (1993), *J. Phys. B* 26:4607
55. P.J.O. Teubner and R.E. Scholten (1992), *J. Phys. B* 25:L301
56. D.H. Madison, K. Bartschat and R.P. McEachran (1992), *J. Phys. B* 25:5199
57. I. Bray and I.E. McCarthy (1993), *Phys. Rev. A* 47:317
58. I. Bray (1994), *Phys. Rev. A* 49:1066
59. K. Bartschat, K. Blum, G.F. Hanne and J. Kessler (1981), *J. Phys. B* 14:3761
60. A. Raeker, K. Blum and K. Bartschat (1993), *J. Phys. B* 26:1491
61. N. Andersen and K. Bartschat (1994), *J. Phys. B* 27:3189; *Corrigendum* (1996), *J. Phys. B* 29:1149
62. N. Andersen, K. Bartschat, J.T. Broad, G.F. Hanne and M. Uhrig (1996), *Phys. Rev. Lett.* 76:208
63. N.S. Scott, P.G. Burke and K. Bartschat (1983), *J. Phys. B* 16:L361
64. K. Bartschat (1989), *Phys. Rep.* 180:1
65. J. Goeke, G.F. Hanne and J. Kessler (1988), *Phys. Rev. Lett.* 61:58
66. J. Goeke, G.F. Hanne and J. Kessler (1989), *J. Phys. B* 22:1075
67. M. Sohn and G.F. Hanne (1992), *J. Phys. B* 25:4627
68. N. Andersen, T. Andersen, J.S. Dahler, S.E. Nielsen, G. Nienhuis and K. Refsgaard (1983), *J. Phys. B* 16:817
69. A.G. Mikosza,R. Hippler, J.B. Wang and J.F. Williams (1994), *Z. Phys. D* 30:129
70. D.T. McLaughlin, B.P. Donnelly and A. Crowe (1994), *Z. Phys. D* 29:259
71. J.B. Wang and J.F. Williams (1995), *Aust. J. Phys.* 49:335
72. A.G. Mikosza and J.F. Williams (1996), *Aust. J. Phys.* 49:375
73. A.G. Mikosza, J.F. Williams and J.B. Wang (1996), Complete description of electron impact excited He(3^1D) state, *PECAM II: Book of Abstracts*, 72
74. R. Schwienhorst, A. Raeker, K. Bartschat and K. Blum (1996), *J. Phys. B* 29:2305
75. P.A. Hayes and J.F. Williams (1996), Simultaneous ionization and excitation to the He^+ 2^2P state, *Phys. Rev. Lett.*: submitted

ELECTRON-HELIUM CORRELATION STUDIES

Albert Crowe

Department of Physics
University of Newcastle
Newcastle upon Tyne
NE1 7RU
United Kingdom

INTRODUCTION

Use of correlation methods over the last quarter of a century has enabled detailed studies of a range of processes involving the impact of electrons on atoms to be made. Studies of single electron excitation in helium have led to the realisation of 'complete experiments' as defined by Bederson[1,2] and exposed severe weaknesses in theoretical approaches to the description of these simple processes.

In this discussion emphasis will be placed on processes which involve production of an atom or ion in an excited state. Depending on the decay mode, the electron-photon angular and polarisation correlation methods and electron-electron (e,2e) technique are used to define the kinematics and probe the dynamics of the process. Figure 1 illustrates schematically the principle of both methods. An incoming electron of energy E_o, momentum $\mathbf{k}_o$ is incident on a helium atom target. In both cases the inelastically scattered electrons with energy E_s are detected at a fixed scattered angle. The vectors $\mathbf{k}_o$, $\mathbf{k}_s$ define a scattering plane. A second outgoing particle is detected in coincidence with the scattered electron. In the case of electron-photon correlation studies, either an angular distribution of the photons from the decay of an excited state is measured for a specific momentum transfer $\mathbf{K} = \mathbf{k}_o - \mathbf{k}_s$ collision or the polarisation of the emitted photons is determined, also for a specific $\mathbf{K}$. In the (e,2e) method, angular correlations between the scattered electrons and electrons ejected, either directly or indirectly, in an ionization process are measured.

Helium provides a convenient target to develop a detailed understanding of excitation from many points of view. Experimentally it is an easily handled source of stable atoms. S-P excitation provides the simplest example of a complete excitation study since the direct and exchange amplitudes cannot be disentangled for the singlet P states while the triplet P states involve only pure exchange amplitudes. The absence of hyperfine structure and fine structure for the singlet states also simplifies both the experiment and data analysis compared with, for example, doublet P excitation in hydrogen.

Helium also provides the simplest examples, experimentally and theoretically, of processes involving two active atomic electrons. Simultaneous excitation-ionization processes can now be probed using both electron-photon and (e, 2e) correlation methods while double excitation-autoionization is best studied using the (e, 2e) technique.

Photon and Electron Collisions with Atoms and Molecules
Edited by Burke and Joachain, Plenum Press, New York, 1997

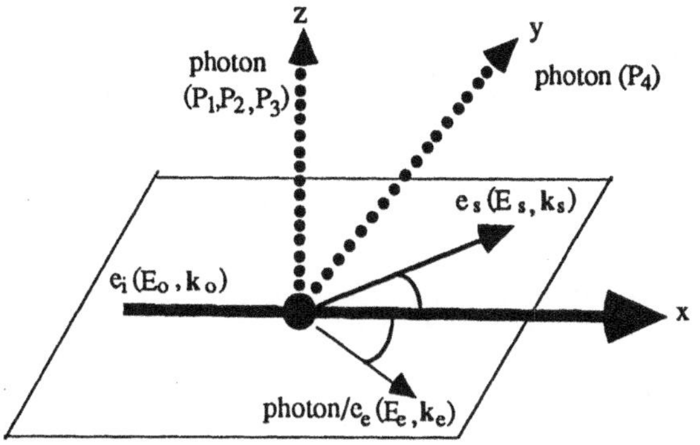

Figure 1. Schematic outline of (e, hv) and (e, 2e) correlation techniques. The electron-photon correlation data presented have been analysed in the natural co-ordinate frame.

SINGLE ELECTRON EXCITATION

Electron-photon angular correlation studies of the 2^1P state were first reported by Eminyan et al[3] and polarisation correlation studies of the 3^1P state by Standage and Kleinpoppen.[4] A comprehensive review of the field up until 1987 has been given by Andersen et al.[5] More recent reviews of aspects of the work have been given by Slevin and Chwirot,[6] Crowe,[7,8] Becker et al,[9] Crowe and Bray[10] and Andersen and Bartschat.[11]

In the present discussion, results are presented in terms of measured Stokes parameters P_1–P_4 describing the polarisation of the emitted radiation and parameters describing the shape and dynamics of the excited state. These are discussed in detail by Andersen et al[5] and only basic definitions are given here. The relevant parameters are defined in the natural co-ordinate frame with the z-axis perpendicular to the (x, y) scattering plane defined by $\mathbf{k}_o$ and $\mathbf{k}_s$. The Stokes parameters P_1 to P_4 are defined by

$$P_1 I_z = I_z(0^\circ) - I_z(90^\circ) \tag{1}$$

where I_z is the total coincident photon intensity in the z-direction and $I_z(\alpha^\circ)$ the intensity transmitted by a polariser with transmission axis making an angle α relative to the incident electron beam direction.

$$P_2 I_z = I_z(45^\circ) - I_z(135^\circ) \tag{2}$$

$$P_3 I_z = I_z(RHC) - I_z(LHC) \tag{3}$$

where *RHC* and *LHC* refer to the handedness of the circular polarisation.

$$P_4 I_y = I_y(0^\circ) - I_y(90^\circ) \tag{4}$$

These yield the parameters P_ℓ, γ describing the shape of the excited state in the collision plane, ρ_{oo}, or alternatively h, the relative height along the z-axis, and $L_\perp$, the expectation value of the angular momentum transferred to the atom, necessarily perpendicular to the scattering plane. Angular correlation methods, in general, give only the shape parameters.

Traditional theoretical methods have struggled to reproduce the results of electron-photon correlation experiments. Even for the simplest dipole allowed excitation to the 2^1P state it was normal that there was good agreement with one, but not both parameters. However the failure of theory was most dramatically demonstrated with availability of data for the 3^1D state. This is vividly illustrated in figure 2 which shows the comparison between theory and experiment for the circular polarisation parameter P_3 at 40 eV and scattering angles up to 60° until mid 1994. In this region three independent sets of experimental data show excellent agreement with each other. In contrast, the four calculations disagree with each other and with the experimental data.

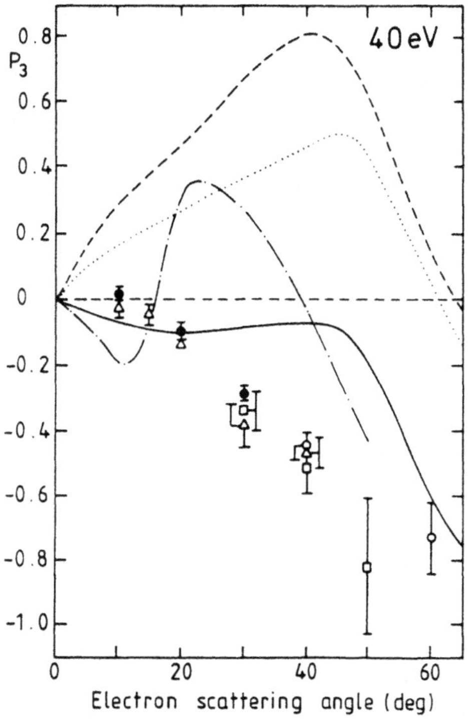

Figure 2. Variation of the Stokes parameter P_3 for the 3^1D state of helium at 40 eV – Status of comparison between theory and experiment – mid 1994. *Circles* (open and closed) – Donnelly and Crowe[12] and private communication; *squares* – Beijers et al;[13] *triangles* – Mikoza et al;[14] *solid line* – Bartschat and Madison[15] – DWBA-EP; *dashed line* – Bartschat and Madison[15] – DWBA-GP; *dotted line* – Cartwright and Csanak[16] – FOMBT; *dash-dot line* – Mansky and Flannery[17] – DMET.

Development of the Convergent Close Coupling (CCC) method and its application to helium, Fursa and Bray,[18] has dramatically transformed this situation. Experimental correlation data for the 3^1D state excitation are well described by this theory over a wide range of electron scattering angles and at a number of energies. Figure 3 shows a comparison between the CCC calculations and experiment for an incident electron energy of 40 eV for the parameters ρ, P_ℓ, $L_\perp$, ρ_{oo}, and the degree of polarisation P. The 75 state CCC differential cross section is also shown.

An interesting aspect of these results is that although the excitation is coherent, the degree of polarisation of the emitted radiation defined by

$$P^2 = P_1^{\,2} + P_2^{\,2} + P_3^{\,2} \tag{5}$$

is almost zero at a scattering angle near 120°. This can be contrasted with the n^1P states where P is unity independent of the kinematics. A detailed discussion of this situation has recently been given by Andersen and Bartschat[11] and can be explained by reference to figure 4 which shows the decay pattern associated with individual sublevels for photon emission perpendicular to the (x, y) scattering plane. Symmetry conditions forbid excitation of the M=±1 states. The D–P transition photons observed in the +z direction arise from two distinct channels, one to the p_{+1} and one p_{-1} level.

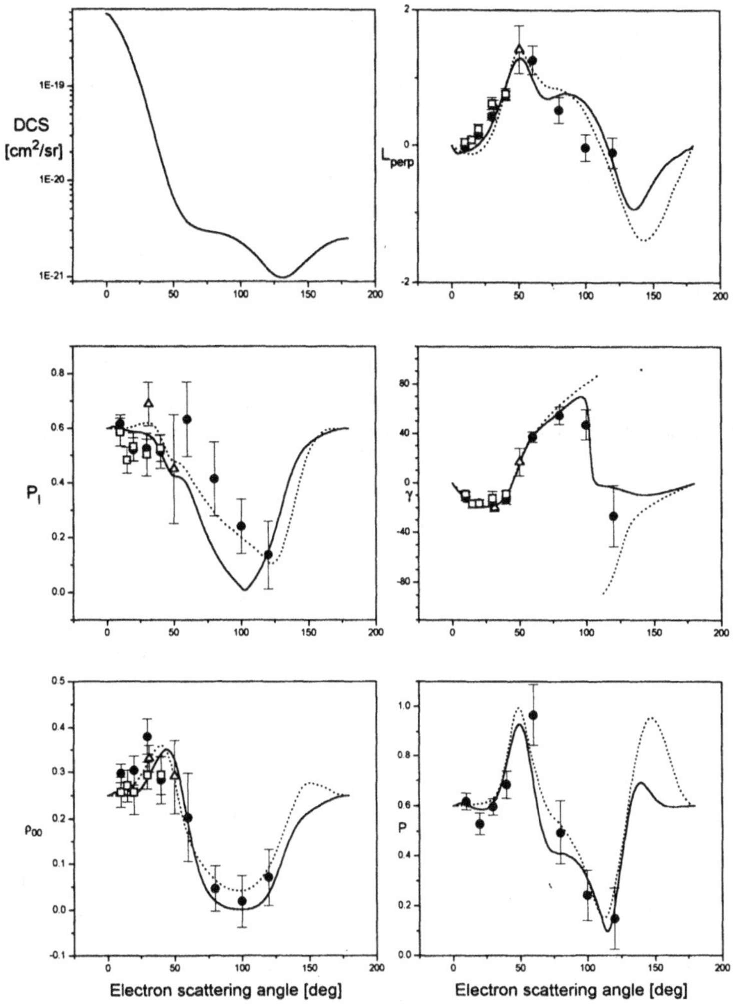

Figure 3. Differential cross section σ, the correlation paramters and the degree of polarisation P for the 3^1D state at 40 eV. ●, McLaughlin et al[19] and unpublished data; ❑, Mikoza et al;[14] Δ, Batelaan et al;[20] ——, 75 state CCC;[18] ········, 62 state CCC.[21]

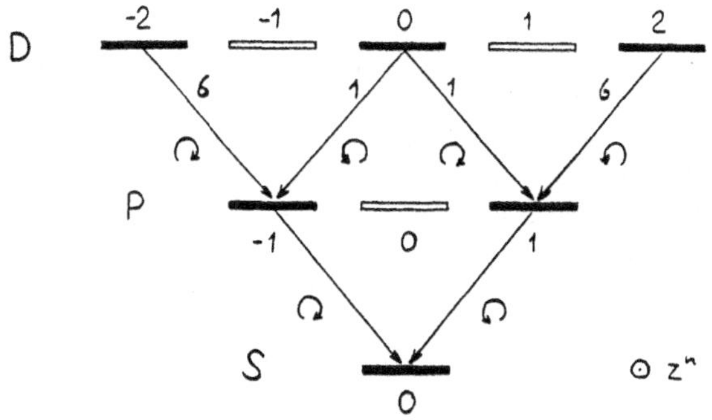

Figure 4. Decay scheme of an atomic D state for photon emission along the z-axis, following electron impact excitation from the ground S state.

Andersen and Bartschat[11] have shown that a triple coincidence experiment involving detection of the scattered electron, the D–P photon and the cascade P–S photon, leads to a complete parameter set $(\sigma, L_\perp^+, L_\perp^-, \gamma^+, \gamma^-)$ for D-state excitation. σ is the differential cross section, and the "+" and "–" refer to the channels involving p_{+1} and p_{-1} respectively. Such experiments are extremely difficult, Wang and Williams,[22] and have yet to be performed. However, $L_\perp^+$ and $L_\perp^-$ can be obtained from existing measured values of P_1, P_3, P_4 using

$$L_\perp^\pm = \pm 2\left(1 - \frac{8\rho_{00}}{3(2 \pm L_\perp)}\right) \tag{6}$$

Values of $L_\perp$, $L_\perp^\pm$ are shown in figure 5 for the 3^1D state at 40 eV. Now it can be seen that although $L_\perp$ is consistent with zero at the larger scattering angles measured, this actually arises from the incoherent sum of $L_\perp^+ \sim 2\hbar$ and $L_\perp^- \sim -2\hbar$.

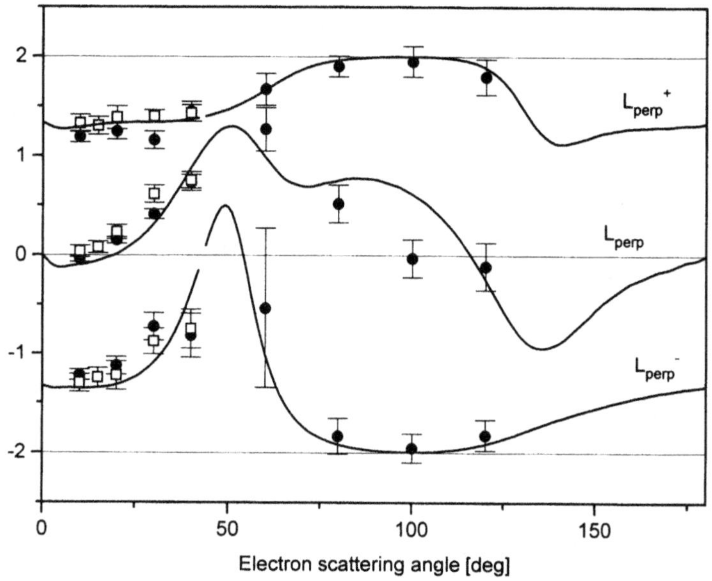

Figure 5. The parameters $L_\perp$, $L_\perp^+$, $L_\perp^-$ for the 3^1D state at 40 eV. Notation as figure 3.

There is a difficulty in extracting the remaining parameters γ^+, γ^- of the complete parameter set from the existing data. In particular there is ambiguity in the **sign** of and angle relating γ and $\gamma^\pm$. However the quality of the CCC calculations enables the ambiguity to be removed. The resulting values of γ^+ and γ^- are shown in figure 6, again for an incident electron energy of 40 eV. Given the unlikelihood of data of good statistical quality from a triple coincidence experiment, the Andersen and Bartschat parameterisation must cast doubt on the value of such experiments. Measurements of correlations between the D–P and P–S photons have enabled Mikoza et al[23] to determine a rank four state multipole and the integrated sub-level cross sections Q_0, Q_1, Q_2 for the 3^1D state.

P_1–P_4 have also been measured for a range of kinematic conditions for the 3^3D state of helium.[24,25] In this case the measured polarisations are reduced due to the fine structure of the excited state. The relationships between the measured Stokes parameters and the parameter set $(P_\ell, \gamma, L_\perp, \rho_{00})$ relating to the excited state produced in the collision have been given by Crowe et al.[24] A comparison between theory and experiment is given in figure 7 for an incident electron energy of 40 eV. The agreement between the experimental data and the CCC calculation is essentially perfect.

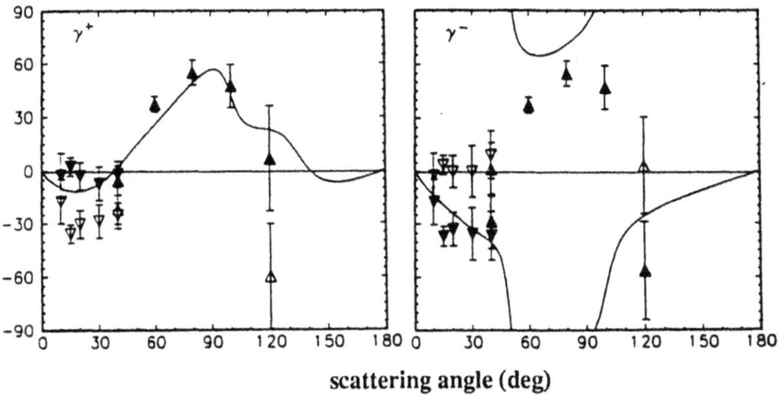

Figure 6. The parameters γ^+, γ^- for 3^1D excitation at 40 eV. The solid triangles represent the real values from the data of McLaughlin et al[19] and Mikoza et al.[14] The open triangles are 'ghost' solutions from this same data. The solid line is the 75 state CCC of Fursa and Bray.[18]

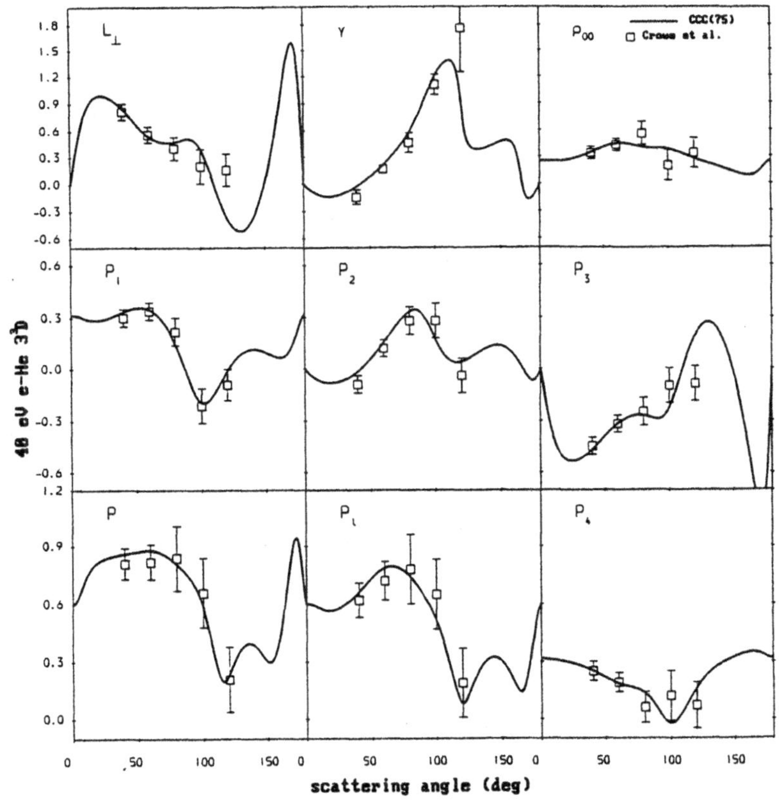

Figure 7. Comparison of the 75 state CCC with experimental data for the 3^3D state of helium at 40 eV.[24] Measured Stokes parameters and derived parameters are shown.

Given the success of the CCC method for D state excitation and the likelihood that new techniques such as the R-matrix with Pseudo States (RMPS), which shows excellent agreement with experiment for the 2^1P state,[26] will be at least equally successful, it would not be unreasonable to conclude that single electron excitation in helium is now well understood. Alternatively, it might be asked if these methods will be as successful for higher L states. Apart from indirect estimates of their total excitation cross sections[27] and lifetimes,[28] no experimental or theoretical data have been available for F states in helium. Compared with L≤2 states, which can be described by a pure Russell-Saunders coupling scheme, the L≥3 states have been shown to be an admixture of singlet and triplet states. Wang et al[29] have recently presented a theoretical description for the coherent excitation of the singlet-triplet mixed 1s4f state of helium, together with Distorted wave Born approximation predictions at an incident electron energy of 100 eV.

Experimentally, the photon wavelengths for transitions to the 3^1D and 3^3D states are 1870 and 1869 nm respectively. These wavelengths are not detectable by single particle detectors necessary for coincidence studies. Nevertheless, detailed information can be obtained from a Stokes parameter analysis of the D–P or P–S cascade photons.[29]

Figure 8 shows preliminary data of Cvejanovic and Crowe[30] for the Stokes parameters P_1 and P_2 measured for the 3^1D–2^1P transition in coincidence with scattered electrons from excitation of the n=4 states. The data have been obtained at an incident electron energy of 29.6eV. It is planned to measure further Stokes parameters and to generalise the formalism of Andersen and Bartschat[11] to F states.

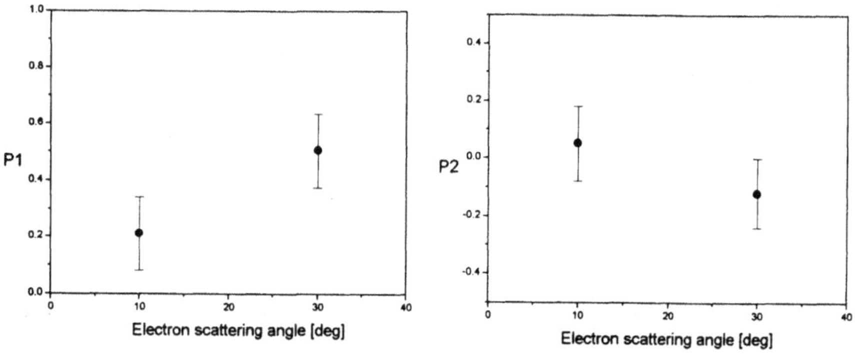

Figure 8. Preliminary values of the Stokes parameters P_1 and P_2 for excitation of the 1s4f states measured for the 3^1D–2^1P transition in coincidence with scattered electrons from excitation of n=4 states.

PROCESSES INVOLVING TWO ACTIVE ATOMIC ELECTRONS

Ionization-Excitation Processes

Processes of the type

$$\begin{array}{l} e + He(1s^2)^1S \rightarrow He^+(n\ell)^2L + e_s + e_e \\ \qquad\qquad\qquad\qquad \downarrow \\ \qquad\qquad\qquad He^+(n'\ell')^2L' + h\upsilon \end{array} \qquad (7)$$

can be studied in detail using both (e, 2e) and electron-photon correlation methods and possibly by a triple coincidence experiment involving both electrons and the photon in the future. Here we concentrate on the production of $He^+(n=2)$ with decay of the $He^+(2p)$ state to the ground ion state $He^+(1s)$ with the emission of 30.4 nm photons.

(e, 2e) dynamics studies have been carried out on the $He^{+}(n=2)$ states by Stefani et al[31] and Dupré et al.[32] Figure 9 shows a comparison between the data of Dupré et al and a calculation by Kheifets et al[33] for ionization to both the n=1 and n=2 states of the ion. The theory is a first Born calculation with an initial correlated multiconfiguration Hartree-Fock helium atom ground state and a coupled-channel final state of the ion and slow ejected electron. For small ejected electron energies, even at an incident energy of 5.5 keV, the triple differential cross sections (TDCS) for n=1 and n=2 are very different in shape and only the 'recoil' ($-\mathbf{q}$) direction is reproduced by the first Born calculations. Kheifets et al associate the peaks in the $\pm\mathbf{q}$ directions with the $He^{+}(2s)$ state and the secondary peaks with the $He^{+}(2p)$ state. At higher ejected electron energies, the n=2 TDCS shapes approach those of the n=1 and agreement with theory is good.

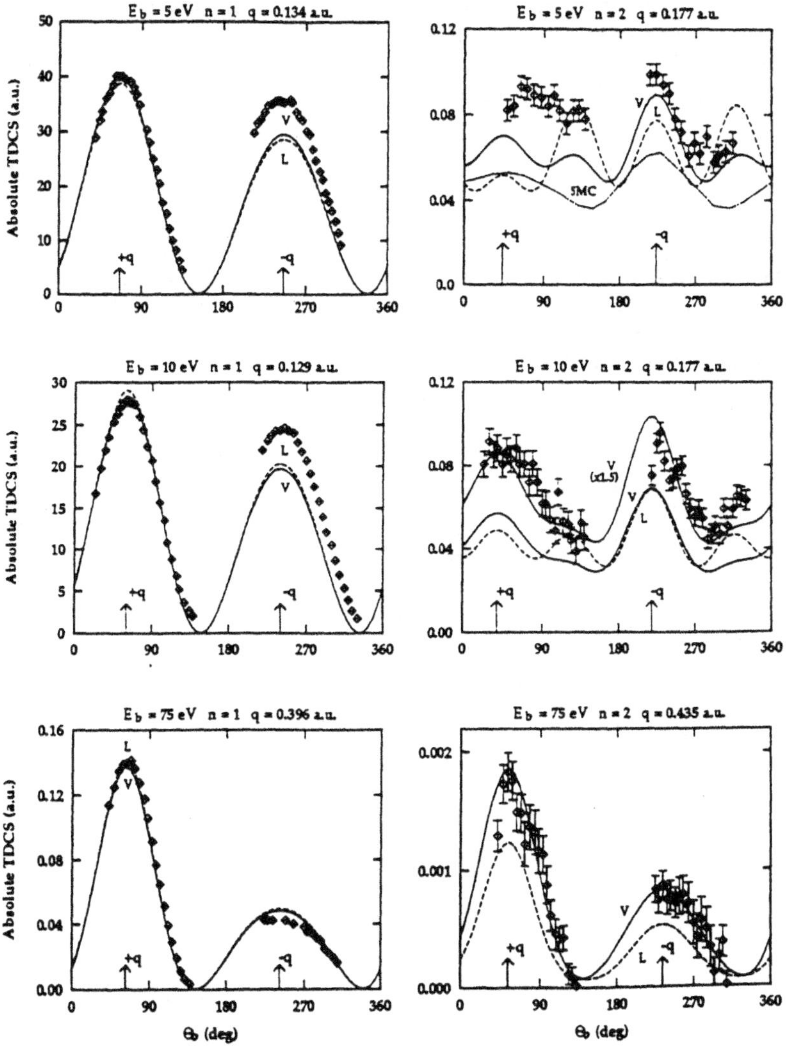

Figure 9. Absolute TDCS for ionization to the n=1 (left) and n=2 (right) ion states by 5.5 keV electrons scattered through 0.35° and for an ejected electron energies of 5, 10 and 75 eV. *Diamonds* – experiment of Dupré et al;[32] 6CC calculations in length (L, *dashed line*) and velocity (v, *solid line*) forms of Kheifets et al;[33] *dotted line* – calculations of Dupré et al.

Studies involving the detection of the emitted photon have two major advantages. They enable study of the $He^+(2p)$ state rather than the unresolved $He^+(2s)$, $He^+(2p)$ states in the (e, 2e) studies. They also provide the possibility of determining the relative phases of scattering amplitudes directly for an ionization process. The inadequacy of present theoretical models in describing the $He^+(2p)$ excitation is already clear from a study of excitation functions. Figure 10 shows a comparison between the measured cross sections of Forand et al[34] and Hayes and Williams[35] and the calculation of Raeker et al.[36] They use a 'distorted wave Coulomb-Born no exchange' method with the incident electron described by distorted waves while an R-matrix expansion is used for the ejected electron and its interaction with the residual ion. Clearly their maximum cross section is a factor of three lower than the experimental values. A close coupling calculation using plane waves for the fast electron by Rudge[37] gives a cross section mid way between the experimental data and the calculations of Raeker et al. It seems unlikely that the influence of cascade and polarisation effects can account for the difference between theory and experiment. The structure around the maximum of the cross section is likely due to the decay of autoionizing states into this channel. A similar explanation has been used to account for structure in the ionization cross section for argon.[38] Clearly more theoretical work is required to arrive at an adequate description of the cross section.

Gotz et al[39] have measured the angular distribution of the 30.4 nm radiation. They show that the angular distribution is given by

$$I(\theta) \sim 1 + \frac{1}{3\sqrt{2}} A_{20}\left({}^2P\right) P_2(\cos\theta) \qquad (8)$$

where A_{20} is an angle integrated alignment parameter and $P_2(\cos\theta)$ the second order Legendre polynomial. Experimentally, in the energy region between the $He^+(n=2)$ and (n=3) thresholds, they find $A_{20} = -0.92 \pm 0.14$ except in the region of the $He(3\ell, 3\ell')$ doubly excited state energies. This compares with their combined DWBA–R-matrix calculation which gives a value of –0.67 almost independent of energy.

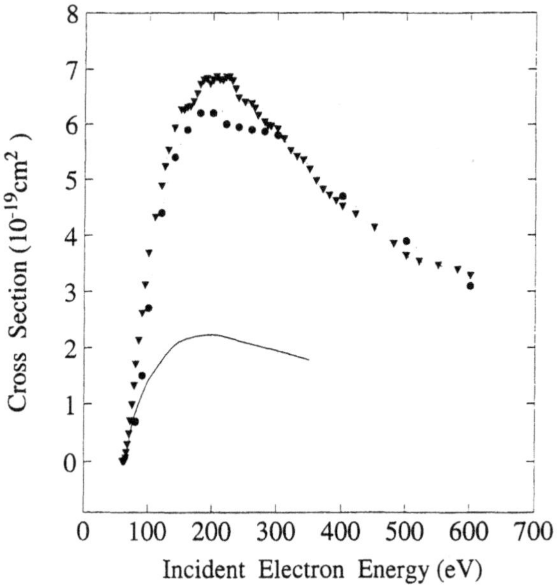

Figure 10. The total cross section for ionization with excitation to the 2p state of the helium ion. *Circles* – Forand et al;[34] *triangles* – Hayes and Williams;[35] *full-line* – theory of Raeker et al.[36]

Very recently attention has been turned both theoretically and experimentally to the study of the process using the electron-photon correlation methods. Schwienhorst et al[40] have developed a general theory and related measurable Stokes parameters and angular correlations to the charge cloud parameters. They also calculate values of the Stokes parameters and charge cloud parameters for an incident electron energy of 200 eV and an ejected electron energy of 1 eV for a scattering angle of 8°. The calculations are based on the R-matrix approach, with the fast scattered electron described by either a plane or distorted wave. Their charge cloud parameters are shown in figure 11.

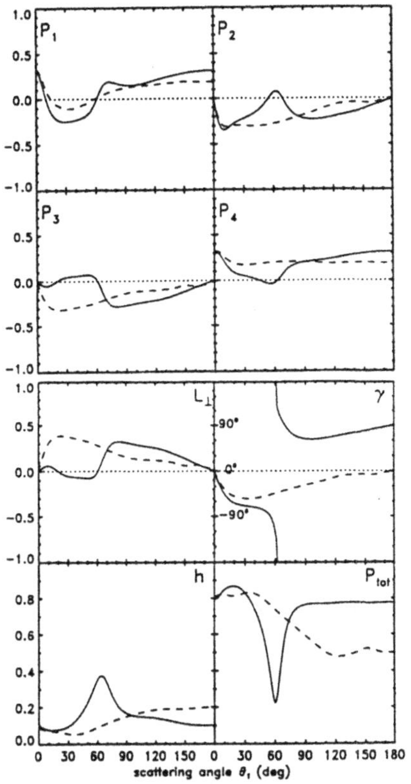

Figure 11. Fine structure depolarised Stokes parameters (top) and charge cloud parameters ($L_\perp$, γ, h) and the degree of reduced (before fine structure depolarisation) polarisation P_{tot} (bottom) of Schwienhorst et al[40] for the He^+(2p) state. The incident electron energy is 200 eV, ejected electron energy is 1 eV and the scattering angle is 8 deg. The *solid line* represents use of a distorted wave to describe the fast electron and the *dashed line*, a plane wave.

Hayes and Williams[35] have recently reported the first experimental electron-photon angular correlation study for He^+(2p). They have measured the angular distribution of 30.4 nm photons in coincidence with scattered electrons having energies of 60 eV and 130 eV, for a scattering angle of 8°. They have analysed their data by expressing the observed coincidence signal N_c in the form

$$N_c = B[1 - A\cos(2(\theta - \gamma))] \tag{9}$$

where B is a normalisation constant, A, a general polarisation, θ is a relative phase and γ the alignment angle of the excited state. Figure 12 shows their values of A and γ as a function of ejected electron energy for the two scattered electron energies. The results are obtained experimentally by fixing conditions as above, and varying the incident electron energy. Within experimental error both parameters remain essentially constant. The value of γ appears consistent with both models of Schwienhorst et al[40] (figure 11).

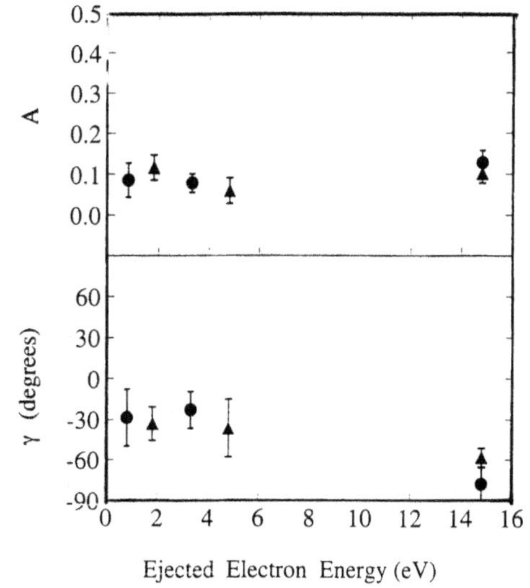

Figure 12. The polarisation A and alignment angle γ for $He^+(2p)$ determined from the angular correlation data of Hayes and Williams.[35] The scattered electrons are scattered through 8° with energies of 60 eV (*triangles*) and 130 eV (*circles*).

Another significant outcome of the work of Hayes and Williams[35] is the role of the doubly excited $3\ell 3\ell'$ states in the $He^+(2p)$ signal. Their double differential cross section for ionization with excitation shows clear interference due to the decay of these states to the $He^+(2p)$ channel.

Double Excitation-Autoionization Processes

Some doubly excited states can decay by photon emission and so in principle could be studied using electron-photon correlation methods. As an example, Westerveld et al[41] have measured the optical excitation function for decay of the $(2p^2)^3P$ state to the $(1s2p)^3P$ singly excited state. However for the n=2 doubly excited states, such transitions lie in the VUV, often close to much stronger lines which can only be resolved using a VUV spectrometer, and have low cross sections. It is unlikely that electron-photon correlations can easily be measured. As a result correlation studies of doubly excited states have only been made on states which undergo autoionization.

The processes of interest are of the type

$$\begin{aligned} e(\mathbf{k}_o) + He\left(1s^2\right)^1 S \rightarrow\ & He(n\ell, n'\ell') + e_s(\mathbf{k}_s) \\ & \downarrow \\ & He^+(n''\ell'') + e_e(\mathbf{k}_e) \end{aligned} \qquad (10)$$

In these cases the (e, 2e) method completely defines the kinematics of the process and provides triple differential cross sections which are a highly sensitive test of theoretical models. So far such studies have been confined to excitation of the $(2s^2)^1S$, $(2s2p)^1P$ and $(2p^2)^1D$ states and their decay to the ground state ion. A major complication in these studies is interference between this channel and the direct ionization channel, i.e.

$$e(\mathbf{k}_o) + He(1s^2)^1S \rightarrow He^+(1s)^2S + e_s(\mathbf{k}_s) + e_e(\mathbf{k}_e) \tag{11}$$

Recent (e, 2e) studies carried out[42–45] yield ejected electron spectra for a specific momentum transfer collision $\mathbf{K} = \mathbf{k}_o - \mathbf{k}_s$. An example of such a spectrum is shown in figure 13 for the $(2s^2)^1S$ state of helium under the kinematic conditions shown. In general, the strong interference seen between the direct and indirect ionization channels is observed over a wide range of conditions. The data can be analysed in terms of either the parameterisations of Fano[46] or of Shore.[47] The Shore parameters have the advantage that they each have units of triple differential cross section (TDCS). The overall TDCS can be written

$$\frac{d^3\sigma}{d\Omega_s d\Omega_e dE} = f(\mathbf{k}_e, \mathbf{K}) + \frac{a(\mathbf{k}_e, \mathbf{K})\varepsilon_r + b(\mathbf{k}_e, \mathbf{K})}{\varepsilon_r^2 + 1} \tag{12}$$

where $\varepsilon_r = (E_e - \overline{E}_r)/\Gamma_r$

E_e and $\overline{E}_r$ are the ejected electron and resonance energies respectively, while Γ_r is the energy full width at half height of the resonance.

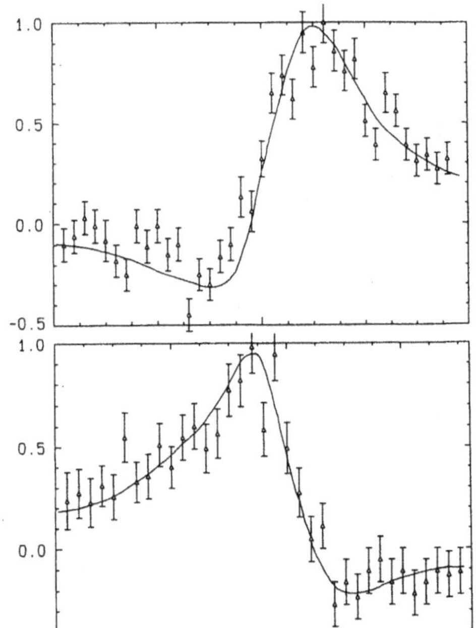

Figure 13. Measured coincidence ejected electron spectra[43] in the region of the $(2s^2)^1S$ state at an incident electron energy of 200 eV and scattering angle $\theta_s = -12°$. Ejected angle $\theta_e = +60°$ (top), $\theta_e = -120°$ (bottom)

$f(\mathbf{k}_e,\mathbf{K})$ is simply the direct ionization cross section in the resonance region.

$a(\mathbf{k}_e,\mathbf{K})$ is an asymmetry parameter originating from interference between the direct and resonance amplitudes in the vicinity of the resonance.

$b(\mathbf{k}_e,\mathbf{K})$ is less straightforward in terms of a simple physical description. Mathematically it consists of two terms. One has a variation with ejected electron angle corresponding to the decay of a pure quasistationary state, i.e. corresponds to the resonance yield over the background direct ionization process. However, the second term can change the form of this angular distribution dramatically, giving either a positive or negative contribution to $b(\mathbf{k}_e,\mathbf{K})$.

The parameters a and b have been determined over a range of kinematic conditions for the $(2s^2)^1S$, $(2s2p)^1P$ and $(2p^2)^1D$ states. Here we concentrate on results at 200 eV incident energy and a scattering angle of −12° (−13°) where the data from this group[43,44] (Flinders[42]) can be compared. Figure 14 shows the a and b parameters for the $(2s^2)^1S$ state under these conditions. In the 'binary' collision region, both experiments show a single positive peak for both a and b close to the momentum transfer direction. In the 'recoil' direction we believe both parameters have a stronger angular dependence than indicated by the Flinders' data. These experiments have led to the development of three independent theoretical models for these processes. Figure 14 also shows the results of calculations by Raeker et al[48] and by Kheifets[49]. Raeker et al again use the distorted wave Coulomb-Born no exchange method, the results of which for simultaneous excitation and ionization were presented earlier. Kheifets used a distorted wave Born approximation with exchange. The results reported here were performed using their distorted wave different field (DWDF) model. This seeks to take account of the fact the scattered electron moves in the field of the $He^+(1s)$ ion for direct ionization and in the field of the excited atom for the resonance excitation. At this energy neither model gives even qualitative agreement with experiment with the exception of the Kheifets' calculation for a.

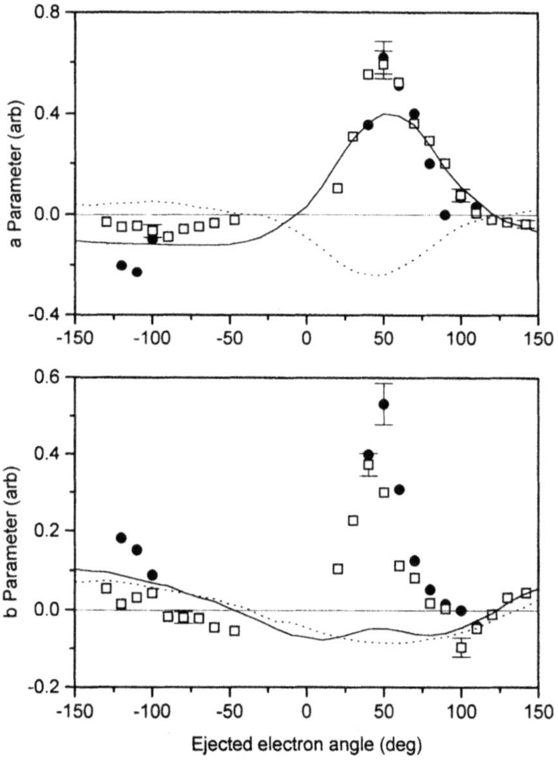

Figure 14. Comparison of experimental and theoretical a and b parameters for the $(2s^2)^1S$ state excited by 200 eV electrons and for a scattering angle of −12 deg[42] (−13 deg)[43]. *Circles* – McDonald and Crowe;[43] *squares* – Lower and Weigold;[42] *solid line* – DWDF theory of Kheifets;[49] *dotted line* – Raeker et al.[48]

For the $(2s2p)^1P$ and $(2p^2)^1D$ states, the experiments show strong oscillatory behaviour as a function of ejected electron angle for both parameters close to the +**K** direction. This is not reproduced by either calculation. McCarthy and Shang[50] have used an equivalent local form of the distorted wave impulse approximation to describe the direct ionization and a six-state momentum coupled channel method for the resonant scattering amplitudes. Their calculated ejected electron spectra show best agreement with experiment under conditions where the direct ionization cross section is small, implying that their model describes the resonant process well.

Recently we have considered the possibility of analysing (e, 2e) experiments in the vicinity of an autoionizing state to obtain parameters relating to the direct and resonance ionization amplitudes. As a first step, the $He(1s^2)^1S - He(2s^2)^1S - He^+(1s)^2S$ double excitation-autoionization process was considered because of its simple symmetries. An angle independent parameter c was defined where

$$[a(\theta_e,\varphi_e)]^2 + \left[b(\theta_e,\varphi_e) - \frac{k}{k_o}|c|^2\right]^2 = 4\frac{k}{k_o} f(\theta_e,\varphi_e)|c|^2 \quad (13)$$

However, the data of McDonald and Crowe[43] and that of Lower and Weigold[42] show a strong dependence of $|c|$ with ejected electron angle. This may be direct evidence for the breakdown of the assumption that the excitation and decay processes can be separated in time.

CONCLUSIONS

It is clear that correlation methods continue to play a major role both in developing a better understanding of the dynamics of various electron-atom collision processes and in the development of accurate theoretical models. This has been illustrated most dramatically for excitation of the 1s3d states in helium. Application of the CCC model to helium over the past two years and its proven ability to predict the measured coherence/correlation parameters over a wide range of kinematic conditions has transformed the accuracy with which single electron excitation cross sections can be predicted. Already a new generation of calculations, RMPS, which builds on the strengths of the CCC method but eliminates its most obvious weakness of treating helium as a one electron system, has been shown to give excellent results for the 2^1P parameters.[26]

In contrast no existing calculations give a good description of experimental data for processes involving both atomic electrons. This is an area now requiring urgent theoretical attention.

ACKNOWLEDGMENTS

Coherence/correlation work in this laboratory is supported by the EPSRC, EU and NATO. I am grateful to D. Cvejanovic, I. Bray, K. Bartschat and J.F. Williams for stimulating discussions in relation to this work and/or data prior to publication.

REFERENCES

1. B. Bederson, *Comments At. Mol. Phys.* 1:41 (1969).
2. B. Bederson, *Comments At. Mol. Phys.* 1:65 (1969).
3. M. Eminyan, K.B. MacAdam, J. Slevin and H. Kleinpoppen, *Phys. Rev. Lett.* 31:576 (1973).
4. M.C. Standage and H. Kleinpoppen, *Phys. Rev. Lett.* 36:577 (1976).
5. N. Andersen, J.W. Gallagher and I.V. Hertel, *Phys. Rep.* 165:1 (1988).
6. J.A. Slevin and S. Chwirot, *J. Phys. B.* 23:165 (1990).

7. A. Crowe, in: *The Physics of Electronic and Atomic Collisions*, T. Andersen et al, eds., A.I.P. Conference Proceedings 295:286 (1993).
8. A. Crowe, in: *Polarized Electron/Polarized Photon Physics*, H. Kleinpoppen and W.R. Newell, eds., Plenum, New York (1995).
9. K. Becker, A. Crowe and J.W. McConkey, *J. Phys. B.* 25:3885 (1992).
10. A. Crowe and I. Bray, in: *Selected Topics on Electron Physics*, D.M. Campbell and H. Kleinpoppen, eds., Plenum, New York (1996).
11. N. Andersen and K. Bartschat, *Adv. At. Mol. Opt. Phys.* (submitted for publication).
12. B.P. Donnelly and A. Crowe, *J. Phys. B.* 21:L637 (1988).
13. J.P.M. Beijers, S.J. Doornenbal, J. van Eck and H.G.M. Heideman, *J. Phys. B.* 20:6617 (1987).
14. A.G. Mikoza, R. Hippler, J.B. Wang and J.F. Williams, *Z. Phys. D.* 30:129 (1994).
15. K. Bartschat and D.H. Madison, *J. Phys. B.* 21:153 (1988).
16. D.C. Cartwright and G. Csanak, *J. Phys. B.* 20:L583 (1987).
17. E.J. Mansky and M.R. Flannery, *J. Phys. B.* 24:L551 (1991).
18. D.V. Fursa and I. Bray, *Phys. Rev. A.* 52:1279 (1995).
19. D.T. McLaughlin, B.P. Donnelly and A. Crowe, *Z. Phys. D.* 29:259 (1994).
20. H. Batelaan, J. van Eck and H.G.M. Heideman, *J. Phys. B.* 24:L397 (1991).
21. I. Bray, D.V. Fursa and I.E. McCarthy, *Phys. Rev. A.* 51:500 (1995).
22. J.B. Wang and J.F. Williams, *Aust. J. Phys.* (in press).
23. A.G. Mikoza, R. Hippler, J.B. Wang and J.F. Williams, *Phys. Rev. A.* 53:3287 (1996).
24. A. Crowe, B.P. Donnelly, D.T. McLaughlin, I. Bray and D.V. Fursa, *J. Phys. B.* 27:L795 (1994).
25. I. Bray, D.V. Fursa, D.T. McLaughlin, B.P. Donnelly and A. Crowe, in: *The Physics of Electronic and Atomic Collisions*, L.J. Dubé et al, eds., A.I.P. Conference Proceedings 360:785 (1995).
26. K. Bartschat, E.T. Hudson, M.P. Scott, P.G. Burke and V.M. Burke, *J. Phys. B.* 29:2875 (1996).
27. R.B. Kay and C.G. Simpson, *J. Phys. B.* 21:625 (1988).
28. G.A. Khayrallah and S.J. Smith, *Phys. Rev. A.* 18:559 (1978).
29. J.B. Wang, J.F. Williams, A.T. Stelbovics, J.E. Furst and D.H. Madison, *Phys. Rev. A.* 52:2885 (1995).
30. D. Cvejanovic and A. Crowe (to be published).
31. G. Stefani, L. Avaldi and R. Camilloni, *J. Phys. B.* 23:L227 (1990).
32. C. Dupré, A. Lahmam-Bennani, A. Duguet, F. Mota-Furtado, P.F. O'Mahony and C. Dal Cappello, *J. Phys. B.* 25:259 (1992).
33. A.S. Kheifets, I. Bray, I.E. McCarthy and B. Shang, *Phys. Rev. A.* 50:4700 (1994).
34. J.L. Forand, K. Becker and J.W. McConkey, *J. Phys. B.* 18:1409 (1985).
35. P.A. Hayes and J.F. Williams (private communication).
36. A. Raeker, K. Bartschat and R.H.G. Reid, *J. Phys. B.* 27:3129 (1994).
37. M.R.H. Rudge, *J. Phys. B.* 21:1887 (1988).
38. A. Crowe, J.A. Preston and J.W. McConkey, *J. Chem. Phys.* 57:1620 (1972).
39. A. Götz, W. Mehlhorn, A. Raeker and K. Bartschat, *J. Phys. B.* (in press).
40. R. Schwienhorst, A. Raeker, K. Bartschat and K. Blum, *J. Phys. B.* 29:2305 (1996).
41. W.B. Westerveld, F.B. Kets, H.G.M. Heideman and J. van Eck, *J. Phys. B.* 12:2575 (1979).
42. J. Lower and E. Weigold, *J. Phys. B.* 23:2819 (1990).
43. D.G. McDonald and A. Crowe, *Z. Phys. D.* 23:371 (1992).
44. D.G. McDonald and A. Crowe, *J. Phys. B.* 26:2887 (1993).
45. O. Samardzic, A.S. Kheifets, E. Weigold, B. Shang and M.J. Brunger, *J. Phys. B.* 28:725 (1995).
46. U. Fano, *Phys. Rev.* 125:1866 (1961).
47. B.W. Shore, *Rev. Mod. Phys.* 39:439 (1967).
48. A. Raeker, R.H.G. Reid, K. Bartschat and R. Schwienhorst, in: *Proc. 19th ICPEAC*, J.B.A. Mitchell et al, eds., 2:508 (1995) and K. Bartschat (private communication).
49. A.S. Kheifets, *J. Phys. B.* 26:2053 (1993).
50. I.E. McCarthy and B. Shang, *Phys. Rev. A.* 47:4807 (1993).
51. V.V. Balashov, S.E. Martin and A. Crowe, *J. Phys. B.* 29:L337 (1996).

IONIZATION OF ATOMS AND IONS BY ELECTRON IMPACT

R. H. G. Reid

Department of Applied Mathematics and Theoretical Physics
The Queen's University of Belfast
Belfast, BT7 1NN, U. K.

INTRODUCTION

We consider some aspects of the theory of the ionization process

$$e^-_{\varsigma_0 \mathbf{k}_0} \quad + \quad X_b \quad \longrightarrow \quad e^-_{\varsigma_1 \mathbf{k}_1} \quad + \quad X^+_\beta \quad + \quad e^-_{\varsigma_2 \mathbf{k}_2} \tag{1}$$

where X_b may be an atom or ion with $N+1$ bound electrons, and b and β are state-labels. The free electrons have spins, wave-vectors and energies ς_i, $\mathbf{k}_i$ and E_i $(i = 0, 1, 2)$, so the energy balance in (1) is:

$$E_0 \quad = \quad E_1 \quad + \quad E_2 \quad + \quad I_{\beta b} \tag{2}$$

where $I_{\beta b}$ is the energy of X^+_β relative to X_b (*i.e.* the "ionization potential").

The simplest quantity associated with process (1) is ***the total cross section***, $\sigma_{\beta b}(E_0)$, where the values of $\{\varsigma_i\}$, $\mathbf{k}_1$ and $\mathbf{k}_2$ are not measured. Data on total cross sections are much in demand for applications such as plasma modelling. The challenge for theory here is to provide reliable computational methods which will work for complicated species X, including when b is an excited state.

Process (1) can also be studied as a means of enhancing our understanding of fundamental physics. Here the quantities of interest include: ***the triple differential cross section***, *i.e.* differential with respect to $\hat{\mathbf{k}}_1$, $\hat{\mathbf{k}}_2$ and E_2 or the energy-loss $\Delta E \equiv E_0 - E_1$; ***spin asymmetry***, when spin-alignment of the incident electron and the target are observed; ***threshold behaviour***, when E_1 and E_2 are both small; ***electron–photon coincidences***, when X^+_β is an excited state that subsequently emits a photon which is observed in coincidence with one of the free electrons. There have been some elegant measurements of these detailed properties, and these provide severe challenges to theory.

It is doubtful if a complete theory of ionization exists for general targets X_b for all energies E_0, E_1 and E_2. The problem lies in the final state comprised of two

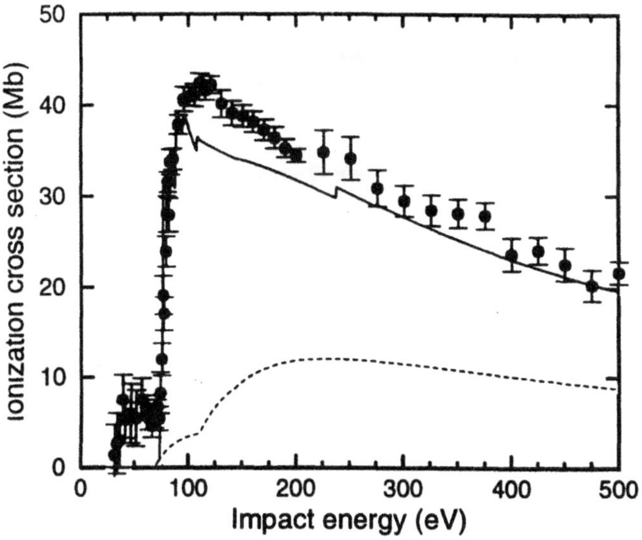

Figure 1. Total cross section for electron-impact ionization of Mo^{5+}. Solid circles: experimental results. Dashed curve: CADW calculation, direct ionization only. Solid curve: CADW calculation, including excitation-autoionization. (From Bannister *et al*[1]).

continuum electrons and N bound electrons. Most theories try to describe those aspects of the physics deemed to be most important for the particular situation in hand. Among the important physics that might be included are: the detailed interaction between the continuum electrons and the initial and final ions, including the effects of this interaction on the ions; exchange effects between the two continuum electrons in the final state; detailed correlated motion of these continuum electrons.

This review will not attempt to cover all aspects of the theory of process (1). Rather, we shall look at some aspects of theory where accurate treatments of *one-continuum-electron problem* have played a prominent part.

TOTAL CROSS SECTIONS

Total cross sections are required for the wide range of ions found in laboratory and astrophysical plasmas. Figure 1 shows the crossed-beam experimental measurements by Bannister *et al*[1] for ionization of Mo^{5+}. This observed cross section is the sum over all the energetically-accessible single-ionization channels β, and is principally from the ground state $Mo^{5+}(4s^2 4p^6 4d)$. However, the non-zero cross section below the threshold at 68.5 eV for producing $Mo^{6+}(4s^2 4p^6)$ indicates the presence of metastable ions in the beam.

Based on the independent-particle model of electronic structure, theorists distinguish the *direct ionization* mechanism. In this, an electron of X_b is excited from a bound orbital to a continuum orbital, with the remaining N electrons playing a passive role. The excitation is caused by the interaction V between the incident electron and the electrons of X. For example, in the $Mo^{5+}(4s^2 4p^6 4d)$ case, the production of $Mo^{6+}(4s^2 4p^6)$ would be due to the removal of the 4d electron. The formulation due to Younger[2], based on the work of Rudge and Schwartz[3], probably represents the direct mechanism to sufficient accuracy. In this, the free electrons are described by distorted Coulomb waves $\{\varphi_{\varsigma \mathbf{k}}\}$, so that the scattering amplitude has the form

$$f = -(2\pi)^{-5/2} \left\langle \Phi_\beta(1 \cdots N)\, \varphi_{\varsigma_2 \mathbf{k}_2}(N+1)\, \varphi_{\varsigma_1 \mathbf{k}_1}(0) \right| V \left| \varphi_{\varsigma_0 \mathbf{k}_0}(0)\, \Psi_b(1 \cdots N+1) \right\rangle \tag{3}$$

where Ψ_b and Φ_β are the wave-functions for the initial and final bound systems. The terms of the bound systems are not distinguished, so the method is known as the *configuration-averaged distorted-wave* (CADW) method. In (3), the continuum functions may be decomposed into partial waves and the interaction into multipoles, so that f is expressed as a sum of radial integrals. Because the two continuum electrons in the final state are treated similarly, no added difficulty arises in the evaluation of the exchange amplitude g derived from f by interchanging these electrons. There remains uncertainty about the relative phases of f and g — Younger adopted the 'maximum interference' approximation. Also, the residual charges seen by the free electrons in the final state is debatable. Finally, to avoid orthogonality problems, the same bound orbitals must be used in Ψ_b and Φ_β, and care must be taken to ensure that the distorted wave of the ejected electron is orthogonal to the bound orbitals of Ψ_b.

The cross section for direct ionization σ^{dir} calculated by Younger's CADW method is shown by the dashed curve in figure 1. The calculation has included direct ionization to the final-ion states $Mo^{6+}(4s^2 4p^6)$, $Mo^{6+}(4s^2 4p^5 4d)$ and $Mo^{6+}(4s4p^6 4d)$, the latter two having ionization potentials 109.8 eV and 142.8 eV, respectively. We see, unequivocally, that direct ionization does not explain the observed cross sections.

An additional mechanism that must be added to the direct mechanism is *excitation-autoionization* (EA). In this, the initial $(N+1)$-electron system is excited to a quasi-bound state X_q lying above the ionization threshold $I_{\beta b}$ — see figure 2.

$$e^-_{\varsigma_0 \mathbf{k}_0} \quad + \quad X_b \quad \longrightarrow \quad e^-_{\varsigma_1 \mathbf{k}_1} \quad + \quad X_q \,. \tag{4}$$

X_q then autoionizes:

$$X_q \quad \longrightarrow \quad X^+_\beta \quad + \quad e^-_{\varsigma_2 \mathbf{k}_2} \,. \tag{5}$$

The quasi-bound state may belong to a Rydberg series: $X_q \equiv [X^+_{\beta'}]n\ell$, where $X^+_{\beta'}$ lies above X^+_β, although the lower quasi-bound states do not have this simple form. If the spin-averaged total cross section for excitation (4) is σ^{ex}_{qb}, then the ionization cross section is given by

$$\sigma_{\beta b}(E_0) \quad = \quad \sigma^{dir}_{\beta b}(E_0) \quad + \quad \sum_q B^a_{\beta q}\, \sigma^{ex}_{qb}(E_0) \tag{6}$$

where $B^a_{\beta q}$ is the branching ratio for decay by autoionization to X^+_β as against radiative decay to a stable state $X_{b'}$. For the dominant, low-lying autoionizing states, $B^a_{\beta q}$ may be taken to be unity, but for higher members of a Rydberg series the autoionizing width decreases as n^{-3} and so radiative stablization becomes relatively more important.

In formula (6), the interference between the direct and EA amplitudes is neglected. This is probably not a major source of error for total cross sections, because, as pointed out by Moores and Reed[4], if the autoionizing resonance in the

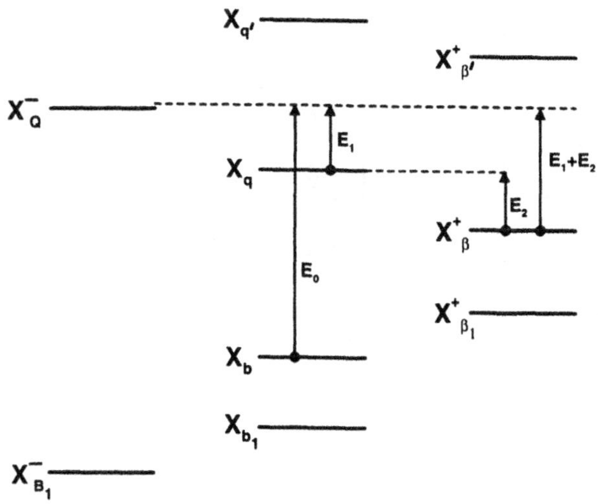

Figure 2. Schematic energy levels of the ions X^-, X and X^+. Bound states are labelled by B, b and β, with ground states B_1, b_1 and β_1. Quasi-bound stated are labelled by Q and q. Vertical arrows denote the energies of the free electrons.

single differential cross section (as a function of energy-loss ΔE) is isolated and narrow, then the interference term is an odd function about the resonance energy and hence makes no contribution to the integral with respect to ΔE that gives the total cross section.

In calculations based on (6), the states X_q are treated as genuinely bound states as far as the calculation of the excitation process (4) is concerned.

The extension of Younger's CADW formulation to include EA was made by Pindzola, Griffin and Bottcher[5]. They represented the quasi-bound states X_q in terms of orbitals similar to those used for the bound state X_b, and calculated the excitation cross section on the same basis as the direct ionization cross section. Also, they neglected radiative stablization. The result of applying their method to Mo^{5+} is shown by the solid curve in figure 1. Ten quasi-bound states were included, the three lowest of which are $Mo^{5+}(4s4p^6 4d^2)$, $Mo^{5+}(4s^2 4p^5 4d4f)$ and $Mo^{5+}(4s^2 4p^5 4d5d)$, with threshold energies 73.1 eV, 75.2 eV and 76.1 eV and excitation cross sections (at threshold) 9.2 Mb, 14.4 Mb and 7.3 Mb, respectively. These three states alone give a rapid increase by 31 Mb in the ionization cross section just above its threshold at 68.5 eV. The inclusion of the EA mechanism dramatically improves the agreement between theory and experiment. In the region of the maximum cross section, the EA contribution is about ten times the direct contribution.

Since EA can be the dominant ionization mechanism, it is important to use accurate methods to calculate the electron-impact excitation process (4). An early example of the use of close-coupled calculations of excitation in the context of ionization is the work of Henry[6]. Examples of the use of R-matrix calculations of excitation are the work of Henry and co-workers[7] and Burke and co-workers[8] — see also the review by Henry and Kingston[9].

An important aspect of the accurate R-matrix treatment of excitation is that it incorporates coherently the *resonant-excitation-double-autoionization* (REDA)

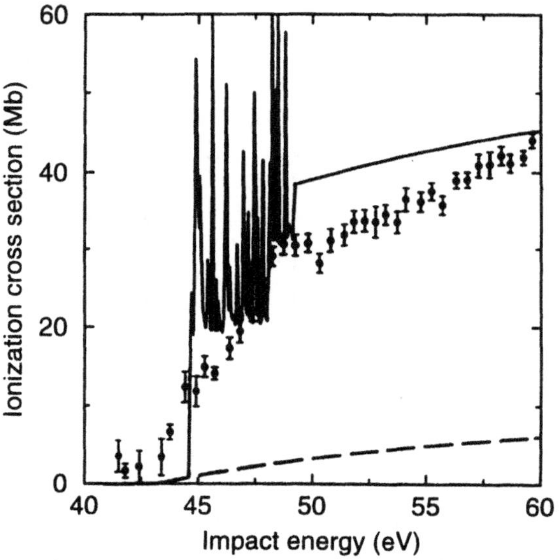

Figure 3. Total cross section for electron-impact ionization of Ti^{3+}. Solid circles: experiment[11,12]. Dashed curve: theory, direct ionization only. Solid curve: theory, including R-matrix calculation of EA and REDA. (From Gorczyca *et al*[13]).

mechanism[10]. According to this, the excitation of the state X_q is enhanced by resonant excitation of (or, more correctly, resonant capture into) a quasi-bound state X_Q^- of the $(N+2)$-electron system lying above X_q (see figure 2):

$$e^-_{\varsigma_0 \mathbf{k}_0} \quad + \quad X_b \quad \longrightarrow \quad X_Q^- \, . \tag{7}$$

X_Q^- then double-autoionizes in two steps:

$$X_Q^- \quad \longrightarrow \quad e^-_{\varsigma_1 \mathbf{k}_1} \quad + \quad X_q \tag{8}$$

followed by the decay of X_q as in (5). Typically, $X_Q^- \equiv \left[X_{q'}\right] n\ell$, where $X_{q'}$ is another quasi-bound state lying above X_q. In the R-matrix treatment of $e^- + X$ collisions, a set of quasi-bound states $\{X_Q^-\}$ shows up automatically by virtue of the accurate treatment of the X^- continuum.

(Another possible decay path of X_Q^- is *auto-double-ionization*, where steps (8) and (5) are combined into a single step. This is the READI mechanism. If the core of X_Q^- were the lowest quasi-bound state above X_β^+ then READI would be the only decay path from X_Q^- to X_β^+.)

As an illustration of the use of R-matrix calculations of excitation, figure 3 shows results for the ionization of $Ti^{3+}(3p^6 3d\,^2D)$. The experimental results are by Falk *et al*[11,12], and the theoretical results by Gorczyca *et al*[13]. The direct ionization contribution was calculated by Younger's CADW method[2], and, as in the example of Mo^{5+}, is completely inadequate to account for the observations. The EA contribution was determined by calculating the excitation cross section by the R-matrix method, using the branching ratios determined by Griffin *et al*[14]. Their results are close to those from the earlier calculation by Burke *et al*[8]. The threshold for ionization to $Ti^{4+}(3p^6\,^1S)$ is 43.3 eV, while ionization with excitation to $Ti^{4+}(3p^5 3d\,^2P^o)$ has a threshold at 77.4 eV. The EA contribution comes from the $^2F^o$, $^2P^o$ and $^2D^o$

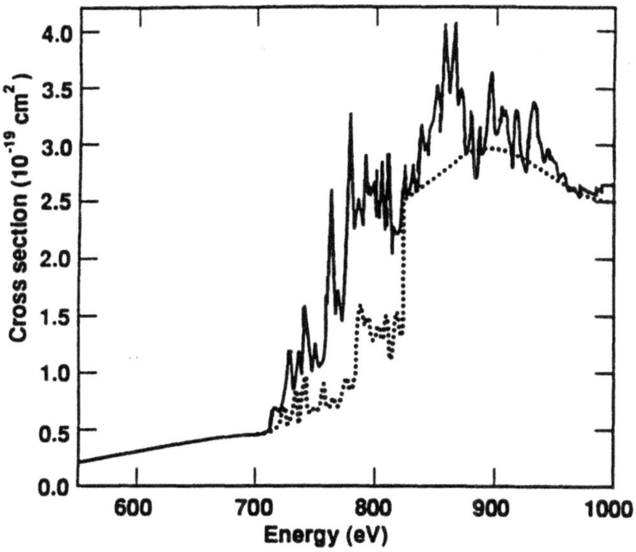

Figure 4. Total cross section for electron-impact ionization of Fe^{15+}. Solid curve: theory, with MCDF calculation of REDA [15]. Dashed curve: theory, with R-matrix calculation of REDA [16]. (From Chen *et al* [15]).

terms of $Ti^{3+}(3p^6 3d)$ which lie above the ionization threshold — the remaining sixteen terms lie below the threshold. A major feature of the excitation calculation is the large amount of configuration-interaction in these terms. The main point to note is the appearance of REDA resonances below the thresholds for excitation of the $^2P^o$ and $^2D^o$ autoionizing states. These resonances are due to quasi-bound states of Ti^{2+} formed by the $(N+2)$-th electron binding to these Ti^{3+} cores. The calculated resonances in EA are too narrow to be resolved experimentally, and, for the sake of comparison with experiment, the theoretical results may be averaged over a distribution in E_0 [13].

Above we have noted that an advantage of using the R-matrix method to calculate the cross section for excitation of $\{X_q\}$ is that series of quasi-bound states $\{X_Q^-\}$ are generated automatically and incorporated coherently in the excitation process. However, the X_Q^- that are generated are limited to those that can be formed in the basis $\{[X_{b'}]\, n\ell\}$, where $\{X_{b'}\}$ is the set of input target states, which is in turn restricted by computational practicalities. It can happen that the automatically-generated set $\{X_Q^-\}$ omits some important states. This is illustrated in figure 4 which shows calculated cross sections for ionization of $Fe^{15+}(1s^2 2s^2 2p^6 3s)$, which was the topic in the pioneering work of LaGutta and Hahn [10]. The solid curve in the figure shows the results of Chen *et al* [15], while the dashed curve shows R-matrix results by Tayal and Henry [16]. The main point to note is the discrepancy by a factor of about two in the energy-range $750-800$ eV. In the work of Chen *et al*, the REDA process (7) was inserted as an independent process, with cross section σ_{Qb}^{cap}, so that the ionization cross section is

$$\sigma_{\beta b}(E_0) \quad = \quad \sigma_{\beta b}^{dir}(E_0) \quad + \quad \sum_q B_{\beta q}^{a}\, \sigma_{qb}^{ex}(E_0) \quad + \quad \sum_Q B_{\beta Q}^{da}\, \sigma_{Qb}^{cap}(E_0) \qquad (9)$$

where $B_{\beta Q}^{da}$ is the branching ratio for double-autoionization to X_β^+. The set $\{X_Q^-\}$ con-

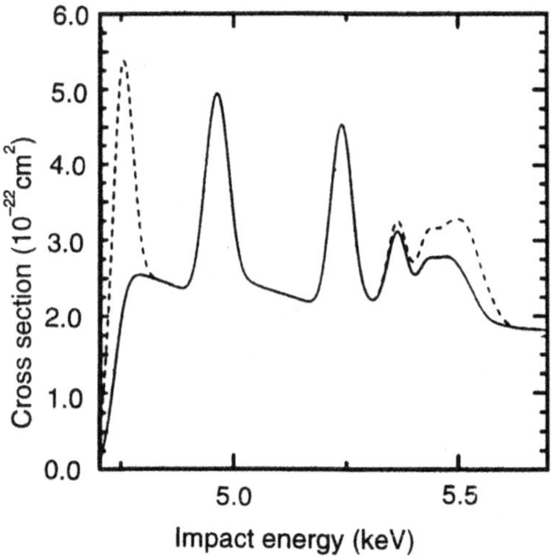

Figure 5. Cross section for $Ti^{20+}\,1s^2\,{}^1S_0 \rightarrow 1s2p\,{}^3P_1$ excitation by electron impact (convoluted with a 50 eV FWHM Gaussian). Solid curve: without radiative damping. Dashed curve: with radiative damping. (From Gorczyca *et al*[18]).

sisted of some 10000 multiconfiguration Dirac–Fock states of Fe^{14+}, with configurations $1s^22s^22p^53s\,3\ell\,n\ell'$, $1s^22s^22p^53s\,4\ell\,n\ell'$ or $1s^22s\,2p^63s\,3\ell\,n\ell'$. The branching ratios $B^{da}_{\beta Q}$ were calculated from the autoionizing and radiative rates computed within the MCDF model, and the capture cross sections σ^{cap}_{Qb} were evaluated from the autoionizing rates by detailed balance. It was found that important contributions to REDA arose from the configurations $Fe^{14+}(1s^22s^22p^53s\,4\ell\,4\ell')$ and $Fe^{14+}(1s^22s\,2p^63s\,3\ell\,n\ell')$. By contrast, the R-matrix calculation by Tayal and Henry[16] included only the $Fe^{14+}(1s^22s^22p^53s\,3\ell\,n\ell')$ configurations, by virtue of the set of Fe^{15+} states included as input. The omission of the configurations that Chen *et al*[15] found to be significant accounts for the shortfall in the R-matrix cross section.

Another short-coming of the R-matrix method, arising from the implicit treatment of the autoionizing states X^-_Q, is the failure to allow for branching in the decay of X^-_Q — equivalent to setting $B^{da}_{\beta Q}$ to unity in (9). This can be overcome by incorporating *radiative damping* in the R-matrix calculation. This has been done by Zhang and Pradhan[17] for excitation of B-like Fe^{21+}, and by Gorczyca *et al*[18] for excitation of He-like Ti^{20+}. In the former the radiative damping was treated by the theory of Davies and Seaton[20], while in the latter an optical-potential method was used. The effects of radiative damping can be significant, as illustrated in figure 5. This shows the results of Gorczyca *et al*[18] for $1s^2\,{}^1S_0 \rightarrow 1s2p\,{}^3P_1$ excitation in Ti^{20+}. The lowest resonant contribution, due to quasi-bound states $Ti^{19+}(1s\,2s\,n\ell')$, is completely removed by radiative damping. On the other hand, for the other resonant contributions which are from states $Ti^{19+}(1s\,3\ell\,n\ell')$, the effect of radiative damping is significant only for $n > 4$.

THE DWBA–R-MATRIX METHOD

As our second example of the use of an accurate treatment of the one-continuum-electron problem in a description of ionization we now consider the DWBA–R-matrix formulation of ionization given by Bartschat and Burke[20]. This is the R-matrix version of the formulation given by Jakubowicz and Moores[21]. The accurate treatment is applied to the final continuum state comprised of the the final ion X_β^+ and the slower of the outgoing electrons ('the ejected electron', with energy E_2). The incident electron and the faster of the outgoing electrons ('the scattered electron', with energy E_1) are described by distorted waves, as in Younger's CADW method[2]. The scattering amplitude (without exchange) thus has the form

$$f = -(2\pi)^{-5/2} \left\langle \Psi^{(-)}_{\beta\varsigma_2\mathbf{k}_2}(1\cdots N{+}1)\, \varphi_{\varsigma_1\mathbf{k}_1}(0) \right| V \left| \varphi_{\varsigma_0\mathbf{k}_0}(0)\, \Psi_b(1\cdots N{+}1) \right\rangle \tag{10}$$

where $\Psi^{(-)}_{\beta\varsigma_2\mathbf{k}_2}$ is the scattering state in which an electron with spin ς_2 and wave-vector $\mathbf{k}_2$ is incident on X_β^+, with incoming scattered waves. Comparing (10) with (3), the accurate scattering function $\Psi^{(-)}_{\beta\varsigma_2\mathbf{k}_2}$ has replaced $\Phi_\beta\, \varphi_{\varsigma_2\mathbf{k}_2}$. In (10), the integral over the coordinates of the scattered electron, $\langle \varphi_{\varsigma_1\mathbf{k}_1} | V | \varphi_{\varsigma_0\mathbf{k}_0} \rangle$, is the effective interaction that causes the bound-to-continuum transition in the $(N+1)$-electron system. As in the CADW formulation, $\langle \varphi_{\varsigma_1\mathbf{k}_1} | V | \varphi_{\varsigma_0\mathbf{k}_0} \rangle$ can be written as a linear combination of operators $\{V_\mu^{(\lambda)}(\ell_1 k_1\, \ell_0 k_0)\}$, where $V_\mu^{(\lambda)}(\ell_1 k_1\, \ell_0 k_0)$ is proportional to the overlap of the 2^λ-pole component of V between the ℓ_0th and ℓ_1th partial waves of the incident and scattered electrons. Also, in the usual way, $\Psi^{(-)}_{\beta\varsigma_2\mathbf{k}_2}$ may be expressed as a linear combination of $\{\psi^{(-)}_{\beta\ell_2 k_2 c}\}$, where $\psi^{(-)}_{\beta\ell_2 k_2 c}$ is the scattering state in which an ℓ_2th partial wave is incident on X_β^+, and which (like Ψ_b) is an eigenstate of angular momentum and parity of the $(N+1)$-electron system, with quantum numbers denoted by c. The scattering amplitude (10) is thus a linear combination of reduced matrix elements

$$\left\langle \psi^{(-)}_{\beta\ell_2 k_2 c} \right\| V^{(\lambda)}(\ell_1 k_1\, \ell_0 k_0) \left\| \Psi_b \right\rangle \ . \tag{11}$$

In the R-matrix formulation, both the initial bound state Ψ_b and the final continuum state $\psi^{(-)}_{\beta\ell_2 k_2 c}$ are expanded in an internal region in terms of the R-matrix basis-fuctions; hence the required quantities are the reduced matrix elements of $V^{(\lambda)}(\ell_1 k_1\, \ell_0 k_0)$ in that basis. Those basis-functions are independent of E_2, and so the major task of evaluating the matrix elements need be performed only once. By contrast, to evaluate (11) in the formulation by Jakubowicz and Moores[21], the function $\psi^{(-)}_{\beta\ell_2 k_2 c}$ is determined by solving coupled integro-differential equations for each value of E_2. Hence the R-matrix formulation is computationally more efficient. Even so, it is computationally intensive because of the large number of operators $\{V^{(\lambda)}(\ell_1 k_1\, \ell_0 k_0)\}$. For example, including multipole components with $\lambda \leq 4$ and partial waves with $\ell_{0,1} \leq 50$ gives 732 operators for each pair of (k_0, k_1) values. In the calculation of total cross sections the rapid variation of $\psi^{(-)}_{\beta\ell_2 k_2 c}$ in the neighbourhood of resonances makes it necessary to consider many thousands of E_2-values. In principle this implies that $V^{(\lambda)}(\ell_1 k_1\, \ell_0 k_0)$ is required at as many E_1-values, since k_1 and k_2 are related through (2). In practice the approximation is made of evaluating $V^{(\lambda)}(\ell_1 k_1\, \ell_0 k_0)$ only at a few 'key' values of k_1.

The major virtue of the DWBA–R-matrix formulation is that the EA mechanism is included implicitly through the accurate treatment of the $(e^- + X^+)$ continuum states. Not only is a set of quasi-bound states $\{X_q\}$ built in to the final continuum states, but the EA process is treated coherently with the direct process, in contrast to the independent-processes picture of formula (6). Hence the DWBA–R-matrix formulation is suitable for studying the effects of the quasi-bound states $\{X_q\}$ on the differential cross sections.

A second virtue of the DWBA–R-matrix method is that the wave-function Ψ_b for the initial bound state is consistent with the final continuum wave-functions $\psi^{(-)}_{\beta\ell_2 k_2 c}$, since they are both represented by R-matrix basis functions in the inner region. At the same time Ψ_b is usually very accurate because of the large basis from which it is constructed. In other formulations, orthogonality problems can arise when 'the best' Ψ_b and 'the best' Φ_β are used.

The DWBA–R-matrix method has shortcomings. Firstly, the distorted-wave treatment of the incident electron means that there is little prospect of representing the REDA mechanism, since this requires a detailed treatment of the mutual interaction between the incident electron and the target ion X. Secondly, the complete dissimilarity in the treatments of the two continuum electrons in the final state makes the evaluation of the exchange amplitude very complicated. So far, in applications of the DWBA–R-matrix method, the exchange amplitude has been ignored. Together, these shortcomings suggest that the method should be more accurate when the incident energy E_0 is well above the ionization threshold and the energy-loss ΔE is small, so that the energy of the scattered electron (E_1) is considerably larger than that of the ejected electron (E_2). Unfortunately, in the calculation of total cross sections these conditions cannot always be met — a point to which we shall return. As a final shortcoming, we repeat the points made in connection with the implicit incorporation of the double-autoionizing states $\{X_Q^-\}$ *via* an R-matrix treatment of the $(e^- + X)$ continuum. Namely, the set of quasi-bound states $\{X_q\}$ that are incorporated is controlled solely by the selection of the input set $\{X_\beta^+\}$, and there is no allowance for radiative stablization of X_q (*i.e.* effectively, $B^a_{\beta q} = 1$). To remedy the latter, radiative damping should be incorporated in the R-matrix calculation.

As an example of the calculation of differential cross sections by the DWBA–R-matrix method, we show in figure 6 the single differential cross section for ionization of $Ar^{9+}(2s^2 2p^5\,{}^2P^o)$ calculated by Laghdas *et al*[22]. In terms of the reduced matrix elements (11) the single differential cross section is given by

$$\frac{d\sigma^c_{\beta b}}{dE}(E_2, E_0) = \frac{16}{E_0 \varpi_b} \sum_{\ell_0 \ell_1 \ell_2 \lambda} \left| \left\langle \psi^{(-)}_{\beta \ell_2 k_2 c} \middle\| V^{(\lambda)}(\ell_1 k_1\, \ell_0 k_0) \middle\| \Psi_b \right\rangle \right|^2 \tag{12}$$

where ϖ_b is the weight of the initial state. We have shown the cross section as a function of E_2 and we have written d/dE for $d/d\Delta E$ or d/dE_2. Expression (12) is the partial cross section that arises *via* continuum symmetry c. The observable cross section would be summed over c, but the autoionizing resonances are best viewed for specific c. The single differential cross sections shown in figure 6 have been summed over the final state β. Five Ar^{10+} states were included in the calculation, namely, $(2s^2 2p^4)\ {}^3P$, ${}^1D$, ${}^1S$ and $(2s\,2p^5)\ {}^3P^o$, ${}^1P^o$ states, with ionization potentials between 17.6 au and 20.9 au. Hence the autoionizing states of Ar^{9+} that have been included automatically by the R-matrix treatment all lie in that narrow energy band. The figure shows the detailed resonant structure due to these states.

The component of the total cross section arising from continuum symmetry c is given by

$$\sigma^c_{\beta b}(E_0) = \int_0^{\overline{E}/2} \frac{d\sigma^c_{\beta b}}{dE}(E_2, E_0)\, dE_2 \tag{13}$$

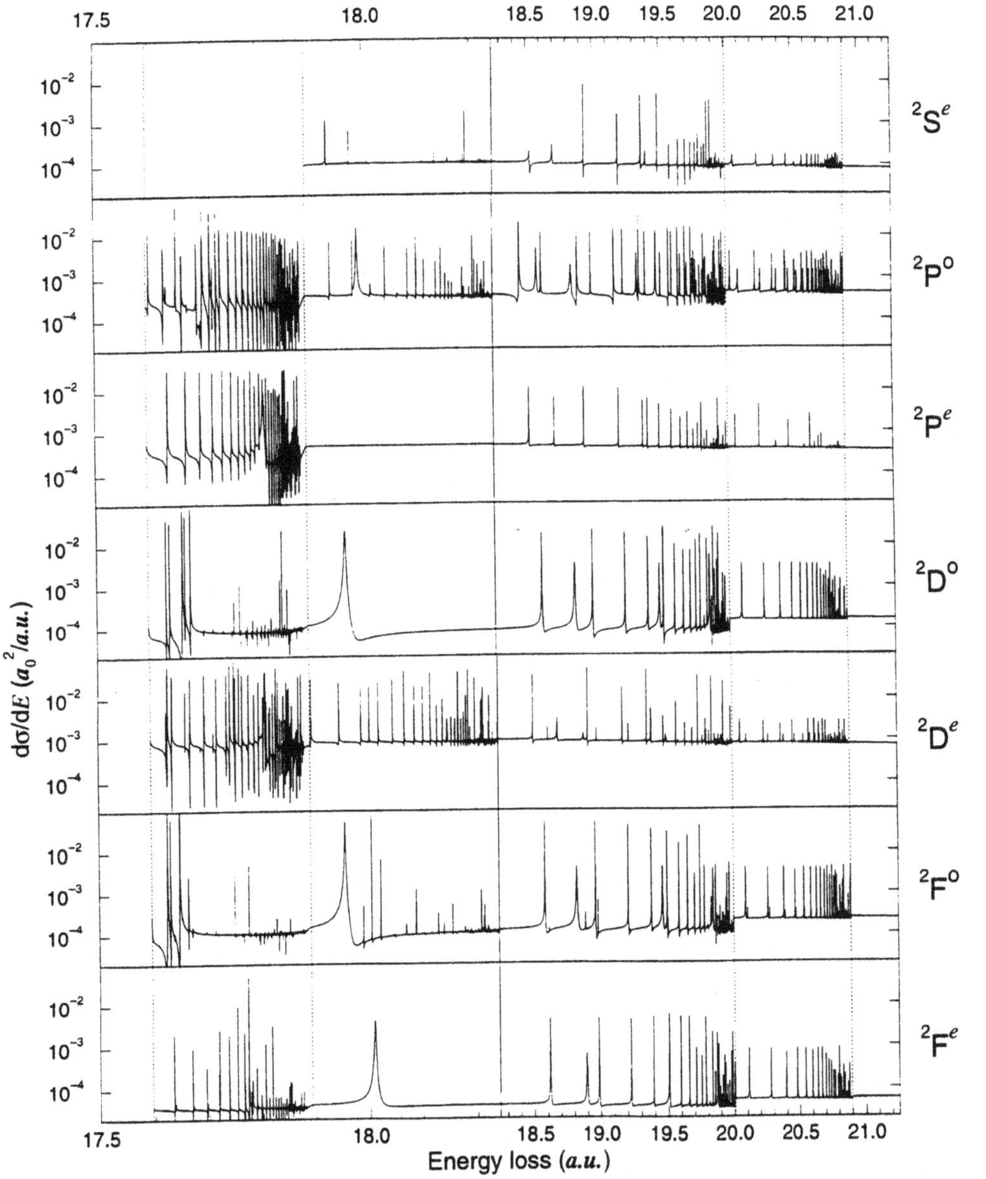

Figure 6. Contributions to the single differential cross section for ionization of Ar^{9+} by electrons with an energy of 40 au. Each contribution is from a particular continuum symmetry of Ar^{9+}, shown at the right margin, and is summed over the final states of Ar^{10+}. The vertical lines are the thresholds for production of each final state of Ar^{10+}, namely, the $(2s^2 2p^4)$ ^3P, ^1D, ^1S and $(2s\,2p^5)$ $^3P^o$, $^1P^o$ states, respectively. Note that the energy scale changes by a factor of four at the $(2s^2 2p^4)$ ^1S threshold. (From Laghdas *et al*[22]).

where

$$\overline{E} \equiv E_1 + E_2 = E_0 - I_{\beta b} . \tag{14}$$

(The observable total cross section is $\sigma_{\beta b} = \sum_c \sigma^c_{\beta b}$). In (13) the upper limit of ejected-electron energy E_2 is taken to be half of the maximum possible value $\overline{E}$, because the case in which $E_2 = x$ and $E_1 = \overline{E} - x$ is physically identical to the case in which $E_2 = \overline{E} - x$ and $E_1 = x$. However this requires that electron-exchange be taken into account in the single differential cross section. If exchange has been ignored, then a threshold-error occurs in (13) for the following reason. If X_q has energy $\mathcal{E}_{qb}$ relative to X_b, then X_q will contribute to ionization if $E_0 \geq \mathcal{E}_{qb}$ (see figure 2 and equation (4)), and the resonance is seen in the non-exchange single differential cross section for $E_1 \approx E_0 - \mathcal{E}_{qb}$. Thus at thresold the resonance appears at $E_1 \approx 0$. On the other hand, the integration range in (13) corresponds to $\overline{E}/2 \leq E_1 \leq \overline{E}$. This implies that, when (13) is evaluated without exchange, the contribution from X_q occurs for $E_0 \geq 2\mathcal{E}_{qb} - I_{\beta b}$, rather than for $E_0 \geq \mathcal{E}_{qb}$. The shift in the threshold, $\mathcal{E}_{qb} - I_{\beta b}$, is the energy of X_q relative to X^+_β. If exchange is taken into account, the single differential cross section is symmetric about $\overline{E}/2$, and the resonance occurs for $E_2 \approx E_0 - \mathcal{E}_{qb}$ through the exchange amplitude.

In the case of Ar^{9+}, the quasi-bound states lie close to the ionization threshold, so that the threshold-error in (13) affects the total cross sections only for a narrow band above threshold. The situation is different when a quasi-bound state has a high excitation energy $\mathcal{E}_{qb}$. We illustrate this by the case of $Ne^{6+}(1s^2 2s^2\,{}^1S)$ where a major contribution to ionization comes from the quasi-bound state $Ne^{6+}(1s\,2s^2 2p\,{}^1P^o)$ whose excitation energy is about four times the ionization potential.

Figure 7 shows the single differential cross section (without exchange) for producing $Ne^{7+}(1s^2 2p\,{}^2P^o)$ calculated by Laghdas *et al*[23] using the DWBA–R-matrix method. Eight Ne^{7+} states were included in the calculation, namely, $1s^2 2s\,{}^2S$, $1s^2 2p\,{}^2P^o$, plus the six $1s\,2\ell\,2\ell'$ doublet states. For the case in figure 7, $I_{\beta b} = 8.2$ au and $E_0 = 35.0$ au. Hence $\overline{E} = 26.8$ au. The figure shows the large resonant contribution due to $Ne^{6+}(1s\,2s^2 2p\,{}^1P^o)$ at energy-loss $\Delta E = 32.9$ au, *i.e.* $E_2 = 24.7$ au. The upper limit in integral (13) is the half-range point $E_2 = \overline{E}/2 = 13.4$ au, and hence the resonance does not contribute to the integral.

Laghdas *et al*[23] have proposed a 'patch' of the threshold-error that is consistent with the level of approximation that led to (13). The integrand in (13), $y(E_2)$, say, is the single differential cross section calculated from the direct amplitude f alone. Then $y(\overline{E} - E_2)$ is the single differential cross section calculated from the exchange amplitude g alone. This is shown by the dashed line in figure 7 for $0 \leq E_2 \leq \overline{E}/2$. In the presence of a strong resonance, as in figure 7, either the direct term or else the exchange term is dominant, and Laghdas *et al* thus proposed that the integrand in (13) should be $\max[y(E_2), y(\overline{E} - E_2)]$. With this approximation, the resonance contributes to the integral at the correct E_0-threshold.

Figure 8 shows the total cross section for ionization of $Ne^{6+}(1s^2 2s^2\,{}^1S)$ calculated by Laghdas *et al*[23]. The theory agrees well with the experimental results by Duponchelle and Defrance[24]. In particular, the location and size of the EA contribution is accounted for satisfactorily.

Resonances in single differential cross sections have been discussed also by Moores and Reed[25] in their study of ionization of Se^{24+} using the DWBA–close-coupled formulation of Jakubowicz and Moores[25]. In contast with the DWBA–R-matrix calculations described above, these authors allowed for radiative decay of the autoionizing states. They did this by identifying, for each prominent resonance

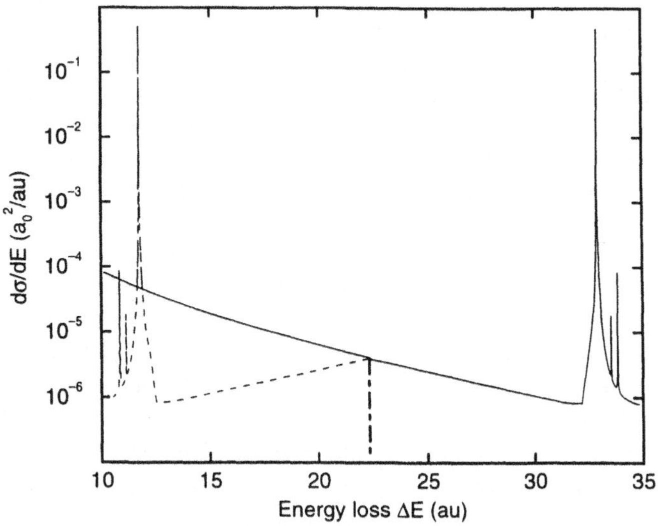

Figure 7. Single differential cross section for ionization of $Ne^{6+}(1s^2 2s^2\ ^1S)$ producing $Ne^{7+}(1s^2 2p\ ^2P^o)$ at impact energy $E_0 = 35$ au. Solid curve: non-exchange term calculated by the DWBA–R-matrix method. Dashed curve: exchange-only term, obtained by reflecting the non-exchange term about the half-range point. (From Laghdas *et al*[23]).

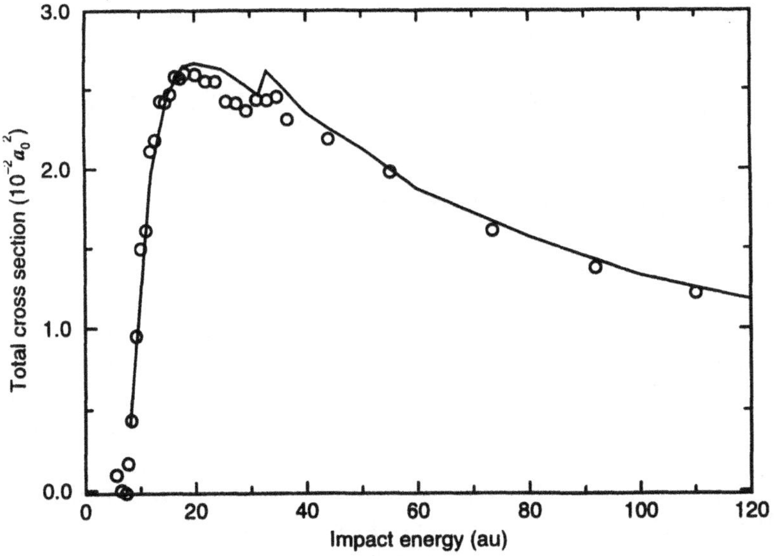

Figure 8. Total cross section for ionization of $Ne^{6+}(1s^2 2s^2\ ^1S)$. Solid curve: DWBA–R-matrix theory. Dots: experiment of Duponchelle and Defrance[24]. (From Laghdas *et al*[23]).

in the single differential cross section, the resonant and non-resonant parts. The resonant part was integrated separately to give the resonant contribution to the total cross section, which could then be multiplied by a branching ratio. Radiative damping reduced their cross section by a factor of about two.

In addition to single differential cross sections, the DWBA–R-matrix method has been applied to double differential cross sections[26], electron-photon coincides[27] and triple differential cross sections[28]. Space-limitations prevent us from discussion of this work. Further details of the DWBA–R-matrix method, complementary to what we have covered here, can be found in the review by Bartschat[29].

DISCRETIZED CONTINUUM METHODS

As our final example of the application of an accurate treatment of the one-continuum-electron problem to ionization, we consider formulations based on the representation of the continuum states of the $(e^- + X^+)$ system by unphysical pseudo-states $\{X_p\}$. The pseudo-states have square-integrable wave-functions $\{\overline{\Psi}_p(1 \cdots \mathcal{N}+1)\}$ and a discrete energy-spectrum $\{\varepsilon_p\}$, where ε_p is relative to $X^+_{\beta_1}$. The set of $\mathcal{N}$ pseudo-states used to represent the target X includes both states with $\varepsilon_p < 0$ as well as states with $\varepsilon_p > 0$. The former may or may not be physical bound states $\Psi_{b'}$ — the only usual requirement is that the initial physical state is accurately represented by a single pseudo-state:

$$\Psi_b \approx \overline{\Psi}_{p_0} \tag{15}$$

The pseudo-states can be expressed in terms of the physical states:

$$\overline{\Psi}_p \quad = \quad \sum_{b'}^{bound} \langle b' | p \rangle \, \Psi_{b'} \quad + \quad \sum_{\beta\varsigma_2} \int d\mathbf{k}_2 \, \langle \beta\varsigma_2 \mathbf{k}_2 | p \rangle \, \rho(k_2) \, \Psi^{(-)}_{\beta\varsigma_2\mathbf{k}_2} \quad + \quad \cdots \tag{16}$$

with the converse relations (which become exact as $\mathcal{N} \to \infty$)

$$\Psi_{b'} \approx \sum_p \langle p | b' \rangle \, \overline{\Psi}_p \qquad \text{and} \qquad \Psi^{(-)}_{\beta\varsigma_2\mathbf{k}_2} \approx \sum_p \langle p | \beta\varsigma_2\mathbf{k}_2 \rangle \, \overline{\Psi}_p \, . \tag{17}$$

The bound–to–bound excitation process involving the $\mathcal{N}$ pseudo-states:

$$e^-_{\varsigma_0\mathbf{k}_0} \quad + \quad X_{p_0} \quad \longrightarrow \quad e^-_{\varsigma_1\mathbf{k}_1} \quad + \quad X_p \tag{18}$$

is a one-continuum-electron problem that may be calculated accurately, and the scattering amplitudes obtained for this excitation process should, in principle, contain the required information on ionization. The issues to decide are: the nature and number $\mathcal{N}$ of the pseudo-states; and the precise methods for extracting the ionization cross sections.

Although there have been several earlier pseudo-state treatments of ionization[30,31,32], we shall start by discussing the calculations on the ionization of atomic hydrogen by Bray and co-workers[33,34,35], using their *convergent close-coupling* (CCC) method[36].

In the CCC method, a Laguerre basis is constructed with radial orbitals

$$\xi_{n\ell}(r) \propto (\lambda_\ell r)^{\ell+1} \exp(-\lambda_\ell r/2) \, L^{2\ell+2}_{n-1}(\lambda_\ell r) \tag{19}$$

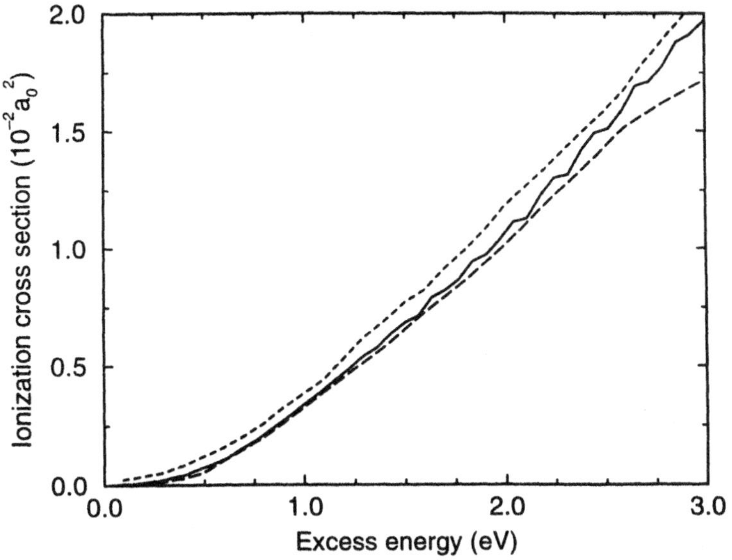

Figure 9. Total cross section for electron-impact ionization of atomic hydrogen. Circles: experiment[37]. Solid curve: CCC calculation[33]. Dashed curve: IERM calulation[38]. (From Bray and Stelbovics[33]).

for $\ell = 0 \cdots \ell_M$, and $n = 1 \cdots \mathcal{N}_\ell$ for each ℓ (so that $\mathcal{N} = \sum_{\ell=0}^{\ell_M} \mathcal{N}_\ell$). The wave-functions $\{\overline{\Psi}_p\}$ and energies $\{\varepsilon_p\}$ of the pseudo-states are found by diagonalizing H_X (the Hamiltonian of X) in this basis. The excitation problem (18) is solved by a close-coupled, momentum-space formulation. Convergence of the results is checked by increasing ℓ_M and $\{\mathcal{N}_\ell\}$.

In their work on the ionization of atomic hydrogen, Bray and Stelbovics[33] took $\lambda_\ell = 2$ for all ℓ, ensuring that the physical 1s-orbital is represented exactly. They calculated the cross section for ionization σ_b (summed over final-ion state β) by subtracting the cross sections $\sigma_{b'b}^{ex}$ for excitation of the physical bound states from the total cross section σ_b^{tot} calculated by the Optical Theorem[31]:

$$\sigma_b(E_0) \quad = \quad \sigma_b^{tot}(E_0) \quad - \quad \sum_{b'}^{bound} \sigma_{b'b}^{ex}(E_0) \; . \tag{20}$$

Using the projection (17), this can be expressed in terms of the cross sections $\sigma_{pp_0}^{ex}$ for pseudo-states excitation (18):

$$\sigma_b(E_0) \quad = \quad \sigma_b^{tot}(E_0) \quad - \quad \sum_p \sigma_{pp_0}^{ex}(E_0) \left(\sum_{b'}^{bound} |\langle p|b'\rangle|^2 \right) \; . \tag{21}$$

The total cross sections calculated by Bray and Stelbovics[33] are shown in figure 9. They are in superb agreement with the experimental results of Shah *et al*[37] over a wide range of impact energies, demonstrating the merit of the CCC method. Bray and Stelbovics also calculated the ionization spin asymmetry and there is good agreement between their results and the experimental measurements by Fletcher *et al*[39] and by Crowe *et al*[40].

The single differential cross section for ionization of hydrogen has been considered by Konovalov *et al*[34]. They extracted the single differential cross section $d\sigma_b/dE$ (summed over final-ion state β) from the total cross section for ionization calculated by the CCC method[33] using an argument due to Bransden and Stelbovics[41]. Firstly they confirmed that, to a good approximation,

$$\sigma_b(E_0) \quad = \quad \sum_{p:\varepsilon_p>0} \sigma^{ex}_{pp_0}(E_0) \; . \tag{22}$$

(This implies that the physical bound states are represented by pseudo-states with $\varepsilon_p < 0$ only, and that the physical continuum states are represented by pseudo-states with $\varepsilon_p > 0$ only.) The excitation cross sections are non-zero only if $\varepsilon_p \leq \overline{E}$, where $\overline{E} \equiv E_0 - I_{\beta_1 b}$, and so the sum in (22) is over p such that $0 < \varepsilon_p \leq \overline{E}$. Hence the sum may be taken to be the integral of a discrete function $F(E_2)$ whose value at E_2 is determined by the ε_p closest to E_2:

$$\sum_{p:\varepsilon_p>0} \sigma^{ex}_{pp_0} \quad = \quad \int_0^{\overline{E}} F(E_2)\, dE_2 \quad = \quad \sum_{p:0<\varepsilon_p\leq\overline{E}} \delta\varepsilon_p F(\varepsilon_p) \tag{23}$$

where $\delta\varepsilon_p$ is the size of the E_2-range lying closest to ε_p. A possible identification is:

$$F(\varepsilon_p) \quad = \quad \frac{1}{\delta\varepsilon_p} \sigma^{ex}_{pp_0}(E_0) \; . \tag{24}$$

Now, by (22) and (23), $\sigma_b = \int_0^{\overline{E}} F(E_2)\, dE_2$. Hence $F(E_2)$ is related to the single differential cross section. An expression with the physically-correct symmetry about $E_2 = \overline{E}/2$ is:

$$\frac{d\sigma_b}{dE}(E_2, E_0) \quad = \quad F(E_2) + F(\overline{E} - E_2) \tag{25}$$

with $F(E_2)$ now regarded as continuous and given by interpolating the values in (24). Adoption of (25) implies that $\sigma_b = \int_0^{\overline{E}/2} (d\sigma_b/dE)\, dE_2$, as should be the case.

The results obtained by Konovalov *et al* agree well with the experimental results of Shyn[42] when the latter are scaled to give the same total cross sections as Shah *et al*[37]. (Konovalov *et al* actually included a factor 1/2 on the right-hand side of (25). This issue is discussed further, and formula (25) given, by Bray and Fursa[43]).

The triple differential cross section for ionization of hydrogen has been calculated by Bray *et al*[35] using the CCC method. Following Curran and Walter[32], they start with an accurate expression for the scattering amplitude for the case of asymmetric energy-sharing with $E_1 > E_2$:

$$f = -(2\pi)^{-5/2} \left\langle \Psi^{(-)}_{\beta\varsigma_2\mathbf{k}_2}(1\cdots N{+}1)\, \varphi_{\varsigma_1\mathbf{k}_1}(0) \right| V \left| \, \Xi^{(+)}_{b\varsigma_0\mathbf{k}_0}(0\cdots N{+}1) \right\rangle \tag{26}$$

where $\Xi^{(+)}_{b\varsigma_0\mathbf{k}_0}$ is the accurate scattering wave-function which is taken to be that for the pseudo-state excitation (18), and here $\varphi_{\varsigma_1\mathbf{k}_1}$ is an undistorted state. The physical continuum state $\Psi^{(-)}_{\beta\varsigma_2\mathbf{k}_2}$ is then projected on to the pseudo-states by (17), to give an expression in terms of the scattering matrix elements for process (18):

$$f = -(2\pi)^{-5/2} \sum_p \langle \beta\varsigma_2\mathbf{k}_2 | p \rangle \left\langle \overline{\Psi}_p(1\cdots N{+}1)\, \varphi_{\varsigma_1\mathbf{k}_1}(0) \right| V \left| \, \Xi^{(+)}_{p_0\varsigma_0\mathbf{k}_0}(0\cdots N{+}1) \right\rangle \; . \tag{27}$$

Bray *et al* present results for coplanar asymmetric geometry for two values of impact energy. At $E_0 = 150\,\text{eV}$, their results are close to the earlier theoretical results of Curran *et al*[44] and are in good agreement with the experimental measurements by Ehrhardt *et al*[45]. At $E_0 = 54.4\,\text{eV}$, again their results agree well with experiment, this time the relative measurements by Brauner *et al*[46].

Following the striking success for ionization of hydrogen, the CCC method has been applied to other systems. Firstly, Bray[47] has calculated total cross sections and spin asymmetry for ionization of sodium, both from the ground state and from the 3P state. The sodium target was treated as an effective one-electron system by adopting a frozen-core Hartree-Fock treatment of the all but the valence electron. The spin asymmetry results (for the 3S initial state) agree well with the experiments of Baum *et al*[48]. For the total cross section, the subsequent measurements by Johnston and Burrow[49] agree with the calculation to within 10%.

Bray[50] has also performed a calculation, similar to that for sodium, on lithium-like ions. However, the results for the total cross sections do not agree with experiment to the spectacular degree found for hydrogen.

As a final application of the CCC method, we report the recent work by Bray *et al*[51,52,53] on the ionization of helium. The target pseudo-states were configuration-interaction two-electron wave-functions constructed from the Laguerre basis (19), with judicious choice of the parameters $\{\lambda_\ell\}$. They found it sufficient to include only configurations with one electron in the lowest orbital, so again this is, effectively, a one-electron target. Their calculated results are again in excellent agreement with experiment: their total ionization cross section agrees well with the measurements by Montague *et al*[54] for $30\,\text{eV} \leq E_0 \leq 500\,\text{eV}$; their single and double differential cross section at $E_0 = 100\,\text{eV}$ agree well with the experiments of Müller-Fiedler *et al*[55]; and their triple differential cross sections agree with the measurements by Röder *et al*[56,57] for asymmetric energy-sharing kinematics at $E_0 = 40\,\text{eV}$ and $E_0 = 50\,\text{eV}$. They further express optimism about the reliability of the CCC results for symmetric energy-sharing.

To conclude this discussion of pseudo-states, we consider how the R-matrix excitation-codes have been adapted to include a pseudo-state representation of the continuum states of the target X, thereby enabling them to be used for ionization.

The original purpose of including pseudo-states in R-matrix calculations was to augment the basis for describing the *bound* states of X. This is necessary because the wave-functions of all the states $\{X_{b'}\}$ being included in the calculation are constructed from a single set of radial orbitals. Pseudo-orbitals are included in this set to allow, through configuration interaction, accurate representation of the bound physical states $\{X_{b'}\}$. In addition, the pseudo-orbitals give rise to pseudo-states, but these were not included in the set of target states, although they show up as pseudo-resonances in the cross sections at high E_0.

One way to adapt the R-matrix excitation-codes so that they can deal with ionization is to change the number and nature of the pseudo-orbitals and to include the pseudo-states, along with the physical bound states, as target states in the excitation calculation. This has been done by Bartschat *et al*[58] in their *R-matrix with pseudo-states* (RMPS) formulation. For their work on hydrogen, the pseudo-orbitals were Sturmian-type functions that were orthogonal to each other and to the physical orbitals. The pseudo-states were obtained by diagonalizing H_X in the basis comprised of the physical and the pseudo orbitals, and thus had radial forms:

$$\phi_{n\ell}(r) = \sum_{i=\ell+1}^{n_{max}} a_{n\ell i}\, r^i e^{-\alpha r} . \tag{28}$$

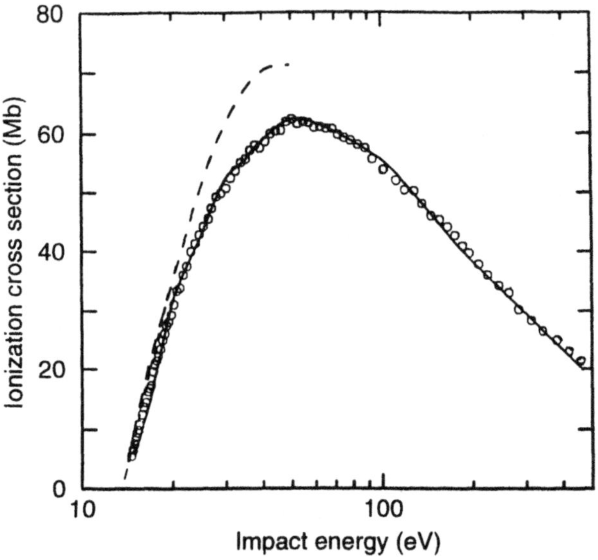

Figure 10. Total cross section for electron-impact ionization of atomic hydrogen. Solid curve: IERM calculation. Dashed curve: RMPS calculation. Dotted curve: CCC calculation. (From Scott *et al*[61]).

The parameter α is arbitrary, and was chosen by comparison with other calculations. The major problem to be overcome was the numerical one of constructing the continuum orbitals (whose purpose is to represent the X^- continuum states in the inner region) so that they are orthogonal to both the physical and the pseudo orbitals.

The RMPS method was applied to ionization of hydrogen by Bartschat and Bray[59]. Their RMPS results for the total cross section and the spin asymmetry are in excellent agreement with the CCC results[33]. In the RMPS calculation, there was a significant contribution to ionization from the excitation of pseudo-states with $\varepsilon_p < 0$, in contrast to the property (22) of the CCC calculation.

An alternative adaptation of the R-matrix method to include a pseudo-state representation of the continuum is the *intermediate-energy R-matrix* (IERM) formulation by Burke *et al*[60]. In this, the continuum orbitals (introduced to represent X^- continuum states) are used as the pseudo-orbitals to describe one of the electrons in X. An early application of the IERM method to the ionization of hydrogen by Scholz *et al*[38] is shown in figure 9. That calculation was less successful than the CCC method for $E_0 > 25\,\mathrm{eV}$. However, an important virtue of the IERM method is that the two continuum electrons are treated identically in the inner region. Also, the density of pseudo-states in the region of the ionization threshold can be made large in the IERM treatment. Hence the IERM method should be useful for near-threshold ionization. Recently Scott *et al*[61] have applied the IERM, the RMPS and the CCC methods to near-threshold ionization of hydrogen, within the Temkin-Poet model. Their results are shown in figure 10. For excess energy $\overline{E}$ less than 2.5 eV, the IERM and the RMPS results are close, while the CCC result is consistently higher.

Acknowledgments

Support from the UK EPSRC (grant GR/K24741) and from the EC–HCM programme (contract ERB CHRX CT 920013) is gratefully acknowledged.

REFERENCES

1. M.E. Bannister, F.W. Meyer, Y.S. Chung, N. Djurić, G.H. Dunn, M.S. Pindzola and D.C. Griffin, Absolute cross sections for the electron-impact single ionization of Mo^{4+} and Mo^{5+} ions, *Phys. Rev. A* 52:413 (1995).
2. S.M. Younger, Electron-impact ionization cross sections for highly ionized hydrogen- and lithium-like atoms, *Phys. Rev. A* 22:111 (1980).
3. M.R.H. Rudge and S.B. Schwartz, Ionization of complex ions by electron impact I. Ionization cross sections for Fe XV and Fe XVI, *Proc. Phys. Soc. London* 88:579 (1966).
4. D.L. Moores and K.J. Reed, Indirect processes in electron-impact of positive ions, *Adv. At. Mol. Opt. Phys.* 34:301 (1994).
5. M.S. Pindzola, D.C. Griffin and C. Bottcher, Electron-impact ionization in the magnesium-isoelectronic sequence, *Phys. Rev. A* 33:3787 (1986).
6. R.J.W. Henry, Cross sections for inner-shell excitation of C^{3+}, N^{4+} and O^{5+} ions, *J. Phys. B: At. Mol. Phys.* 12:L309 (1979).
7. S.S. Tayal and R.J.W. Henry, Contributions of excitation autoionization to electron-impact ionization of Mg-like Al^{+}, S^{4+}, Cl^{5+} and Ar^{6+} ions, *Phys. Rev. A* 33:3825 (1986).
8. P.G. Burke, W.C. Fon and A.E. Kingston, Electron impact ionization of Ti^{3+}, *J. Phys. B: At. Mol. Phys.* 17:L733 (1984).
9. R.J.W. Henry and A.E. Kingston, Electron impact excitation: V. Inner shell excitation-autoionization, *Adv. At. Mol. Phys.* 25:267 (1988).
10. K.J. LaGattuta and Y. Hahn, Electron impact ionization of Fe^{15+} by resonant excitation double Auger ionization, *Phys. Rev. A* 24:2273 (1983).
11. R.A. Falk, G.H. Dunn, D.C. Griffin, C. Bottcher, D.C. Gregory, D.H. Crandall and M.S. Pindzola, Excitation-autoionization contributions to electron-impact ionization, *Phys. Rev. Lett.* 47:494 (1981).
12. R.A. Falk, G.H. Dunn, D.C. Gregory and D.H. Crandall, Measurement of the contribution of excitation autoionization to electron-impact ionization of ions: Ti^{3+}, Zr^{3+}, Hf^{3+} and Ta^{3+}, *Phys. Rev. A* 27:762 (1983).
13. T.W. Gorczyca, M.S. Pindzola, D.C. Griffin and N.R. Badnell, Configuration-interaction effects on the electron-impact ionization of Ti^{3+}, *J. Phys. B: At. Mol. Opt. Phys.* 27:2399 (1994).
14. D.C. Griffin, C. Bottcher and M.S. Pindzola, Theoretical calculations of the contribution of excitation autoionization to electron-impact ionization in ions of the transition series of elements, *Phys. Rev. A* 25:1374 (1982).
15. M.H. Chen, K.J. Reed and D.L. Moores, Contributions of resonant excitation double autoionization to the electron-impact ionization of Fe^{15+}, *Phys. Rev. Lett.* 64:1350 (1990).
16. S.S. Tayal and R.J.W. Henry, Role of indirect-ionization processes in the electron-impact ionization of Fe^{15+}, *Phys. Rev. A* 39:3890 (1989).
17. H.L. Zhang and A.K. Pradhan, Relativistic and radiative damping effects in electron-impact excitation of highly charged ions, *J. Phys. B: At. Mol. Opt. Phys.* 28:L285 (1995).
18. T.W. Gorczyca, F. Robicheaux, M.S. Pindzola and N.R. Badnell, Radiative damping and interference in the resonance structure of the electron-impact excitation cross section for Ti^{20+}, *Phys. Rev. A* 52:3852 (1995).

19. P.C.W. Davies and M.J.Seaton, Radiative damping in the optical continuum, *J. Phys. B: At. Mol. Phys.* 2:757 (1969).

20. K. Bartschat and P.G. Burke, The R-matrix method for electron-impact ionization, *J. Phys. B: At. Mol. Opt. Phys.* 20:3191 (1987).

21. H. Jakubowicz and D.L. Moores, Electron-impact ionization of Li-like and Be-like ions, *J. Phys. B: At. Mol. Phys.* 14:3733 (1981).

22. K. Laghdas, R.H.G. Reid, C.J. Joachain and P.G. Burke, Electron-impact ionization of Ar^{9+}, *J. Phys. B: At. Mol. Opt. Phys.* 28:4811 (1995).

23. K. Laghdas, R.H.G. Reid, C.J. Joachain and P.G. Burke, Electron-impact ionization of Ne^{6+}, *J. Phys. B: At. Mol. Opt. Phys.* Submitted.

24. M. Duponchelle, Experimental studies of ionization of the ions Ne^{q+} (q=5,6,8) by electron impact, *PhD Thesis,* Université Catholique de Louvain (1993).

25. D.L. Moores and K.J. Reed, Electron-impact ionization of Se^{24+}, *Phys. Rev. A* 39:1747 (1989).

26. R. Schwienhorst, A. Raeker, R.H.G. Reid and K. Bartschat, R-matrix calculations for double-differential cross sections in electron-impact ionization of helium *J. Phys. B: At. Mol. Opt. Phys.* 28:4651 (1995).

27. R. Schwienhorst, A. Raeker, K. Bartschat and K. Blum, Electron-photon coincidences in electron impact ionization-excitation *J. Phys. B: At. Mol. Opt. Phys.* 29:2305 (1996).

28. R.H.G. Reid, K. Bartschat and A. Raeker, Initial-state, final-state and higher-order effects in electron-impact ionization of helium atoms *J. Phys. B: At. Mol. Opt. Phys.* Submitted.

29. K. Bartschat, Simultaneous ionization and excitation of atoms by electron impact, in: *Proceedings – Electron Impact Symposium,* R.A. Phaneuf, ed., University of Nevada-Reno (1996).

30. D.F. Gallagher, Ioization of atomic hydrogen by an incident electron in a pseudo-state approximation, *J. Phys. B: At. Mol. Phys.* 7:362 (1974).

31. J. Callaway and D.H. Oza, Ionization of hydrogen atoms by electron impact, *Phys. Lett.* 72A:207 (1979).

32. E.P. Curran and H.R.J. Walters, Triple differential cross sections for electron-impact ionization of atomic hydrogen — a coupled pseudo-state calculation, *J. Phys. B: At. Mol. Phys.* 20:337 (1987).

33. I. Bray and A.T. Stelbovics, Calculation of the total ionization cross section and spin asymmetry in electron-hydrogen scattering from threshold to 500 eV, *Phys. Rev. Lett.* 70:746 (1993).

34. D.A. Konovalov, I. Bray and I.E. McCarthy, Convergent close-coupling calculation of singly differential cross sections in the ionization of atomic hydrogen by electron impact, *J. Phys. B: At. Mol. Opt. Phys.* 27:L413 (1994).

35. I. Bray, D.A. Konovalov, I.E. McCarthy and A.T. Stelbovics, Calculation of triple-differential cross sections in electron scattering on atom hydrogen, *Phys. Rev. A* 50:R2818 (1994).

36. I. Bray and A.T. Stelbovics, Convergent close-coupling calculations of electron-hydrogen scattering, *Phys. Rev. A* 46:6995 (1992).

37. M.B. Shah, D.S. Elliott and H.B. Gilbody, Pulsed crossed-beam study of the ionization of atomic-hydrogen by electron-impact, *J. Phys. B: At. Mol. Phys.* 20:3501 (1987).

38. T.T. Scholz, H.R.J. Walters and P.G. Burke, The total cross-section for electron-scattering by atomic-hydrogen, *J. Phys. B: At. Mol. Opt. Phys.* 23:L467 (1990).

39. G.D. Fletcher, M.J. Alguard, T.J. Gay, V.W. Hughes, P.F. Wainwright, M.S. Lubell and W. Raith, Experimental-study of spin-exchange effects in elastic and ionizing collisions of polarized electrons with polarized hydrogen-atoms, *Phys. Rev. A* 31:2854 (1985).

40. D.M. Crowe, X.Q. Guo, M.S. Lubell, J. Slevin and M. Eminyan, Spin-tagged electron hydrogen scattering - new measurements of ionization asymmetries from threshold to 500 eV,

J. Phys. B: At. Mol. Opt. Phys. 23:L325 (1990).

41. B.H. Bransden and A.T. Stelbovics, An L^2 approach to the solution of coupled-channel scattering equations, *J. Phys. B: At. Mol. Phys.* 17:1877 (1984).
42. T.W. Shyn, Doubly differential cross-sections of secondary electrons ejected from atomic-hydrogen by electron-impact, *Phys. Rev. A* 45:2951 (1992).
43. I. Bray and D.V. Fursa, Calculation of singly differential cross-sections of electron-impact ionization of helium at 100 eV, *J. Phys. B: At. Mol. Opt. Phys.* 28:L435 (1995).
44. E.P. Curran, C.T. Whelan and H.R.J. Walters, On the electron-impact ionization of H(1s) in coplanar asymmetric geometry, *J. Phys. B: At. Mol. Opt. Phys.* 24:L19 (1991).
45. H. Ehrhardt, K. Jung, G. Knoth and P. Schlemmer, Differential cross-sections of direct single electron-impact ionization, *Z. Phys. D* 1:3 (1986).
46. M. Brauner, J.S. Briggs, H. Klar, J.T. Broad, T. Rosel, K. Jung and H. Ehrhardt, Triply differential cross-sections for ionization of hydrogen-atoms by electrons - the intermediate and threshold energy regions, *J. Phys. B: At. Mol. Opt. Phys.* 24:657 (1991).
47. I. Bray, Calculation of electron-impact total, ionization, and nonbreakup cross-sections from the 3S and 3P states of sodium, *Phys. Rev. Lett.* 73:1088 (1994).
48. G. Baum, M. Moede, W. Raith and W. Schröder, Measurement of spin asymmetries in the electron-impact ionization of alkali atoms, *J. Phys. B: At. Mol. Phys.* 18:531 (1985).
49. A.R. Johnston and P.D. Burrow, Electron-impact ionization of Na, *Phys. Rev. A* 51:R1735 (1995).
50. I. Bray, Calculation of electron-impact ionization of lithium-like targets, *J. Phys. B: At. Mol. Opt. Phys.* 28:L247 (1995).
51. I. Bray, D.V. Fursa and I.E. McCarthy, Calculation of electron-helium scattering at 40 eV, *Phys. Rev. A* 51:500 (1995).
52. D.V. Fursa and I. Bray, Calculation of electron-helium scattering, *Phys. Rev. A* 52:1297 (1995).
53. I. Bray and D.V. Fursa, Convergent close-coupling method - a "complete scattering-theory"?, *Phys. Rev. Lett.* 76:2674 (1996).
54. R.G. Montague, M.F.A. Harrison and A.C.H. Smith, A measurement of the cross section for ionization of helium by electron impact using a fast crossed beam technique, *J. Phys. B: At. Mol. Phys.* 17:3295 (1984).
55. R. Müller-Fiedler, K. Jung and H. Ehrhardt, Double differential cross sections for electron-impact excitation of helium, *J. Phys. B: At. Mol. Phys.* 19:1211 (1986).
56. J. Röder, H. Ehrhardt, I. Bray, D.V. Fursa and I.E. McCarthy, Absolute triple differential cross-section for electron-impact ionization of helium at 50 eV, *J. Phys. B: At. Mol. Opt. Phys.* 29:L67 (1996).
57. J. Röder, H. Ehrhardt, I. Bray, D.V. Fursa and I.E. McCarthy, Absolute triple differential cross-section for electron-impact ionization of helium at 40 eV, *J. Phys. B: At. Mol. Opt. Phys.* 29:2103 (1996).
58. K. Bartschat, E.T. Hudson, M.P. Scott, P.G. Burke and V.M. Burke, Electron-atom scattering at low and intermediate energies using a pseudo-state/R-matrix basis, *J. Phys. B: At. Mol. Opt. Phys.* 29:115 (1996).
59. K. Bartschat and I. Bray, Electron-impact ionization of atomic-hydrogen from the 1S and 2S states, *J. Phys. B: At. Mol. Opt. Phys.* 29:L577 (1996).
60. P.G. Burke, C.J. Noble and M.P. Scott, R-matrix theory of electron-scattering at intermediate energies, *Proc. R. Soc. A* 410:289 (1987).
61. M.P. Scott, P.G. Burke, K. Bartschat and I. Bray, Near-threshold electron-impact ionization of atomic hydrogen within the Temkin-Poet model, *J. Phys. B: At. Mol. Opt. Phys.* In preparation.

R-MATRIX THEORY OF ELECTRON-MOLECULE COLLISIONS

Lesley A. Morgan

Computer Centre, Royal Holloway, University of London, Egham, Surrey TW20 0EX, England

1. INTRODUCTION

The R-matrix method has been used extensively in recent years, primarily by groups in the UK and Germany, to obtain high quality cross sections for low energy electron scattering by diatomic molecular targets. The first calculations for polyatomic targets are now being attempted and results for several molecules have now been published by the Bonn group. This article will review the progress that has been made in the two years that have elapsed since the EC Network meeting in Rennes.

2. NEW METHODS

The multi-centre R-matrix method, as applied to diatomic, targets has been described in detail by, for example, Gillan *et al*[1] and Morgan[2]. The application to polyatomic targets is a new development. The approach being used by the Bonn group is described in some detail by Pfingst *et al*[3]. The UK group is also well advanced in developing a polyatomic code, though using a slightly different approach. Many of the fundamental problems are common to both and these will be discussed later in this section.

In all calculations, configuration space is partitioned into two regions, bounded by the R-matrix sphere. The scattering wavefunction is expanded in the inner region on a basis ψ_k where

$$\psi_k = \mathcal{A}\sum_{ij}\Phi_i\eta_j\alpha_{ijk} + \sum_j \chi_j\beta_{jk} \tag{1}$$

where $\mathcal{A}$ is the usual antisymmetrisation operator, Φ_i are N electron target wavefunctions, η_j are single particle continuum functions, χ_j are N+1 electron correlation/polarization functions and α_{ijk} and β_{jk} are expansion coefficients.

The calculations now being carried out for diatomic targets routinely use CI target wavefunctions Φ_i and frequently include several electronic states in the sum over i. The UK code is based on the Alchemy quantum chemistry package[4, 5] which uses Slater type orbitals (STOs). These are only really suitable for linear molecules and integrals involving them must be evaluated using numerical quadrature. We are heavily reliant on quantum chemists to provide us with good quality target wavefunctions and most recent calculations, even for diatomics, use Gaussian type orbitals (GTOs). This is due to the ease in which the required integrals can be evaluated from closed form expressions. Modern wavefunctions routinely involve up to a million configurations and when used in a scattering calculation, these will be multiplied by typically 50 - 100 continuum functions η_j, thus increasing the size of the problem by up to two orders of magnitude.

The first problem to be tackled, regardless of which type of orbital is used, is to find an efficient way of building the Hamiltonian matrix when very large target CI wavefunctions are used. Serious bottlenecks already occur when using 'small' wavefunctions comprised of 100 - 200 configuration state functions (CSFs). Another problem, which affects all methods based on a wavefunction expansion of the type given in equation (1), is that of enforcing a consistent phase convention in all stages of its construction. Jonathan Tennyson[6] has recently developed a new algorithm which addresses these problems. It is similar in philosophy to the new RMATRX II atomic scattering code[7] but the methods employed are, of necessity, very different. The algorithm exploits the fact that the wavefunction is ultimately a sum over products of target wavefunction and continuum orbital, rather than a sum over all of the individual CSFs multiplied by all of the continuum orbitals, which then have to be explicitly evaluated and summed before the Hamiltonian construction can be initiated. It is based on the identification of reusable 'prototype' angular momentum integrals and the utilization of the fact that the target wavefunctions diagonalize the target-only part of the Hamiltonian.

A more fundamental problem, which is independant of the type of basis function used, is the choice of 'correlation/polarization functions' χ_j. When CI target wavefunctions are used, care must be taken not to introduce any additional correlations into the parts of these functions which describe the target electrons, but which are not already included in the target wavefunctions Φ_i . It is equally important not to undercorrelate the scattering state, relative to the target, since this will also lead to inaccuracies in the energies of resonances and (N+1) electron bound states. In extreme cases, this can lead to unphysical effects due to the incorrect ordering of thresholds. If accurate experimental data is available, a small amount of adjustment of thresholds is possible, though since the R-matrix method is entirely *ab initio* the scope for such adjustments is limited. To date we do not have a satisfactory algorithmic approach to the construction of correlation/polarization functions, though work is underway to test some possible models.

If Gaussian type orbitals are used, two further problems arise. The first is that the analytic expressions used to evaluate the integrals only apply over the range $0-\infty$. The R-matrix method requires integrals evaluated over the finite range $0-r_a$ where r_a is the R-matrix radius. The most straightforward way to tackle the problem is to use existing codes to evaluate integrals over the infinite range, then subtract the contribution from $r_a-\infty$. The R-matrix sphere must be chosen to be large enough to contain the target charge distribution, hence the only integrals which have a non-zero contribution from the outer region will be those involving a continuum function. These will be relatively few in number and can be evaluated numerically.

The second problem is to obtain an adequate representation of the continuum basis in terms of functions which do not require major modification to existing quantum chemistry codes. This means, in practice, representing Bessel or Coulomb functions as linear combinations of GTOs. The Bonn group have pioneered work in this area[8] but problems still remain with linear dependance when diffuse orbitals are also included in the target basis.

Another development, of more general interest, is an improved method for locating and fitting the broad resonances often found in molecular scattering. Conventional fitting procedures break down when the non-resonant background is itself rapidly varying, a common problem in polar molecules, or the resonance is cut off by a threshold, also a frequent problem due to the plethora of electronically excited states which often lie only a few eV above the ground state of the target. Jonathan Tennyson[9] has adapted the time-delay matrix method, first developed by Felix Smith[10] for use in these situations. It also gives a means of calculating branching ratios in cases where a resonance can decay into the continua associated with more than one target state.

3. NEW RESULTS

3.1. Diatomic Targets

Recent calculations have mostly focussed on open shell systems, such as O_2[11], NO[12], CH[13] and OH[14], but there is also new work on NO^+[15]. and previous work on H_2[16], H_2^+[17] and N_2[18] has been reassessed and extended.

The most extensive calculations have been for the low energy scattering of electrons by O_2[11]. The basic model includes the nine target states which have the configuration $[\text{core}]1\pi_u^4 1\pi_g^2$ or $[\text{core}]1\pi_u^3 1\pi_g^3$. These were represented by rather small (~ 30 CSFs) CI wavefunctions. Three virtual orbitals, $4\sigma_g$, $1\delta_g$ and $1\delta_u$ are also included in the basis and are used to represent correlation and polarization effects. These effects, both in the target and the negative ion, are not well represented by such a small basis. Minor adjustments were therefore made to the energy levels used in the scattering calculations so that the various thresholds were in agreement with experiment.

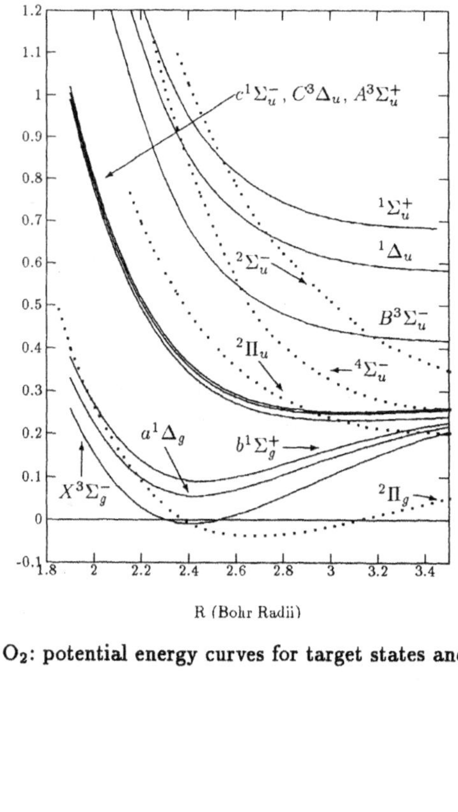

Figure 1. O_2: potential energy curves for target states and resonances

Low energy scattering in this system is dominated by the O_2^-, $^2\Pi_g$ state. This is only bound at bond lengths greater than about $2.5a_0$ and appears as a resonance at smaller internuclear separations. Vibrationally resolved cross sections for scattering by the electronic ground state were obtained using a non-adiabatic model of vibrational motion[19]. These were found to consist entirely of a sequence of sharp peaks, superimposed on an effectively zero background, and correspond to the vibrational levels of the temporary anion. The positions and relative magnitudes of these peaks are in very good agreement with the experimental measurements of Field *et al*[20].

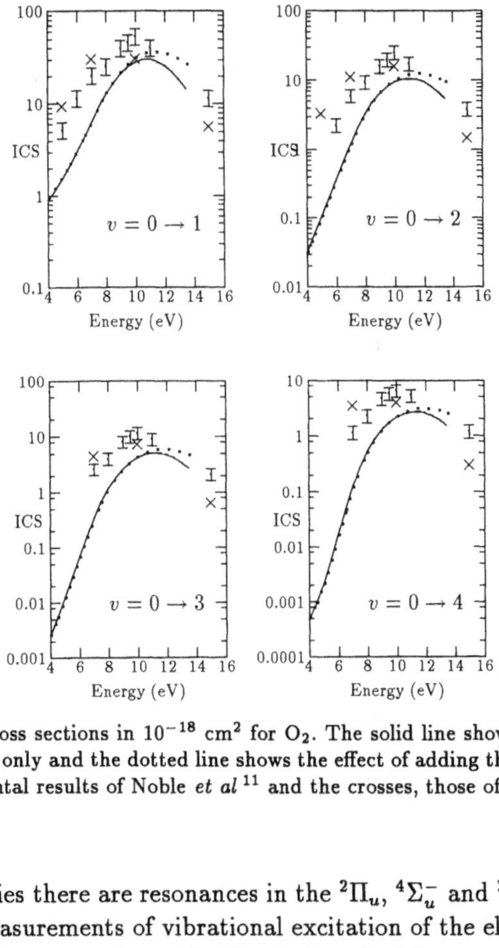

Figure 2. Integrated cross sections in 10^{-18} cm^2 for O_2. The solid line shows the theoretical results from the $^4\Sigma_u^-$ symmetry only and the dotted line shows the effect of adding the $^2\Sigma_u^-$ contribution. The bars show the experimental results of Noble *et al*[11] and the crosses, those of Shyn and Sweeney[21]

At higher energies there are resonances in the $^2\Pi_u$, $^4\Sigma_u^-$ and $^2\Sigma_u^-$ symmetries. Earlier experimental measurements of vibrational excitation of the electronic ground state showed a broad maximum at about 10 eV and there had been some debate over which of the resonances were responsible. The dominant configurations of each of the two Π resonances correspond to the addition of an electron to one of the open π orbitals of the target, whereas the Σ resonances correspond to the additional electron being temporarily captured into the first unoccupied, $3\sigma_u$, orbital. The higher resonances are short lived and hence an adiabatic approximation to nuclear motion is valid. The shapes of the resultant integrated cross sections agree well with experiment. They are found to be dominated by the $^4\Sigma_u^-$ resonance with the $^2\Sigma_u^-$ only significant at energies above 10 eV. The two Π resonances, which are core-excited resonances, have been found in earlier work to dominate the electronic excitation process, but, apart from the spikes in the low energy cross section mentioned above, play an insignificant role in the electronically elastic scattering.

Another open shell system of current interest is NO. The ground state has symmetry $^2\Pi$ and has one electron in a π orbital outside closed shells. Experimental results[22, 23] show a complicated resonance structure. By analogy with O_2, which is iso-electronic with NO^-, these will be vibrational series having symmetry $^3\Sigma^-$, $^1\Delta$ or $^1\Sigma^+$. The problem is to identify the individual peaks. Early work by Tennyson and Noble[24] found that, at equibrium geometry, the $^3\Sigma^-$ state was weakly bound and that the other two negative ion states appeared as resonances. New calculations[12] are currently in progress for this system. These include up to 6 target configurations, represented by CI wavefunctions obtained in a CASCI model, where the active space consists of the 4σ, 5σ, 6σ, 1π and 2π orbitals. The experimental results all indicate that the electron affinity of NO^- is in the range 24 – 50 meV whereas our calculations currently give a value close to 1 eV. This is clearly caused by including correlation/polarization effects in the negative ion wavefunction which are missing from those used for the neutral target. The $^1\Delta$ and $^1\Sigma^+$ curves are similarly too low. A variety of approaches to the problem of obtaining a representation of polarization effects, which are consistent for both target and anion, are currently being investigated.

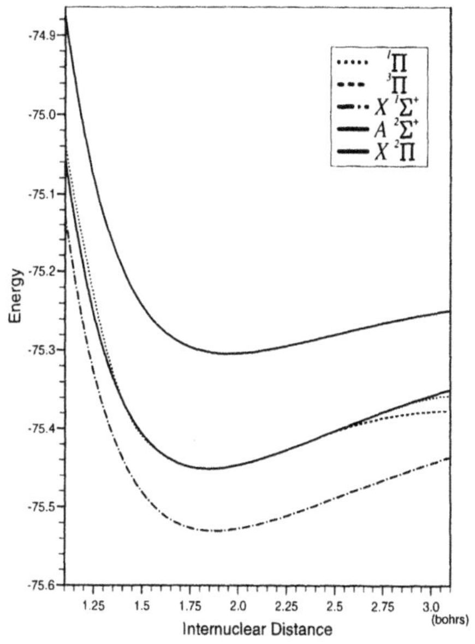

Figure 3. Potential energy curves for OH and OH^-

A calculation nearing completion is a two state model of electron scattering by OH[14]. This is also an open shell system with a $^2\Pi$ ground state. There are no experimental results, but several bound state calculations have been carried out for the negative ion states. Again we use a CASCI representation of the target, but due to the smaller number of electrons involved, we need only freeze the 1σ orbitals to get a manageable size wavefunctions (160 and 108 CSFs respectively). In this case, our value for the electron affinity of OH^- is slightly less than the best previous calculation by Werner *et al*[25], which, in turn. is slightly less than experimentally derived values. In this case, our model would appear to be underestimating correlation effects. This is in contrast to the NO results, but can possibly be explained by the fact that the OH model is nearly a complete CI and therefore polarization and correlation effects in the target and anion are treated at the same consistent level of approximation. Some *ab initio* calculations by Tellinghuisen *et al*[26] suggest that the lowest $^3\Pi$ and $^1\Pi$ states of the anion are also bound, at least at large bond lengths. Our calculations show both states to be bound at all but very short internuclear distances, though the binding is extremely weak at all but large bond lengths. The binding mechanism is probably the strong permanent dipole of the neutral parent.

The R-matrix method can also be used to obtain accurate energies for the Rydberg states of molecules. The system is treated in the inner region of configuration space as electron scattering by the ionic core, but in the outer region, the appropriate negative energy boundary conditions are imposed. This technique has recently been used to study the Rydberg states of NO[15]. The model included the lowest four states ($X^1\Sigma^+$,$a^3\Sigma^+$, $b^3\Pi$ and $w^3\Delta$) of NO^+. Quantum defect theory was used to classify the large number of bound states and, above the NO^+ threshold, the series of resonances converging to the three excited states. For the bound states, it was found that most of the Rydberg series show a linear relationship between their quantum defects and the energy of the Rydberg electron. At $n = \infty$ a surprisingly large p – d mixing was observed. Satisfactory agreement was obtained with experimental data[27, 28] for the resonance positions, except for the d series, where the theoretical results appear to be too high. The original work was only carried out at the equilibrium bond length of NO^+ but it is now being extended to cover a range of geometries and uses an improved target basis.

A similar calculation has also been carried out for the super-excited states of H_2[17]. In this case, the target state expansion included the lowest three states of H_2^+ and results are tabulated for 12 symmetries and 13 geometries. Positions and widths of the Feshbach resonances converging to both the $^2\Sigma_u^+$ and $^2\Pi_u$ states of the ion were calculated. When these resonances lie above the first excited state, they can decay either to it or to the ground state. The time-delay matrix method[9, 10] was used to calculate the corresponding autoionization branching ratios.

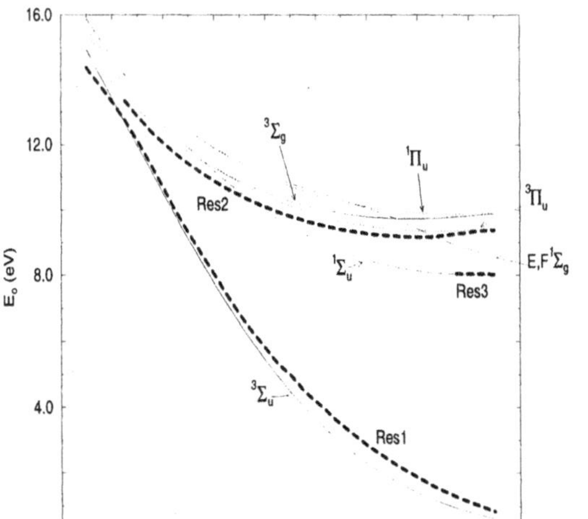

Figure 4. Potential energy curves for H_2 and low energy resonances in the $^2\Sigma_g^+$ symmetry

Stibbe and Tennyson[16] have also used the time-delay method to re-evaluate earlier work on H_2[29]. The original model was improved significantly, by reoptimizing the target wavefunctions, increasing the number of continuum basis functions used, and by eliminating phase inconsistencies between the target and scattering wavefunctions. The fact that there are only two resonances of symmetry $^2\Sigma_g^+$ below 12.3 eV was confirmed and more accurate values for their positions and widths were obtained. An avoided crossing between the two was observed at small (R $\sim$ 1.2a_0) bond lengths. At larger bond lengths the lower resonance can be clearly identified with the parent state $(1\sigma_g, 1\sigma_u^2)$ $b^3\Sigma_u^+$. This resonance is well known from experimental data where it appears as a peak at about 10 eV. The parentage of the second, 12 eV, resonance is not so obvious but the authors conclude that it is predominantly the $a^3\Sigma_g^+$, contrary to previous classifications. The branching ratios and, to a lesser extent, the resonance width, shown in Figure 5, show quite dramatically the effects of the avoided crossing and the changes in the decay paths as more states become energetically accessible.

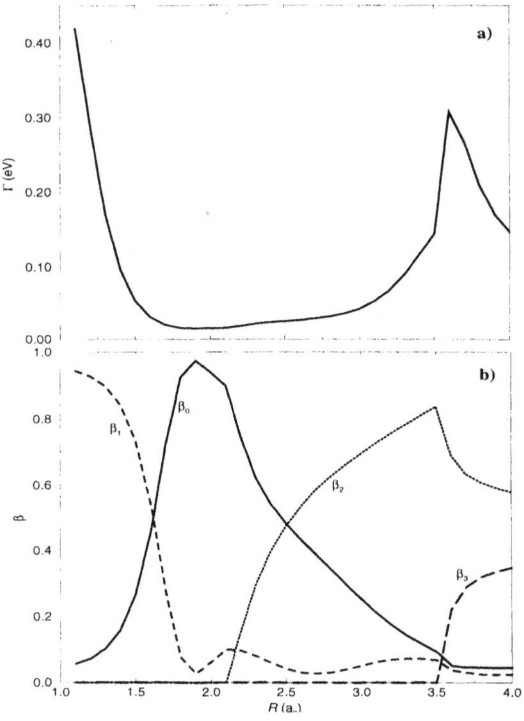

Figure 5. H_2: $^2\Sigma_g$ total symmetry: a) resonance width (Γ) and b) branching fractions (β) for res2 (see Figure 4) as a function of bondlength (R). β_0 decays into the $X^1\Sigma_g$ state, β_1 into the b $^3\Sigma_u$, β_2 into the B $^1\Sigma_u$ and β_3 into the E,F $^1\Sigma_g$

Earlier R-matrix calculations have been also been improved and extended for electron impact excitation of N_2[18]. The new calculations include the ground and lowest seven valence states of the target. More sophisticated target wavefunctions were employed and a variety of models were investigated for the inclusion of polarization effects. Only one geometry was considered and no attempt was made to include vibrational motion. This restriction was due to the large number of crossings between the potential energy curves of the excited states in the Franck-Condon region of the ground state.

There are still more crossings with states not included in the expansion at both large and small bond lengths. The integrated cross sections for the excitation of the $A^3\Sigma_g^+$ state were relatively insensitive to the model employed and and are fairly good agreement with the measurements of Trajmar*et al*[30]. The differential cross sections, however, have a completely different shape at both 10 eV and 15 eV. At the higher energy, the R-matrix results are in better accord with the more recent measurements of Brunger and Teubner[31]. Integrated cross sections for the excitation of the $B^3\Pi_g$, $W^3\Delta_u$ and $B'^3\Sigma_u^-$ states show somewhat more sensitivity to the model employed but, in contrast to earlier work, are in reasonable agreement with experiment.

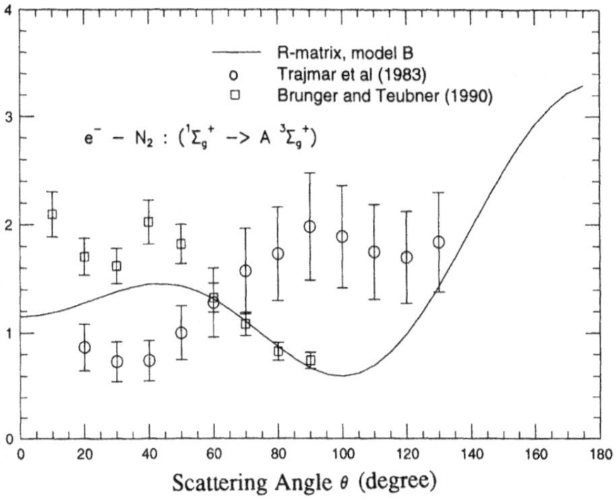

Figure 6. N_2: differential cross sections at 15 eV for the excitation of the A $^3\Sigma_u^+$ state. The solid curve is theory[18]; the circles, the experimental data of Trajmar*et al*[30] and the squares the experimental data of Brunger and Teubner[31].

3.2. Polyatomic Targets

The application of the R-matrix method to polyatomic systems is being tackled by both the UK group and the group in Bonn. The approaches are broadly similar in that both groups are using existing quantum chemistry codes with minimal modification, though there are significant differences in, for example the way in which the structure of the scattering wavefunction as given in equation (1) is imposed when using a code, which in its original form, has no mechanism for distinguishing between target and continuum orbitals. To date, all of the published results are from the Bonn group.

The first results published for a polyatomic target were for low energy electron scattering by CH_4[32]. A major difference between the this calculation, and the diatomic calculations carried out by the UK group, is the use of the 'Selected State R-matrix Method' [3] which allows individual R-matrix poles to be evaluated at different levels of approximation. The R-matrix is given in terms of eigenvalues and boundary amplitudes, E_k, w_{ik} and E'_k, w'_{ik}, obtained in the two approximations by

$$R_{ij} = \sum_{k=1}^{n} \frac{w_{ik}w_{jk}}{E_k - E^N_{ref} - \mathcal{E}} + \sum_{k=n+1}^{K} \frac{w'_{ik}w'_{jk}}{E'_k - E'^N_{ref} - \mathcal{E}} \tag{2}$$

where E^N_{ref} is the target energy relative to which the scattering energy $\mathcal{E}$ is defined. The advantage of this method is that not all of the eigenvalues and eigenvectors of the

Hamiltonian used in the first approximation need be extracted. Large scale CI models for the scattering state will lead to Hamiltonian matrices of large dimension and full diagonalization will not be practicable. In such a case, only the first few eigenvalues and their corresponding eigenvalues need be obtained, the rest of the R-matrix being built from those from a smaller Hamiltonian corresponding to a simpler model.

In the methane calculation the lower poles were obtained using static-exchange plus polarization (SEP) model and the higher ones using the simple static-exchange (SE) model. The SEP model is obtained by adding to the SE, two-particle one-hole terms constructed using the valence orbitals of the target. Integral and differential cross sections were obtained for $^2A_1$, $^2T_2$ and ^2E scattering symmetries. The shape of the integrated cross section is in good agreement with experiment[33, 34, 35]. The magnitude in the vicinity of the Ramsauer minimum is rather too low in comparison with experiment, but is in good agreement with the Complex Kohn results of Lengsfield *et al* [36].

Electron scattering by O_3[37] has also been studied, but only at the static exchange level of approximation. Two shape resonances were found below 20 eV, neither of which had previously been observed. The early experimental data[38] was limited to a few widely spaced energies, as was the only previous low energy calculation[39]. Recent measurements [40] have confirmed the existance of these resonances, but at somewhat lower energies. This is consistent with general experience with the static exchange approximation which, though it can predict shape resonances, always gives too high an energy due the omission of any correlation in its description of the scattering state.

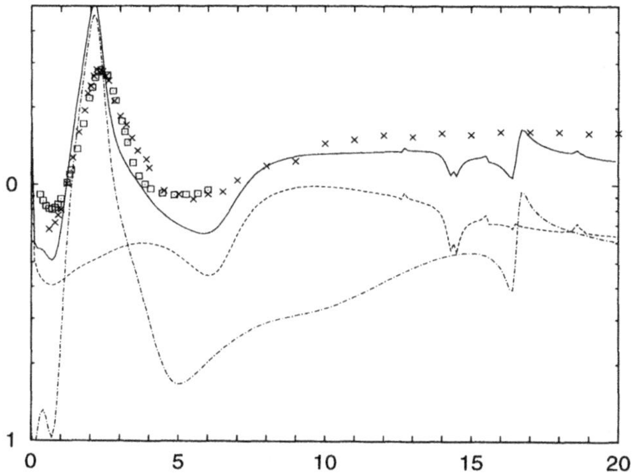

Figure 7. Calculated cross sections for elastic scattering by N_2O, where the $^2\Sigma$ symmetry (broken curve) is from the SEP calculations and the $^2\Pi$ symmetry (chain curve) is from the CI calculations. The solid line is the sum. The experimental data of Szmytkowski *et al* [42] are shown as crosses and those of Strakeljahn *et al* [43] are shown as squares.

The group have recently published results for N_2O[43]. These are the first polyatomic calculations to use CI target wavefunctions. Three levels of approximation were considered, the SE and SEP models described above, using an SCF target, and a CI model where an attempt was made to construct both the target and scattering wavfunctions in a consistent manner. Only the linear, equilibrium, geometry was considered. As in previous work on diatomic systems, polarization effects were found to influence the position of the dominant shape resonance in the $^2\Pi$ symmetry and only the SEP and CI models were in good accord with experiment. Polarization effects were found to have a dramatic effect on the magnitude and shape of the cross section for the $^2\Sigma$ symmetry. The best agreement with experiment was obtained by using the SEP approximation for this symmetry. This would appear to be another manifestation of the problem of obtaining a reliable prescription for the addition of polarization terms to a model which uses a trunctated CI expansion for the target wavefunction. The structure observed at higher energies may not be physical since channels corresponding to electronic excitation have not been included in the model.

4. SUMMARY

The calculations described above are representative of work currently being carried out with the UK and Bonn R-matrix codes. Recent developments will enable us to use considerably better target wavefunctions but problems still remain with the inclusion of the large number of, often diffuse, states which lie close above the lowest valence states of most molecules. The consistent representation of correlation and polarization effects remains the most challenging theoretical problem. Calculations for other processes, such as photoionization, dissociative scattering, dissociative recombination/attachment and scattering from molecules adsorbed on to surfaces are in progress.

References

[1] C.J. Gillan, O. Nagy, P.G. Burke, L.A. Morgan and C.J. Noble, J. Phys. B: At. Mol. Phys. **20** 4583 (1987)

[2] L.A. Morgan in "Computational Methods for Electron Molecule Collisions" ed. W.M. Huo and F.A. Gianturco, (New York: Plenum) (1995)

[3] K. Pfingst, B.M. Nestmann and S.D. Peyerimhoff, J. Phys. B: At. Mol. Opt. Phys. **27** 2283 (1994) and in "Computational Methods for Electron Molecule Collisions" ed. W.M Huo and F A Gianturco, (New York: Plenum) (1995)

[4] A.D. McLean " Conference on Potential Energy Surfaces in Chemistry ", W.A. Lester Jr., ed., IBM Research Laboratory, San Jose, p87 (1971).

[5] C.J. Noble, Daresbury Laboratory Technical Memorandum DL/SCI/TMT33T (1982).

[6] J. Tennyson, J. Phys. B: At. Mol. Opt. Phys. **29** 1817 (1996)

[7] P.G Burke, V.M. Burke and K.M. Dunseath, *J. Phys. B: At. Mol. Opt. Phys.* **27**, 5341 (1994).

[8] B.M. Nestmann and S.D. Peyerimhoff, J. Phys. B: At. Mol. Opt. Phys. **23** L773 (1990)

[9] J. Tennyson, submitted to *Comput. Phys. Commun.*,

[10] F.T. Smith, *Phys. Rev.* **114**, 349 (1960).

[11] K. Higgins, C.J. Noble and P.G. Burke, J. Phys. B: At. Mol. Opt. Phys. **27** 3203 (1994); K. Higgins, C.J. Gillan, P.G. Burke and C.J. Noble, J. Phys. B: At. Mol. Opt. Phys. **28** 3391 (1995); G. Wošte, K. Higgins, P. Duddy, C.M. Fullerton and D.G. Thompson, J. Phys. B: At. Mol. Opt. Phys. **29** 2553 (1996); C.J. Noble, K. Higgins, G. Wošte, P. Duddy, P.G. Burke, P.J.O. Teubner, A.G. Middleton and M.J. Brunger, submitted to Phys. Rev. Lett.

[12] L.A. Morgan, in preparation.

[13] B.M. McLaughlin, C.J. Gillan and P.G. Burke, in preparation.

[14] X. Chen and L.A. Morgan, in preparation.

[15] I. Rabadan and J. Tennyson, submitted to *J. Phys. B: At. Mol. Opt. Phys.*

[16] D. Stibbe and J. Tennyson, submitted to *J. Phys. B: At. Mol. Opt. Phys.*

[17] J. Tennyson, submitted to *Atomic Data and Nuclear Data Tables*

[18] C.J. Gillan, J. Tennyson, B.M. McLaughlin and P.G. Burke, J. Phys. B: At. Mol. Opt. Phys. **29** 1531 (1996)

[19] B.I. Schneider, M. Le Dourneuf and P.G. Burke, *J. Phys. B: At. Mol. Phys.* **12**, L365 (1979).

[20] D. Field, G. Mrotzek, D.W. Knight, S.L. Lunt, and J.P.Ziesel, *J. Phys. B: At. Mol. Opt. Phys.* **21**, 171 (1988).

[21] T.W. Shyn and C.J. Sweeney, *Phys. Rev.* **A 48**, 1214 (1993).

[22] J. Randell, S.L. Lunt, G. Mrotzek, D. Field and J.P.Ziesel, *Chem. Phys. Lett.* **252**, 253 (1996).

[23] D.T. Alle, M.J. Brennan, and S.J. Buckman, *J. Phys. B: At. Mol. Opt. Phys.* **29**, L277 (1996).

[24] J. Tennyson and C.J.Noble, *J. Phys. B: At. Mol. Phys.* **19**, 4025 (1986).

[25] H-J. Werner, P. Rosmus and E-A. Reinsch, J. Phys. B: At. Mol. Opt. Phys. **29** 1531 (1996)

[26] J. Tellinghuisen and C.S. Ewig *Chem. Phys. Lett.* **165**, 355 (1990).

[27] E. Miescher and K.P. Huber, *International Review of Science (Physical Chemistry Series Two)* volume 3 (London, Butterworth) (1964) *J. Phys. B: At. Mol. Opt. Phys.*

[28] S. Fredin, D. Gauyacq, M. Horani, C. Jungen, G. Lefevre and F. Masnou-Seeuws, Mol. Phys. **60** 825 (1987) *J. Phys. B: At. Mol. Opt. Phys.*

[29] S.E. Branchett and J. Tennyson *Phys. Rev. Lett.***64**, 2889 (1990).

[30] S. Trajmar, D.F. Register, and A. Chutjian, *Phys. Rep.***7**, 219 (1983)

[31] M.J. Brunger and P.J.O. Teubner *Phys. Rev.***A 41**, 1413 (1990)

[32] B.M. Nestmann, K.Pfingst and S.D. Peyerimhoff, J. Phys. B: At. Mol. Opt. Phys. **27** 2297 (1994)

[33] J. Ferch, B. Granitza and W Raith, *J. Phys. B: At. Mol. Phys.* **18**, L445 (1985).

[34] B. Lohmann and S.J. Buckman, *J. Phys. B: At. Mol. Phys.* **19**, 2565 (1986).

[35] W. Sohn, K.H. Kochem, K-M. Scheuerlein, K. Jung and H Ehrhardt, *J. Phys. B: At. Mol. Phys.* **19**, 3625 (1986).

[36] B. H. Lengsfield, T.N. Rescigno and C.W. McCurdy, *Phys. Rev.* **A 44**, 4296 (1991).

[37] B K Sarpal, K Pfingst, B M Nestmann and S D Peyerimhoff, Chem. Phys. Lett. **230** 231 (1994)

[38] T.W. Shyn and C.J. Sweeney *Phys. Rev.* **A 47**, 2919 (1993).

[39] Y. Okamoto and Y. Itikawa, *Chem. Phys. Lett.* **203**, 61 (1993).

[40] N.J. Mason, J.M. Gingell, J.A. Davies, H. Zhao, I.C.Walker and M.R.F. Siggel, J. Phys. B: At. Mol. Opt. Phys. **29** 3075 (1996)

[41] C. Szmytkowski, G Karwasz and K Maciag Chem. Phys. Lett. **107** 481 (1984)

[42] G. Strakeljahn, J. Ferch and W Raith, *5th European Conf. on Atomic and Molecular Physics (Edinburgh) Abstracts* 542 (1995)

[43] B.K. Sarpal, K. Pfingst, B.M. Nestmann and S.D. Peyerimhoff, J. Phys. B: At. Mol. Opt. Phys. **29** 857 (1996)

ELECTRON COLLISIONS WITH MOLECULES ON METAL SURFACES

J.P. Gauyacq, B.Bahrim, A.Djabri, V.Djamo and D. Teillet-Billy

Laboratoire des Collisions Atomiques et Moléculaires

(Unité associée au CNRS 281),

Bât. 351, Université Paris-Sud

91405 Orsay Cédex, France.

1. INTRODUCTION

Inelastic scattering of low energy electrons is a widely used method in surface analysis. Indeed, the High Resolution Electron Energy Loss Spectroscopy (HREELS) efficiently allows to recognize the molecules adsorbed on a surface, the vibrational energy loss being characteristic of a given molecule. The high resolution achieved in HREELS (below 1 meV now[1]) also allows to analyse the vibrational movement of the adsorbate with respect to the substrate, as well as to give informations on the state of the adsorbed molecule (for example, the O_2 and O_2^- molecules have rather different vibrational frequencies and can then be recognized). More recently, the inelastic electron-adsorbed molecule collision processes have received a more direct interest, experimental and theoretical studies being devoted to the dynamics of the energy transfer between a low energy electron and molecules adsorbed on a surface. Similary to the case of collisions with free molecules, it was quickly recognized that resonances were playing a key role in the vibrational excitation of adsorbed molecules. Indeed, the large mass ratio between the electron and the nuclei makes the vibrational excitation process very weak in general and the presence of a resonance which increases the collision time strongly enhances the vibrational excitation process in a well defined energy region. The above argument holds both for free and adsorbed molecules since it is only based on mass ratios and collision times. The first experimental observations of the importance of resonant scattering were reported in 1981 by Sanche and Michaud[2] and Demuth et al[3] for simple molecules. Since then, numerous studies have been devoted to resonant scattering by adsorbed molecules (see e.g. the reviews by Sanche[4] and Palmer and Rous[5]). Resonance were shown to dominate vibrational excitation for low energy electron -adsorbed molecule collisions and also to induce electronic excitation[6] as well as to excite the molecule movement with respect to the substrate[7].

Photon and Electron Collisions with Atoms and Molecules
Edited by Burke and Joachain, Plenum Press, New York, 1997

Usually, adsorbed molecules are classified in physisorbed and chemisorbed molecules according to the strength of the molecule surface bond. In the case of chemisorption, the bond is very strong, possibly associated with a charge transfer between the molecule and the surface. In the case of physisorption, the molecule-surface bond is weaker and the molecule is only weakly perturbed by the neighbouring surface, so that one can expect that the resonances present in the free molecule will also influence the scattering in the case of physisorbed molecules. In the present paper, we will only discuss the case of physisorbed molecules and by concentrating on a few examples of theoretical studies on simple molecules (N_2, O_2 and H_2) we will try to illustrate some qualitative features of the inelastic resonant electron scattering by molecules physisorbed on metals and their comparison with the case of free molecules. Before describing theoretical results obtained with the Coupled Angular Mode method (parts 3-5), we will analyze what can be the leading effects introduced by the presence of the surface.

(i) First of all, with the case of a resonant process :

$$e^- + AB \rightarrow AB^{-*} \rightarrow e^- + AB^*$$

the intermediate state is a charged species. At large distances, the interaction between a charge and a surface can be described as the image charge interaction and from this one can deduce a few qualitative features : the resonance energy should be shifted down by an amount of the order of $1/4Z$ when Z is the molecule surface distance ; the image interaction being attractive, it should pull the electron towards the surface thus modifying the angular distributions of scattered electrons and decreasing the resonance lifetime.

(ii) A second effect arises from the anisotropy of space that is introduced by the surface : the collisional electron can go into (come from) the metal or the vacuum and the energetics of these two sides are different, leading to strong effects on the excitation processes.

(iii) A third aspect is connected with the crystal nature of the surface : electrons are diffracted by the metal crystal which thus presents a reflectivity that depends on the electron energy and angle. This reflectivity thus can influence the decay of the intermediate negative ion. In addition, adsorbed molecules are usually not unique and can present a well structured layer on the metal surface. Scattering of the electron inside this layer could also influence the inelastic processes.

2. THEORETICAL APPROACH : THE COUPLED ANGULAR MODE (CAM) METHOD

Following the first experimental observations of resonant vibrational excitation, a few theoretical model studies were performed on the N_2/Ag system in order to analyze the various features of the process[8,9] ; their main conclusion was that reproducing the experimental results for adsorbed molecules requires the lowering of the resonance energy and the decrease of its lifetime compared to the case of the free molecule. In a few other studies, a description of the electron surface interaction was added to that of the electron-free molecule interaction and this was used to study the position and width of resonant intermediates[10-13]. In some of these studies, a crystallographic structure was introduced which concluded on its effect at relatively high energy and not at low collision energy in the few eV range. Below, we will present results obtained in the Coupled Angle Mode (CAM) method. This can study both the static problem (position and width of the resonant intermediate) and the dynamic problem (vibrational excitation probabilities and cross sections). This method will be only briefly presented here and more detailed accounts can be found elsewhere[14-15]. Basically, this method studies the scattering of an electron by the superposition of an electron-surface and an electron-molecule interactions which are supposed to be additive. These two interactions have completely different symmetries and thus the metal surface mixes the various molecular symmetries. The electron surface interaction is described as a local potential only function of z, the electron-surface distance. This corresponds to a jellium approximation for a free electron metal[16] and does not contain any crystallographic aspects. In practice we used potentials given by Jennings et al[17]. The electron molecule interaction is described in the Effective Range Approximation (ERT) which was developped to treat low energy inelastic electron scattering by free molecules[18]. In this method, the space for the collisional electron is cut into two regions. In the inner region ($r > r_c$, r is the electron molecule distance), the electron interacts with the molecule via a local potential. The inner region ($r < r_c$) is represented by a boundary condition on the electron wave function expanded on uncoupled angular modes at $r = r_c$. This boundary condition is assumed to be independent of the electron energy in the limit of low energy collisions. The boundary condition in $r = r_c$ can either be extracted from ab initio calculations on the e^-- AB system, or adjusted so that the ERT results reproduce ab initio results on electron scattering or experimental results on electron scattering. The collisional system is then described by an expansion over spherical harmonics centered on the molecule $Y_{\ell m}$ and over the target molecule vibrational wavefunctions $\chi_v(R)$:

$$\psi = \sum_{v} \sum_{\ell m} \frac{1}{r} F^{v}_{\ell m}(r)\, Y_{\ell m}\, \chi_v(R) \tag{1}$$

This expansion is brought into the Schroedinger equation to yield a set of coupled equations for the radial wavefunction $F^{v}_{\ell m}(r)$:

$$-\frac{1}{2}\frac{d^2F^{v}_{\ell m}}{dr^2} + V^{ext}_{\ell m}\, F^{v}_{\ell m} + \sum_{\ell'} < \ell m \left| V_{es} \right| \ell' m)\, F^{v}_{\ell' m} = EF^{v}_{\ell m} \tag{2}$$

where $V^{ext}_{\ell m}$ is the e^--molecule interaction in the external region ; V_{es} is the e^--surface interaction potential.

The $F^{v}_{\ell m}$ radial wavefunctions also have to fulfill the ERT boundary conditions. The solution of the above coupled equations yields the S matrix from which we compute summed cross sections. The expansion over $Y_{\ell m}$ in (1) is well adapted for e^--molecule interaction and not for the e^--surface interaction. This leads to difficulties in the computation of differential cross sections with a finite number of $Y_{\ell m}$ terms and so we only computed summed cross sections which are, in a way, the equivalent of total cross sections for non-isotropic environments. They are defined as the integral of the differential cross section over the exit angle, averaged over the incident angle, the integrations being restricted to a half space : either the metal or the vacuum half space. One thus defines the following summed cross sections : σ_{VV} when both the incident an exit electrons are in the vacuum side, this corresponds to a usual scattering experiment ; σ_{VM} when the electron comes from the vacuum and goes into the metal, this process is present in a scattering experiment but not observed ; σ_{MV} and σ_{MM} which are processes involving electrons coming from the metal (hot electrons or photo-electrons).

3. VIBRATIONAL EXCITATION OF N_2 PHYSISORBED ON Ag

The N_2 molecule presents around 2 eV a $^2\Pi_g$ resonance associated with a $d\pi_g$ wave which leads to a very strong excitation process, exhibiting the well known "boomerang" oscillations[19-20]. This resonance has been observed in molecules adsorbed on a few substrates[3,21,22]. We present below the results obtained with the CAM method[11,15] using an ERT modelling of the e⁻-free N_2 interaction adjusted on the study of Dubé and Herzenberg[23].

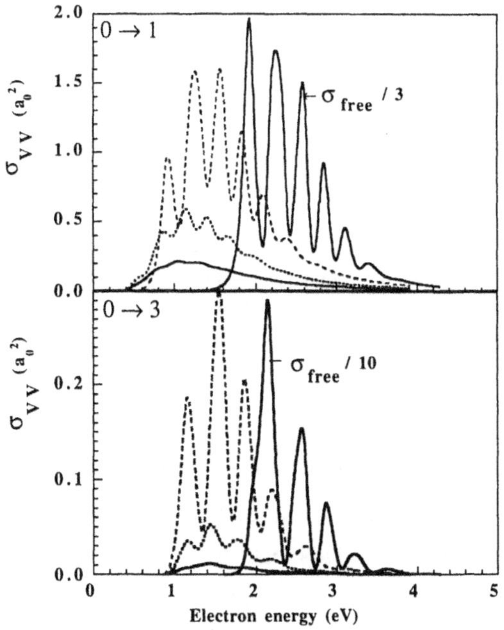

Figure 1. Summed cross sections σ_{VV} for vibrational excitation of N_2 in the VV process as functions of the electron incident energy (v=0 → v=1,3 excitations). Full line with label : σ_{VV} for the free molecule (total excitation cross section divided by 4) ; dashed line : σ_{VV} for a molecule-surface distance Z of 7 a_0 ; dotted line : σ_{VV} for Z=5 a_0 ; full line (unlabeled) : σ_{VV} for Z=4 a_0. (Reproduced with permission from Djamo et al[15]).

Figure 1 presents the results of the σ_{VV} summed cross section for various vibrational excitations in the case of a molecule standing upright on the surface at various adsorption heights ($4a_O$, $5a_O$, 7ao and infinity). The metal surface is Ag for which the adsorption distance is around $5a_O$[24] ; studies of the electron scattering process with different molecule-surface distances illustrate the effect of the surface on the excitation process. As the main results, when the molecule approaches the surface the excitation cross section shifts to lower energies, it decreases and the boomerang oscillations are progressively damped. This is consistent with the energy position of the resonance being lowered and the resonance lifetime shortened due to the image charge interaction[11]. In particular, the boomerang oscillations come from interferences due to the vibrational motion of the intermediate N_2^- ion ; the increase of the resonance width is rather moderate (around 50 % between the free molecule and the distance $Z = 5a_O$) but nevertheless, it is enough to almost completely remove the oscillations, the molecule being unable to vibrate during the collision. Figure 2 presents a comparison of the CAM results (σ_{VV}, $Z = 5\ a_O$, molecule perpendicular to the surface) with the experimental results of Demuth et al[3]. The latter are relative values of the differential cross section at a fixed angle. The agreement between the two energy dependences is quite satisfactory.

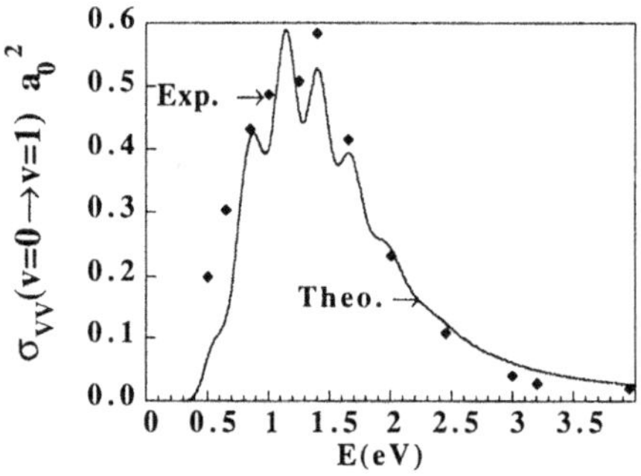

Figure 2. Comparison of the CAM theoretical result for the summed σ_{VV} cross section (v=0 → v=1) excitation) with the experimental results of Demuth et al[3] (black diamonds) as functions of the incident electron energy. The absolute vertical scale corresponds to the present results and the experimental results have been scaled to it. (Reproduced with permission from Djamo et al[15]).

Figure 3 presents the same comparison for the overtone excitation i.e. the ratio between the vibrational excitation cross section for 0 → v and that for 0 → 1. The CAM overtone excitation is smaller than for the free molecule and it decreases very rapidly when v increases in good agreement with experiment. This feature could be attributed to the increase of the resonance width. However, the situation is more complex as can be seen on figure 4 which shows a comparison of σ_{VV} and σ_{VM} for $Z = 5\ a_O$ and the 0 → 1 and 0 → 3 excitations. The σ_{VM} cross section is much larger than the σ_{VV} cross section, so in a scattering experiment, one only looks at a small fraction of the excitation. In addition, they display different shapes as functions of the collision energy. This is due to the difference in energetics between the two processes : indeed, in the VM process, the outgoing electron goes into the metal, thus gaining some energy and then the vibrational excitation does not present any threshold in the studied energy range. This is quite visible on the high v excitations, for which the σ_{VM} cross sections maximize around the N_2^- resonance position.

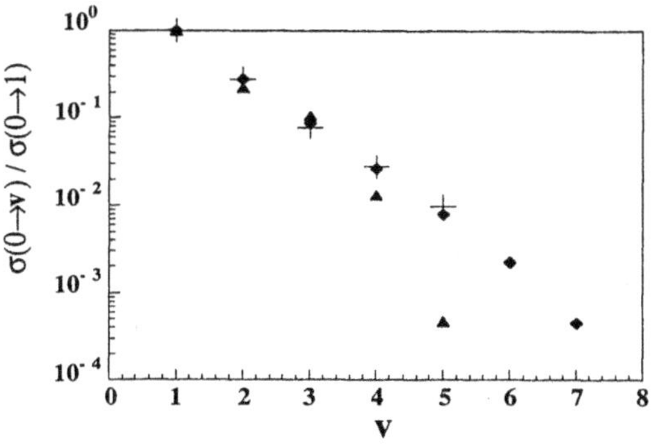

Figure 3. Comparison of the CAM theoretical result for the overtone excitation ratio with the experimental results of Demuth et al[3] (crosses). Theoretical results : black diamonds : ratio between the excitation summed cross sections (VV process) at their maxima ; black triangles : ratio between the excitation summed cross sections at the fixed energy 1.5 eV. (Reproduced with permission from Djamo et al[15]).

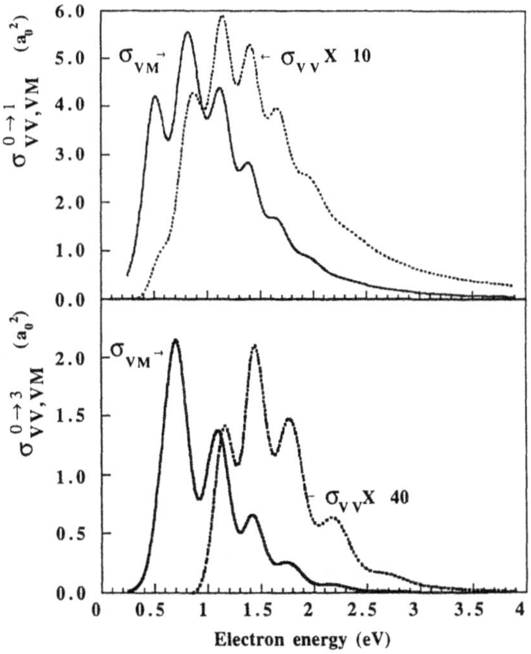

Figure 4. Comparison between the summed cross sections (v=0 → v=1,3 excitation) for the VV (dotted lines) and VM (full lines) processes as functions of the incident electron energy. Note the multiplicative factors on the σ_{VV} cross sections. (Reproduced with permission from Djamo et al[15]).

In contrast for the VV process, the N_2^- resonance is below the excitation threshold and the σ_{VV} cross section maximizes in the wing of the resonance and consequently with a smaller value. The decrease of the VV overtone excitation thus appears as also due to the shift of the resonance below to the excitation threshold and not only to the increase of the resonance width. This effect is a consequence of the existence of two very different half-spaces introduced by the surface which modifies the energetics of the problem. In the present case, it is rather strong because the resonance is at low energy. In the case of the N_2^- ($^2\Pi_g$) resonance, the vibrational excitation process thus appears to be much modified by the adsorption. As we will see below, the situation can be different in other systems.

4 ELECTRONIC EXCITATION OF O_2 PHYSISORBED ON A JELLIUM METAL

The electronic excitation of the low lying states of O_2 :

$$e^- + O_2\ (\pi_u^4\,\pi_g^2,\ ^3\Sigma_g^-) \rightarrow e^- + O_2\ (\pi_u^4\pi_g^2,\ a^1\Delta_g,\ b^1\Sigma_g^+)$$

is a spin forbidden transition and it implies an exchange process. In has been shown [25-28], that for low energy collision on a free molecule, this process is dominated by the low lying O_2^- ($\pi_u^4\pi_g^3\ ^2\Pi_g$) resonance. In this case, the O_2^- resonance is located below the excitation threshold and the excitation process only occurs in the high energy wing of the resonance. This process has been studied using the CAM method with the e^- - O_2 interaction taken from the ab initio study of Teillet-Billy et al [26].

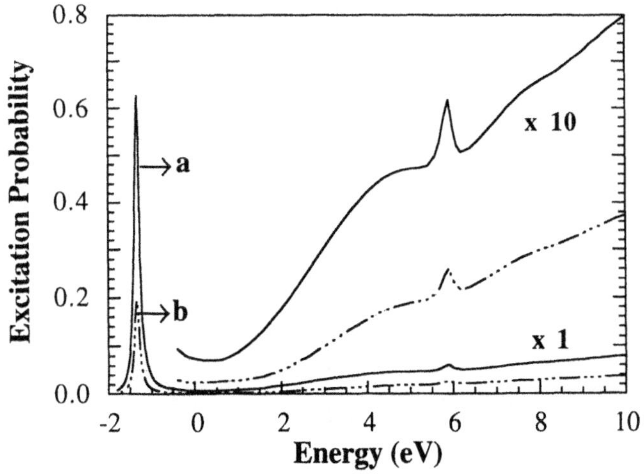

Figure 5. Excitation probability of the O_2 molecule via the $^2\Pi_g$ resonance as function of the incident electron energy. Full line : excitation to the $a^1\Delta_g$ state ; dashed-triple dotted line : excitation to the $b^1\Sigma^+_g$ state. The O_2 molecule is perpendicular to the metal surface at a distance 5 a_0. (Reproduced with permission from Bahrim et al[29]).

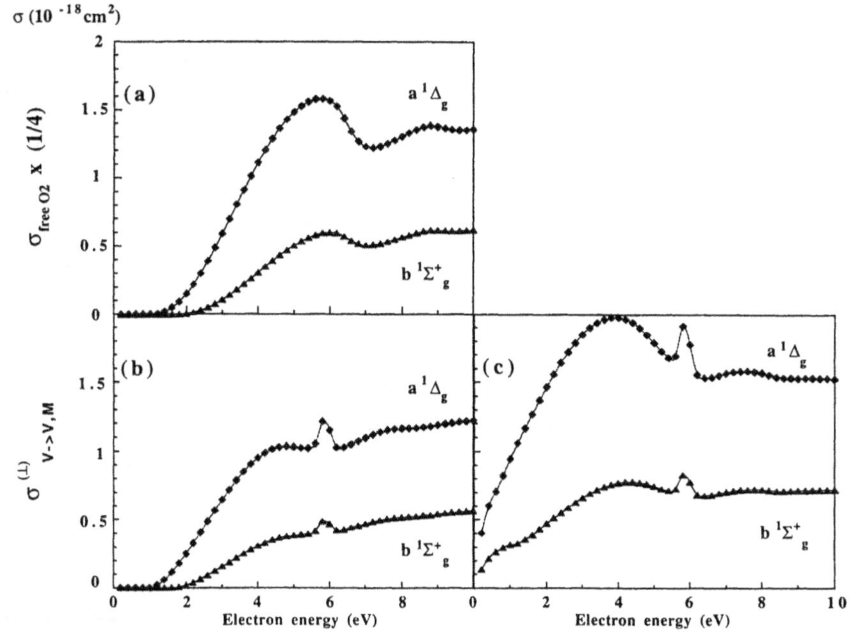

Figure 6. Summed excitation cross sections for the O_2 molecule as functions of the incident electron energy. Black diamonds : excitation of the $a^1\Delta_g$; black triangles : excitation of the $b^1\Sigma^+{}_g$ state. a) : the O_2 molecule is free ; b) : VV process for the O_2 molecule perpendicular to the metal surface at a distance 5 a_o ; c) : VM process for the O_2 molecule perpendicular to the metal surface at a distance 5 a_o. (Reproduced with permission from Bahrim et al[29]).

Figure 5 shows the excitation probabilities ($X \rightarrow a$) and ($X \rightarrow b$) as functions of the collision energy for an O_2 molecule standing upright at a distance $Z = 5a_o$ on a jellium metal surface with the Al characteristics[29]. For this study, only the $^2\Pi_g$ O_2^- intermediate is considered in order to analyze how this resonant excitation involving a resonance hidden below the threshold is modified in our physisorbed molecule. For positive energies, the excitation probabilities are slowly rising as functions of the collision energy in a way very similar to the case of the free molecule. (Note that the peak around 6 eV should be considered as an artefact in the calculation, it stands at the position of the $^2\Pi_u$ O_2^- resonance, however the contributions to the a and b excitation through this resonance are not properly introduced in the calculation). For negative energies, the excitation probabilities present a sharp peak at the position of the O_2^- ($^2\Pi_g$) resonance ; this peak is indeed absent in calculations for the free molecule. It thus appears that putting the O_2 molecule in front of a metal allows to probe the electron -molecule scattering at negative energies and thus to uncover hidden resonances. The figure 5 thus displays two different resonant contributions : one at negative energies with the usual more or less lorentzian shape peaked at the resonance position and a second one corresponding to transitions in the wings of the resonance (the latter is associated with a very small time delay and is mainly a way of recoupling the various angular momenta in the collision , see discussion in[26]). Summed cross sections are presented on figure 6 for an O_2 molecule perpendicular to the surface at a distance Z = $5a_0$[29]. Figure 6 displays the a and b excitation summed cross sections for the free molecule and for the VV and VM processes. As the main feature, the excitation cross sections for the adsorbed molecules are not much different from those for the free molecule : the VV (resp. VM) cross sections are smaller (resp. larger) than the free molecule ones, with a shift toward low energies. It is noteworthy that these changes due to adsorption are much weaker than the ones observed in the case of N_2. This is attributed to the peculiar resonant character of the O_2 excitation process : it involves the wing of the resonance, far from its center, where the time delay is very small ; such a process will have a very little sensitivity on the characteristics of the resonance (energy position and time delay at its center) and thus on the effect of a neighbouring surface. Indeed, close to the energy threshold, the sharing of the excitation between the Vacuum and Metal sides is very effective and leads to sizable differences between the VV and VM processes.

5. VIBRATIONAL EXCITATION OF H_2 MOLECULES PHYSISORBED ON A JELLIUM METAL.

The H_2 molecule presents at low energy a $^2\Sigma_u^+$ resonance associated with a pσ wave which leads to an important vibrational excitation in the few eV energy range. This resonance has been observed in adsorbed molecules in a few systems[3,7]. It has been studied using the CAM method and an ERT modelling of the e^--free H_2 interaction ($^2\Sigma_u$ symmetry) adjusted on the results of Berman et al[30].

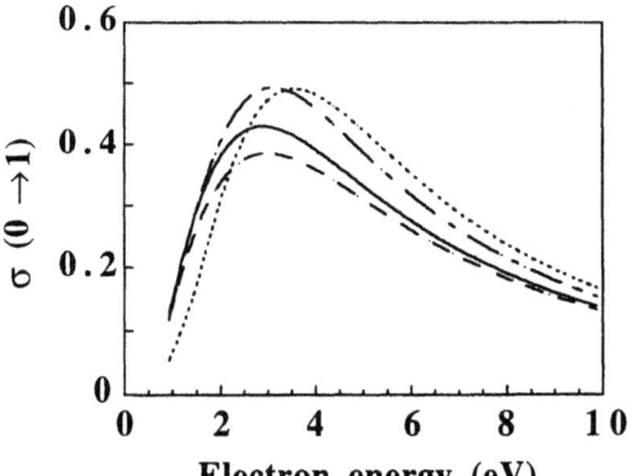

Figure 7. Summed vibrational excitation cross section (v=0 → v=1, VV process, units of 10^{-16} cm^2) for the H_2 molecule perpendicular to the metal surface as a function of the incident electron energy. Dotted line : free H_2 molecule ; dashed-dotted line : H_2 molecule at a distance Z=10 a_0 from the surface ; full line : H_2 molecule at Z=5 a_0 ; dashed line : H_2 molecule at Z=4 a_0.

Figure 7 presents preliminary results[31] for the VV summed cross section for the $0 \rightarrow 1$ vibrational excitation in the case of an H_2 molecule physisorbed on a jellium metal (Al characteristics) at different distances ($Z = 4a_o$, $5a_o$, $10a_o$ and infinity). These preliminary results use a simplified description of the e^--H_2 interactions and treat the vibrational motion of the molecule in the energy modified adiabatic approximation[32]. Only the vibrational excitation occuring through the $^2\Sigma_u^+$ resonance is taken into account. The VV summed cross section appears to be only slightly modified by the adsorption in contrast with the similar result for the N_2^- ($^2\Pi_g$) resonance. When the adsorption height decreases, the cross section is shifted toward low energy. However, this shift is smaller than what can be expected from an image charge shift. The size of the VV cross section slightly decreases when the adsorption height decreases, by 12 % between infinity and $Z = 5a_o$. This is much smaller than the factor 6 reduction observed in the case of the N_2^- ($^2\Pi_g$) resonance. Similar results are obtained for the $0 \rightarrow 2$ vibrational excitation.

This very weak perturbation of the $^2\Sigma_u^+$ resonant vibrational excitation of H_2 due to the presence of the surface is attributed to the very large width of the H_2^- ($^2\Sigma_u^+$) resonance. Although difficult to determine because of the very short lifetime, the $^2\Sigma_u^+$ width is usually assumed to be larger than a few eV at the H_2 equilibrium distance. In such a case, the perturbation introduced by the surface cannot significantly modify the resonance width ; in addition, the low energy part of the excitation cross section is determined as much by the threshold law as by the resonance characteristics and as a consequence, such a cross section is only weakly sensitive to the presence of the metal surface. This is quite consistent with the experimental results of Demuth et al[3] who observed very similar energy dependences of the excitation cross section for the free and physisorbed molecule.

6. CONCLUSIONS

We have presented recent theoretical results obtained with the CAM method on the electron impact excitation of N_2, O_2 and H_2 molecules physisorbed on jellium metals. The resonances that are very efficient for the free molecule excitation are also found to play a very important role in the physisorbed molecules. In this approach with a jellium description of the metal surface, the effect of the surface is mainly to modify the resonance characteristics (lowering of the energy, increase of the width) and to introduce a strong anisotropy in space (Vacuum and Metal). Depending on the system, this leads to more or less important modifications of the strength of the excitation process. The relative orientation of the surface normal and of the molecular axis has also been shown to lead to significant effects[33].

Although significant progress has been achieved in the past year on this problem of electron impact excitation of physisorbed molecules a few questions still require further work :

- The effect of the crystallography : it has been shown to influence the resonance characteristics for high lying resonances, but not for low lying ones[13,34], however, the effect on the excitation cross sections has still to be studied,
- The effect of multiple scattering : the adsorbed molecules form a layer on the metal through which the electron can be multiply scattered[35] and this could influence the excitation process,
- The excitation of molecule-surface vibrational modes : it has been shown experimentally[7] that the resonant scattering was not only resulting in the excitation of the intramolecular vibration but also of the relative molecule-surface motion,
- Differential scattering : the above results only concern summed cross sections. The differential inelastic cross sections are directly linked to the symmetry of the involved resonance and to the geometry of the adsorption. Recently, it has been shown that the knowledge of the inelastic differential cross section for the free molecule could be used to analyse the electron angular distribution for physisorbed molecules and so to provide a surface analytical tool[36,37].

REFERENCES

1. H. Ibach, *Surf. Sci* 299/300:116 (1994).
2. L. Sanche and M. Michaud, *Phys. Rev. Lett.* 47:1008 (1981).
3. J.E. Demuth, D. Schmeisser and Ph. Avouris , *Phys. Rev. Lett.* 47:1166 (1981)
   D. Schmeisser, J. E. Demuth and Ph. Avouris, *Phys. Rev B* 26:2457 (1982).
4. L. Sanche, *J. Phys. B* 23:1597 (1990).
5. R. Palmer and P.J. Rous, *Rev. Mod. Phys.* 64:383 (1992).
6. P.J. Rous, E.T. Jensen and R.E. Palmer, *Phys. Rev. Lett.* 63:2496 (1989).
7. M. Gruyters and K. Jacobi, *Chem. Phys. Lett.* 225:309 (1994).
8. J.W. Gadzuk, *J. Chem. Phys.* 79:3982 (1983).
9. A. Gerber and A. Herzenberg, *Phys. Rev. B* 31:6219 (1985).
10. D. Teillet-Billy and J.P. Gauyacq, *Nucl. Inst. Meth. B* 58:393 (1991).
11. D. Teillet-Billy, V. Djamo and J.P. Gauyacq, *Surf. Sci.* 269/270:425 (1992).
12. P.J.. Rous, *Surf. Sci.* 260:361 (1992)
13. P.J. Rous, *Surf. Sci.* 279:L191(1992)
14. D. Teillet-Billy and J.P. Gauyacq, *Surf.Sci.* 239:343 (1990).
15. V. Djamo, D. Teillet-Billy and J.P. Gauyacq, *Phys. Rev. Lett* 71:3267 (1993).
    V. Djamo, D. Teillet-Billy and J.P. Gauyacq, *Phys. Rev. B* 51:5418 (1995).
16. M.C. Desjonquères and D. Spanjaard, *Concepts in Surface Physics*, Springer, Berlin, Heidelberg (1993)
17. P.J. Jennings, R.O. Jones and M. Weinert, *Phys. Rev. B* 37:6113 (1988).
18. J.P. Gauyacq, *Dynamics of Negative Ions*, World Scientific, Singapore (1987).
    D. Teillet-Billy and J.P. Gauyacq, *J. Phys. B* 17:4041 (1984).
19. G.J. Schulz, *Rev. Mod. Phys.* 45:423 (1973).
20. D.J. Birtwistle and A. Herzenberg, *J. Phys. B* 4:53 (1971).

21. L. Sanche and M. Michaud, *Phys. Rev. B* 27:3856 (1983).
22. K. Jacobi, C. Astaldi, P. Geng and M. Bertolo, *Surf. Sci.* 223:569 (1989).
23. L. Dubé and A. Herzenberg, *Phys. Rev. A* 20:194 (1979).
24 G. Corey and M. Alexander, *J. Chem. Phys.* 87:4937 (1987).
25. C.J. Noble and P.G. Burke, *J. Phys. B* 19:L35 (1986).
26. D. Teillet-Billy, L. Malegat and J.P. Gauyacq, *J. Phys B* 20:3201 (1987).
27. C.J. Noble and P.G. Burke, *Phys. Rev. Lett.* 68:2011 (1992).
28. A.G. Middleton, P.J.O. Teubner and M.J. Brunger, *Phys. Rev. Lett.* 69:2495 (1992).
29. B. Bahrim, D. Teillet-Billy and J.P. Gauyacq, *J. Chem. Phys.* 104:10014 (1996).
30. M. Berman, C. Mündel and W. Domcke, *Phys. Rev. A* 31:641 (1985).
31. A. Djabri, D. Teillet-Billy and J.P. Gauyacq *work in progress.*
32. R.K. Nesbet, *Phys. Rev. A* 19:551 (1979).
33. V. Djamo, D. Teillet-Billy and J.P. Gauyacq, in *Electron Collisions Molecules, Clusters and Surfaces*, L. Morgan and H. Elrhardt Eds , Plenum, New York (1994) p. 227.
V.Djamo, D. Teillet-Billy and J. P. Gauyacq, *Surf. Sci.* 346:253 (1996).
34. P.J. Rous, *Phys. Rev. B* 53:11076 (1996).
35. P.J. Rous, R.E. Palmer and R.F. Willis, *Phys. Rev.* 39:7552 (1989).
36. D. Teillet-Billy and J.P. Gauyacq *Nucl. Inst. Meth. B* 101:88 (1995).
37. K.B.K. Tang, R.E. Palmer, J. Villette, D. Teillet-Billy and J.P. Gauyacq, *Surf. Sci*, in print (1996).

RECENT MEASUREMENTS OF DIFFERENTIAL CROSS SECTIONS FOR ELECTRON-MOLECULE COLLISIONS AT LOW ENERGIES

Michael Allan

Institute of Physical Chemistry
University of Fribourg
CH-1700 Fribourg, Switzerland

INTRODUCTION

Absolute differential cross sections for electron-atom and electron-molecule scattering have been measured for a long time,[1,2] but there is still need for improvement of the techniques. The measurements of elastic differential cross sections (DCSs) are in general easier than those of inelastic DCSs. The reasons are higher signal intensity, but mainly the possibility to calibrate the instrumental response function on the well known elastic cross section of helium. The available data for inelastic scattering is often less satisfactory than for elastic cross sections, in particular where electronic excitation and thus large energy-losses are involved. The aspects, where improvement of the inelastic measurements is desirable, are:

- Extension of scattering angle. Large scattering angles are not accessible to instruments using electrostatic hemispherical energy analyzers, the largest attainable angle being between 120° and 156° (136° for the instrument in Fribourg). Measurements at very small angles are difficult because the main beam enters the analyzer where it often causes large stray electron background.
- Improvement of reliability. Apart from the obvious problems like the notorious drifts of contact potentials encountered in the presence of the sample gases, low signal and substantial background often encountered with the small inelastic cross sections, there is the serious problem of determining the instrumental response function, more problematic for inelastic scattering because, unlike for elastic scattering, there are no reliable standards for calibration.
- Improved performance near threshold.

The present contribution will discuss some aspects of the above problems, present new instrumental developments and illustrate them with some examples. In particular a novel 'electron-mirror spectrometer' is presented, a modification of the magnetically focused trochoidal electron spectrometer (TES), capable of measuring the ratio of the 0° and 180° inelastic DCSs. The second spectrometer in Fribourg, with hemispherical energy analyzers, can measure absolute inelastic DCSs in the range 0° to 136°. Combining the absolute value at 0° with the ratio measured by the first instrument yields absolute DCS at 180°. The two instruments together thus provide absolute DCS from 0° to 136° continuously, plus one more point at 180°.

EXPERIMENTAL

Measurement of inelastic cross sections at 180°: the electron-mirror spectrometer

The trochoidal electron spectrometer, utilizing an axial magnetic field to focus the electrons, has been shown[3] to detect both forward and backward inelastically scattered electrons. The latter are reflected into the forward direction by a potential barrier which they encounter after leaving the target chamber, as shown schematically in figure 1. The instrument could thus be called an 'electron mirror spectrometer'. (The electrons scattered elastically into 180° are not reflected because their energy is higher than the potential barrier, and re-enter the monochromator. They have been collected and used by Burrow and Sanche[4] to measure the *elastic* cross sections at 180°.) The reflected electrons re-enter the target chamber, traverse it, at a sufficiently low pressure, without a significant chance of a second collision and enter the analyzer together with the forward scattered electrons.

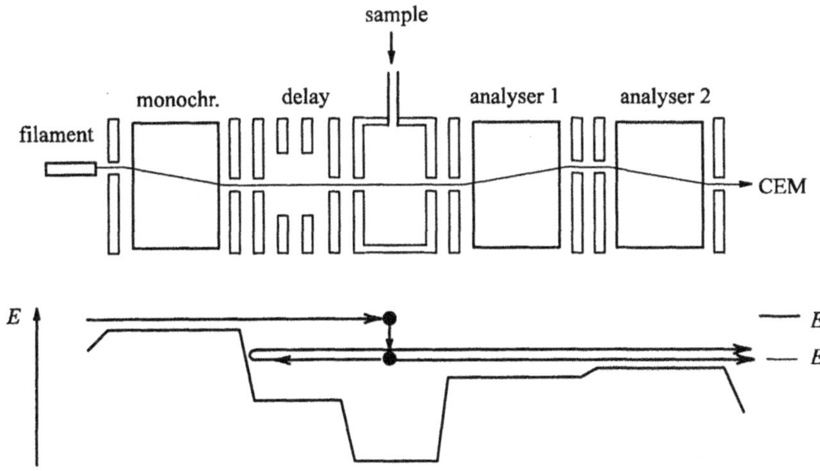

Figure 1 Schematic diagram of the electron mirror spectrometer.

The forward and backward scattered electrons are distinguished using a pulsed electron beam and a time-of-flight detection scheme.[5] Both forward and backward scattered electrons have a given range of arrival times, primarily because of the finite length of the target chamber, along which the collisions take place, causing an overlap of the arrival times.[3] An 'electron delay line' has therefore been developed,[5] delaying the backward scattered electrons before they re-enter the target chamber. The delay line consists of a short (1 cm) path which the electrons have to traverse with a constant and low (≈0.1 eV) energy.

The backward scattered electrons travel a relatively long distance and the alignment of the magnetic field is critical to achieve equal collection efficiency for the 0° and 180°

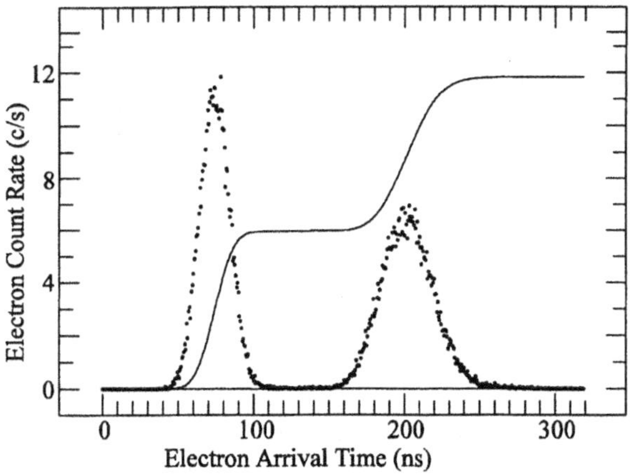

Figure 2. Illustration of the pulsed mode of operation of the trochoidal electron spectrometer. The figure shows the yield of electrons having excited the $\upsilon = 5$ level of N_2 at an incident energy of 2.4 eV as a function of their arrival time at the detector. The first band corresponds to the forward scattered electrons, the second to the backward scattered electrons. The integral, shown by a solid line, indicates that the areas under both bands are equal, that is an equal collection efficiency for electrons scattered into 0° and into 180°.

electrons. It is tested on the vibrational excitation in N_2 in the 2-4 eV energy range. The inelastic scattering in N_2 in this energy range is dominated by the $^2\Pi_g$ resonance.[2] The angular distribution is therefore symmetrical around 90° and the DCS is the same at 0° and 180° making this process suitable to test the collection efficiency in the 0° and the 180° directions. A properly adjusted instrument yields equal time integrated signal for the forward and the backward electrons, as shown in figure 2.

The instrument can be used to measure the ratio of differential cross sections at 0° and 180° at a given incident energy and a given energy-loss, by recording the time histogram of the electron arrival times as shown in figure 2. The ratio of the forward and backward cross sections is equal to the ratio of the areas under the two bands. The instrument can further measure the dependence of the 0° or 180° DCS as a function of incident electron energy by defining discriminator levels passing only pulses within a window corresponding to either forward or backward scattered electrons.

The hemispherical analyzer spectrometer

This instrument has been described previously.[6-8] It is an essentially standard electrostatic spectrometer using hemispherical deflectors for electron energy selection. The analyzer can be rotated from -3° to 135° with respect to the incident beam. The angular dependence spectra are measured by repetitively scanning the analyzer position with a stepping motor. Absolute cross sections are determined using the relative flow technique.[9] The error of the absolute values is dominated by the uncertainties in determining the instrumental response function. This error is larger at low incident energies and near threshold. As a general guide, the present data is believed to be accurate within ±35%. The energy scale is calibrated on the 19.366 eV resonance in helium and is accurate within ±30 meV. The stray electron rejection of this instrument is sufficiently high to permit the measurement of inelastic cross section even at 0°. This value is then used to normalize the data from the electron mirror spectrometer, yielding absolute data at 180°.

A small Wien filter, incorporated between the second hemispherical analyzer and the channel electron multiplier, was used where needed to distinguish between scattered electrons and stable fragment ions from dissociative attachment.

Central and critical part of the measurement is the determination of the instrumental response (or sensitivity) as a function of the incident and residual energies and of scattering angle.

Determination of the instrumental response for elastic scattering does not pose major problems: The helium elastic signal is divided by the calculated elastic DCS of Nesbet[10] to yield the sensitivity as a function of either the electron energy or the scattering angle. Examples of the latter at different energies are given in figure 3.

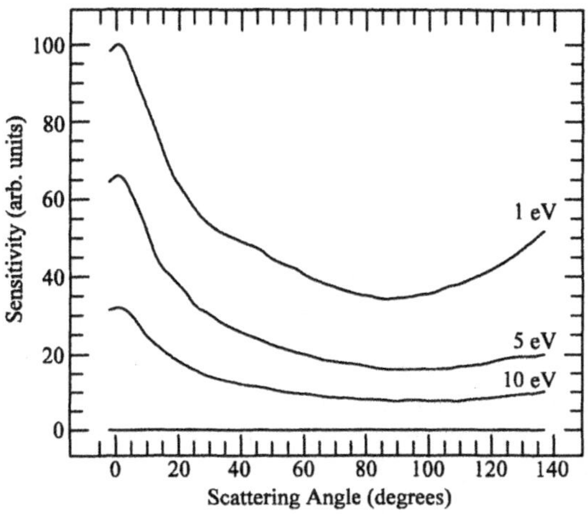

Figure 3. Sample instrumental response curves for elastic scattering as a function of scattering angle.[18] (The elastic signal can be recorded down to angles of about 5°, the response curves were extrapolated below this angle using a numerical simulation as a qualitative guide.)

Determination of the instrumental response for inelastic scattering is more difficult, in particular when the residual, or both the incident (E_i) and the residual (E_r) energies are low (below about 3 eV), because there are no generally accepted standards. High quality calculations of the inelastic cross sections in helium are now becomming available[11] and promise to provide a suitable standard, but they have not (yet) been used as standard in the present work.

We used the accepted (details and original references can be found in Ref. 6) procedure of calibrating the sensitivity of the analyzer $SA_\theta(E_r)$, a function of the residual energy E_r and the scattering angle θ, on the near threshold ionization continuum of helium, which provides a standard in the form of a uniform distribution of secondary electrons. When the incident energy is high (above about 6 eV), providing a constant profile of the incident beam, or when the incident energy is constant, it suffices to divide the observed spectrum point-by-point by $SA(E_r)$ to correct it for the variation of the instrumental response. When E_i is low, then the variation of the incident beam profile (and intensity) with energy come into play and a more elaborate procedure is necessary. To correct spectra recorded at a constant angle θ, we used the simplifying assumption that the instrumental (total) sensitivity $ST(E_i,E_r)$ at any combination of E_i and E_r can be expressed as a product $ST(E_i,E_r) = SM(E_i) \times SA(E_r)$, where $SM(E_i)$ is the monochromator sensitivity determined on elastic scattering.[6] Results from the electron-mirror spectrometer indicate, however, that the ionization continuum of helium does not provide a suitable standard at small and large scattering angles, necessitating additional assumptions at scattering angles far from 90°.

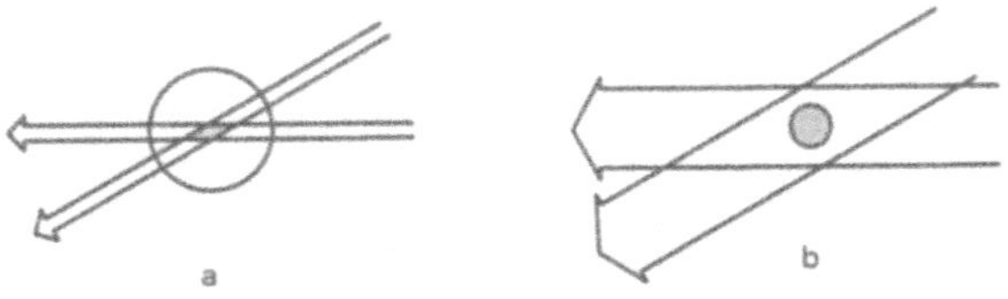

Figure 4. Schematic diagram of the interaction region in two extreme situations. Left: narrow, well defined beams, obtained at higher energies. Right: diffuse beams, at very low energy. The circle indicates the molecular beam (which has the same size in both cases; the left picture is shown enlarged). The effective interaction volume is highlighted.

To justify these assumptions we analyse in more detail the response of the instrument for elastic signal as a function of scattering angle. The response is given by the overlap of the incident electron beam, the analyzer acceptance cone, and the effusive molecular beam as shown schematically in figure 4. It is not possible to calculate the overlap because the exact beam profiles are not known. In particular at low energies the incident beam and the analyzer acceptance cones become more diffuse because of uncontrolled contact potential variations ('patches') on the surface of the electron optics elements. This effect varies slowly with time and depends on the gas being measured. The beam profile can not be measured precisely at low energies, because any 'probe' (such as a small electron collector) introduced into the collision region distorts the fields by its own contact potential, affects the beam profiles and distorts the measurement.

The behavior of the instrument can be understood qualitatively, however, by observing the two extreme situations shown qualitatively in figure 4. When the electron beam and analyzer acceptance cone are narrow with respect to the molecular beam (situation (a)), the response function varies with scattering angle as $1/\sin\theta$ (except at very low scattering angles, where the finite size of the molecular beam causeS a 'cut-off'). This situation is encountered in the experiment at high energies, when the 'patch effect' is negligible (see figure 3). The response function would not depend on the scattering angles at all in the extreme case of very wide electron beam and acceptance cone (figure 4, situation (b)). The experimental sensitivity curves become flatter at low energies, indicating more diffuse electron beam and acceptance cone, that is departure from the situation (a) towards situation (b). The strategy applied in this work is to optimize the beam profiles through computer control of the focussing and steering voltages such as to preserve the $1/\sin\theta$ - like shape of sensitivity down to low energies (it is reasonably preserved even at 1 eV in figure 3) and then use the average of the angular response curves to correct the angular distribution of inelastic signal. (i.e. angular distribution of inelastic signal recorded at E_i and E_r is corrected using an average of the angular sensitivity curves determined on elastic helium signal at $E = E_i$ and $E = E_r$)

Finally the analyzer and monochromator sensitivities $SA(E_r)$ and $SM(E_i)$ at low scattering angles and at low energies, where the calibration on the ionization continuum of helium is no longer possible, were based on the curves determined at 90°, modified in such a way as to be consistent with the 'average rule' used for angular distributions.

The DCSs for the excitation of the $n = 2$ states of helium were measured as a function of scattering angle at 22, 25, and 29.6 eV.[12] The results for the 2^3S state are shown and compared with the theoretical results of Fon *et al.*[11] in Fig. 5. The agreement is satisfactory as far as the general shape is concerned. There are substantial differences in the absolute values at certain angles and energies, however, in particular at 180° and 22 eV, and at all angles at the highest energy, 29.6 eV. More work will be needed to reconcile the details of the experimental and theoretical data.

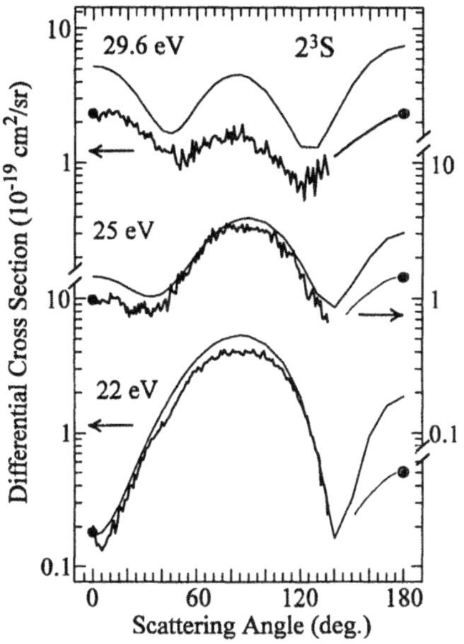

Figure 5. Differential cross sections for the excitation of the 2^3S state of helium. Heavy line: present data from the hemispherical analyzer instrument; filled circles at 0° and 180°: results from the electron-mirror instrument (with visual interpolations indicated by thin lines connected to the circles at 180°); thin line: calculation of Ref.[11] Horizontal arrows point to appropriate DCS scale.

The DCS for the excitation of the n = 2 states of helium were further measured as a function of incident energy at 0°, 30° and 180° and sample results are shown in Fig. 6. The DCS at 0° can be measured both with the electron mirror spectrometer and with the hemispherical analyzer spectrometer, providing a test of mutual consistency of the instrumental response correction procedure. The agreement of the 0° spectra from the two instruments can be considered as very satisfactory when the higher energy and angular resolutions of the hemispherical analyzer instrument are taken into account in the comparison.

Finally, yields of secondary electrons resulting from near-threshold ionization have been recorded as a function of their energy at 0° and 180°.[12] They have shown that whereas such an electron distribution is uniform (i.e. the intensity is nearly constant for all electron energies) at 90° even at energies as high as 3 eV or 5 eV above threshold (where it is used to calibrate the response function of electrostatic instruments), it is not uniform at 0° or 180°. The faster secondary electrons are found preferentially in the forward direction, the slower secondary electrons in the backward direction.

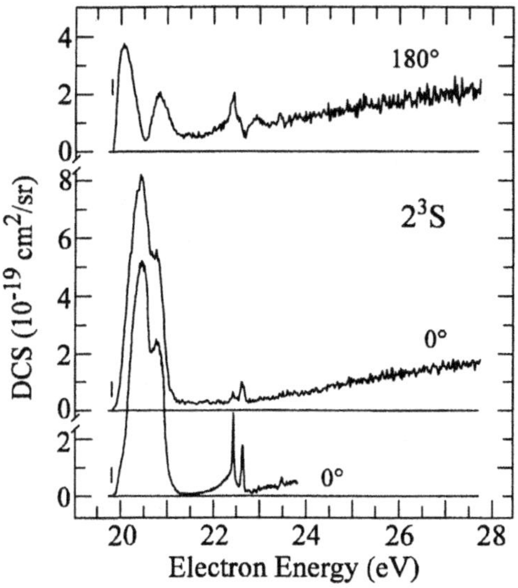

Figure 6. DCS for exiting the 2^3S state in helium at 0° and 180°. The lowest curve has been recorded with the hemispherical analyzer instrument, the top two curves with the electron mirror instrument.

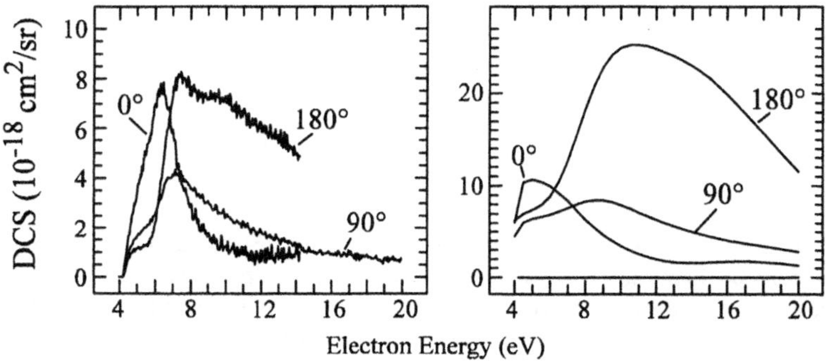

Figure 7. Differential cross sections for the excitation of the first triplet state of ethene. On the left are the present results, on the right results of a Schwinger multichannel calculation. [14,15]

ETHENE

The DCS for the excitation of the lowest triplet state of ethene at 90° have been published.[13] These measurements have recently been extended to 0° and 180° using the new techniques.[12] The DCS are shown in figure 7 as a function of the incident electron energy. The measured curves agree in the essential qualitative features with the Schwinger multichannel calculation of Sun, Winstead, McKoy, and Lima, shown for comparison in figure 7.[14,15] The cross sections consist of a shoulder extending from threshold to about 5-6 eV, and assigned to the π^* shape resonance, followed by a broad peak, assigned to the (π,π^{*2}) core excited shape resonance, followed by a gradual decrease characteristic for excitation of spin-forbidden transitions. An interesting and somewhat unexpected trend, observed in the experimental DCS and qualitatively well reproduced by theory, is a strong dependence of the width and position the (π,π^{*2}) resonance peak on scattering angle. The peak is much narrower and closer to threshold at 0° and becomes gradually wider and its maximum moves to higher energy with increasing angle.

OZONE

Ozone, apart from its environmental and technological significance which need not be repeated here, is a very interesting molecule. It has an unusually low separation of the highest occupied and the lowest unoccupied molecular orbitals (HOMO-LUMO gap), resulting in very low-lying electronically excited states, and a biradicaloidal character, responsible for high chemical reactivity. The dissociation energy into O and O_2 is only 1.05 eV, much less than the typical strength of a covalent bond. These properties put ozone somewhere between a stable molecule and a reactive intermediate.

An extensive study of electron-ozone collisions, combining the mobile ozone production and handling system of University College London[16] and the Fribourg spectrometers, has recently been performed[17,18,19] and a few excerpts are shown here.

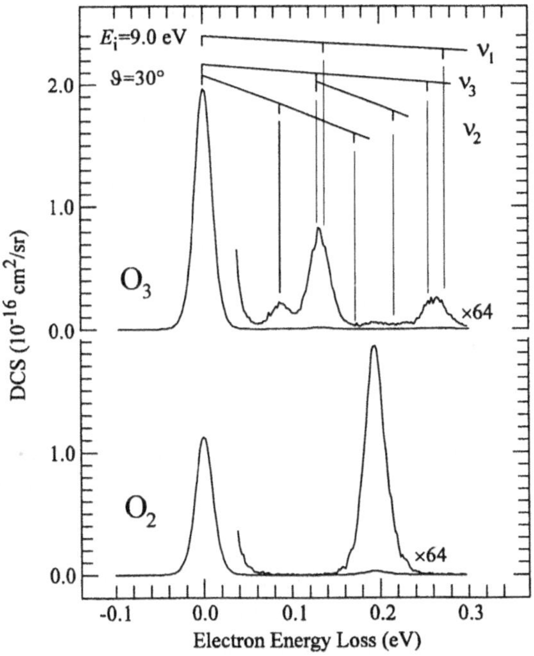

Figure 8. Electron-energy-loss spectra illustrating the determination of the oxygen content of the ozone sample.[18] The upper spectrum indicates that the ozone sample contained less than 3% of oxygen.

The measurement of ozone is complicated by its reactive nature, which causes contact potential drifts to be more severe than with the majority of less reactive samples. Ozone decomposes to oxygen in contact with metals, and the sample consequently contains some oxygen. Vibrational energy-loss spectra at 9 eV proved to be a convenient means of *in-situ* determination of the ozone content, as shown in Fig. 8. Up to 97% ozone was obtained for periods of several hours after prolonged passivation of the inlet system.

Ozone also decomposes in the capacitance manometer and the manifold used to measure the sample flow. A somewhat modified procedure was therefore used.[18]

Figure 9 shows the absolute elastic DCS in comparison with the measurements of Shyn and Sweeney[20] and the calculations of Okamoto and Itikawa.[21] The agreement of the two sets of experimental data can be regarded as satisfactory in view of the additional sources of error caused by ozone's reactivity.

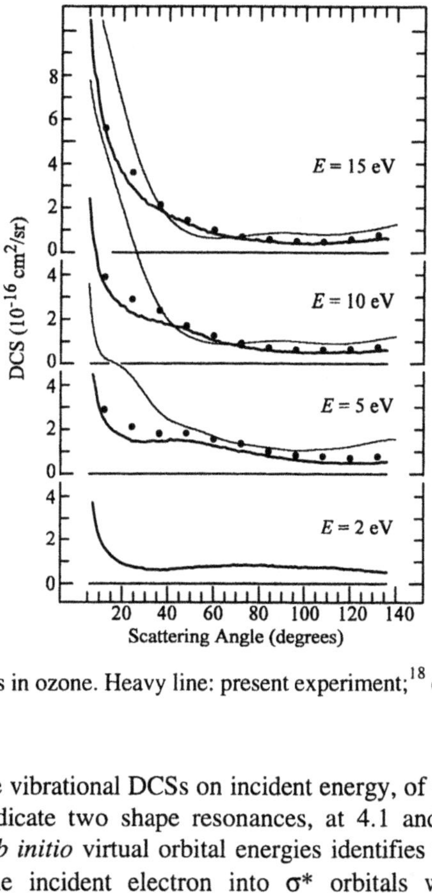

Figure 9. Elastic DCSs in ozone. Heavy line: present experiment;[18] dots: Ref. 20; thin line: calculation of Ref. 21.

The dependence of the vibrational DCSs on incident energy, of which two sample curves are shown in Fig. 10, indicate two shape resonances, at 4.1 and 6.6 eV. Application of Koopmans' theorem to *ab initio* virtual orbital energies identifies them as A_1 and B_2, with temporary capture of the incident electron into σ^* orbitals which are very strongly antibonding with respect to the O-O bond length.[18] The selectivity of the vibrational excitation has two remarkable aspects. Vibrational energy-loss spectra[18] show that all three vibrations are active, in particular that the antisymmetric stretch vibration ν_3 (b_2) is excited, apart from the expected totally symmetric stretch and bending vibrations ν_1 and ν_2. We propose that the antisymmetric stretch vibration is excited by vibronic coupling between the A_1 and B_2 shape resonances.

A second remarkable selectivity can be discerned in Fig. 10: the lower-lying A_1 resonance excites both the stretch and bending vibrations, whereas the higher-lying B_2 resonance does not excite the bending vibration. This finding in well rationalized by the potential curves obtained by the application of the Koopmans' theorem and an empirical scaling equation, shown in figure 11.[22] This treatment is oversimplified in many respect, but it provides a simple estimate of the real part of the resonant potential surfaces and is suitable to rationalize qualitative observations. The lower resonant potential curve is seen to have a much larger slope with respect to the bending normal coordinate than the upper resonance, thus explaining the observation.

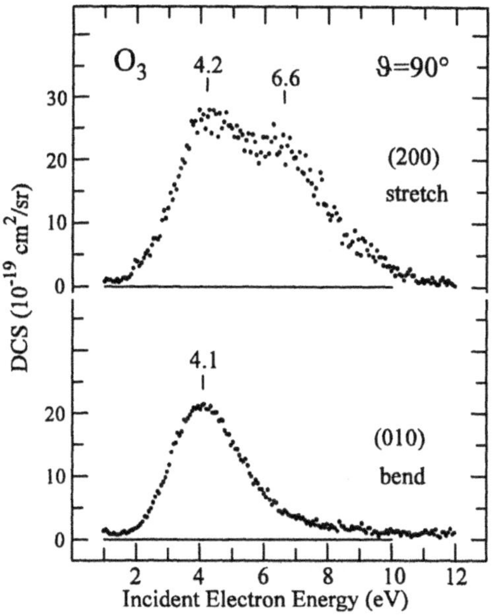

Figure 10. DCSs for exciting the specified vibrations in ozone.

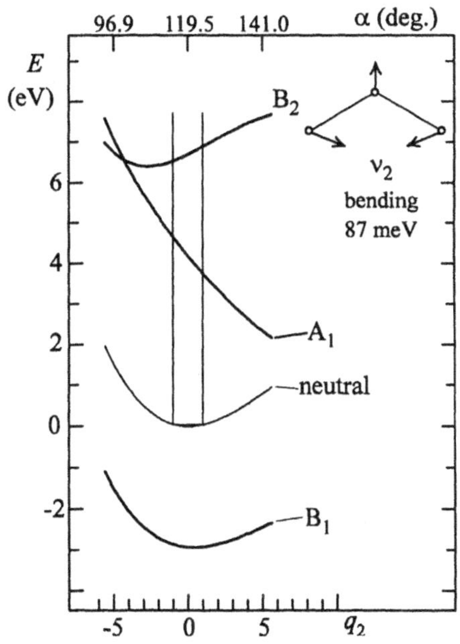

Figure 11. Potential energy curves for ozone and the two shape resonances as a function of the bending normal coordinate q_2 (of neutral ozone). The potential curve for the ground state of the ozonide negative ion is also included. The potential energy of neutral ozone shows the SCF energy calculated with the 6-31G basis set. The negative ion energies were calculated from virtual orbital energies using the Koopmans' theorem and an empirical scaling relation as explained in the text.

The low dissociation energy of ozone leads to low threshold for dissociative attachment. Dissociative electron attachment below 5 eV yields O^- and O_2^- with considerable intensity.[23,24] Four resonant peaks are observed in the dissociative attachment spectra, at 0.4 eV, 1.3 eV, 3.2 eV, and 7.5 eV.[18] The ions are formed with moderate kinetic energies for the first, nearly zero kinetic energy for the second, and with large kinetic energies for the third resonance. O_3^- formation through three-body attachment or ion-molecule reactions is also observed.

Attachment of electrons with energies of 1.2 to 2 eV to ozone was found to lead to the production of O and vibrationally excited O_2^-, the latter subsequently lose electrons by vibrational autodetachment.[17] This type of electron scattering is intermediate between inelastic electron scattering and dissociative attachment. It is similar to the formation of autodetaching N^- in electron collisions with N_2, discovered nearly simultaneously in Nebraska[25] and in Paris[26] in the late 70's, except that here the fragment anion is molecular, giving the detached electron spectra rich vibrational structure, which can be assigned to the individual O_2^- / O_2 vibronic transitions, providing detailed information on the product state distribution (figure 12).

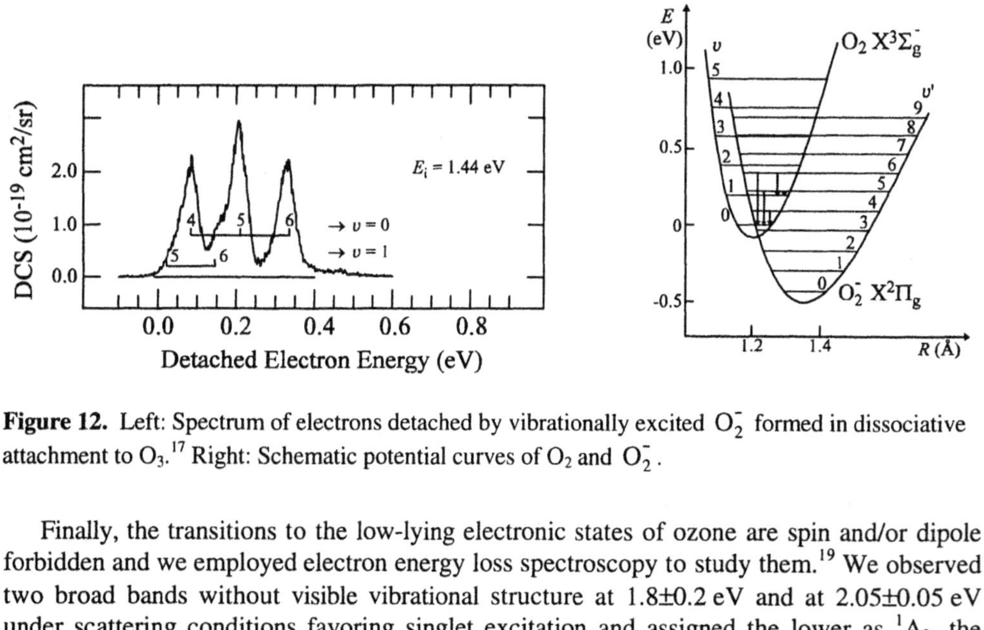

Figure 12. Left: Spectrum of electrons detached by vibrationally excited O_2^- formed in dissociative attachment to O_3.[17] Right: Schematic potential curves of O_2 and O_2^-.

Finally, the transitions to the low-lying electronic states of ozone are spin and/or dipole forbidden and we employed electron energy loss spectroscopy to study them.[19] We observed two broad bands without visible vibrational structure at 1.8±0.2 eV and at 2.05±0.05 eV under scattering conditions favoring singlet excitation and assigned the lower as $^1A_2$, the higher as $^1B_1$ (Chapuis band).

Bands with rich vibrational structure were observed under scattering conditions favoring triplet excitation, as shown in figure 13. At least two vibrational progressions can be discerned, the first with an origin at 1.30 eV, the second either at 1.53 eV, or at 1.45 eV.

Information on the low-lying electronic states has been available from photoabsorption spectroscopy[27] (the spin-forbidden transitions gaining very weak intensity from spin-orbit coupling) and from anion photoelectron spectroscopy of O_3^-.[28] The dependence of the excitation DCSs on the scattering angle and electron energy brought important evidence about the spin-forbidden nature of these states.

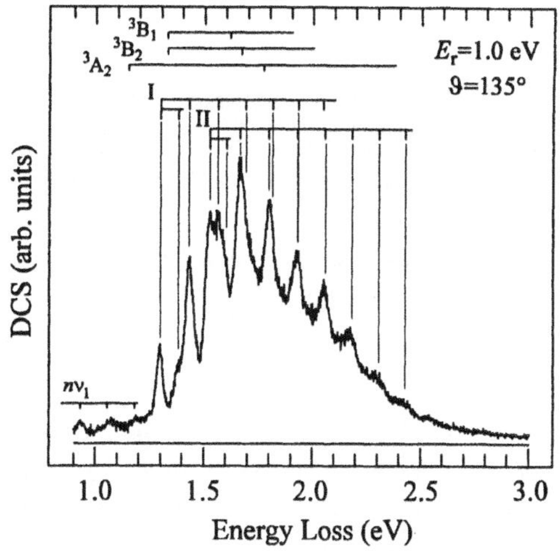

Figure 13. Electron-energy-loss spectrum of the low-lying excited states of ozone.[19]

The well known Hartley band and a number of other singlet and triplet excited states were observed at higher energy-losses. Excitation functions and angular distributions of the triplet band at 1.30 eV and of the Hartley band were presented.[19] The absolute value of the differential cross section for excitation of the Hartley band was measured and it compares favorably with the recent work of Sweeney and Shyn.[29]

CONCLUSIONS

A progress has been made in the measurement of absolute differential cross section for inelastic scattering by:

- Extending the angular range of the measurements down to 0° and, most importantly, obtaining one more point in the angular distribution, and an excitation function, at 180°.
- Increasing the sensitivity and decreasing the background of the measurements, permitting for example recording the cross sections for the higher channels in helium.

A serious factor limiting the accuracy of the measurements are the uncertainties in the determination of the instrumental sensitivity as a function of the incident and residual energies, and the scattering angle, for inelastic signal. High quality calculations of the inelastic cross sections in helium promise to provide a suitable standard in the future.

Encouraging agreement between experiment and recent high quality calculations of the inelastic cross sections in helium has been obtained, but sizeable differences remain at some scattering angles and energies, requiring more work to reconcile the details of the experimental and theoretical data.

Acknowledgment

This work is part of project 2028-040398.94/1 of the Swiss National Science Foundation.

REFERENCES

1. H. S. W. Masey, *Negative Ions*, Cambridge University Press, Cambridge (1976).
2. G. J. Schulz, Resonances in electron impact on diatomic molecules, *Rev. Mod. Phys.* 45:423 (1973); Resonances in electron impact on atoms, *Rev. Mod. Phys.* 45:378 (1973).
3. M. Allan, Study of triplet states and short-lived negative ions by means of electron-impact spectroscopy, *J. Electr. Spectr.* 48:219-351 (1989).
4. P. D. Burrow and L. Sanche, Elastic scattering of low-energy electrons at 180° in CO_2, *Phys. Rev Lett.* 28:333 (1972)
5. K. R. Asmis, *Applications of electron energy-loss spectroscopy; development of an instrument to measure differential cross sections for inelastic electron-molecule scattering at 0° and 180°*, Ph. D. thesis, University of Fribourg (1996).
6. M. Allan, Measurement of differential cross sections for excitation of helium by electron impact within the first 4 eV above threshold, *J. Phys. B: At. Mol. Opt. Phys.* 25:1559-1575 (1992).
7. M. Allan, Experimental differential cross sections for the electron-impact excitation of the $a^1\Delta_g$, $b^1\Sigma_g^+$, and the 6 eV states of O_2, *J. Phys. B: At. Mol. Opt. Phys.* 28:4329-4345 (1995).
8. M. Allan, Measurement of absolute differential cross sections for vibrational excitation of O_2 by electron impact *J. Phys. B: At. Mol. Opt. Phys.* 28:5163 (1995).
9. J. C. Nickel, P. W. Zetner, G. Shen, and S. Trajmar, Principles and procedures for determining absolute differential electron-molecule (atom) scattering cross sections, *J. Phys. E: Sci. Instrum.* 22:730 (1989).
10. R. K. Nesbet, Variational calculations of accurate e^--He cross sections below 19 eV, *Phys. Rev. A* 20:58 (1979).
11. W. C. Fon, K. P. Lim, K. Ratnavelu, and P. M. J. Sawey, Differential cross sections for the electron impact excitation of the $n^{3,1}S$ (n = 2, 3, and 4) states of helium at low energies, *J. Phys. B: At. Mol. Opt. Phys.* 27:1561 (1994).
12. K. R. Asmis and M. Allan, in preparation.
13. M. Allan, Measurement of absolute differential cross section for the excitation of the lowest triplet state of ethene, *Chem. Phys. Lett.* 225:156 (1994).
14. Q. Sun, C. Winstead, V. McKoy, and M. A. P. Lima, Low-energy electron-impact excitation of the $\tilde{a}\,^3B_{1u}\,(\pi\rightarrow\pi^*)$ state of ethylene, *J. Chem. Phys.* 96:3531 (1992).
15. C. Winstead, V. McKoy, private communication
16. W. M. Johnstone, N. J. Mason, W. R. Newell, P. Biggs, G. Marston, and R. P. Wayne, Differential oscillator strengths produced by electron impact energy-loss spectroscopy of ozone, *J. Phys. B: At. Mol. Opt. Phys.* **25**:3873 (1992)
17. M. Allan, K. R. Asmis, D. B. Popovic, M. Stepanovic, N. J. Mason and J. A. Davies, Production of vibrationally autodetaching O_2^- in low-energy electron impact on ozone, *J. Phys. B: At. Mol. Opt. Phys.* 29 (1996), in press.
18. M. Allan, K. R. Asmis, D. B. Popovic, M. Stepanovic, N. J. Mason and J. A. Davies, Resonances in collisions of low-energy electrons with ozone: Experimental elastic and vibrationally inelastic differential cross sections and dissociative attachment spectra *J. Phys. B: At. Mol. Opt. Phys.* 29 (1996), in press.

19. M. Allan, N. J. Mason and J. A. Davies, Study of electronically excited states of ozone by electron-energy-loss spectroscopy, *J. Chem. Phys.* (1996), in press.

20. T. W. Shyn and C. J. Sweeney, Angular distribution of electrons elastically scattered from ozone, *Phys. Rev. A* 22:2916 (1993).

21. Y. Okamoto and Y. Itikawa, Elastic scattering of electrons by ozone molecules, *Chem. Phys. Lett.* 203:61 (1993).

22. M. Allan and D. b. Popovic, in preparation.

23. R. K. Curran, Negative ion formation in ozone, *J. Chem. Phys.* 35:1849 (1961).

24. J. D. Skalny, S. Matejcik, A. Kiendler, A. Stamatovic, and T. D. Märk, Dissociative electron attachment to ozone using a high-resolution crossed beam tecnique, *Chem. Phys. Lett.* 255:112 (1996).

25. D. Spence and P. D. Burrow, Resonant dissociation of N_2 by electron impact, *J. Phys. B: At. Mol. Opt. Phys.* 12:L179 (1979).

26. J. Mazeau, F. Gresteau, R. I. Hall, and A. Huetz, Energy and width of N^- (^3P) from observation of its formation by dissociative attachment to N_2 and NO, *J. Phys. B: At. Mol. Opt. Phys.* 11:L557 (1979).

27. S. M. Anderson and K. Mauersberger, Ozone absorption spectroscopy in search of low-lying electronic states, *J. Geophys. Res.* 100:3033 (1994).

28. D. W. Arnold, C. Xu, E. H. Kim, and D. M. Neumark, Study of low-lying electronic states by anion photoelectron spectroscopy of O_3^-, *J. Chem. Phys.* 101:912 (1994).

29. C. J. Sweeney and T. W. Shyn, Measurement of inelastic electron impact on ozone: Absolute differential cross sections for excitation of the Hartley band, *Phys. Rev.* A 53:1576 (1996).

DISSOCIATIVE RECOMBINATION : NEW DEVELOPMENTS STIMULATED BY ION STORAGE RING EXPERIMENTS

A. Suzor-Weiner[1-2], I. F. Schneider[2-3] and L. Carata[3]

[1]Laboratoire de Chimie Physique, Université Paris VI, 75231 Paris Cedex 05, France
[2]Laboratoire de Photophysique Moléculaire, Bat.213, Université Paris XI, 91405 Orsay
[3]Laser Department, Institute of Physics and Technology of Radiation Devices, 76900 Bucharest-Magurele, Romania

INTRODUCTION

The dissociative recombination (DR) of a molecular ion AB^+ :

$$AB^+ + e \dashrightarrow A + B^*$$

is an important process for determining the electronic densities and the molecular abundances in the ionosphere or in the interstellar medium. A new generation of experiments, very promising as a source of astrophysical data, is being developped since a few years (Tanabe et al 1993, Forck et al 1993, Larsson et al 1993) : Several experimental groups using ion storage rings to study atomic ions recently turned to store molecular ions (TARN at Tokyo, TSR at Heidelberg, CRYRING at Stockholm, ASTRID at Aarhus). The residence time of the ions may reach several tens of seconds, allowing relaxation of all the electronic and (at least for heteronuclear ions with a permanent dipole) vibrational excited states. Additional phase-space cooling is achieved using an electron cooler device, which also provides the recombining electrons. The DR cross section is then measured as a function of the relative velocity between the electron and ion merged beams. As an example, Figure 1 shows the device used at CRYRING. The reader can find detailed experimental information in the review papers by Forck et al (1993) and by Larsson (1995).

The large energy range covered in these new experiments require several extensions to the theoretical treatments of DR, both in the mechanisms involved and in the number of molecular states to be included in the calculations. The simplest molecular ions H_2^+, D_2^+ and HD^+ have been extensively studied as test cases. It is crucial to achieve there a good

Photon and Electron Collisions with Atoms and Molecules
Edited by Burke and Joachain, Plenum Press, New York, 1997

agreement between theory and experiment, in particular to check the efficiency of ion cooling and the calibration of the rate coefficient absolute measurements. Then, one will be able to use safely the experimental results for more complex ions, in particular large polyatomic ions, very important for atmospheric or interstellar equilibrium but almost untractable theoretically.

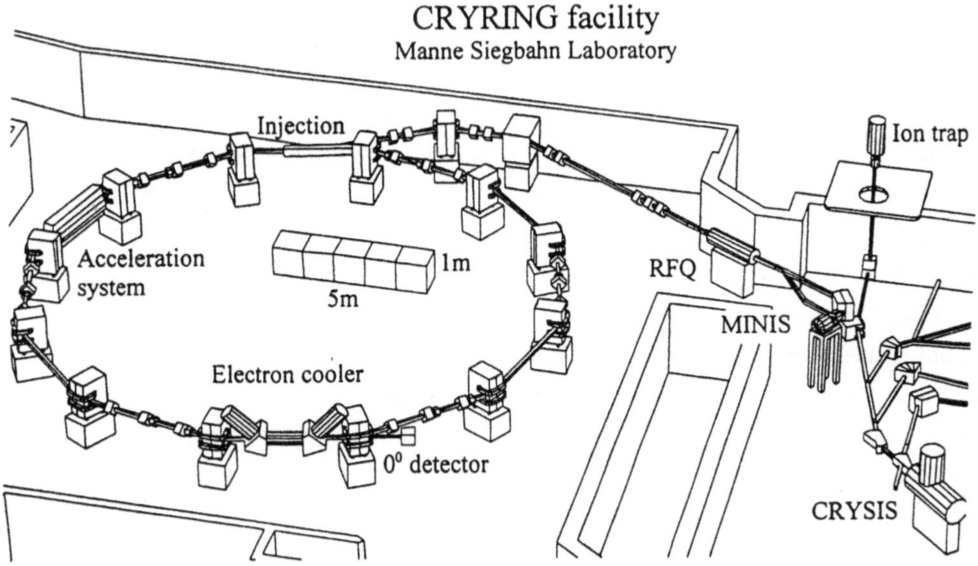

Figure 1 : The CRYRING facility at the Manne Siegbahn Laboratory at Stockholm University. The circumference of the ring is 52m. The electron-ion recombination takes place in the electron cooler, and the DR neutral products are detected in the 0°-detector arm (from Larsson 1995).

MECHANISMS AND THEORETICAL TREATMENT OF DR

Several review papers describe the basic mechanisms and the main theoretical treatments of dissociative recombination (Giusti-Suzor 1990; Guberman 1993; Takagi 1993b; Giusti-Suzor, Schneider and Dulieu 1993), with many references therein. Briefly, dissociative recombination may be viewed as a *multichannel reactive collision* involving two types of "channels": *ionization channels* (electron-molecular ion scattering) on one hand, among which the entrance channel, and *dissociation channels* (atom-atom scattering) on the other hand. An ionization channel with threshold above the total energy available in a specific recombination process is termed "closed" and introduces in the calculation a series of

Rydberg states differing only by the principal quantum number of the external electron. At high energy, some open ionization channels become also dissociative, when the total energy is higher than the ion state dissociation limit. This gives rise to *dissociative excitation*, a competitive process to dissociative recombination.

Dissociative recombination, as well as associative ionization (the inverse process, see e.g. Weiner et al 1988) results from various couplings between these two types of channels. The dominant interaction is usually the *electronic coupling* between the Rydberg manifold or adjoigning ionization continuum (singly excited configurations) and the dissociative valence states (doubly excited configurations) whose potential curves cross the ion ground state in the Franck-Condon region. This interaction allows *direct* electron capture into the dissociative channels, measured by the so-called "electronic width" Γ(R), with competing autoionization back to the open ionization channels. Another type of interchannel couplings, the*vibrational* and*rotational* interactions between alternative ionization channels due to non-Born Oppenheimer effects, may cause electron capture into ro-vibrationally excited Rydberg levels (pertaining to the closed ionization channels), which are then predissociated due to electronic coupling with the same dissociative states as for the direct process previously described. This second process has been termed *indirect* DR (Bardsley 1968), as contrasted to the purely electronic direct one. It is responsible for narrow resonances (mostly in form of dips) in the DR cross-sections, as observed in the merged beam experiments at low energy (Van der Donk et al 1991). When no doubly excited dissociative state favourably crosses the ion ground state, these ro-vibronic interactions might nevertheless induce fast DR via low-lying Rydberg states, as for the case of HeH^+ (Guberman 1994 ; Sarpal, Tennyson and Morgan 1995) and H_3^+.

The first requirement for a complete study of any DR process is the determination of potential curves (or surfaces) for the dissociative states and the evaluation of their autoionization widths, i.e. of their electronic or non-adiabatic couplings to the entrance ionization channel. Besides using spectroscopic or ab initio data in the region of Rydberg-valence states mixing, which may be then extrapolated above the ionization threshold, several methods from electron-molecule scattering theory are used and developed for this step : the R-matrix method (Tennyson and Noble 1985 ; Shimamura, Noble and Burke 1990 ; Tennyson 1996), the linear algebraic method (Collins, Schneider and Noble 1993,1995), the complex Kohn variational method (Orel et al 1991), the Ritz variational method with static approximation for the continuum state (Sato and Hara 1986).

This progress report will focus on diatomic molecules where at least one dissociative curve corresponding to a doubly excited state favorably crosses the potential curve of the ion ground state. The whole process is thus dominated by strong *electronic* interactions, which induce various resonance structures depending on the energy range. We review below the recent developments or challenges induced by the ongoing experimental studies on various storage rings.

LOW ENERGY REGION : ROLE OF ROTATIONAL COUPLINGS

Only recently did the ion storage rings provide DR cross sections for very slow electrons (1meV to 1 eV), with high enough resolution to show the resonance structure near threshold. In this energy range, below the ion dissociation threshold, electron capture into bound levels of Rydberg monoexcited states induces a typical resonant structure in the cross section, the signature of interferences between the direct and indirect processes.

Most of previous DR calulations at low energy, either for H_2^+ (Giusti-Suzor et al 1983; Nakashima et al 1987; Schneider et al 1991, 1994) or for O_2^+ (Guberman and Giusti-Suzor 1991), N_2^+ (Guberman 1991) and CH^+ (Takagi and Ledourneuf 1994), used the two-step MQDT approach (Giusti 1980) with an electronic reaction matrix calculated at first or second order. Vibrational couplings among singly excited neutral states were included via the R-dependence of the electronic quantum defect, but rotational interactions among Rydberg states were neglected (only rotational initial excitation was considered, together with the corresponding centrifugal distorsion of the molecular potential curves). In more recent work on H_2^+ and HD^+, Takagi included the rotational interactions (Takagi 1993, Tanabe et al 1995) through the same frame-transformation as in the original MQDT approach for photoionization (Jungen and Dill 1980).

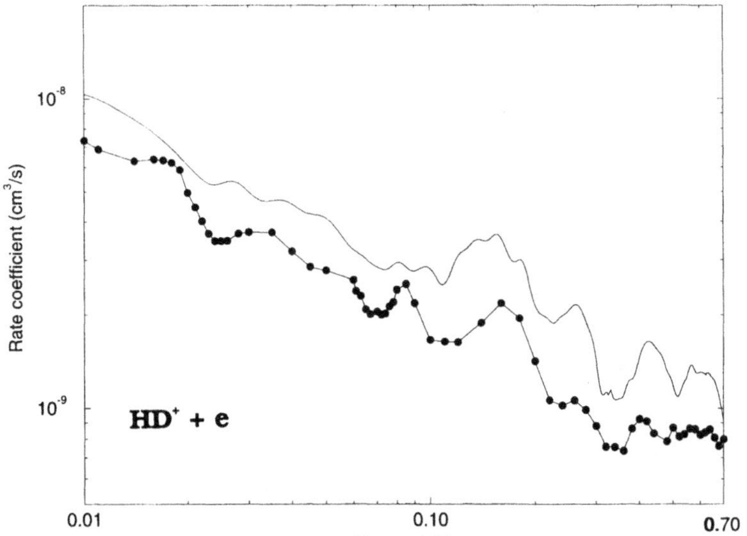

Figure 2. Low-energy cross-section for the dissociative recombination of HD^+. The dots show the absolute cross-section as measured at CRYRING (Strömholm et al 1995). The full line represents the T=300K Bolzman average of our theoretical results for ion initial levels v=0, N=0-10. The calculations (Schneider et al 1996) have been performed in the first order of the reaction K-matrix, taking into account the contribution of "d" and "s" partial waves of the incident electron and simultaneous electronic, vibrational and rotational interactions.

Following the same approach (but with two Rydberg series - dσ and sσ - included instead of only dσ in Takagi's calculations) we have performed calculations in the low energy region for the case of HD^+ initially in its ground vibrational level, taking into account both *vibrational* and *rotational* interactions, besides the dominant electronic coupling (Schneider et al 1996). Before comparing to the experimental results, two kinds of average have to be performed. First, one must average over the velocity distribution of the electrons in the electron cooler, assumed to be an anisotropic Maxwell distribution (Larsson 1995). Then, the ions may be initially in various rotational states, since their relaxation time often exceeds the storage time. A separate calculation is performed for each initial rotational level and the final result is obtained by averaging the different cross sections over a Boltzman distribution of rotational levels corresponding to the room temperature of 300K. In spite of the large number of resonances due to the ro-vibrational structure of the Rydberg states, these two averages lead to a relatively smooth cross section, in good agreement with experiments (see figure 2).

INTERMEDIATE ENERGY REGION : CORE EXCITED RESONANCES

Two experiments performed on the test storage ring (TSR) in Heidelberg have revealed strong resonances around 1eV in the DR cross sections of CD^+ (Forck et al 1994) and OH^+ (Amitay et al 1995), broader and more prominent than the low-energy resonances discussed above. They could result from a new kind of indirect dissociative recombination process, with temporary capture in a core-excited (instead of only vibrationnally excited) Rydberg states. Many molecular ions (except for H_2^+ and its isopotomers HD^+ and D_2^+) have low lying *bound* excited states giving rise to series of bound core-excited Rydberg states in the neutral molecule, e.g. in CH, OH, N_2. Such doubly excited Rydberg states are electronically coupled both to the initial continuum (incoming electron + ground state ion) and to the dissociative states. They might thus induce the observed resonances in the DR cross sections.

No calculations including this mechanism have been published yet, but some studies are underway, in particular by S. Guberman (1995) for OH^+ and N_2^+ and in our group (Carata et al 1996) for CH^+ and CD^+. The MQDT formalism can easily include the core-excited closed channels, as was done previously in molecular photoionization studies to account for electronic autoionization (Giusti-Suzor and Lefebvre-Brion 1980, 1984). The most difficult part is the calculation of the numerous electronic couplings between the different types of channels (Rydberg-Rydberg with different cores, Rydberg-dissociative states). Spectroscopic analysis of molecular spectra can help for this point.

Figure 3 shows the result of a preliminary calculation around 1eV in CD^+, where such a prominent resonance has been observed (Forck et al 1994, Fig.3(a)). Besides the closed ionization channels (or Rydberg series) with an excited vibrational level of the ion ground state as a threshold, we have included in the MQDT treatment additional closed channels with a

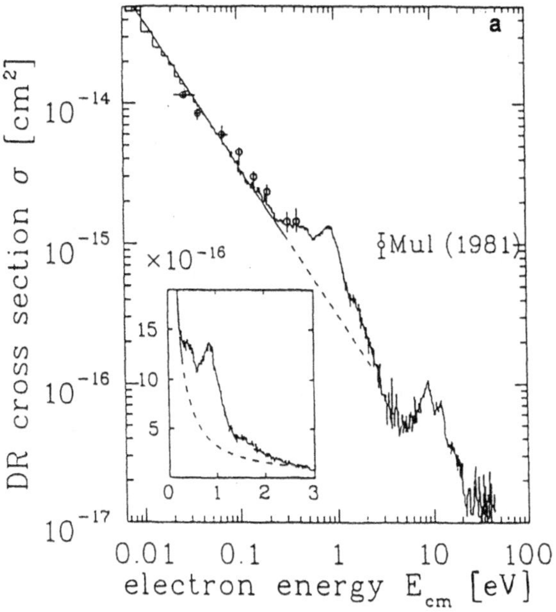

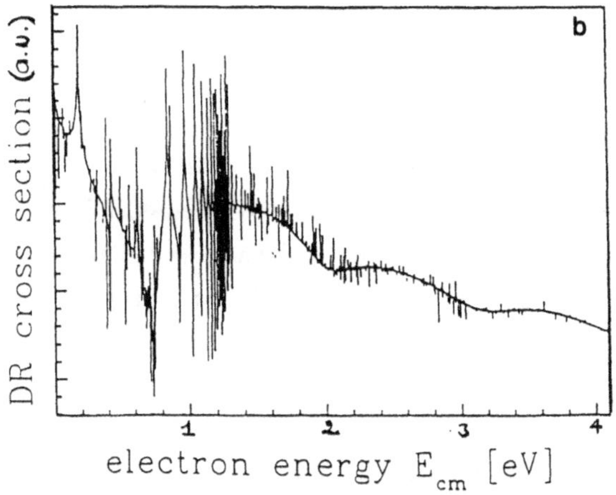

Figure 3. (a) Dissociative recombination cross section of CD^+ measured at TSR by Forck et al, normalized to the results (dots) of Mul et al (1981). The $E^{-0.9}$ dependence of the direct process is indicated by the dashed line and the inset is a linear scale enlargement of the cross section around the core-excited resonance. (b) Result of an exploratory MQDT calculation with closed channels (see text) corresponding both to the $X^1\,\Sigma^+$ and to the $a^3\,\Pi$ ion cores. (Carata et al 1966).

bound *electronically excited* ion state as threshold, namely the $a^3\Pi$ ion state with low vibrational quantum numbers (v = 0 to 5). The ion potential curves and the electronic couplings between mono- and diexcited configurations have been guessed mainly from the CH^+ paper by Takagi et al (1991). A complex resonance structure is obtained in the calculated cross section between 0.8 and 1.2 eV (Fig. 3b). The broad peak corresponds to the v=4 level of the $[a^3\Pi(4p\sigma)]\ ^2\Pi$ Rydberg state of CD, which interacts electronically both with the ionization continuum of the $X^1\Sigma^+$ ground state of CD^+ (the entrance channel) and with the lowest dissociative $^2\Pi$ valence state. The narrow resonances on both sides of this relatively broad structure are due to temporary capture (through vibrational interactions) into excited vibrational levels of $^2\Pi$ Rydberg states $[X^1\Sigma^+(np\pi)]$ with ground state ion core. Such interfering resonance pattern closely reminds observations reported in photoionization experiments, e.g. in N_2 by Dehmer et al (1984) where vibrational and electronic autoionization processes interfer in the same region of the spectrum. The MQDT approach used to analyse this "complex resonance" feature (Giusti-Suzor and Lefebvre-Brion 1980, 1984) should also lead to a quantitative explanation of this new type of resonances observed in DR cross sections, in CD^+ or other molecules, once more precise data for the molecular interactions involved are known.

HIGH ENERGY REGION : SERIES OF DISSOCIATIVE STATES, DISSOCIATIVE EXCITATION

Rydberg series of dissociative states

At still higher energy, above the dissociation limit of the bound ion states, electron capture into bound Rydberg levels is no more possible and DR proceeds through direct electronic recombination into doubly excited dissociative states. More and more dissociative states contribute with increasing energy, and most of them (except for the lowest ones) may be considered as the successive n* members of Rydberg series with a given dissociative ion core state and various molecular symmetries. A good picture of this scenario may be found for the case of H_2 in Fig.3 of Guberman (1983) reproduced here as Fig.4 : Two sets of dissociative Rydberg states, the Q_1 states with $2p\sigma_u$ core, and the Q_2 ones with $2p\pi_u$ core, clearly appear below each ion excited state.

These series of dissociative states give rise to very broad resonances in the cross sections, as seen for example in the HD^+ results around 9 eV (Q_1 series) and then about 16 eV (Q_2 series) in different storage ring experiments (Forck et al 1993, Tanabe et al 1995, Strömholm et al 1995), or in the HeH^+ measurements of Sundström et al (1993) and Tanabe et al (1993) near 20 eV. The analysis of such broad composite resonances is a theoretical challenge: how many dissociative states are involved, of which symmetry?

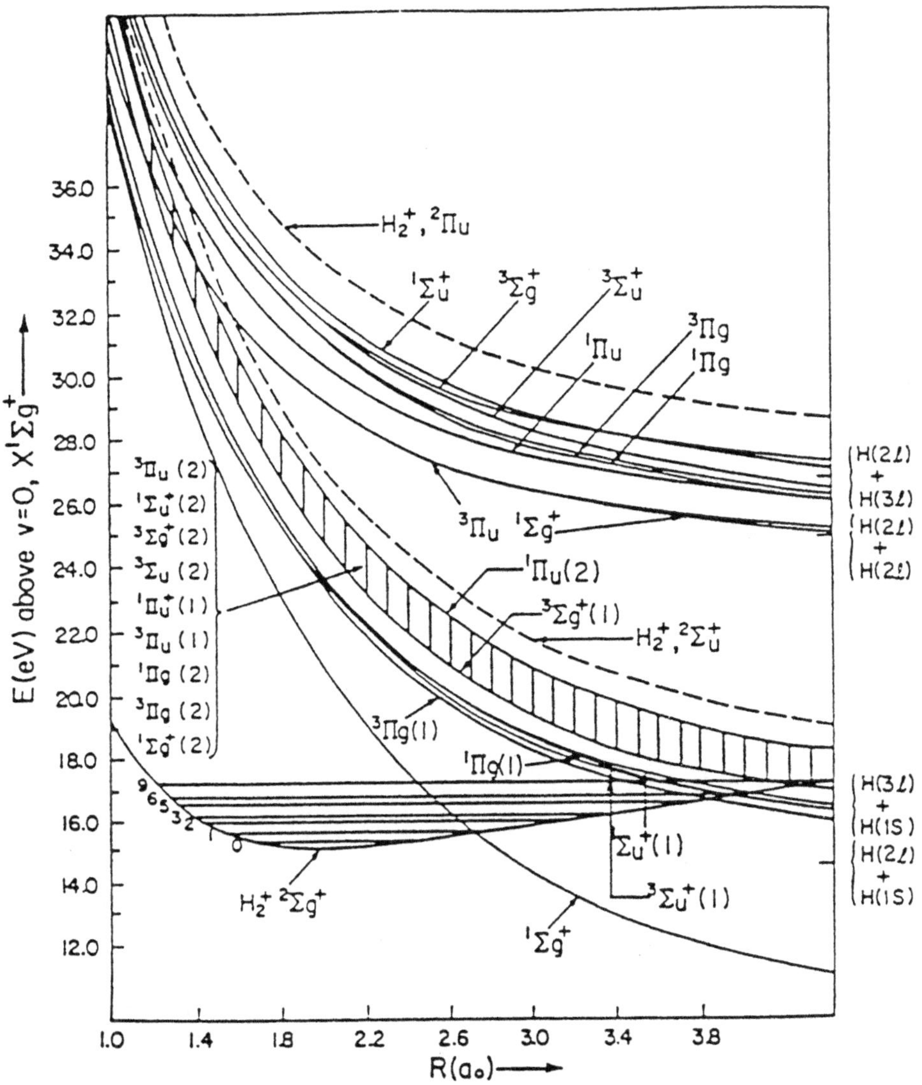

Figure 4 : Potential energy curves for H_2 and H_2^+, from S.Guberman (1983). The dashed curves correspond to the two first excited states of H_2^+ ($2p\sigma_u$ and $2p\pi_u$), to which many Rydberg series of H_2 dissociative states converge.

symmetry should be included in order to reproduce the correct *shape* and *magnitude* of the observed composite resonance (see figure 5). This reminds the situation encountered in atomic dielectronic recombination, where entire series of core excited Rydberg states contribute to the broad autoionization resonances observed in the storage ring experiments (see e.g. Kilgus et al 1992).

Unfortunately, ab initio structure or collisional calculations become more and more difficult with increasing energy because the electron orbitals involved become very diffuse and large electronic as well as internuclear distances must be explored. Precise molecular data are thus lacking except for the lowest members of each series. Extrapolation procedures based on Rydgerg series scaling laws may be used. In particular, the higher members of each series are very close in energy and may be represented by a single effective state with a total width obtained by summing the $n \geq n_0 \approx 6$ individual widths.

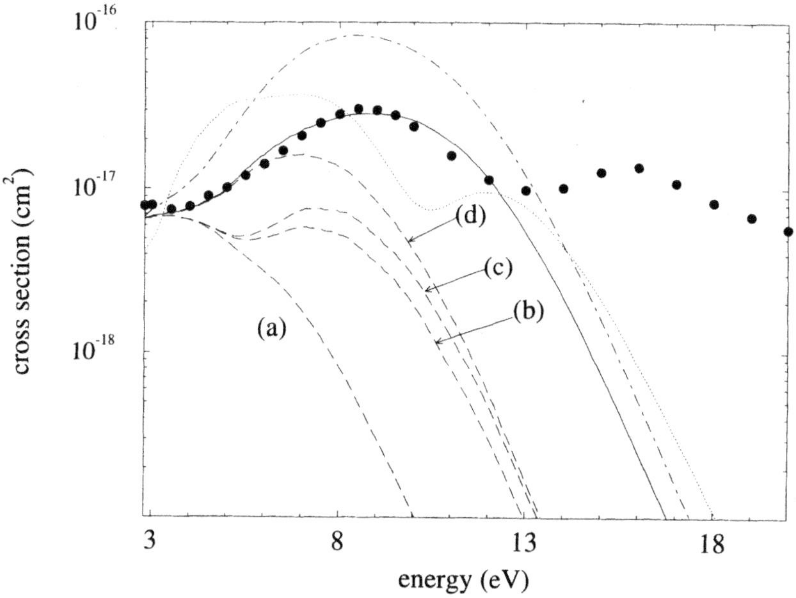

Figure 5 : High energy absolute cross section measured for HD^+ at CRYRING compared with theoretical MQDT results (for initial v=0 ion level) including 4 series of dissociative states converging to the $2p\sigma_u$ state of H_2^+ (Strömholm et al 1995). The dashed lines represent the cumulative contributions of the lowest state $^1\Sigma_g^+$ (a), added to the lowest $^1\Pi_g$ state (b), then to the lowest $^1\Sigma_u^+$ (c), and finally to the lowest $^3\Pi_g$ state (d). The dot-dashed line is when the entire series of each symmetry is included, and the full line is when dissociative autoionisation is taken into account. The dotted line shows the result of the same global calculation for initial ion level v=1.

Extensive calculations for the higher series of dissociative Rydberg states (Q_2 series with $2p\pi_u$ ion core) have just been performed by Tennyson (1996) using the R-matrix method. Not only the total autoionization width but also the partial widths for autoionization into the $1s\sigma_g$

or $2p\sigma_u$ electronic continua have been evaluated. These new data should allow to reproduce the second broad peak observed around 16 eV in the DR cross section of HD^+ (e.g. by Strömholm et al 1995).

Dissociative excitation and autoionization

Besides these infinite series of dissociative states, another feature specific to the high energy region is autoionization not only into the bound vibrational levels of the ion ground state but also into its *vibrational continuum* :

$$(AB^+)_v + e \longrightarrow AB^{**} \longrightarrow \begin{cases} A + B^* & \text{dissociation} \\ (AB^+)_{v'} + e & \text{autoionisation to bound ion levels} \\ [A + B^+]_{\text{ion ground state}} + e & \text{dissociative autoionisation} \end{cases}$$

At still higher energy the direct *dissociative excitation* process:

$$(AB^+)_v + e \longrightarrow [A + B^+]_{\text{ion excited state}} + e$$

also begins to compete with dissociative recombination, as observed in the experiment at TSR on HD^+ (Forck 1994). As for *dissociative photoionization* (see e.g. the recent experiments on H_2 and D_2 by Geddes et al 1994, Latimer et al 1995), we face here the problem of double continuum (electronic and nuclear), or three-particle fragmentation whose exact treatment requires integration over correlated electron and nuclei kinetic energies. The method proposed by Kanfer and Shapiro (1983) for dissociative photoionization, in which they solve coupled equations for the nuclear motion after having discretized the electronic continuum, could be adapted to dissociative recombination with little changes. Until now, only approximate first-order calculations of dissociative autoionization in HD^+ have been performed (Takagi 1995, Schneider et al 1995) but these preliminar results are in encouraging agreement with the dissociative excitation measurements of Forck (1994). They also lead to corrected DR cross sections in good quantitative agreement with the *absolute values* measured on CRYRING (Strömholm et al 1995, see figure 5), assuming complete vibrational relaxation of the stored ions. We stress the importance of accurate calculations in this energy range because they can help to calibrate the experimental results, both for absolute values and for the efficiency of internal cooling. Besides the HD^+ case, complete calculations have been recently performed by Orel, Kulander and Rescigno (1995) for HeH^+ in the same high energy region (10-30 eV), both for the electronic collision and the multichannel dissociation dynamics, leading to good agreement with the experimental results of Sundström et al (1993).

Long-range dynamics and branching ratios

Besides the rates for dissociative recombination, it is crucial for the study of equilibrium in atmospheric or interstellar media to know the outcome of a given recombination process, especially in the case of polyatomic ions. Recent experimental progress (Zajfmann et al 1995, van der Zande et al 1995, Semaniak et al 1996) are very promising. They yield not only the energy distribution of fragments but also their angular distribution, related to the symmetry of the dissociative states involved. Theoretical support is clearly needed to help the analysis, but very little has been done for the calculation of branching ratios up to now. This aspect involves a "half-collision" between heavy particles, atomic or molecular partners, and requires quite different theoretical tools (e.g. close-coupling methods) than for the short range calculation of total cross sections. The dissociative molecular states often undergo severe configuration mixing on their way to dissociation, as well as non-adiabatic couplings, and their evaluation is the major bottleneck to accurate calculations of branching ratios. The long range dynamics have been well studied in simple cases like H + H, H^+ + H^- collisions (see e.g. Sidis et al 1983) at high energy. This type of studies provides enough data to "branch" a close-coupling atomic collision calculation to the MQDT treatment of the first steps of DR, as was done already for the inverse process of associative ionization (Urbain 1990; Urbain et al 1991). This aspect of DR studies is certainly worth theoretical efforts in the coming years.

ACKNOWLEDGMENTS

Support for travel and communication related to this work has been provided by EC within the HCM network no.ERB CHRX CT920013, in particular through the PECO addition to the network. One of us (L.C.) was partly supported by a grant from the EC TEMPUS program. The Bucharest team also acknowledges the financial and technical support of the International Atomic Energy Agency in Vienna, through Contract no. 6642/RB, and of the SOROS fundation.

REFERENCES

Amitay Z. et al 1996, *Phys. Rev.***A 53**, R644

Bardsley J.N. 1968, *J.Phys. B* **1**, 365

Carata L., Schneider I. F. and Suzor-Weiner A. 1996, in preparation

Collins L.A., Schneider B.I. and Noble C.J.1992, *Phys. Rev.* **A45**, 4610; 1995, *Phys. Rev. A* **52**, 3010

Forck P. et al 1993, *Phys. Rev. Lett.* **70**, 426

Forck P. et al 1993, *Nucl. Instrum. Methods Phys. Rev. B* **79**, 273

Forck P. et al 1994, *Phys. Rev. Lett.* **72**, 2002

Forck P. 1994, Ph.D. Thesis, Universität Heidelberg (and Fork P. et al, to be published)

Geddes J. et al 1994, *J.Phys. B* **27**, 2961

Giusti A. 1980, *J.Phys. B:***13**, 3867

Giusti-Suzor A. and Lefebvre-Brion H. 1980, *Chem. Phys. Lett.* **76**, 132; 1984 *Phys. Rev.A* **30**, 3057

Giusti-Suzor A., Bardsley J.N. and Derkits C. 1983, *Phys. Rev.A* **28**, 682

Giusti-Suzor A.1989, in "*Dissociative Recombination:Theory, Experiments and Applications*", Mitchell J.B. and S.L. Guberman eds, World Scientific, p.14

Giusti-Suzor A., Schneider I.F. and Dulieu O.1993, in "*Dissociative Recombination:Theory, Experiments and Applications*", Rowe B.E.,Mitchell J.B. and Canosa A. eds, NATO ASI series **B313**, p.11

Guberman S.L. 1983, *J. Chem. Phys.* **78**, 1404

Guberman S.L. and Giusti-Suzor A.1991, *J. Chem. Phys.* **95**, 2602

Guberman S. L. 1991, *Geophys. Res. Lett.* **18**, 1051

Guberman S.L. 1993, in "*Dissociative Recombination: Theory, Experiments and Applications*", Rowe B.,Mitchell J.B. and Canosa A. eds, NATO ASI series **B313**, p.47

Guberman S.L. 1994, *Phys. Rev.A* **49**, R4277

Jungen Ch. and Dill D. 1980, *J. Chem. Phys.* **73**, 3338

Kanfer S. and Shapiro M. 1993, *Laser Chem.* **3**, 73

Kilgus G. et al 1992, *Phys. Rev.A* **46**, 5730

Larsson M. et al 1993, *Phys. Rev. Lett.* **70**, 430

Larsson M. 1995, *Int. J. Mass Spectr. Ion Proc.* **149/150**, 403

Latimer C. J. et al 1995, *J. Chem. Phys.* **102**, 722

Mul P.M. et al 1981, *J. Phys.B* **14**, 1353; see also Mitchell J.B.A. 1990, *Phys. Rep.* **186**, 215

Nakashima K., Takagi H. and Nakamura H. 1987, *J. Chem. Phys.* **86**, 726

Orel A., Rescigno T.N. and Lengsfield III B.H.1991, *Phys. Rev. A* **44**, 4328

Orel A. and Kulander K.C. 1993, *Phys. Rev. Lett.* **71**, 4315

Orel A., Kulander K.C. and Rescigno T.N.1995, *Phys. Rev. Lett.* **74**, 4807

Sarpal B.K., Tennyson J. and Morgan L.A. 1995, *J. Phys. B* **27**, 5943

Sato H. and Hara S. 1986, *J.Phys. B* **19**, 2611

Schneider I. F., Dulieu O. and Giusti-Suzor A. 1991, *J.Phys. B* **24**, L289

Schneider I. F., Dulieu O. E, Giusti-Suzor A. and Roueff E.1994, *Ap. J.* **424**, 983

Schneider I. F. et al 1996, to be sumitted to *J. Phys.B*

Semaniak et al 1996, submitted to Phys. Rev.A

Shimamura I., Noble C.J. and Burke P.G. 1990, *Phys. Rev. A* **45**, 4610

Strömholm C. et al 1995, *Phys. Rev.* **A52**, R4320

Sundström G. et al 1993, *Phys. Rev. A* **50**, R206

Takagi H., Kosugi N. and Le Dourneuf M. 1991, *J.Phys. B* **24**, 711

Takagi H. 1993, *J.Phys.B* **26**, 4815

Tanabe T. et al 1993, *Phys. Rev. Lett.* **70**, 422

Tanabe T. et al 1995, *Phys. Rev. Lett.* **75**, 1066

Tennyson J. and Noble C.J. 1985, *J. Phys. B* **18**,155

Tennyson J. 1996, *Atomic Data and Nuclear Data Tables* (in press)

Urbain X. 1990, thesis, Université de Louvain-la Neuve, Belgium, unpublished

Urbain X., Cornet A., Brouillard F. and Giusti-Suzor A., 1991, *Phys. Rev. Lett.* **66**, 1685

Van der Donk P., Yousif F.B., Mitchell J.B.A and Hickman A.P. 1991, *Phys. Rev. Lett.* **67**, 42

Van der Zande W.J. et al 1996, in preparation

Weiner J., Masnou-Seeuws F. and Giusti-Suzor A. 1989, *Adv. At. Mol. Opt. Phys.* **26**, 209

Zajfman D. et al 1995, *Phys. Rev. Lett.* **75**, 814

ONE- AND TWO-ELECTRON RESONANCES IN MOLECULAR PHOTOIONIZATION

Robert W. Zurales, Eric Stratmann, Shaleen Botting,
and Robert R. Lucchese

Department of Chemistry
Texas A&M University
College Station, TX 77843-3255
USA

INTRODUCTION

Significant progress has been made in recent years in the ability to compute accurate electron-molecule scattering wave functions and the corresponding observables such as scattering and photoionization cross sections. It is now possible to consider the electron-molecule scattering problem using correlated targets with an accurate treatment of the wave function representing the scattered electron. In particular we can now compute molecular photoionization cross sections where the target electrons are described by thousands of configuration state functions (CSFs) and the continuum is accurately described on a numerical grid.[1]

The main features in the photoionization cross sections of small molecules are due to the various scattering resonances. In the fixed-nuclei approximation, these resonances can be characterized either by the mechanism of decay or by the orbital occupation relative to the initial state of the molecule. Resonances can be classified as being either a one-electron or a two-electron resonance based on the decay mechanism. We will give a review of some recent results obtained using the Schwinger variational method with Padé approximant corrections[1, 2] to illustrate the qualitative features of these resonant processes.

TREATMENT OF THE MANY-ELECTRON PROBLEM

Expansion in Product Basis

We treat the electron scattering problem in photoionization in an asymmetric manner. The N-electron wave function is written as a linear combination of CSFs constructed from one-electron functions. In the asymptotic region of the wave function, where one electron is far from the nuclei, the wave function approaches an N–1 electron target wave function coupled with the continuum electron represented as a Coulomb scattering function leaving

Photon and Electron Collisions with Atoms and Molecules
Edited by Burke and Joachain, Plenum Press, New York, 1997

in some fixed direction. In this part of configuration space, an appropriate representation of the target is an expansion in N-1 electron CSFs where the one-electron functions are written as a sum of square-integrable functions. Following the usual close-coupling approach,[3] we then write the full wave function as a linear combination of the antisymmeterized and spin adapted products of the asymptotic target states and continuum functions, with a unique continuum function for each target state. The resulting wave function is then inserted in the Demkov variational expression[4] to yield a set of coupled integro-differential equations for the continuum functions.

Overcompleteness of the Close-Coupling Expansion

If each of the continuum functions in the close-coupling expansion is unrestricted, the full expansion will be an overcomplete representation of the wave function.[5] This can be seen easily in the case where the continuum functions are expanded in the same set of one-electron functions, ϕ_α, which were used to construct the N-1 electron target states and where the target states are represented by single spin-adapted Gelfand CSFs[6], ψ_i. In an expansion for two electrons, the term $\mathcal{A}_2[\phi_1(\vec{r}_1)\phi_2(\vec{r}_2)]$, where $\mathcal{A}_2$ is the two electron antisymmeterizer, will appear two times, once when the target state is ϕ_1 and once when the target state is ϕ_2. This redundancy leads to the overcompleteness of the full unrestricted close-coupling expansion.

It is possible to remove the redundancy in the close-coupling expansion by imposing appropriate restrictions on the orbitals representing the Nth electron, i. e. the continuum electron. If the N–1 electron target is represented by the CSF ψ_i and the Nth electron is represented by the function ϕ_α, the antisymmeterized spin-adapted N-electron state constructed from these two functions will be written as $\psi_i(\phi_\alpha)$. Then if the orbitals ϕ_α form an ordered set, the redundancy is removed if the index is restricted to only certain ranges of values based on the occupation of the original N–1 electron CSF ψ_i. Assume that the highest occupied orbital in ψ_i has index β. If this orbital is a doubly occupied orbital then the continuum function must have index $\alpha > \beta$. If the highest orbital is singly occupied then two subcases must be considered. Assuming that the spin quantum of the final N-electron state is S, if the N–1 electron state has spin S+1/2 then we need to constrain the added electron to have index $\alpha \geq \beta$. Alternatively, if the N–1 electron state has spin S–1/2 then we need to constrain the added electron to have index $\alpha > \beta$. Note that this last case will only occur for $S > 0$. If the one-electron basis set is ordered by eigenvalue with respect to some effective one-electron Hamiltonian, then the restricted expansion can be seen to correspond qualitatively to requiring the continuum electron to have an energy greater than that of any of the occupied orbitals of the target.

This approach can be generalized to treat the case where the target states are represented by multiconfigurational wave functions. In this case one must make a one-to-one correspondence between the multiconfigurational target states and their principle configurations. Assume that the target states are written as a linear combination of CSFs

$$\Phi_p = \sum_i \psi_i C_{ip} \,. \tag{1}$$

Then the mapping can be constructed in a unique fashion by making a list of all the expansion coefficients C_{ip} for all the states ordered by the absolute values of the coefficients. The indices of the expansion coefficient with the largest absolute value is then the first element of the one-to-one mapping, i. e. $i \leftrightarrow p$. The list is then modified by removing all coefficients which contain either the CSF index i or the total N–1 state index p. The largest remaining coefficient is then used to identify the next element of the one-to-one mapping. This process can be repeated until the full mapping is completed. With this mapping we can then construct the N electron states by constraining the index of the added orbital to have the relationship as discussed above with respect to the orbitals occupied in the principle CSF of the N–1 electron state as defined by the mapping.

Thus we can see that for the close-coupling expansion to be not overcomplete, we must constrain the continuum orbitals in some fashion. Note that if a full close-coupling expansion is used the overcompleteness does not cause any essential problem, although it can lead to numerical instabilities in a claculation. However, when a limited expansion of correlated target states is used, one can encounter unphysical pseudoresonances at low energy which are due to the overcompleteness of the expansion.[7, 8] In practice, there are two ways to avoid these unphysical pseudoresonances in the close-coupling method: either constrain the continuum functions appropriately or include all open channels which can be obtained from a complete active space configuration interaction (CASCI) type calculation of the ion state.[9] The overcompleteness of the close-coupling expansion also causes problems at intermediate energies, i.e. above the first ionization potential of the target, for any form of the target wave functions.[10]

Treatment of the One-Electron Problem

The use of the close-coupling expansion reduces the equations which one must solve to predict electron-scattering cross sections or photoionization cross sections to a set of constrained coupled integro-differential equations for the three dimensional scattering continuum orbitals. We solve these equations for the photoionization problem using the Schwinger variational principle with Padé approximant corrections.[2] The appropriate Lippmann-Schwinger equation for the vector of one-electron scattering functions ψ_S is of the form

$$\left|\psi_S\right\rangle = \left|\psi_S^0\right\rangle + \mathbf{GV}\left|\psi_S\right\rangle , \tag{2}$$

where $\psi_S$is a vector of the channel solutions, $\psi_S^0$is a vector of the homogeneous solutions (i.e. solutions of the Schrödinger equation in the absence of $\mathbf{V}$), $\mathbf{V}$ is the potential matrix[3, 11] and $\mathbf{G}$ is the matrix of channel Green's functions for the Hamiltonian in the absence of the potential $\mathbf{V}$. For electron scattering from neutral molecules, $\mathbf{G}$ is then the free particle Green's function and ψ_S^0 are free particle scattering states, e. g. plane waves or spherical Bessel functions. In the case of photoionization or scattering from an ion, $\mathbf{G}$ is the Coulomb Green's function and ψ_S^0 are Coulomb waves. The potential $\mathbf{V}$ has the form of a Phillips-Kleinman pseudopotential which incorporates the various orthogonality constraints required to yield a nonredundant close-coupling expansion. The exact form of this potential has been discussed previously.[12, 13]

In general, we do not need to compute directly the solutions to the Lippmann-Schwinger equation given in Eq. (2), we only need to compute matrix elements of the scattering wave function of the form

$$M\left(\chi_R^0, \psi_S^0\right) = \left\langle \chi_R^0 \middle| \psi_S \right\rangle, \tag{3}$$

where χ_R^0 is a second function which depends on the physical observable which one is predicting. Thus if one is computing the probability for a scattering process, elements of the S matrix are obtained as $M(\mathbf{V}\psi_{S'}^0, \psi_S^0)$. Photoionization cross sections are obtained from dipole matrix elements using the form $M(\vec{\mu}\phi, \psi_S^0)$ where ϕ is the appropriate vector of occupied orbitals.

In our present implementation of this approach, all of the required matrix elements are computed using numerical techniques based on single-center expansions.[14-16] This approach yields the exact result for the matrix elements of the scattering function within the truncated partial-wave and target-state expansions used. In the iterative procedure we have implemented to obtain the Padé sequence, the asymmetry of the required matrix elements leads to a great reduction in computational effort.[5] In particular, given that in the Padé approximant approach we compute the matrix elements for photoionization of the form

$$\langle \vec{\mu}\phi | \psi_S \rangle = \sum_{n=0}^{\infty} \langle \vec{\mu}\phi | (\mathbf{GV})^n | \psi_S^0 \rangle, \tag{4}$$

we compute the sequence of terms $\langle \vec{\mu}\phi | (\mathbf{GV})^n |$ for the one initial state and take simple overlaps with the many homogeneous scattering states $| \psi_S^0 \rangle$.

Types of Resonances

A resonance in electron-molecule scattering is usually a process involving a state whose wave function is very similar to a bound state, in that it is localized in the region of the molecule. Such a state is not an eigenstate of the total Hamiltonian, however, so that when the system is prepared in a resonant state, it slowly decays into the continuum allowing one or more electrons to escape from the system.

Within the fixed-nuclei approximation, such resonances are characterized in two ways. The first way one can describe the resonance is in terms of the one-electron orbitals used in the expansion of such a state.[17] Thus there are valence-Rydberg states where the resonant state can be characterized by a single electron in a Rydberg-like orbital and where an electron has been removed from one of the valence orbitals. There are a class of resonant states which can be described as valence-valence states, where an electron has been removed from one valence orbital and placed in a previously unoccupied or unfilled valence orbital. This type of excitation can be further subdivide into V type excitations where the two orbitals involved in the excitation process form a bonding-antibonding pair, thus they occupy the same region of space and differ from each other by a single nodal surface between nuclei, e. g. a $\sigma \rightarrow \sigma^*$. If the orbitals are not related in this fashion, e.g. a $\sigma \rightarrow \pi^*$ excitation, the transition is classified as an A type valence-valence excitation. The V type excitations usually have much higher oscillator strengths and are often more difficult to describe theoretically since the electron-electron correlation effects are larger when the electrons occupy the same region of space. In addition, there are resonant states which are described in the simplest picture as doubly excited states. Of course all of these description are somewhat simplified and a given state may be a superposition of two or more of the different types of states described here. This mixing is important since the excitation of a doubly excited state would not have any oscillator strength without configuration mixing.

Resonances in electron scattering can be also classified by the mechanism by which they decay into the continuum. If the essential mechanism for the decay is an exchange of energy between electrons, i.e. an inelastic electron-electron scattering process, then we will describe the resonance as being a two-electron resonance. Such a resonance can also be variously described as a configuration mixing, correlation, or a $1/r_{12}$ resonance. Alternatively, if the decay mechanism involves an electron elasticly scattering from the other electrons in the system or from the nuclei, then we will describe the resonance as a one-electron resonance. Such a resonance can also be described as a conjugate mixing, potential barrier penetration, overlap, or shape resonance.

ONE-ELECTRON RESONANCES

We will restrict our discussion of one-electron resonances to scattering processes which induce a time delay in the scattering event. The time delay is the delay relative to the motion of a free particle in the electron-neutral molecule scattering case and relative to the motion of a particle scattering from a pure Coulombic potential in the case of electron scattering from an ion or in photoionization of a neutral molecule. Such a time delay is characterized theoretically by a rising eigenphase sum, and can be observed experimentally as either a peak or minimum in the corresponding scattering cross section.[18] The rising eigenphase sum is usually due to the presence of a pole in the *S* matrix near the real energy axis on the unphysical sheet.

We note that the interaction potential between an electron and a molecule is almost everywhere attractive in the region near the molecule. This is true for electrons scattering from neutral molecules and for electrons scattering from positive ions when only the forces in addition to those due to the long-rage Coulombic potential are considered. The essential reasons for this are that the other electrons in the system do not shield the nuclei perfectly so the static interaction is attractive. The other interactions in the one-electron model, exchange and polarization, also are attractive interactions. The lack of a repulsive potential in electron scattering then implies that any time delay cannot be due to a slowing down of the particle due to repulsive interactions.[19]

Thus, for a resonance to be evident in an experiment or a calculation, the time delay must then be in excess of the time advancement due to the attractive forces which increase the momentum of the scattering electron in the vicinity of the molecule. This leaves only two mechanisms for the time delay in the scattering of electrons. Either there is a time delay due to multiple scattering between centers, i. e. the electron is scattered back and forth between centers thus delaying the scattered electron relative to the electron which is not scattered, or there is a dynamic trapping of the electron, i. e. the energy of the electron is trapped in a mode which does not couple to the mode for leaving the system. The classic example of this type of dynamical trapping is found in resonances with angular momentum barriers where the energy is temporarily trapped in the angular motion of the electron and is only slowly converted to radial motion via tunneling through an angular momentum barrier.

The concept of the motion of a particle being principally in a mode other than the mode corresponding to the continuum is essential to an understanding of the mechanism for trapping. As has been shown by Langhoff and coworkers,[20] it is possible to correlate the energies of shape resonances with the length of a simple one-dimensional attractive square well potential. However, scattering from such a potential will never exhibit time delays. In particular, the characteristic transmission resonances found in such one-dimensional systems do not correspond to time delays, just scattering in different directions, i. e. backward scattering versus forward scattering.

Although multiple scattering resonances can occur in extended systems, none of the isolated small molecular systems we have examined have had resonances which can only be explained by such a mechanism.

Asymptotically, for small molecular systems, a natural coordinate system is the spherical coordinate system. Thus at least the outer part of any resonant wave function can be described by a particular partial wave, which then leads to a description of the trapping by a given angular momentum barrier.[21] Although the angular momentum picture provides the correct picture for the asymptotic part of the wave function, in the region of the molecule, the linear combination of atomic orbital (LCAO) provides the most compact description of electronic wave functions. Thus it has been observed that the unoccupied (or virtual) molecular orbitals found in a minimum basis set (MBS) (one atomic centered basis function for each occupied valance or core atomic orbital) calculation of the electronic structure often correspond to resonant states.[22, 23] However, there is not a strict one-to-one correspondence between resonant states and virtual MBS orbitals. For example, some virtual orbitals of C_6H_6 do not correspond to electron scattering shape resonances[21] and there is a non-MBS resonant state predicted in the photodetachment spectrum of F_2^-.[24, 25]

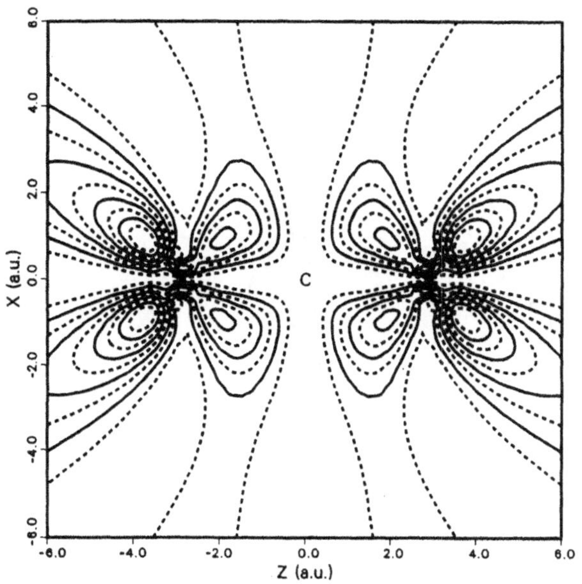

Figure 1. The probability density of the resonant π_g state in the photoionization of CS_2 leading to the $(5\sigma_u)^{-1}$ $B\ ^2\Sigma_u^+$ state of CS_2^+ from Ref. (26).

One example which illustrates both the fact that resonant states often look like unoccupied valence states and that the inner and outer regions can have very different angular momentum states is found in the photoionization of CS_2.[26] In Fig. 1 the probability density is shown of the one-electron resonant state found in the photoionization of CS_2 in the channel leading to the $(5\sigma_u)^{-1}$ $B\ ^2\Sigma_u^+$ state of CS_2^+. This resonance orbital is of π_g symmetry and occurs with an asymptotic energy of 4.9 eV and a width of $\Gamma = 5.5$ eV. Qualitatively, this state can be seen to be the anti-bonding combination of the unoccupied valence *d* orbitals on the S atoms. Thus this state has a strong $l = 2$ component near each of the centers. However, by extending the nodal surfaces to the asymptotic region, we can see that this orbital connects to an $l = 4$, $m = 1$ state. This then provides a sufficient angular momentum barrier to resonantly trap the photoelectron.

Behavior of One-Electron Resonances Above the Angular Momentum Barrier

An interesting and very practical question concerns the location and density of poles in the *S* matrix for realistic potentials. Consider the simple spherically symmetric piecewise model potential shown in Fig. 2 which can be used to model e–C_6H_6 scattering in the e_{2g} symmetry.[27] For $l = 2$, the barrier in the effective radial potential is ~0.06 au. By varying the well depth in the potential between an R of 2.0 and 4.0 au, the resonant state in this system can be moved above or below the barrier. In Fig. 3 the phase shifts for several well depths are given. We can see that as the well is deepened the position of the rise in phase shift is shifted to an energy which is below the energy of the barrier. For energies below the barrier, the rise in the phase shift is fairly rapid, eventually approaching a rise in π as the resonance approaches the threshold. Note however, even when the resonance is above the barrier, there is a rising phase shift, although the total rise is much less than π. Also note how the phase shift near threshold does not approach the characteristic form of δ

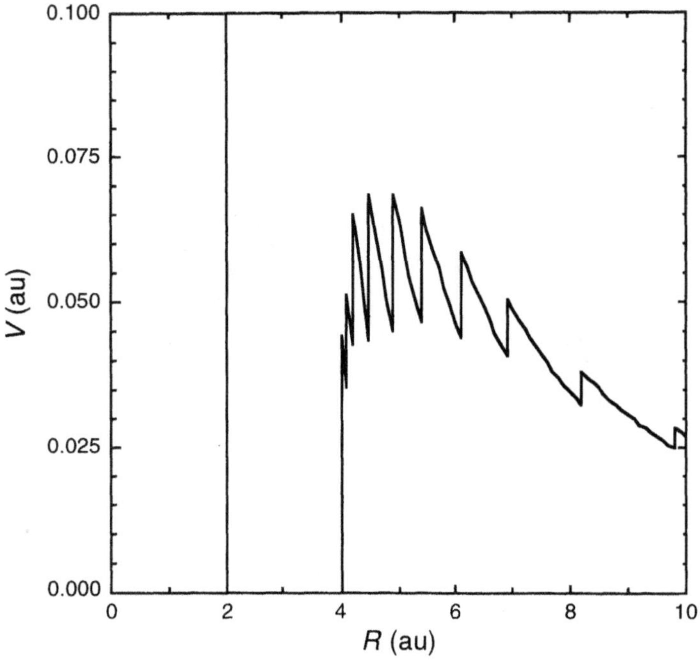

Figure 2. Model potential constructed from a piece-wise constant spherical potential and an l = 2 angular momentum barrier. The value of the potential between 2.0 au and 4.0 au -2.2 to -1.0 au.

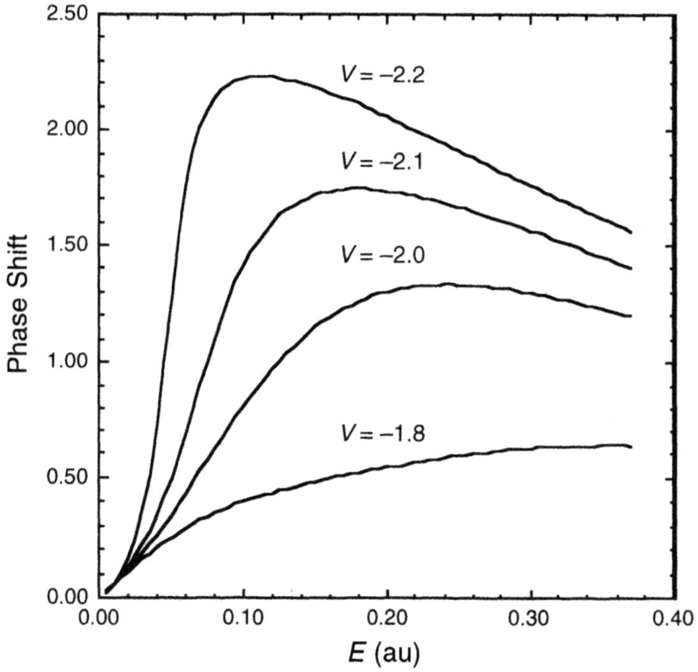

Figure 3. Behavior of the l = 2 phase shift as a function of the well depth V (in au) where V is the value of the potential between R = 2.0 and 4.0 au

$\sim E^{l+1}$ for the energies plotted here. The nearly linear rise in the phase shift is characteristic of potentials with a r^{-4} spherical component as in the model potential used here.

In Fig. 4 we plot the location of the poles in the *S* matrix as a function of the well depth. At well depths where the resonance is below the barrier, the pole in the *S* matrix corresponding to the resonance is near the real axis and is isolated from the poles which combine to yield the background scattering in this system. Then, as the well is made shallower, the resonance pole moves towards the line of background scattering poles. This continues until the poles form a somewhat distorted sequence with the resonance pole being indistinguishable from the other poles in the system. At this point, it would be difficult to conclude that there is a distinct resonance even though there is a rising phase shift in the system. With an even shallower well the line of poles again becomes a nearly straight line of evenly spaced poles. Thus we can see that in even slightly complicated systems, there is not a sharp transition between a pole corresponding to a resonance and a pole corresponding to background scattering. Indeed, there is a slowly varying continuum of possible resonance types.

Correlation in One-Electron Resonances

Although a one-electron resonance can be qualitatively described using the single-electron picture, there are usually correlation effects which are important in describing the details of the energy, width and decay mechanism of the resonance. The importance of correlation is most apparent in electron-neutral scattering where polarization effects must be included in order to get the correct energy and consequently the correct width of the resonance. In photoionization, interchannel coupling is one significant correlation effect in shape resonances. The effects of the interchannel coupling are most prominent in the non-resonant channel, where interchannel coupling can lead to enhanced cross sections due to

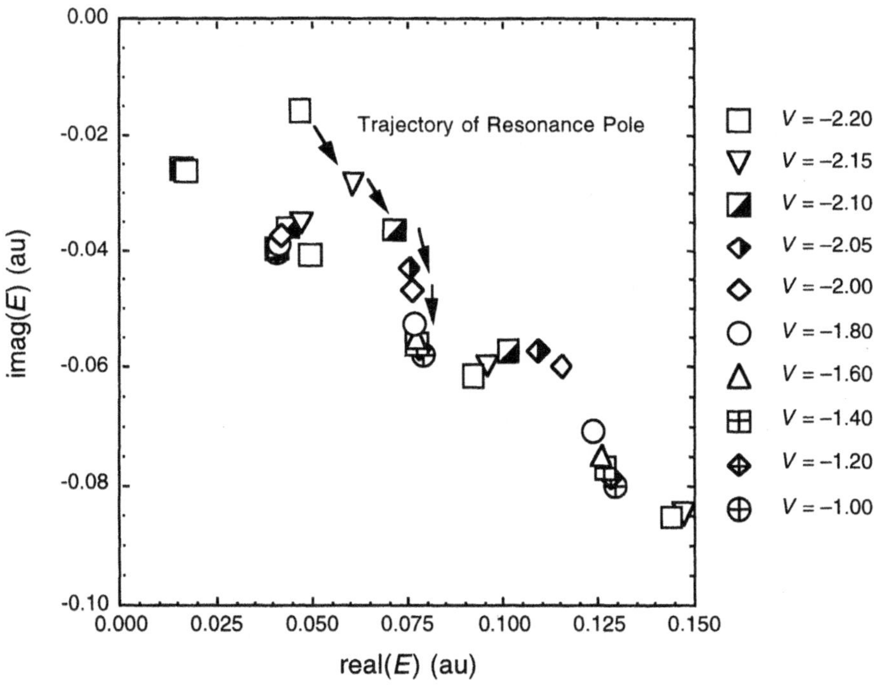

Figure 4. Motion of the poles of the *S* matrix in the complex energy plane for the $l = 2$ model potential.

intensity borrowing or coupling can lead to dramatic modifications in photoelectron angular distributions.[28, 29]

Another correlation effect in interchannel coupling can occur in shape resonant states in the photoionization of open-shell molecules. An example is found in the photoionization of NO.[8] In this system there is a σ^* shape resonance in the channels leading to the two $(5\sigma)^{-1}$ hole states of the NO^+ ion. The open shell part of the electronic configuration of the resonant state is $5\sigma\, 2\pi\, 6\sigma^*$, where the $6\sigma^*$ orbital is the shape resonant state which is qualitatively described as the σ anti-bonding orbital which asymptotically is an $l = 3$ angular state. This electronic configuration leads to two linearly independent CSFs of $^2\Pi$ symmetry which is one of the allowed symmetries for photoionization from the initial $X\,^2\Pi$ state of NO. Asymptotically, the correct coupling scheme is to couple the target 5σ and 2π electrons to form a $^1\Pi$ and a $^3\Pi$ target state and then couple the $6\sigma^*$ continuum electron to form the $^2\Pi$ final states. This description of the scattering has been called the multiplet specific form since in a single-channel picture each resonance is governed by a different multiplet specific potential.[30, 31] In the multiplet specific picture the resonance in both channels would have comparable intensities and be located at the energies shifted by the different potentials and thresholds for the two targets.

In Fig. 5 the various cross sections for ionization of NO leading to the two $(5\sigma)^{-1}$ hole states are shown. The coupled channel results were obtained in the multichannel complete-active-space configuration-interaction (MCCASCI) approximation, where several channels are included in the close-coupling expansion and the target states are represented by valence complete-active-space configuration-interaction (VCASCI) wave functions. It is clear in the coupled channel calculations that the resonance occurs at the same energy and

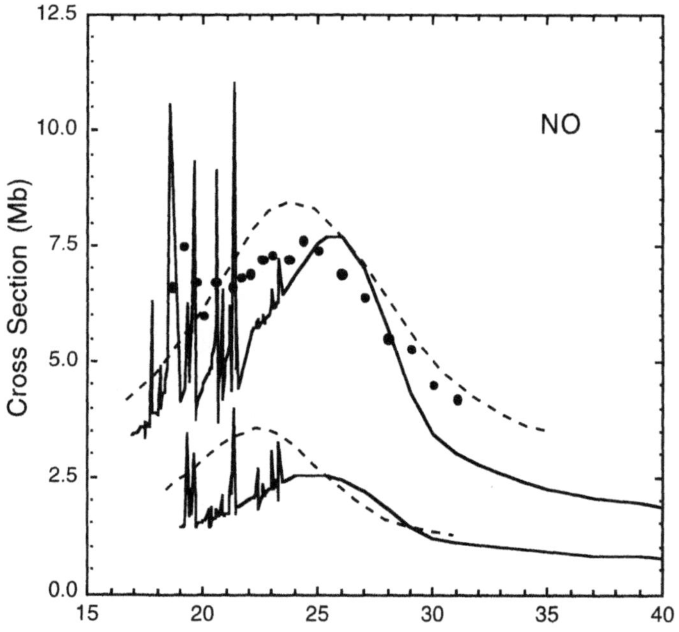

Figure 5. Photoionization cross sections of NO leading to the $b\,^3\Pi$ and $A\,^1\Pi$ states of NO^+ ——— calculated cross section in the MCCASCI approximation[8]; — — — single channel cross section using the multiplet specific potential, computed by Herrman et al.[32] Cross sections in megabarns (Mb) = 10^{-18} cm^2.

not at two different energies as predicted by the single-channel multiplet-specific calculation of Hermann *et al.*[32] The photoionization of NO is an example of a one-electron resonance being strongly affected by coupling between the asymptotic channels.

Analyzing the MCCASCI results, we found that at energies in the resonance region the conjugate relationship between the 5σ and $6\sigma^*$ orbitals makes the interaction between these orbitals much stronger than the interaction between the 5σ and 2π orbitals. Thus a much better single-configuration description of the resonance is to first couple the 5σ and $6\sigma^*$ orbitals to form $^1\Sigma^+$ and $^3\Sigma^+$ states. Then the 2π orbital is coupled to these states to form the $^2\Pi$ final states. In the single configuration picture, this second description would lead to all the intensity going to the state constructed from the $^1\Sigma^+$ (5σ $6\sigma^*$) sub-system with no intensity going to the other configuration. When this coupling is used in a single channel calculation, it is referred to as the multiplet averaged approach since the resonant electron only feels the potential found in the $^1\Sigma^+$ (5σ, $6\sigma^*$) state which in fact is equal to a weighted average of the two potentials found in the multiplet specific form.[31, 33]

TWO-ELECTRON RESONANCES

We have seen that one-electron resonances can be described by simple valence molecular orbitals which are temporarily bound by an angular momentum barrier. In a similar fashion two-electron resonances often can be described in terms of valence molecular orbitals in a single (or few) CSF picture. From this perspective, the *N*-electron states which can be obtained in a VCASCI calculation including all valence orbitals (and perhaps a few low lying Rydberg orbitals) are candidates for two-electron resonant states. These resonant states can be grouped into three categories: states with appreciable oscillator strength for direct excitation, states with little oscillator strength but which strongly couple to the ionization channel being investigated, and finally states which neither couple to the measured ionization channel nor have direct oscillator strength. The states with direct oscillator strength can significantly increase the total oscillator strength leading to a particular final ion state, whereas the resonances with no direct oscillator strength can only redistribute the oscillator strength as a function of energy.[34] For closed shell targets in the single CSF approximation (i.e. the Hartree-Fock approximation) only excited states which are singly excited with respect to the initial state can have direct oscillator strength. For this reason these singly excited states, whether *A* type or *V* type, are the most prominent in the photoionization continuum. Most doubly excited states have little oscillator strength and thus are less important. However, there are a few examples of doubly excited states with significant direct oscillator strength due to mixing with a singly excited state and/or initial state correlation.

Rydberg Autoionization

The simplest autoionizing resonance to describe is the autoionization of a Rydberg level leading to an excited state of the molecule. A Rydberg state is a one-electron state which is very similar to a large principle quantum number hydrogenic state. Such resonances correspond to single excitations from the ground state but often are fairly weak due to the small dipole matrix element coupling the resonance to the initial state. They are also usually fairly narrow due to the weak coupling to the continuum.

An example of such resonances is the series of levels leading up to the $(2\sigma_u)^{-1}$ $B\ ^2\Sigma_u^+$ excited state of N_2^+. In Fig. 6 we show the resonances which correspond to the excitation of the $n = 3$ and $n = 4$ Rydberg states.[35] The calculated results were obtained from a four-coupled-channel calculation where the ion states were represented by single CSFs, referred to as the multichannel frozen-core Hartree-Fock (MCFCHF) approximation. The energy of

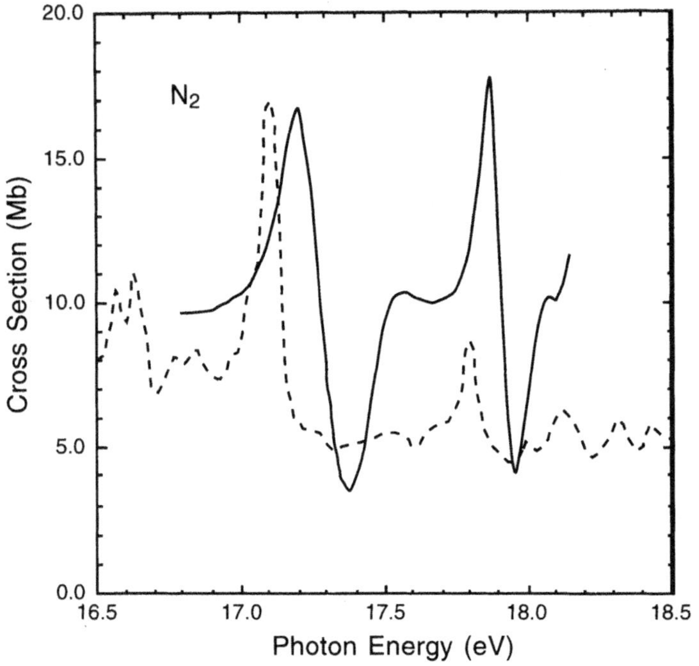

Figure 6. Photoionization cross section of N_2 leading to the $X\ ^2\Sigma_g^+$ state of N_2^+ in the region of autoionization below the $B\ ^2\Sigma_u^+$ state: ——— MCFCHF in the mixed form [35]; — — — relative experimental results of Plummer *et al.*[36]

the resonance relative to the ionization potential (IP) for the excited ion state can be represented by the quantum defect in that channel. The quantum defects are usually easier to calculate than the absolute value of the resonance energy. Thus in order to get good agreement between experiment and *ab initio* theory for these resonances, it is important to obtain accurate IPs. In the results shown here we have not attempted to obtain the IPs in an *ab initio* calculation but have instead used the experimentally determined values. These results indicate that this fairly simple calculation can in favorable cases obtain a reasonable estimate of the width and line-shape for these resonances.

Valence Autoionization

Most prominent valence resonances are also single excitations. An example of this is found in the low energy photoionization of C_2H_2. By examining the VCASCI of C_2H_2 we identified that the major resonant states in the low energy spectrum of C_2H_2 are the two singly excited states formed by the $1\pi_u \rightarrow 1\pi_g$ *V* type excitation and the $2\sigma_u \rightarrow 1\pi_g$ *A* type excitation. In Fig. 7 we see the predicted photoionization cross section compared to relative absorption measurement of Ibuki *et al.*[37] The theoretical curve has been shifted by –0.8 eV to bring the theoretical and the experimental IPs into better agreement. There are two very broad features in the cross section. Between 12 and 15 eV the broad background is found which corresponds to the $1\pi_u \rightarrow 1\pi_g$ excitation. This is actually an open channel one-electron resonance which is very broad, probably due to the fact that it is above the angular momentum barrier which is very low for $l = 2$ in electron-ion scattering. The other broad feature in the photon energy range from 15 to 16 eV corresponds to the $2\sigma_u \rightarrow 1\pi_g$ state which is decaying by a two-electron transition since the $(2\sigma_u)^{-1}$ channel is closed at

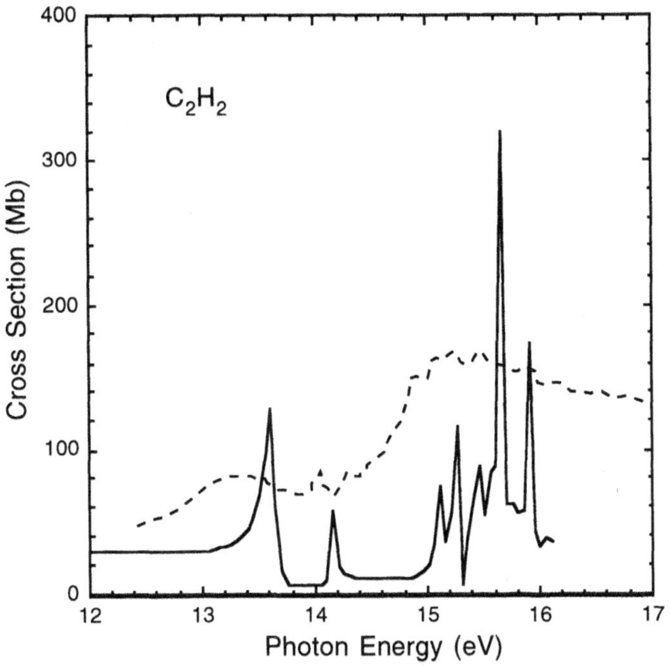

Figure 7. Photoionization of C_2H_2 leading to the $X\ ^2\Pi_u$ state of $C_2H_2\ ^+$: ———— MCCASCI mixed form results; — — — relative experimental cross sections of Ibuki *et al.*[37]

this energy. On top of these broad valence features are a set of two-electron resonances due to a series of autoionizing Rydberg levels. This figure clearly illustrates that the resonance structure in photoionization cross sections are in general a fairly complex combination of broad one-electron resonances and both narrow and broad two-electron resonances.

Participator Auger Resonance

Another type of resonance in photoionization occurs when core electrons are excited at a photon energy which is insufficient to form the core-hole ion state but is at an energy which is resonant with an excitation of the core electron to a valance one-electron state. An example of this type of resonance is found in the $C1s \rightarrow 2\pi^*$ excitation of CO which occurs at about 292.4 eV,[38, 39] just below the energy needed to form the $C1s$ hole state of CO^+ at 296.3 eV.[40] This resonant state can decay through autoionization into the many ion states with IPs less than 292.4 eV. These autoionization processes are referred to as participator Auger decays. We computed the branching ratios for the various possible channels by performing a series of two-channel calculations including the $(C1s)^{-1}$ hole state as a closed channel and one additional open ion state channel. The width of the resonance in this two-state approximation was assumed to give a good indication of the bound continuum coupling so that the ratio of the widths would correspond to the ratios of the probabilities for decay into the respective channels.

In Table 1 the results of a series of such two-state calculations is given. The decay pattern is very different from the branching ratios which would be observed in a direct ionization process. There are a number of strong decays to valance shake-up states of CO^+ where the $2\pi^*$ orbital remains occupied while a two electron Auger-like decay fills the core level and create two holes in the valance shell. In addition to these decay channels, the low

Table 1. Computed IPs of the states of CO^+ and partial widths for decay to the ion state from the $C1s \rightarrow 2\pi^*$ resonant state of CO.

Configuration	Symmetry	IP (eV)	Width (meV)
$(4\sigma)^{-1}(5\sigma)^{-1}(2\pi)^{+1}$	$^2\Pi$	22.6	15.9
$(1\pi)^{-1}$	$^2\Pi$	12.6	9.8
$(5\sigma)^{-1}(3\sigma)^{-1}(2\pi)^{+1}$	$^2\Pi$	41.5	4.6
$(5\sigma)^{-1}(1\pi)^{-1}(2\pi)^{+1}$	$^2\Delta$	20.7	4.5
$(4\sigma)^{-1}(5\sigma)^{-1}(2\pi)^{+1}$	$^2\Pi$	23.9	4.2
$(5\sigma)^{-1}(1\pi)^{-1}(2\pi)^{+1}$	$^2\Sigma$	19.2	4.2
$(5\sigma)^{-1}(1\pi)^{-1}(2\pi)^{+1}$	$^2\Delta$	17.8	3.5
$(5\sigma)^{-1}$	$^2\Sigma$	8.5	3.1
$(5\sigma)^{-1}(1\pi)^{-1}(2\pi)^{+1}$	$^2\Sigma$	21.2	2.5
$(4\sigma)^{-1}$	$^2\Sigma$	15.3	2.0
$(5\sigma)^{-1}(1\pi)^{-1}(2\pi)^{+1}$	$^2\Sigma$	17.7	1.3
$(4\sigma)^{-1}(1\pi)^{-1}(2\pi)^{+1}$	$^2\Delta$	28.1	1.0
$(4\sigma)^{-1}(1\pi)^{-1}(2\pi)^{+1}$	$^2\Sigma$	30.1	0.8
$(3\sigma)^{-1}(1\pi)^{-1}(2\pi)^{+1}$	$^2\Delta$	45.7	0.8
$(4\sigma)^{-1}(1\pi)^{-1}(2\pi)^{+1}$	$^2\Sigma$	30.1	0.8
$(3\sigma)^{-1}$	$^2\Sigma$	33.9	0.7
$(5\sigma)^{-1}(1\pi)^{-1}(2\pi)^{+1}$	$^2\Sigma$	45.8	0.6
$(4\sigma)^{-1}(1\pi)^{-1}(2\pi)^{+1}$	$^2\Sigma$	28.4	0.5
Total Width			62

lying single-hole valence states of CO^+ are also populated through a decay which removes the electron from the $2\pi^*$ orbital. In direct photoionization the single hole state intensities are much stronger than any of the shake-up states found here since than can be populated directly in the independent particle picture.

Resonant Doubly-Excited States

As discussed above, a doubly excited valence state has no oscillator strength for excitation from the ground state in the independent particle picture, and thus can only be excited through various electron-electron correlation effects. One example of such a state occurs in the photoionization of N_2. In this system, there is a low-lying shake-up state of the ion, $C\,^2\Sigma_u^+$, which occurs experimentally at 23.6 eV.[40] This state can be qualitatively described by the electronic configuration $(3\sigma_g)^{-1}(1\pi_u)^{-1}(1\pi_g)^{+1}$ relative to the ground state of N_2. Below the threshold for the production of N_2^+ in the $C\,^2\Sigma_u^+$ state there is a resonant state of overall $^1\Pi_u$ symmetry which can be described by the configuration $(3\sigma_g)^{-1}(1\pi_u)^{-1}(1\pi_g)^{+2}$ and thus in part corresponds to an electron in the $1\pi_g$ orbital in the field of the $C\,^2\Sigma_u^+$ ion state[5, 9] although penetration terms from other ion states also contribute to the $^1\Pi_u$ resonance. In the independent particle picture, this state corresponds to a doubly excited state relative to the $X\,^1\Sigma_g^+$ state of N_2.

In Fig. 8 the computed photoionization cross section leading to the $B\,^2\Sigma_u^+$ of N_2^+ is given.[9] There is a strong resonant enhancement of this cross section at approximately 20.5 eV which is due to the $(3\sigma_g)^{-1}(1\pi_u)^{-1}(1\pi_g)^{+2}$ resonant state. One source of the intensity of this transition is the strong correlation effects in the $X\,^1\Sigma_g^+$ state of N_2 which contains approximately a 5% contribution from the $(1\pi_u)^{-2}(1\pi_g)^{+2}$ configuration. The resonant state is only a single excitation $(3\sigma_g \rightarrow 1\pi_u)$ from this configuration and thus the dipole operator can couple the resonance and the ground state through this initial state correlation effect. In addition, a VCASCI calculation of the resonant state indicates a 13% mixing with the

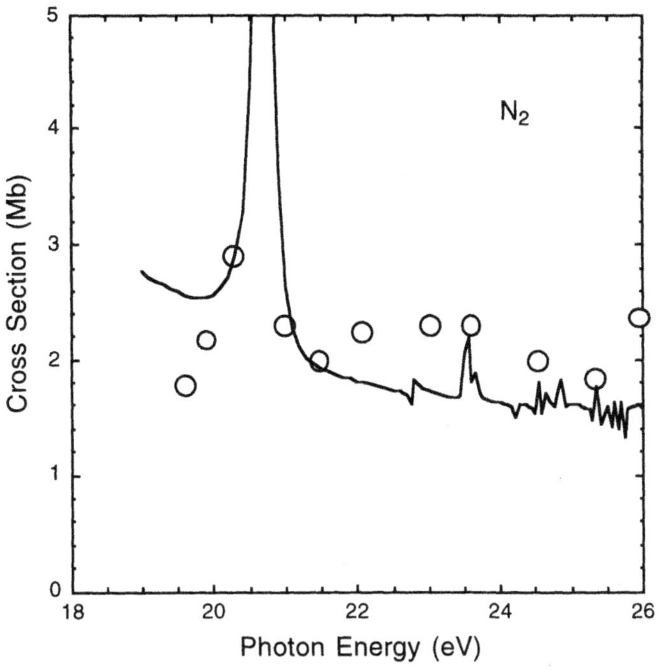

Figure 8. Photoionization of N_2 leading to the $(2\sigma_u)^{-1}$ B $^2\Sigma_u^+$ state of N_2^+; ——— MCCASCI calculations[9]; **O** experimental results of Plummer *et al.*[36]

lower-lying $(2\sigma_u)^{-1}$ $(1\pi_g)^{+1}$ $^1\Pi_u$ configuration[5] which has a non-zero dipole matrix element with the principal configuration of the initial state. Thus this final state correlation also enhances the oscillator strength for excitation of the $^1\Pi_u$ resonance in the photoionization of N_2.

CONCLUSIONS

The calculation of molecular photoionization within the fixed nuclei approximation has reached a fairly sophisticated level. We can include extensive correlation effects in both initial and final states. We have discussed how an analysis of the form of the expansion and the removal of terms corresponding to an overcomplete expansion can lead to reliable results with limited expansions. We have further demonstrated that, lacking actual repulsive interactions, angular momentum barriers provide the main mechanism for resonant trapping of electrons in one-electron resonances. We have also seen that, when variations of the interaction potentials are considered, a one-electron resonance does not immediately disappear when the energy of the resonance is moved above the angular momentum barrier, but the time delay slowly lessens as the energy gets further above the confining barrier. In terms of the poles of the *S* matrix, this corresponds to the resonant pole slowly becoming part of the series of poles which combine to make up the background scattering continuum. We have also seen that valance two-electron resonances can be described with small VCASCI wave functions. These approximate wave functions can then be used as a guide to the calculation of the photoionization cross section within a close-coupling expansion.

The level of sophistication in the current electronic structure calculations is such that the largest source of disagreement between theory and experiment is most likely due to the neglect of nuclear motion. Work is currently underway to consider the main effects of nuclear motion on the photoabsorption and photoionization cross sections.

ACKNOWLEDGMENTS

The partial support of the Robert A. Welch Foundation (Houston, Texas) under grant A-1020 is gratefully acknowledged. This work has also been supported by the NATO organization under the Collaborative Research Grant n. 950552.

REFERENCES

1. R. E. Stratmann and R. R. Lucchese, A graphical unitary group approach to study multiplet specific multichannel electron correlation effects in the photoionization of O_2, *J. Chem. Phys.* 102:8493 (1995).
2. R. R. Lucchese and V. McKoy, Padé-approximant corrections to general variational expressions of scattering theory: Application to 5σ photoionization of carbon monoxide, *Phys. Rev. A* 28:1382 (1983).
3. R. K. Nesbet, *Variational Methods in Electron-Atom Scattering Theory*, Plenum Press, New York (1980).
4. Y. N. Demkov, *Variational Principles in the Theory of Collisions*, MacMillan, New York (1963).
5. R. W. Zurales, Ph. D. thesis, Texas A&M University, (to be published).
6. J. Paldus, Group theoretical approach to the configuration interaction and preturbation theory calculations for atomic and molecular systems, *J. Chem. Phys.* 61:5321 (1974).
7. B. H. Lengsfield III and T. N. Rescigno, Electron-molecule close coupling with correlated target wave functions: Application to impact dissociation of F_2, *Phys. Rev. A* 44:2913 (1991).
8. R. E. Stratmann, R. W. Zurales, and R. R. Lucchese, Multiplet-specific multichannel electron-correlation effects in the photoionization of NO, *J. Chem. Phys.* 104:8989 (1996).
9. R. E. Stratmann, G. Bandarage, and R. R. Lucchese, Electron-correlation effects in the photoionization of N_2, *Phys. Rev. A* 51:3756 (1995).
10. R. W. Zurales and R. R. Lucchese, Pseudostate expansions for two-electron systems, *Phys. Rev. Lett.* submitted for publication.
11. R. R. Lucchese, Multichannel variational expressions of scattering theory, *Phys. Rev. A* 33:1626 (1986).
12. R. R. Lucchese, G. Raseev, and V. McKoy, Studies of differential and total photoionization cross sections of molecular nitrogen, *Phys. Rev. A* 25:2572 (1982).
13. J. D. Weeks, A. Hazi, and S. A. Rice, On the use of pseudopotentials in the quantum theory of atoms and molecules, in: *Advances in Chemical Physics*, I. Prigogine and S. A. Rice ed., Interscience, New York (1969).
14. R. R. Lucchese, Effects of interchannel coupling on the photoionization cross sections of carbon dioxide, *J. Chem. Phys.* 92:4203 (1990).
15. B. Basden and R. R. Lucchese, High-order Newton-Cotes integration methods in scattering theory, *J. Comput. Phys.* 77:524 (1988).
16. R. R. Lucchese, K. Takatsuka, and V. McKoy, Applications of the Schwinger variational principle to electron-molecule collisions and molecular photoionization, *Phys. Rep.* 131:147 (1986).
17. M. B. Robin, *Higher Excited States of Polyatomic Molecules*, Academic Press, New York (1985).
18. J. R. Taylor, *Scattering Theory*, Wiley, New York (1972).
19. C. R. Natoli, Near edge absorption structure in the framework of the multiple scattering model. Potential resonance or barrier effects?, in: *EXAFS and Near Edge Structure*, A. Bianconi, L. Incoccia and S. Stipcich ed., Springler-Verlag, Berlin (1983).
20. J. A. Sheehy, T. J. Gil, C. L. Winstead, R. E. Farren, and P. W. Langhoff, Correlation of molecular valence- and *K*-shell photoionization resonances with bond lengths, *J. Chem. Phys.* 91:1796 (1989).
21. R. R. Lucchese and F. A. Gianturco, One-electron resonances in electron scattering from polyatomic molecules, *Int. Rev. Phys. Chem.* 15:429 (1996).

22. P. W. Langhoff, Molecular photoionization resonances. A theoretical chemist's perspective, in: *Resonances In Electron-Molecule Scattering, van der Waals Complexes, and Reactive Chemical Dynamics*, D. G. Truhlar ed., American Chemical Society, Washington, D. C. (1984).
23. F. A. Gianturco, C. Guidotti, and U. Lamanna, Electronic properties of sulphur hexafluoride. II. Molecular orbital interpretation of its X-ray absorption spectra, *J. Chem. Phys.* 57:840 (1972).
24. R. W. Zurales and R. R. Lucchese, Separable coordinates in the description of resonances in electron-molecule scattering, (to be published).
25. W. Sun, R. M. Pitzer, and C. W. McCurdy, Photodetachment cross section for $3\sigma_u$ channel of F_2^- in the static-exchange approximation, *Phys. Rev. A* 40:3669 (1989).
26. R. E. Stratmann and R. R. Lucchese, Resonances and the effects of interchannel coupling in the photoionization of CS_2, *J. Chem. Phys.* 97:6384 (1992).
27. R. R. Lucchese and F. A. Gianturco, Elastic scattering cross sections and resonances in electron scattering from benzene, (to be published).
28. J. A. Stephens and D. Dill, Shape-resonant effects mediated by channel interaction: Angular distributions of N_2 $2\sigma_u$ photoelectrons, *Phys. Rev. A* 31:1968 (1985).
29. B. Basden and R. R. Lucchese, Angular distributions of N_2 $(2\sigma_g)^{-1}$ photoelectrons including the effects of coupling to the N_2 $(3\sigma_g)^{-1}$ channel, *Phys. Rev. A* 34:5158 (1986).
30. M. R. Hermann, S. R. Langhoff, T. J. Gil, and P. W. Langhoff, Theoretical studies of photoionization in nitric oxide: Multiplet-specific effects in vibrationally resolved $5\sigma^{-1}$ resonant channels, *Chem. Phys. Letts.* 125:336 (1986).
31. M. E. Smith, V. McKoy, and R. R. Lucchese, Multiplet-specific shape resonant features in photoionization of NO, *J. Chem. Phys.* 82:4147 (1985).
32. M. R. Hermann, C. W. Bauschlicher Jr., W. M. Wuo, S. R. Langhoff, and P. W. Langhoff, Theoretical studies of valence-shell photoionization and ionization cross sections in nitric oxide, *Chem. Phys.* 109:1 (1986).
33. S. Wallace, D. Dill, and J. L. Dehmer, Shape resonant features in the photoionization spectra of NO, *J. Chem. Phys.* 76:1217 (1982).
34. U. Fano, Effects of configuration interaction on intensities and phase shifts, *Phys. Rev.* 124:1866 (1961).
35. R. R. Lucchese and R. W. Zurales, Comparison of the random-phase approximation with the multichannel frozen-core Hartree-Fock approximation for the photoionization of N_2, *Phys. Rev. A* 44:291 (1991).
36. E. W. Plummer, T. Gustafsson, W. Gudat, and D. E. Eastman, Partial photoionization cross sections of N_2 and CO using synchrotron radiation, *Phys. Rev. A* 15:2339 (1977).
37. T. Ibuki, Y. Horie, A. Kamiuchi, Y. Morimoto, M. C. K. Tinone, K. Tanaka, and K. Honma, Radiative dissociation of C_2H_2, C_2HD, and C_2D_2 superexcited at 50-90 nm region, *J. Chem. Phys.* 102:5301 (1995).
38. S. J. Osborne, A. Ausmees, S. Svensson, A. Kivimäki, O.-P. Sairanen, A. Naves de Brito, H. Aksela, and S. Aksela, The vibrationally resolved participator Auger spectra of selectively excited C 1*s* $(2\sigma)^{-1}$ $(2\pi)^{+1}$ vibrational states in carbon monoxide, *J. Chem. Phys.* 102:7317 (1995).
39. M. Neeb, J.-E. Rubensson, M. Biermann, and W. Eberhardt, Coherent excitation of vibrational wave functions observed in core hole decay spectra of O_2, N_2 and CO, *J. Electron Spectrosc. Relat. Phenom.* 67:261 (1994).
40. K. P. Huber and G. Herzberg, *Molecular Spectra and Molecular Structure IV. Constants of Diatomic Molecules*, Van Nostrand Reinhold Co., New York (1979).

DOUBLE PHOTOIONIZATION OF HELIUM

Marcel Pont and Robin Shakeshaft

Physics Department,
University of Southern California,
Los Angeles, California 90089-0484, USA

INTRODUCTION

The differential cross section for double photoionization of helium provides a sensitive tool for exploring electron-electron correlation since the absorption of one photon cannot lead to double escape without the two electrons interacting. However, differential cross sections for double photoionization are neither easy to calculate nor easy to measure. Moreover, while a great deal has been learned about the dynamics of double photoionization since the pioneering work by Wannier[1] in 1953, a complete understanding has yet to be achieved. Over the last several years we have developed a method,[2,3] based on the earlier work of Rudge and Seaton,[4] for calculating differential cross sections for double photoionization of helium. In this paper we present some of our results and we compare them with selected data from recent measurements in the energy range 0 to 80 eV above threshold.

The dipole matrix element, $f(\mathbf{k}_1, \mathbf{k}_2)$, for two electrons to be ejected from an atom by one photon, and to emerge with momenta $\mathbf{k}_1$ and $\mathbf{k}_2$ and energies $E_1 \equiv k_1^2/2$ and $E_2 \equiv k_2^2/2$, is

$$f(\mathbf{k}_1, \mathbf{k}_2) = \langle \Psi^{(-)}_{\mathbf{k}_1,\mathbf{k}_2} | V | \Psi_i \rangle,$$

where Ψ_i and $\Psi^{(-)}_{\mathbf{k}_1,\mathbf{k}_2}$ are the initial- and final-state wavefunctions and where V is the atom-photon interaction. The direct evaluation of $f(\mathbf{k}_1, \mathbf{k}_2)$ is complicated by the difficulty of calculating $\Psi^{(-)}_{\mathbf{k}_1,\mathbf{k}_2}$. It is a formidable task to exactly solve the homogeneous Schrödinger equation subject to the exit-channel boundary condition describing three unbound particles moving with specified velocities. Nevertheless, Maulbetsch and Briggs[5] have successfully calculated angular distributions by approximating $\Psi^{(-)}_{\mathbf{k}_1,\mathbf{k}_2}$ as a product of 3 Coulomb continuum wavefunctions which take final-state correlation explicitly into account. However, rather than evaluate $f(\mathbf{k}_1, \mathbf{k}_2)$ directly, we follow Rudge and Seaton[4] and approximate $\Psi^{(-)}_{\mathbf{k}_1,\mathbf{k}_2}$ as a product of 2 *screened* Coulomb wavefunctions employing effective charges. This approximate final-state wavefunction, which we call the 2SC wavefunction, and denote by $\chi^{(-)}_{\mathbf{k}_1,\mathbf{k}_2}$, satisfies the Schrödinger equation

$$(E - H_{\mathrm{a}}) | \chi^{(-)}_{\mathbf{k}_1,\mathbf{k}_2} \rangle = W | \chi^{(-)}_{\mathbf{k}_1,\mathbf{k}_2} \rangle,$$

where H_a is the Hamiltonian of the bare atom, E is the excess energy above threshold, i.e.

$$E \equiv E_1 + E_2,$$

and W is a sum of screened Coulomb potentials, i.e.

$$W = \frac{1}{|\mathbf{r}_1 - \mathbf{r}_2|} - \frac{\Delta_1}{r_1} - \frac{\Delta_2}{r_2},$$

with the screening parameters Δ_1 and Δ_2 chosen so that W vanishes along the classical asymptotic trajectories, $\mathbf{r}_1 \approx \mathbf{v}_1 t$ and $\mathbf{r}_2 \approx \mathbf{v}_2 t$, of the departing electrons. The 2SC wavefunction is substituted into the flux formula

$$f(\mathbf{k}_1, \mathbf{k}_2) = \langle \Psi^{(-)}_{\mathbf{k}_1,\mathbf{k}_2} | (H_a^\dagger - H_a) | \mathcal{F}^+ \rangle,$$

where $\mathcal{F}^+$ represents the response of the initial state to the photon-atom perturbation and satisfies the inhomogeneous Dalgarno-Lewis equation

$$(E - H_a)|\mathcal{F}^+\rangle = V|\Psi_i\rangle.$$

In principle, though this is not done in practice, the flux formula can be transformed to a surface integral whose value is determined (using the method of stationary phase — see Rudge and Seaton[4]) by the final-state wavefunction along the classical asymptotic trajectories of the electrons. Since the 2SC wavefunction satisfies $H_a|\Psi^{(-)}_{\mathbf{k}_1,\mathbf{k}_2}\rangle = E|\Psi^{(-)}_{\mathbf{k}_1,\mathbf{k}_2}\rangle$ along these trajectories, the burden of the calculation is shifted from the final-state wavefunction to the response function $\mathcal{F}^+$, and in this respect our approach is similar to the one taken earlier by Dalgarno *et al.*[6] The advantage of shifting the burden onto $\mathcal{F}^+$ is two-fold. First, $\mathcal{F}^+$ satisfies an inhomogeneous equation, subject to the simple entrance-channel boundary condition that the electrons are initially bound and, after photoabsorption, move radially outwards. This boundary condition is already incorporated through the inhomogeneous term, aside from the stipulation that the electrons move radially outwards, rather than inwards. Second, since $\mathcal{F}^+$ satisfies an outgoing-wave boundary condition, in contrast to the standing-wave boundary condition satisfied by $\Psi^{(-)}_{\mathbf{k}_1,\mathbf{k}_2}$, it can be represented on a set of basis functions having outgoing-wave character. Furthermore, Ψ_i can also be represented on this basis provided that the basis functions have *damped* outgoing-wave character. In any given region of space it is as easy, if not easier, to calculate $\mathcal{F}^+$ as it is to calculate Ψ_i, although inevitably one needs $\mathcal{F}^+$ over a much larger region of space in using the flux formula than one needs Ψ_i in using the direct formula. We choose a basis composed of products of one-electron complex radial Sturmian functions and spherical harmonics.

After substituting $\chi^{(-)}_{\mathbf{k}_1,\mathbf{k}_2}$ into the flux formula, and taking note of the Dalgarno-Lewis equation for $\mathcal{F}^+$, we arrive at

$$f(\mathbf{k}_1, \mathbf{k}_2) = \langle \chi^{(-)}_{\mathbf{k}_1,\mathbf{k}_2} | V | \Psi_i \rangle + \langle \chi^{(-)}_{\mathbf{k}_1,\mathbf{k}_2} | W | \mathcal{F}^+ \rangle,$$

where the second term on the right-hand side of this last equation is the correction compensating for the error in $\chi^{(-)}_{\mathbf{k}_1,\mathbf{k}_2}$. Formally, this correction is a divergent volume integral since W decreases at large distances only as a Coulomb potential (except along the classical asymptotic trajectories), but the integral can be given meaning in the Lebesgue sense. In practice, since the response function is represented on a discrete basis, the correction is a divergent sum, which can be rendered convergent by standard means, e.g. Padé summation.

We use atomic units unless stated otherwise.

RESULTS

Triply Differential Cross Section

The triply differential cross section for the electrons to emerge into solid angles $d\Omega_1$ and $d\Omega_2$, with one electron having energy E_1, is

$$\frac{d^3\sigma^{2+}}{dE_1 d\Omega_1 d\Omega_2} = \frac{4\pi^2}{\omega c} k_1 k_2 |f(\mathbf{k}_1, \mathbf{k}_2)|^2,$$

where ω is the photon frequency; the prefactor depends on the normalization of the final-state wavefunction, and we have chosen to normalize on the momentum scale, and to symmetrize with respect to the electrons.

Since the atom absorbs only one unit of angular momentum, the amplitude $f(\mathbf{k}_1, \mathbf{k}_2)$ has the geometrical decomposition

$$f(\mathbf{k}_1, \mathbf{k}_2) = f(k_1, k_2, k)\hat{\mathbf{k}}_1 \cdot \hat{\mathbf{e}} + f(k_2, k_1, k)\hat{\mathbf{k}}_2 \cdot \hat{\mathbf{e}},$$

where $\hat{\mathbf{e}}$ is the unit (linear) polarization vector, $\hat{\mathbf{k}}_1$ and $\hat{\mathbf{k}}_2$ are unit vectors along $\mathbf{k}_1$ and $\mathbf{k}_2$, and

$$\mathbf{k} = \mathbf{k}_1 - \mathbf{k}_2,$$

the relative momentum of the electrons, and $k = |\mathbf{k}|$. The dynamical components $f(k_1, k_2, k)$ and $f(k_2, k_1, k)$ peak when the relative momentum of the electrons has its maximum maginitude $k = k_1 + k_2$, for then $\hat{\mathbf{k}}_1 = -\hat{\mathbf{k}}_2$ and the electrons are moving in opposite directions thereby minimizing their Coulomb repulsion. In contrast, $f(k_1, k_2 = k_1, k = 0)$ vanishes due to the strong Coulomb repulsion. Following e.g. Huetz *et al*[11] it is useful to reexpress the dynamical components so as to directly display the identity of the electrons; thus we write

$$f(\mathbf{k}_1, \mathbf{k}_2) = f^+(k_1, k_2, k)(\hat{\mathbf{k}}_1 + \hat{\mathbf{k}}_2) \cdot \hat{\mathbf{e}} + f^-(k_2, k_1, k)(\hat{\mathbf{k}}_1 - \hat{\mathbf{k}}_2) \cdot \hat{\mathbf{e}},$$

where

$$f^\pm(k_1, k_2, k) = (1/2)[f(k_1, k_2, k) \pm f(k_2, k_1, k)],$$

with

$$f^\pm(k_1, k_2, k) = \pm f^\pm(k_2, k_1, k).$$

Note that

$$f^-(k_1, k_2 = k_1, k) = 0,$$

so that if $k_1/k_2 \approx 1$ the component $f^-(k_1, k_2, k)$ is small relative to $f^+(k_2, k_1, k)$. Since $f^\pm(k_1, k_2, k)$ peak when $\hat{\mathbf{k}}_1 = -\hat{\mathbf{k}}_2$ we expect that if k_1/k_2 differs significantly from unity the triply differential cross section has a strong maximum at $\hat{\mathbf{k}}_1 = -\hat{\mathbf{k}}_2$ due to the term $f^-(k_2, k_1, k)(\hat{\mathbf{k}}_1 - \hat{\mathbf{k}}_2) \cdot \hat{\mathbf{e}}$ in $f(\mathbf{k}_1, \mathbf{k}_2)$. The factor $(\hat{\mathbf{k}}_1 + \hat{\mathbf{k}}_2) \cdot \hat{\mathbf{e}}$, and hence the term $f^+(k_2, k_1, k)(\hat{\mathbf{k}}_1 + \hat{\mathbf{k}}_2) \cdot \hat{\mathbf{e}}$, vanish at this maximum. While $(\hat{\mathbf{k}}_1 + \hat{\mathbf{k}}_2) \cdot \hat{\mathbf{e}}$ peaks when $\hat{\mathbf{k}}_1 = \hat{\mathbf{k}}_2$, the dynamical component $f^+(k_2, k_1, k)$ is small for this configuration, and so the term $f^+(k_2, k_1, k)(\hat{\mathbf{k}}_1 + \hat{\mathbf{k}}_2) \cdot \hat{\mathbf{e}}$ gives its maximum contribution when the angle between $\hat{\mathbf{k}}_1$ and $\hat{\mathbf{k}}_2$ is significantly different from 0 and π.

In Fig. 1 we show the angular distribution of electron 2 for the case where electron 1 emerges along the major polarization axis of the light with energy $E_1 = 5$ eV, and electron 2 has energy 47.9 eV. The polarization of the light is specified by the Stokes parameter S_1, whose value is $S_1 = 0.99$. We show theoretical results from our calculations and those of Maulbetsch and Briggs, and we compare with the measured data of

Schwarzkopf *et al.*[7] We have rescaled all results so that they agree at the maximum of the distribution; the overall agreement is rather good. The angular distribution has a strong maximum at $\hat{\mathbf{k}}_1 = -\hat{\mathbf{k}}_2$ due to the term $f^-(k_2, k_1, k)(\hat{\mathbf{k}}_1 - \hat{\mathbf{k}}_2) \cdot \hat{\mathbf{e}}$ in $f(\mathbf{k}_1, \mathbf{k}_2)$. The small wings, which peak when the angle between $\hat{\mathbf{k}}_1$ and $\hat{\mathbf{k}}_2$ is roughly $\pi/2$ or $3\pi/2$, receive a significant contribution from the term $f^+(k_2, k_1, k)(\hat{\mathbf{k}}_1 + \hat{\mathbf{k}}_2) \cdot \hat{\mathbf{e}}$.

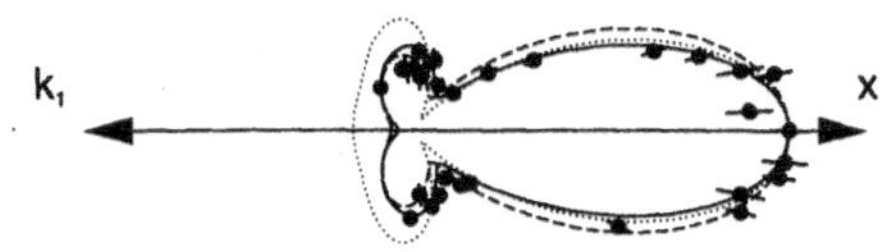

Fig. 1. Polar plots of the (relative) triply differential cross section when electron 1 emerges along the major polarization axis (x-axis) with energy $E_1 = 5$ eV and electron 2 emerges with energy $E_2 = 47.9$ eV. The experimental data are from Ref. 7. The dashed and dotted lines are from the velocity and length gauge versions, respectively, of the theory of Maulbetsch and Briggs[5], and the solid line represents our theoretical results.[3]

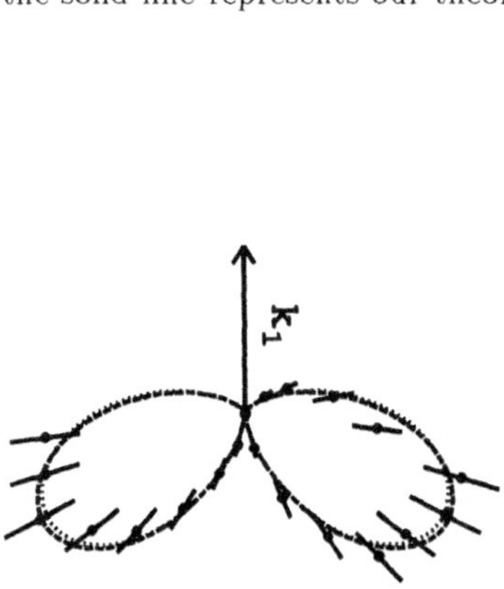

Fig. 2. Polar plots of the coplanar triply differential cross section on an absolute scale. Electron 1 emerges along the major polarization axis, and $E_1 = E_2 = 10.0$ eV. The experimental data are from Ref. 8, but with the original data, and its fit, multiplied by 2.2 (see text); the broken line is a fit to this data, and the dotted line represents our theoretical results[10].

Almost all measurements of the triply differential cross section up to now have been relative measurements, but an absolute measurement was recently made by Schwarzkopf and Schmidt[8] for the case where electron 1 emerges along the major polarization axis of the light with energy $E_1 = 10$ eV, and electron 2 has the same energy. (The Stokes parameter $S_1 = 0.59$.) The data, as originally published,[8] differed from our results by a factor of about two,[4,10] but recently the measured data were revised upwards by a factor of two[9]; in Fig. 2 we have multiplied the original data, and the analytical fit to this data, by a factor of 2.2 (see Ref. 10 for further discusssion of this factor) which gives almost perfect agreement with our results. Since $k_1/k_2 = 1$ the component

$f^-(k_1, k_2, k)$ vanishes, and the angular distribution peaks where the contribution from the term $f^+(k_2, k_1, k)(\hat{\mathbf{k}}_1 + \hat{\mathbf{k}}_2) \cdot \hat{\mathbf{e}}$ is maximum. The angular distribution vanishes when $\hat{\mathbf{k}}_1 = \hat{\mathbf{k}}_2$ since $f^\pm(k_1, k_2 = k_1, k = 0)$ vanish, and it vanishes also when $\hat{\mathbf{k}}_1 = -\hat{\mathbf{k}}_2$ since the term in $f^+(k_2, k_1, k)$ vanishes.

Energy Distribution and Total Cross Section

Integrating the triply differential cross section over the solid angles Ω_1 and Ω_2 yields the singly differential cross section $d\sigma^{2+}/dE_1$. To perform this integration we make use of the preceding geometrical decomposition of $f(\mathbf{k}_1, \mathbf{k}_2)$. Defining μ by $\mathbf{k}_1 \cdot \mathbf{k}_2 \equiv k_1 k_2 \mu$ it is straightforward to show that[12]

$$\frac{d\sigma^{2+}}{dE_1} = \frac{2^6\pi^4}{3\omega c} k_1 k_2 \int_{-1}^{1} d\mu \, [(1+\mu)|f^+(k_1, k_2, k)|^2 + (1-\mu)|f^-(k_1, k_2, k)|^2].$$

To obtain the total cross section, σ^{2+}, for double photoionization we integrate $d\sigma^{2+}/dE_1$ over the range $0 \leq E_1 \leq E/2$. The photoelectron energy distribution is symmetric about the midpoint $E/2$ and we must integrate over only half the allowed range of E_1 to avoid double counting; we cannot distinguish between electron 1 emerging with energy E_1 or $E - E_1$. Note that there is no contribution to σ^{2+} from the interference of $f^+(k_1, k_2, k)$ and $f^-(k_1, k_2, k)$; this is expected since under the interchange of E_1 and E_2, the terms $f^+(k_1, k_2, k)(\hat{\mathbf{k}}_1 + \hat{\mathbf{k}}_2) \cdot \hat{\mathbf{e}}$ and $f^-(k_2, k_1, k)(\hat{\mathbf{k}}_1 - \hat{\mathbf{k}}_2) \cdot \hat{\mathbf{e}}$ merely change sign, and if they were to interfere the energy distribution would be asymmetric about $E/2$.

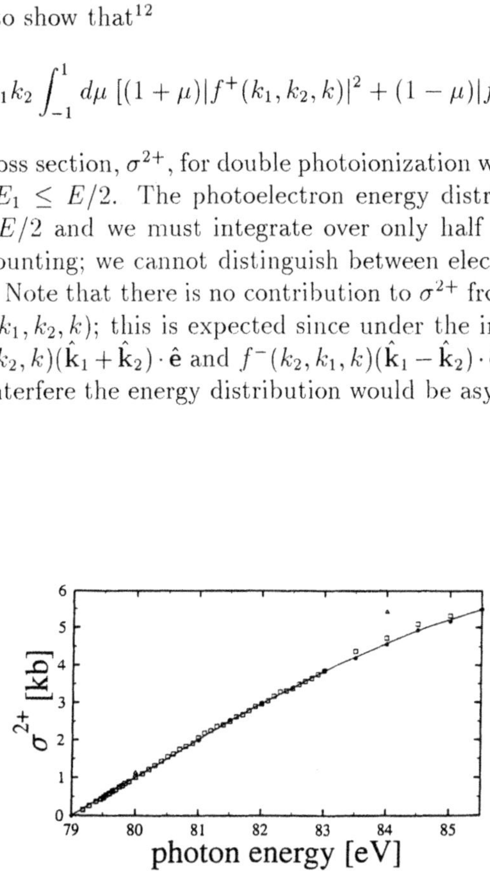

Fig. 3 Integrated cross section for double photoionization of He *vs* photon energy in the near-threshold region. The solid line is a fit through our calculated points (solid circles).[12] The open squares and open triangles are, respectively, data from experiments of Kossmann *et al*[13,14] and Bizau and Wuilleumier[15].

In Fig. 3 we show our estimates[12] of the total cross section, $\sigma^{2+}(E)$ in the near-threshold region, and we compare with the absolute data from the experiments of Kossmann *et al*[13,14] and Bizau and Wuilleumier[15]. In our calculations we were able to obtain good convergence with respect to increasing basis size down to an excess energy E of 2eV; by studying the rate of convergence with respect to increasing basis size we estimate the relative error in our values for $\sigma^{2+}(E)$ to be typically 2%. For E below 2eV convergence was much more difficult to achieve, and rather than attempt to achieve convergence we have interpolated instead over the interval $0 < E < 2$ eV using a spline fit to our results for $E \geq 2$ eV and to the accepted result $\sigma^{2+}(0) = 0$. The agreement between our results and the experimental data is very good. For excess energies close to threshold the Wannier theory yields the law

$$\sigma^{2+}(E) = \sigma_0^{2+} E^m,$$

where the exponent m has the value $m = 1.056$. Kossmann *et al* measured the value $m = 1.05(2)$, and our calculations yield $m = 1.06$, in good agreement with the Wannier theory. The Wannier theory does not give the value of the coefficient σ_0^{2+}, but Kossmann *et al* measured the value $\sigma_0^{2+} = 1.02(4)$ kb, and our calculations yield the fairly similar value $\sigma_0^{2+} = 0.97$ kb.

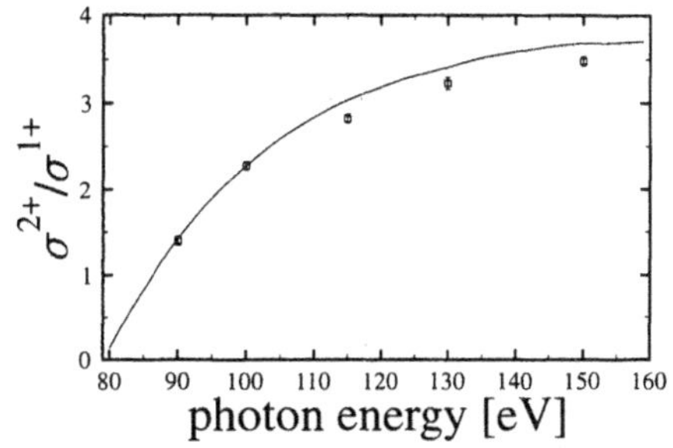

Fig. 4 Ratio of cross sections for double photoionization and single photoionization of He *vs* photon energy. The solid line is a fit through our calculated points,[12] and the open squares are data from the experiment of Dörner *et al.*[16]

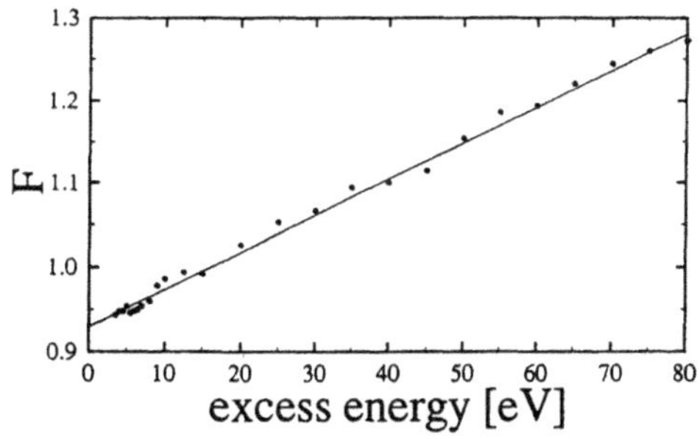

Fig. 5. Curvature function $F(E)$ *vs* excess energy E. The solid line is a linear fit.

In Fig. 4 we show the ratio of cross sections for double and single photoionization of He over a photon energy range extending up to 160 eV. We compare our results[12] with the data from the experiment of Dörner *et al.*[16] The agreement is good, although at the higher photon energies of Fig. 4 the data of Dörner *et al* lie slightly below our results whereas the data of Kossmann *et al*[13,14] and Bizau and Wuilleumier[15] (not shown) lie significantly above our results (by about 20% or so). At asymptotically large photon energies (well above the 160 eV considered here) the ratio is 1.66, a value first obtained in the pioneering calculation by Byron and Joachain.[17]

It is useful to introduce a function which provides a measure of the curvature of the energy distribution. Thus we define the dimensionless "curvature" function, $F(E)$, through the following relation between the total cross section and the singly differential cross section at the midpoint of the energy distribution:

$$\sigma^{2+}(E) \equiv \frac{E}{2} F(E) \frac{d\sigma^{2+}(E/2, E)}{dE_1}.$$

If the energy distribution were flat, $F(E)$ would be equal to unity; on the other hand, if the energy distribution were to have positive (negative) curvature, $F(E)$ would be greater (smaller) than unity. We show our estimates[12] of the curvature function in Fig. 5 over the energy range $2 \leq E \leq 80$ eV. We see that $F(E)$ is approximately linear in E over this range, and that the curvature of the energy distribution changes from positive to negative as E falls below about 17 eV. Although we cannot calculate $F(E)$ below 2 eV, we note that Read[18] has carried out classical calculations which clearly demonstrate that the energy distribution has negative curvature in the near-threshold region, a finding supported by the experiment of Hammond *et al.*[19]

Momentum Distribution and Asymmetry Parameter of the Nucleus

Dörner *et al*[20] recently used the technique of Cold Target Recoil Ion Momentum Spectroscopy to measure the asymmetry parameter and the momentum distribution of the recoiling nucleus when a helium atom undergoes double photoionization. The nucleus recoils with momentum $-\mathbf{K}$ where

$$\mathbf{K} \equiv \mathbf{k}_1 + \mathbf{k}_2$$

is the center-of-mass momentum of the two-electron subsystem (the dielectron). Defining $E_K \equiv K^2/2$ and $E_k \equiv k^2/2$ we have

$$E_k + E_K = 2E.$$

The triply differential cross sections for double electron emission and nuclear recoil are related:

$$\frac{d^3\sigma^{2+}}{dE_K \Omega_{\mathbf{K}} d\Omega_{\mathbf{k}}} = \left(\frac{kK}{4k_1k_2}\right) \frac{d^3\sigma^{2+}}{dE_1 d\Omega_1 d\Omega_2},$$

where $\Omega_{\mathbf{K}}$ and $\Omega_{\mathbf{k}}$ are solid angles containing $\mathbf{K}$ and $\mathbf{k}$. As noted above, the singly differential cross section $d\sigma^{2+}/dE_1$ is symmetric about $E/2$, and if we were to integrate over the full range $0 \leq E_1 \leq E$ we would obtain $2\sigma^{2+}$. However, E_K can vary from 0 to $2E$, and $d\sigma^{2+}/dE_K$ is *a*symmetric about E; we are obliged to integrate over the full range $0 \leq E_K \leq 2E$. Hence we include a factor of 1/2 in defining the doubly differential cross section for the nucleus to recoil into the solid angle $d\Omega_{\mathbf{K}}$ with momentum K:

$$\frac{d^2\sigma^{2+}}{dE_K d\Omega_{\mathbf{K}}} = \frac{kK}{8} \int d\Omega_{\mathbf{k}} \left(\frac{1}{k_1k_2}\right) \frac{d^3\sigma^{2+}}{dE_1 d\Omega_1 d\Omega_2}.$$

To perform the integration over $\Omega_{\mathbf{k}}$ it is useful to first reexpress the amplitude $f(\mathbf{k}_1, \mathbf{k}_2)$ as

$$f(\mathbf{k}_1, \mathbf{k}_2) = g^+(k_1, k_2, k)\mathbf{K} \cdot \hat{\mathbf{e}} + g^-(k_2, k_1, k)\mathbf{k} \cdot \hat{\mathbf{e}},$$

where

$$g^{\pm}(k_1, k_2, k) \equiv \frac{1}{2}\left[\frac{f(k_1, k_2, k)}{k_1} \pm \frac{f(k_2, k_1, k)}{k_2}\right],$$

Defining $\bar{\mu}$ by $\mathbf{k} \cdot \mathbf{K} \equiv kK\bar{\mu}$ and introducing the auxilliary functions

$$I_l^{\pm}(K) = \frac{1}{2}\int_{-1}^{1} d\bar{\mu}\ |g^{\pm}(k_1, k_2, k)|^2 P_l(\bar{\mu}),$$

$$J_l(K) = \int_{-1}^{1} d\bar{\mu}\ \mathrm{Re}\{g^+(k_1, k_2, k)[g^-(k_1, k_2, k)]^*\}P_l(\bar{\mu}),$$

we arrive at

$$\frac{d^2\sigma^{2+}}{dE_K d\Omega_{\mathbf{K}}} = \frac{1}{4\pi}\frac{d\sigma}{dE_K}[1 + \beta_n(K)P_2(\cos\theta_K)],$$

where θ_K is the angle of nuclear recoil relative to the polarization axis, where $\beta_n(K)$ is the asymmetry parameter for the nucleus, which can be expressed as

$$\beta_n(K) = 2\frac{K^2 I_0^+(K) + k^2 I_2^-(K) + kKJ_1(K)}{K^2 I_0^+(K) + k^2 I_0^-(K) + kKJ_1(K)},$$

and where $d\sigma^{2+}/dE_K$ can be expressed as

$$\frac{d\sigma^{2+}}{dE_K} = kK\left(\frac{8\pi^4}{3\omega c}\right)[K^2 I_0^+(K) + k^2 I_0^-(K) + kKJ_1(K)].$$

Switching variables from E_K to K, the recoil momentum distribution is $d\sigma^{2+}/dK = Kd\sigma/dE_K$.

In Figs. 6 and 7 we compare our results[21] for the momentum distribution of the nucleus, and the asymmetry parameter, with the data from the experiment of Dörner *et al* at photon energies of 85, 99, and 159 eV — or, equivalently, at excess energies $E = 6$, 20, and 80 eV. The horizontal scale is the dimensionless variable $K/K_{\max}$, where $K_{\max} \equiv 2\sqrt{E}$ is the maximum value of K.

In Fig. 6 we show the momentum distribution; the agreement with experiment is rather good. We see that $d\sigma^{2+}/dK$ vanishes at both $K = 0$ and $K = K_{\max}$. The vanishing of $d\sigma^{2+}/dK$ at $K = 0$ is due to both the phase space factor K^2 contained in $d\sigma^{2+}/dE_K$, and exchange and inversion symmetry, which precludes the two electrons from emerging with equal and opposite momenta (as happens when $K = 0$). The reason $d\sigma^{2+}/dK$ vanishes at $K = K_{\max}$ is that, when $K = K_{\max}$, the two electrons emerge together, with equal momenta, which is impossible since the electrons repel each other. When $K \approx k$ we have $|\mathbf{k}_1 \cdot \mathbf{k}_2| \ll E$, that is, either one electon emerges with a very small velocity relative to the other, or the two electrons emerge in perpendicular directions. As the photon energy increases it is increasingly likely that one electron emerges slowly and the other rapidly owing to the difficulty of imparting large momentum to both electrons. Appreciable momentum can be imparted to both electrons, however, through a binary encounter collision in which one electron absorbs the photon and subsequently collides with the other (more or less stationary) electron; in this event, the two electrons emerge in perpendicular directions. Hence, noting that $K = k$ implies that $K = \sqrt{2E} = K_{\max}/\sqrt{2}$, we expect the momentum distribution to peak at a value of K that approaches $K_{\max}/\sqrt{2}$ as the photon energy becomes sufficiently high; we see that confirmed in Fig. 6.

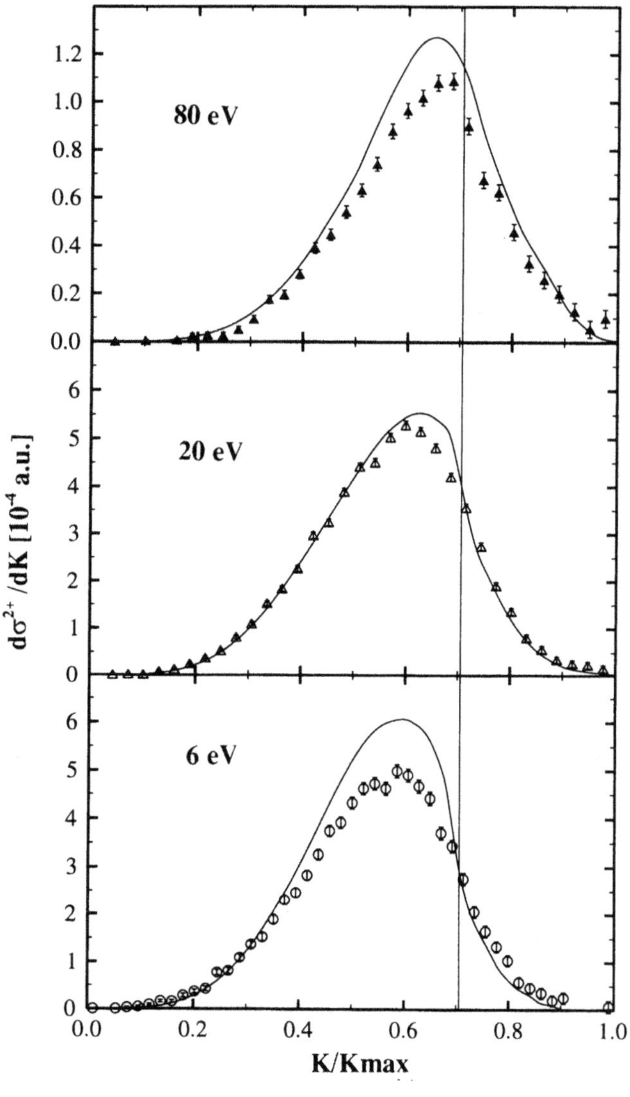

Fig. 6. Singly differential cross section, $d\sigma^{2+}/dK$, versus rescaled momentum of recoiling nucleus when helium undergoes double ionization by one linearly polarized photon of energy 85, 99, or 159 eV. The excess energy, E, is 6, 20, or 80 eV. The abscissa is $K/K_{\rm max}$, where K is the momentum of the nucleus and $K_{\rm max} = 2\sqrt{E}$. The vertical solid line passes through point $K = K_{\rm max}/\sqrt{2}$. The solid lines are our theoretical results[21] and the experimental data are from Ref. 20

Turning to the asymmetry parameter, we see that $\beta_n(K)$ approaches 2 as K approaches K_{max}. For $K \approx K_{max}$ the two electrons emerge as if they are a single particle, a dielectron, and the process resembles photoionization of hydrogen; there is only one preferred axis, the electric field axis, and the angular distribution is dipolar, corresponding to an asymmetry parameter of 2. Moreover, if we let k vanish (corresponding to $K = K_{max}$) on the right side of our expression for $\beta_n(K)$ we obtain $\beta_n(K) = 2$. [Note that neither $I_l^{\pm}(K)$ nor $J_l(K)$ are infinite when $K = K_{max}$ since $g^+(k_1, k_2 = k_1, k)$ is finite above threshold and $g^-(k_1, k_2 = k_1, k) = 0$.] Since $\beta_n(K) \leq 2$ for all K, we may infer that $\beta_n(K)$ either remains flat or begins to decrease as K begins decreasing below K_{max}. We see, in fact, that as K decreases $\beta_n(K)$ is nearly flat at first, it then drops sharply in the neighborhood of the maximum of the momentum distribution, exhibits a shallow miminum, and flattens out again as K approaches 0. Inspection of of our expression for $\beta_n(K)$ reveals that $\beta_n(K)$ deviates from 2 only to the extent that $I_2^-(K)$ in the numerator and $I_0^-(K)$ in the denominator differ from either zero or from each other. The denominator is proportional to $d\sigma^{2+}/dK$, and this was seen in Fig. 6 to be a smooth function with a peak at approximately $K_{max}/\sqrt{2}$. It is reasonable to assume that the numerator has a shape that is not too dissimilar from the denominator, but it is unlikely to peak at exactly the same value of K; consequently the ratio of numerator and denominator varies rapidly in the vicinity of the peak of either. Note that $\beta_n(0)$ is about 1.42 when $E = 6$ eV, in good agreement with the value 7/5 calculated by Feagin[22] using a different approach. Note also that $\beta_n(0)$ rises as E increases, and that $\beta_n(K)$ tends to become a flatter function of K as E rises. In fact, we expect $\beta_n(K)$ to approach 2 for (almost) all K as E increases, for the following reason. At large E, one electron is most likely to take away most of the linear momentum, and so the angular pattern of the nucleus mirrors the angular pattern of the fast electron. Since the fast electron can most easily acquire its linear momentum from a zero-angular-momentum component of the ground state, it experiences only one preferred axis, the electric field axis, and so its angular pattern is dipolar; hence $\beta_n(K) \approx \beta_e(E) \approx 2$.

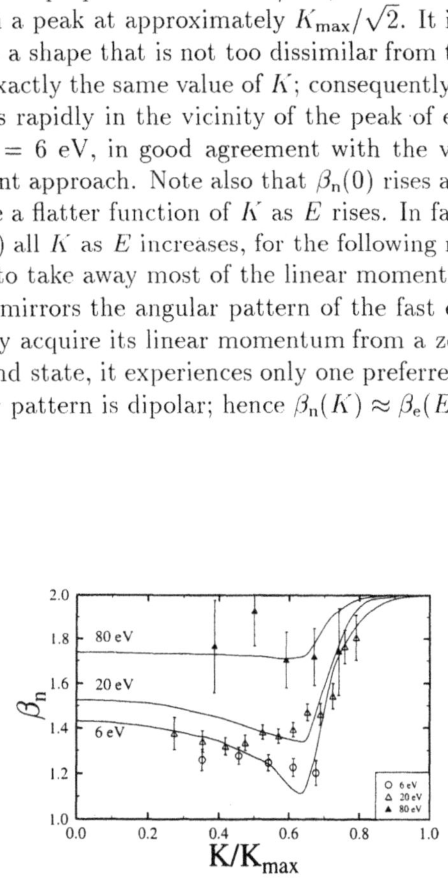

Fig. 7. Same as Fig. 6 but for asymmetry parameter, $\beta_n(K)$, of recoiling nucleus. The curves, from lower to upper, correspond to $E = 6$, 20, and 80 eV.

If, following e.g. Huetz *et al*,[11] we adopt a model wherein we neglect $f^-(k_1,k_2,k)$ and we approximate $k_1k_2|f^+(k_1,k_2,k)|^2$ by a Gaussian in $\theta_{12} \equiv \cos^{-1}(\mu)$ which is peaked at π, i.e. we put $k_1k_2|f^+(k_1,k_2,k)|^2 \propto \exp[-(\theta_{12}-\pi)^2/(\bar{\theta})^2]$, where the width $\bar{\theta}$ is an E-dependent parameter, we can easily evaluate $d\sigma/dK$ and $\beta_n(K)$. We find that the Gaussian fits to $d\sigma/dK$ are very good even at quite large E, and, furthermore, the renormalized $d\sigma/dK$ is almost invariant on the scale $K/K_{\rm max}$, at least over the energy range considered. (For E less than about 2 eV the width $\bar{\theta}$ is strongly dependent on E, and scales with E as $E^{1/4}$, so that the renormalized $d\sigma/dK$ is not invariant.) The Gaussian model is presumably exact in the limit $E \to 0$, and within this model we can evaluate $\beta_n(0)$ exactly; we find $\beta_n(0) = 7/5$, independently of the value of $\bar{\theta}$, in perfect agreement with the value calculated by Feagin.[22] However, for the three energies considered in Fig. 7 we find the fits to $\beta_n(K)$ are not as good as they are for $d\sigma^{2+}/dK$.

SUMMARY

To summarize, we have developed a method, inspired by the earlier work of Rudge and Seaton,[2] which is capable of reproducing to high accuracy most of the experimental data which has appeared during the last few years for photon energies in the range 80 to 160 eV. However, some discrepancies remain; most notably there are significant unexplained differences between our results for the the triply differential cross section and the measurements of Lablanquie *et al.*[23] Nevertheless, the success of the method does open the possibility of theoretically investigating *two*-photon double ionization of helium. This process may be of particular interest when one of the photons (produced by a synchrotron, say) resonantly excites an autoionizing state, while the other photon (produced by a laser, say) may be used, by varying its frequency, to control the branching ratio for double to single ionization.

ACKNOWLEDGMENTS

This work was supported by the DOE, Division of Chemical Sciences, and by the NCSA at Urbana-Champaign.

REFERENCES

1. G. H. Wannier, Phys. Rev. **90**, 817 (1953).
2. M. R. H. Rudge and M. J. Seaton, Proc. Roy. Soc. London A **283**, 262 (1965).
3. D. Proulx and R. Shakeshaft, Phys. Rev. A **48**, R875 (1993).
4. M. Pont and R. Shakeshaft, Phys. Rev. A **51**, R2676 (1995).
5. F. Maulbetsch and J. S. Briggs, J. Phys. B **26**, 1679 (1993).
6. A. Dalgarno, H. Doyle, and M. Oppenheimer, Phys. Rev. Lett. **29**, 1051 (1972).
7. O. Schwarzkopf, B. Krässig, V. Schmidt, F. Maulbetsch, and J. S. Briggs, J. Phys. B **27**, L347 (1994).
8. O. Schwarzkopf and V. Schmidt, J. Phys. B **28**, 2847 (1995).
9. V. Schmidt, J. Phys. B **29**, 1877 (1996).
10. M. Pont and R. Shakeshaft, Phys. Rev. A **53**, 3671 (1996).
11. A. Huetz, P. Selles, D. Waymel, and J. Mazeau, J. Phys. B **24**, 1917 (1991).
12. M. Pont and R. Shakeshaft, J. Phys. B **28**, L571 (1995).
13. H. Kossmann, V. Schmidt, and T. Andersen, Phys. Rev. Lett. **60**, 1266 (1988).
14. H. Kossmann, *Ph D Thesis, Freiburg University* (1992).
15. J. M. Bizau and F. J. Wuilleumier, J. Electron Spectrosc. Relat. Phenom., **71**, 205 (1995).

16. R. Dörner, T. Vogt, V. Mergel, H. Khemliche, S. Kravis, C. L. Cocke, J. Ullrich, M. Unverzagt, L. Spielberger, M. Damrau, O. Jagutzki, I. Ali, B. Weaver, K. Ullmann, C. C. Hsu, M. Jung, E. P. Kanter, B. Sonntag, M. H. Prior, E. Rotenberg, J. Denlinger, T. Warwick, S. T. Manson, and H. Schmidt-Böcking, Phys. Rev. Lett. **76**, 2654 (1996).
17. F. W. Byron and C. W. Joachain, Phys. Rev. **164**, 1 (1967).
18. F. H. Read, J. Phys. B **17**, 3965 (1984).
19. P. Hammond, F. H. Read, S. Cvejanovic, and G. C. King, J. Phys. B **18**, L141 (1985).
20. R. Dörner, J. Feagin, C. L. Cocke, H. Bräuning, O. Jagutzki, M. Jung, E. P. Kanter, H. Khemliche, S. Kravis, V. Mergel, M. H. Prior, H. Schmidt-Böcking, L. Spielberger, J. Ullrich, M. Unverzagt, T. Vogt, Phys. Rev. Lett. in press (1996).
21. M. Pont and R. Shakeshaft, Phys. Rev. A in press (1996).
22. J. Feagin, J. Phys. B in press (1996).
23. P. Lablanquie, J. Mazeau, L. Andric, P. Selles, and A. Huetz, Phys. Rev. Lett. **74**, 2192 (1995).

MULTICOLOR PHOTOIONIZATION OF ATOMS WITH AN IR LASER AND ITS HIGHER UV HARMONICS.

Valérie Véniard, Richard Taïeb and Alfred Maquet.

Laboratoire de Chimie Physique-Matière et Rayonnement
Université Pierre et Marie Curie
11, Rue Pierre et Marie Curie
75231 Paris Cedex O5 FRANCE.

1. INTRODUCTION

The recently observed generation, by samples of atomic rare gas, of high-order harmonics of the fundamental frequency of an infrared laser opens the possibility to have at one's disposal a table-top source of coherent and pulsed VUV, (or even soft X-ray) radiation[1-3]. The main motivation of the present paper is to address the possibility to perform a new class of (N+1)-color ($N\geq 1$) photoionization experiments, which fully exploit the unique characteristics of high-order harmonic radiation. Among these characteristics is the fact that the harmonic emission spectrum is constituted of coherent UV radiation lines with equally spaced frequencies and comparable intensities, the spacing between two lines being twice the frequency ω_L of the infrared laser which has been used to generate the harmonics. The idea is to expose simultaneously an atomic target to the fields of a high order harmonic radiation with frequency ω_H and the fundamental of the laser ω_L. Then, if the harmonic order is high enough, i.e. $\omega_H > E_I$, where E_I is the ionization energy of the atom, Laser-Assisted-Single-Photon-Ionization (LASPI) can be observed. Experimental observations have already been reported[4,5].

A simplified picture of LASPI is that of an atom being first brought in the continuum via the absorption of the single high-frequency photon, followed by the exchange of one or several photon of the laser fields, as shown in Fig. 1.The physics of the process is however more complex and approximate treatments based on this scheme do not account for the photoelectron spectra obtained from a complete time-dependent calculation.

One of the main outcome of our simulations is to evidence the prominent role of the harmonic phase. Indeed, the phase difference between the laser and its harmonic is one of the key parameters which govern the relative intensities of the photoelectron peaks in the 2-color spectra[6,7]. In multicolor spectra, involving several harmonics the phase difference between harmonics becomes also important. Furthermore, one of the objective of this paper is to show that conversely, a detailed analysis of multicolor photoionization spectra, involving high harmonics and the (attenuated) laser field can provide interesting informations on the phase differences between successive harmonics[8].

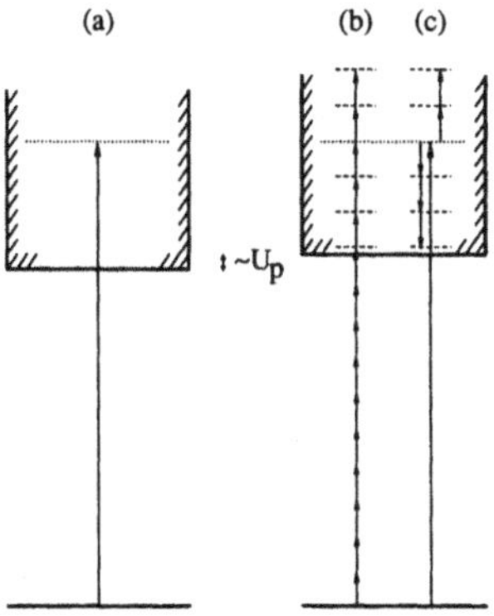

Figure 1. Energy diagrams schematically representing photoionization processes involving respectively: (a) single-photon ionization through the absorption of one-high frequency photon; (b) multiphoton ATI; (c) laser-assisted single-photon ionization in the simultaneous presence of the IR laser and of its high-order harmonic. The shift (~ U_p) of the ionization threshold in the presence of the (strong) laser is indicated.

2. MODEL AND THEORY

The model we have chosen in order to illustrate the general properties of this class of processes is the one of a hydrogen atom, initially in its ground state, submitted to a pulse of radiation containing N+1 frequencies, associated to N higher harmonics together with the laser which has been used to generate them. Unless otherwise stated, the pulse shape is trapezoidal, its duration is eight laser cycles with one-cycle linear turn-on and turn-off. The polarization for both fields is assumed to be linear and parallel to the quantization axis. We have solved the time-dependent Schrödinger equation for this model:

$$i\frac{\partial\psi(\mathbf{r},t)}{\partial t}=\left[H_{at}+H_L(t)+\sum_{i=1}^{N}H_{H_i}(t)\right]\psi(\mathbf{r},t)$$

where H_{at} is the atomic hamiltonian for hydrogen and $H_L = -z\,F_L\,(t)\,\sin(\omega_L t)$ and $H_{Hi} = -z\,F_{Hi}\,(t)\,\sin(\omega_{Hi}t+\varphi_i)$, i=1...N, are the interaction hamiltonians for the laser field and one of its harmonic (written in the length gauge) with time-dependent amplitudes $F_L(t)$ and $F_{Hi}(t)$, the indexes L and H_i labelling the laser and its harmonics respectively. The angles φ_i denotes the phase differences between the two fields at time t=0. Note that we have also used the velocity gauge in our computations. The computation is conducted along a line similar to the one developped by Kulander *et al* [9]. The computations have been performed on a radial grid from r=0 to r=1250 a.u. with steps δr=0.25 a.u. and we have included up to l_{max}=30 angular momenta in the partial wave expansion when the computation was performed in the velocity gauge. The photoelectron spectra have been determined with the help of a spectral analysis of the atomic wave function, as obtained directly after the pulse[10].

3. TWO-COLOR PHOTOIONIZATION

For the sake of illustration, we have considered here the case of a laser frequency equal to that of a Ti:Sapphire operated at ω_L=1.55 eV. We present in this section photoelectron spectra for two-color processes, the high frequency field consisting of the 13th harmonic of the laser, i.e. ω_H=20.15 eV.

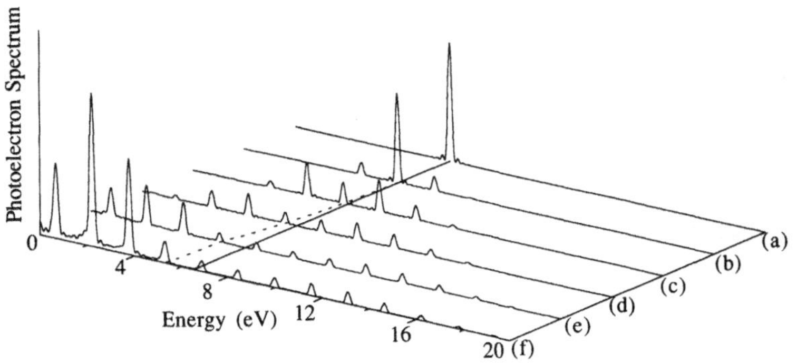

Figure 2. Effect of the laser intensity on the two-color photoelectron spectra, for radiation pulses containing the fundamental frequency of a Ti:Sapphire laser, ω_L = 1.55 eV and its 13th harmonic with a fixed intensity I_H = 3. 10^8 W/cm^2. (a) : I_L=0 (single-photon ionization); (b): I_L= 5. 10^{11} W/cm^2 ; (c): I_L=3. 10^{12} W/cm^2; (d): I_L= 8. 10^{12} W/cm^2 ; (e): I_L= 1.75 10^{13} W/cm^2 ; (f): I_L= 2. 10^{13} W/cm^2. Note that the peaks are shifted towards the lower energies due to the ponderomotive shift.

We illustrate in Fig. 2 the influence of the laser intensity I_L on the angle-integrated photoelectron spectra. To this end, we have kept the (13th-) harmonic intensity fixed at the value I_H=3.10^8 W/cm^2. Note that such an intensity is attainable, after focalisation, by currently developed harmonic sources. On the other hand, we have varied the intensity of the fundamental radiation I_L between 0, see Fig. 2a and 2.10^{13} W/cm^2, see Fig. 2f. In the following, we consider only the case of fields with the same phase at time t=0, i.e. φ=0.

The curve (a) represents the single-photon ionization spectrum obtained with the harmonic radiation alone. The maximum of the photoelectron peak is located at the position given by energy conservation $E_0 = \omega_H + E_{1s}$ = 6.55 eV. LASPI begins to show up at a relatively moderate laser intensity, see Fig. 2b corresponding to I_L=5.10^{11} W/cm^2. Two satellite peaks are clearly visible on each side of the main peak, located at E_0. At such a low laser intensity, the two satellites can be unambiguously associated to two-photon transitions, implying the absorption of a high-frequency photon ω_H and the absorption (or emission) of a low frequency one ω_L[11].

At higher laser field intensities, the number of satellite peaks increases, as can be seen in Figs. 2c and 2d, corresponding to I_L=3. 10^{12} W/cm^2 and 8. 10^{12} W/cm^2 respectively. We are clearly in a nonperturbative regime as large numbers of photons are exchanged with the atomic system. However, for the short pulse considered here, the laser itself does not contribute to the ionization process. This can be checked by verifying that the total area under the photoelectron peaks is the same as in the laser-free case (Fig. 2a). In these cases, the role of the laser is to redistribute the photoeletrons into continuum dressed states.

Interferences between Above-threshold-ionization (ATI) and LASPI are observed at higher laser field intensity, as shown in Figs. 2e and 2f, corresponding to I_L=1.75 10^{13} W/cm^2 and 2. 10^{13} W/cm^2 respectively. Two important features can be noted in these cases. First, one can see an important growth of the low-energy peaks, which can be ascribed to the opening of new channels, associated to ATI, leading directly to the ionization of the target. Here, the laser radiation contributes to the ionization yield as the relative importance of ATI increases with the laser intensity. Then, one can note also the extension of the spectra towards high energies, corresponding to an increase of the number of satellites.

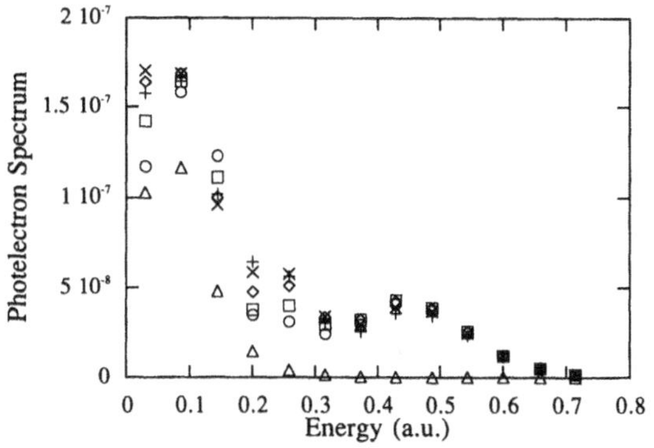

Figure 3. Phase dependence of two-color photoelectron spectra for radiation pulses containing the fundamental of a Ti-Sapphire laser, ω_L=1.55 eV, I_L=1.75 10^{13} W/cm^2 and its 13th harmonic with a fixed intensity I_H=3 10^8 w/cm^2. ATI (I_H=0): Δ, Δφ=0°: O,Δφ=45°: ❒, Δφ=90°: +, Δφ=135°: ×, Δφ=180°: ◊.

A remarkable result of the present simulation is that the change in the spectra between the regime in which ATI is almost not visible and the one in which ATI is dominant in the low photoelectron energy range takes place in a narrow laser intensity range : here it occurs between $I_L=10^{13}$ W/cm^2 and I_L=2. 10^{13} W/cm^2. This is precisely in this range that quantum interferences can be identified. In this situation, the relative phase φ between the two fields plays an important role and governs the relative heights of the photoelectron peaks when both processes (ATI and LASPI) have comparable probabilities. In spite of the fact that the question of the control of the relative phase between the laser field and one of its higher harmonic is still an open problem for the currently developed harmonic sources[1-3], we have illustrated this point in Fig. 3, where we have schematically represented the magnitudes of the photoelectron peaks for different values of φ. The interesting point is that in the low energy part of the spectrum, i.e. when ATI and LASPI amplitudes have comparable magnitude, one observes important changes in the heights, depending on the relative phase as indicated. Note that these changes are notable in spite of the fact that, in contrast with most previous studies on phase effects, the field strengths are quite different, since here the ratio $I_H/I_L \approx 2.\ 10^{-5}$. They open the possibility to realize a, at least partial, coherent control of the photoelectron current[12,13].

4. PHASE DEPENDENCE OF (3+1) COLOR PHOTOIONIZATION

We now address the case in which three harmonics with unequal intensities are used. This situation approximatly models preliminary experimental attemps to observe the process in helium, using UV harmonics of a femtosecond Ti:Sapphire laser at the facility in Palaiseau, in which a multilayer mirror was used to select the needed frequencies[4]. As a result of the finite frequency bandwidth of the mirror used, the photoelectron spectrum, obtained with the sole harmonics, then consists in three peaks with relative magnitudes as shown in Fig. 4.

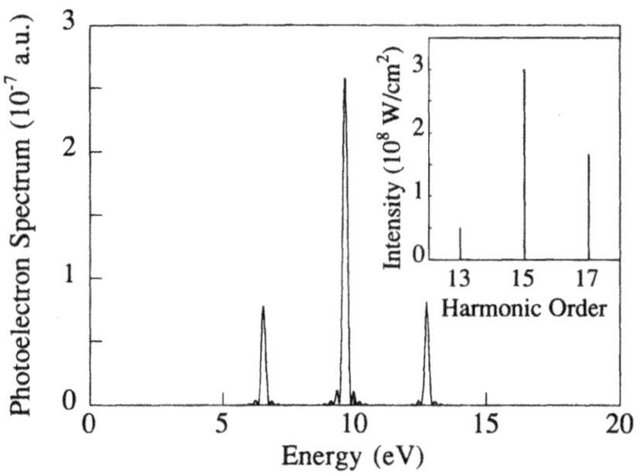

Figure 4. Simulated three-color photoelectron spectrum, for a radiation pulse containing the 13th, 15th and 17th harmonics of a Ti:Sapphire laser, ω_L=1.55 eV. The inset shows the corresponding harmonic intensities needed to reproduce the spectrum.

The spectrum shown has been simulated in hydrogen, with the harmonics 13, 15 and 17 of the laser frequency ω_L = 1.55 eV. We have chosen these frequencies in order to reproduce a photoelectron spectrum with the same characteristics as the one obtained in the above mentioned experiment. Note that we have checked, by comparing with a time-dependent Hartree-Fock calculation for He, that the main conclusions which can be drawn from our study in hydrogen are not altered when considering a more sophisticated model, at least in the low-intensity regime considered here. We have kept with the calculation in hydrogen for the sake of computational convenience. Then, in order to reproduce the spectrum, the intensities of each harmonic component must be as shown in the inset. In fact, any departure from these relative intensities would notably modify the photoelectron spectrum, a feature which provides a quite stringent test on the relative intensities of the harmonics and, consequently, on the frequency dependence of the reflectivity of the mirrors. Note that in the course of the reflection process the phases of the harmonics are changed. The corresponding changes can be estimated and we shall not discuss this point here.

Once these relative intensities are determined, one can consider the situation in which one adds a fourth color, namely the laser radiation with a moderate intensity I_L=8. 10^{11} W/cm^2. This corresponds to an intensity regime in which lowest-order perturbative estimates are expected to account fairly for the processes. Then, as shown in Figs. 5, some new photoelectron lines do appear resulting predominantly of 2-photon processes involving the exchange of one laser photon in addition to the absorption of the UV photons[11]. Shown in

Fig 5a, is the photoelectron spectrum for the test case in which the three harmonics have the same phase , i.e. $\Delta\varphi_{15,13} = \Delta\varphi_{17,15} = 0$, where $\Delta\varphi_{i,j} = \varphi_i - \varphi_j$.

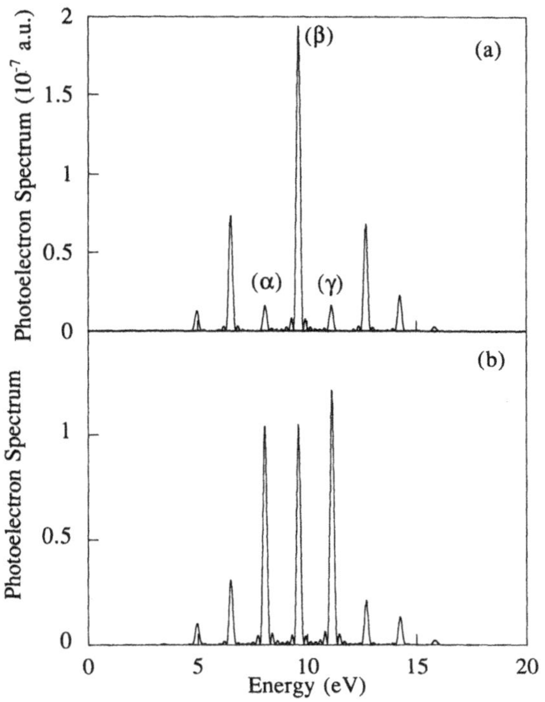

Figure 5. Effect of the harmonic phases on the (3+1)-photoelectron spectrum with ω_L=1.55 eV, I_L=8. 10^{11} W/cm^2, the 3-color spectrum being shown in Figure 4 ; (a): $\Delta\varphi_{15,13} = \Delta\varphi_{17,15} = 0$; (b): $\Delta\varphi_{15,13} = \Delta\varphi_{17,15} = \pi$.

One observes the presence of new, relatively small, satellite peaks. Comparing with the laser-free spectrum displayed in Fig. 4, one notes also that the new peaks have grown at the expense of the original ones which are slightly depleted. This was expected since the overall ionization probability is unchanged, the laser itself being not intense enough to contribute to the ground state depletion, for such a short pulse. Changing the phases can dramatically modify the photoelectron spectrum. For instance, if the phases differences are $\Delta\varphi_{15,13} = \Delta\varphi_{17,15} = 180°$, the spectrum changes as shown in Fig. 5b (note the change of scale). We have introduced the phase differences between successive harmonics because it appears that they are the relevant parameters for discussing these effects. For this choice of phases, one observes that the satellites are much bigger: Their magnitudes are now comparable to the ones of the original peaks which, in turn, are significantly depleted. The strong dependence of the sizes of the satellite peaks on these phases results from the fact that each of them can be reached via, (at least two) distinct quantum paths. It indicates also that the amplitudes associated to each of these paths have comparable magnitudes, so that strongly constructive or destructive interferences can occur.

We show now an example of the feasibility of determining the relative phases of the harmonics, starting from an experimental spectrum. The main idea is to monitor the variations of the relative heights of the peaks in terms of the phase differences between the harmonics and then to find the optimal combination needed to fit the spectrum. More precisely, for the above set of fields intensities, we have monitored the heights of the peaks labelled (α), (β) and (γ) in Fig. 5a. We only present here the variation of the peak labelled (α). Note that these peaks can be associated with the following net absorption of energies: $E_\alpha=14\ \omega_L$; $E_\beta=15\ \omega_L$; $E_\gamma=16\omega_L$, and that the peaks (α) and (γ) are reached through the exchange of at least one laser photon, in addition to the absorption of one UV (odd)-harmonics. These peaks are obviously the most sensitive to interference effects.

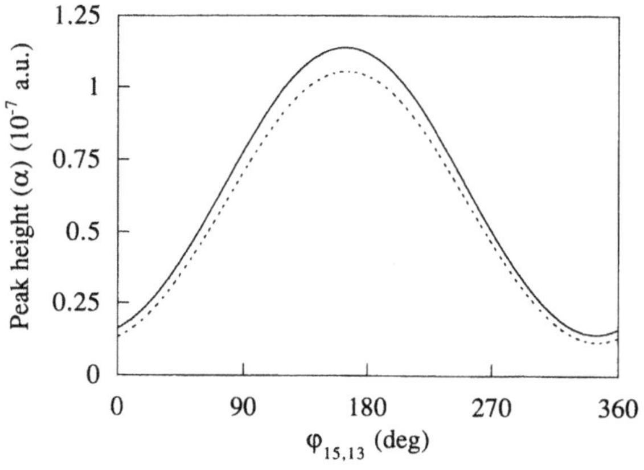

Figure 6. Peak height (α) of the photoelectron spectrum as a function of the phase difference $\Delta\varphi_{15,13}$, $\Delta\varphi_{17,15}$ being kept fixed at the indicated value: full line $\Delta\varphi_{17,15}$=0°, dashed line $\Delta\varphi_{17,15}$=180°.

In Fig. 6 are reported the variations of the height of the peak (α), in terms of the phase difference $\Delta\varphi_{15,13}$, the other difference $\Delta\varphi_{17,15}$ being kept fixed at the indicated values. Several interesting conclusions can be drawn from the perusal of this figure. First, the peak (α) is more sensitive to the phase difference $\Delta\varphi_{15,13}$ .This results from the fact that it can be reached via two dominant quantum paths involving either the absorption of the 13rd harmonic and of one laser photon, or the absorption of the 15th harmonic and the (stimulated-) emission of one laser photon. In such a case, the resulting probability amplitude depends strongly on the phase difference between the two harmonics. One observes that the maximum height for (α) is reached for values of $\Delta\varphi_{15,13}$ slightly larger than π. The interesting point is that the variations of this peak are linked to the ones of the peak (β), in the sense that any growth of (α) occurs at the expense of (β). This is a direct consequence of the overall conservation of the ionization probability. However, one has to remember that (β) is depleted also when transitions leading to the peak (γ) take place. These are however less affected by changes in $\Delta\varphi_{15,13}$. Performing a similar analysis for the influence of $\Delta\varphi_{17,15}$, one can univocally find the set of phases differences which permit to fit the experimental spectrum ($\Delta\varphi_{15,13}=\Delta\varphi_{17,15}$= 90° in the present case).

5. (N+1)-COLOR PHOTOIONIZATION

Harmonic spectra generated by atoms submitted to an intense laser pulse can be schematically represented in Fig. 7[14]. The harmonics within the plateau have approximately the same intensities and a subset of them can be idealized as a Dirac comb. One can use such a finite subset to perform single-photon ionization experiments in atoms, provided the frequencies are larger than the atomic ionization frequency.

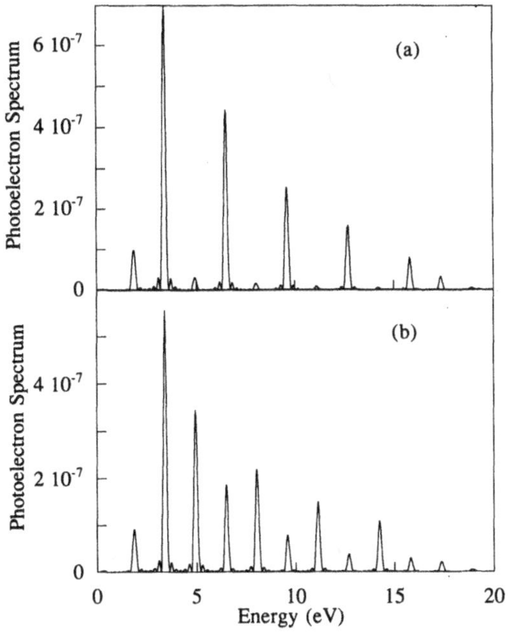

Figure 7. (a) Schematic representation of an harmonic spectrum; (b) enlargement of the spectrum showing the "Dirac comb" of harmonic frequencies used in the calculation.

Starting from a comb with five harmonics and adding a sixth color, namely the IR laser radiation leads to photoelectron spectra such as the ones displayed in the Figures 8. We will exemplify the influence of the harmonic phases in this set of Figures, in which the respective intensities of the laser and of the harmonics are kept fixed as follows : $I_L = 8.\ 10^{11}$ W/cm^2; $I_j = 3.\ 10^8$ W/cm^2, where here j = 11, ..., 19.

First is shown in Fig. 8a, the case in which the harmonics have all the same phase, i.e. $\Delta\varphi_{ij} = 0$, where the indices i,j label consecutive (odd-) harmonics. Quite surprisingly, no new satellite peaks show up between the original ones, at least with the linear scale chosen here. In fact their magnitudes differ by almost two orders of magnitudes from the ones originating from the absorption of the harmonics. There is however a notable exception for the two small peaks located on each side of the original distribution. These peaks are the only ones which cannot be reached via interfering paths, at least in the low intensity regime considered here. This indicates that, if the harmonics had the same phase, the presence of the laser would almost not change the photoelectron spectrum. This results from strong destructive interferences which systematically take place for the peaks associated to laser-assisted transitions which can occur via distinct quantum paths. Moreover, the strong cancellations observed are the signature of the fact that the amplitudes associated to the interfering paths have almost equal magnitudes.

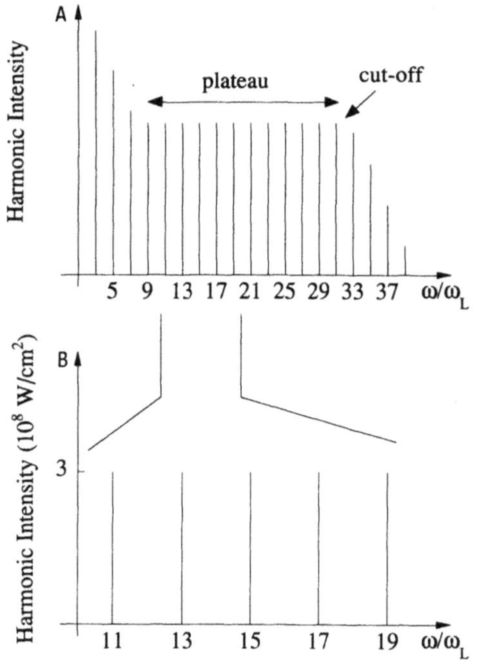

Figure 8. Effect of the harmonic phases on the (5+1)-photoelectron spectrum, with ω_L=1.55 eV, $I_L=8.10^{11}$W/cm^2. (a) : $(\varphi_{11},\varphi_{13},\varphi_{15},\varphi_{17},\varphi_{19}) = (0,0,0,0,0)$; (b) : $(\varphi_{11},\varphi_{13},\varphi_{15},\varphi_{17},\varphi_{19}) = (0,\pi,0,\pi,0)$.

We turn now to another extreme case in which the successive harmonics have phases which differ from π, i.e. $\Delta\varphi_{ij} = \pi$, nothing else being changed with respect to the preceding case. The photoelectron spectrum is now dramatically changed, see Fig. (8b). The satellite peaks become dominant, the original ones being notably depleted. This clearly indicates that constructive interferences have now taken place. Comparing the set of Figures 8, helps to evidence the strong influence of the harmonic phases and also hints at the possibility to determine them from the analysis of multicolor photoelectron spectra.

6. CONCLUSION

We have presented a computation of (N+1)-color atomic photoionization spectra, (N≥1), as they should be observed when using a radiation pulse containing a Dirac comb of N harmonic UV frequencies, together with the one of the IR laser which has been used to generate them. A remarkable result of our analysis is that the satellite peaks have notable magnitudes, even at moderate laser intensities. In fact, even in the regime in which lowest-order perturbation theory applies, the satellite peaks can have magnitudes comparable to the ones of those resulting from the absorption of harmonic photons. This results from the strong free-free radiative coupling induced by the IR laser field, between continuum states. One observes further that the satellite peaks grow at the expense of the original ones, the total ionization probability not being significantly changed by the presence of the laser in the short pulses considered here. Even more interesting is the strong sensitivity of the magnitudes of the photoelectron peaks on the phases of the harmonics. We have shown that extreme situations can be envisioned, such as the satellite peaks being almost negligible or, on the contrary, being dominant. The relevant parameter to control the relative heights of the peaks is the phase difference between successive harmonics. As a consequence, we believe that the study of (N+1)-color photoionization spectra, should allow one to determine the relative phases of successive harmonics, an intrinsic quantity of interest from both the experimental and theoretical points of views.

Acknowledgments

The Laboratoire de Chimie Physique-Matière et Rayonnement is a Unité de Recherche Associée au CNRS, URA 176.This work was partially supported by the EC contracts ERB CHRXCT 920013 and 940470 . Parts of the computations have been performed at the Centre de Calcul pour la Recherche (CCR, Jussieu, Paris) and at the Institut du Développement et des Ressources en Informatique Scientifique (IDRIS).

REFERENCES

1. A. L'Huillier, L. A. Lompré, G. Mainfray and C. Manus in *Atoms in Strong Fields*, edited by M. Gavrila, Adv. Atom. Molec. Opt. Phys. Suppl. 1, Acad. Press, San Diego, p 139 (1992).
2. R. Haight and D. R. Peale, Phys. Rev. Lett. **70**, 3979 (1993); R. Haight and P. F. Seidler, Appl. Phys. Lett. **65**, 517 (1994).
3. J. Larsson, E. Mevel, R. Zerne, A. L'Huillier, C-G. Wahlström and S. Svanberg, J. Phys. B **28** L51 (1995).
4. J. M. Schins, P. Breger, P. Agostini, R. C. Constantinescu, H. G. Muller, A. Bouhal, G. Grillon, A. Antonetti and A. Mysyrowicz, J. Opt. Soc. Am. B **13**, 197 (1996).
5. T. E. Glover, R. W. Schoenlein, A. H. Chin and C. V. Shank, Phys. Rev. Lett. **76**, 2468 (1996).
6. V. Véniard, R. Taïeb and A. Maquet, Phys. Rev. Lett. **74**, 4161 (1995).
7. R. Taïeb, V. Véniard and A. Maquet, J. Opt. Soc. Am. B **13**, 363 (1996).
8. V. Véniard, R. Taïeb and A. Maquet, Phys. Rev. A **54**, 721 (1996).
9. K. C. Kulander, K. J. Schafer and J. L. Krause in *Atoms in Strong Fields*, edited by M. Gavrila, Adv. Atom. Molec. Opt. Phys. Suppl. 1, Acad. Press, San Diego, p 247 (1992).
10. K. J. Schafer, Comp. Phys. Comm. **63**, 427 (1991).
11. A. Cionga, V. Florescu, A. Maquet and R. Taïeb, Phys. Rev. A **47**, 1830 (1993).
12. M. Shapiro and P. Brumer, Int. Rev. Phys. Chem. **13**, 187 (1194) and references therein.
13. K. J. Schafer and K. C. Kulander Phys. Rev. A **45**, 8026 (1992).
14. J. J. Macklin, J. D. Kmetec and C. L. Gordon III, Phys. Rev. Lett. **70**, 774 (1993).

STABILIZATION OF ATOMS BY ULTRA-STRONG LASER FIELDS

Mihai Gavrila

FOM-Institute for Atomic and Molecular Physics
Amsterdam, Netherlands, and
Institute for Theoretical Atomic and Molecular Physics
at the Harvard-Smithsonian Center for Astrophysics
Cambridge, Massachussets, USA

1. Intense-field ionization

The incessant advance of laser technology has pushed the peak electric-field strengths produced by lasers to well above the atomic unit, 5.14 10^9 V /cm [1], and there appears to be no limit in sight. Under these circumstances, the amplitude of the oscillating field is many times larger than the average electrostatic field experienced by an electron in the ground state of the hydrogen atom. Obviously, this changes drastically the laser-atom interaction phenomena as known at low intensities, and in particular ionization [2]. Indeed, in superintense fields ionization acquires a striking feature, sometimes termed as "counter-intuitive": stabilization. In the following we present an overview of ultra-strong field stabilization, theoretical and experimental [3], [4]. In this Introduction, we start by placing the phenomenon into perspective.

At low radiation intensities, ionization is tantamount to *one-photon ionization*, a phenomenon extensively studied over the years. The condition for one-photon ionization is that the frequency be larger than the threshold imposed by Einstein's equation. However, the advent of the laser, has made possible the discovery of *multiphoton ionization* (MPI): ionization can occur, with no limitation on the frequency, by the absorption of several photons, provided that enough intensity is available (Delone and Voronov, 1965; Hall, Robinson, and Branscomb, 1965 ; [5]). Following the extensive studies of MPI in the 1970', it was discovered that the

Photon and Electron Collisions with Atoms and Molecules
Edited by Burke and Joachain, Plenum Press, New York, 1997

ionized electron could in fact absorb more than the minimal number of photons needed to get it to the continuum (Agostini, Fabre, Mainfray, Petite, and Rahman, 1979; [6]). This is the phenomenon of *excess-photon ionization (EPI)*, or above-threshold ionization (ATI), extensively studied in the 1980'.

The traditional framework for describing radiation-atom interactions has been *perturbation theory* with respect to the intensity of the radiation, in practice limited to its lowest non-vanishing orders (LOPT). The forms of ionization involving several photons had been predicted within this framework well in advance of their experimental discovery (Göppert-Mayer, 1931; [7]). A characteristic feature is that their rates are proportional to powers of the intensity I^n, where n is the number of photons absorbed. It was to be expected that LOPT would fail at some intensity. Signs of this starting to happen were encountered in EPI studies at about 10^{13} - 10^{14} W / cm^2, when the electron spectra of noble gas atoms showed "peak suppression" for the first peaks, and too slow a decrease in the magnitude of the following ones (Kruit, Kimman, Muller, van der Wiel, 1983, [8]). Moreover, it was realized that, although in principle perturbation theory could be extended to include higher orders, this was in practice unwieldy. Thus, one was lead to develop *nonperturbative methods* for describing laser-atom interactions, based on the direct solution of the Schrödinger equation. Such methods require large scale computations, and their progress has been made possible over the last decade due to advances in hardware, and programming. There are mainly two nonperturbative approaches to consider: the steady-state Floquet theory, and wave-packet dynamics (WPD). Both theories proceed from the familiar Schrödinger equation for the laboratory frame (in the momentum, or length gauges), or from the "space-translated" form of the equation (see Eq.(1) further on), pertaining to an oscillating frame. The latter was discovered and rediscovered over the years by Pauli and Fierz (1938; [9a]), Kramers (1947; [9b]), Henneberger (1968; [9c]), see also Faisal (1973; [9d]).

Floquet theory describes a regime of steady-state ionization that can be characterized by constant ionization rates. Most frequently, the field was considered to be monochromatic with constant amplitude, although the theory has been extended successfully to the case of several frequencies. Time can be eliminated from the problem by Fourier analysis, whereby one has to solve a system of coupled partial differential equations for the "Floquet components" of the wave function. The specific physical conditions are introduced via boundary conditions for this system. They are usually chosen of the "resonance" (Gamow-Siegert) type, although scattering boundary conditions have also been used. The first alternative (the only one we shall consider) leads to an eigenvalue problem, with complex eigenenergies, the "quasienergies": $E = W - (i / 2) \Gamma$. W is interpreted as the average binding energy in the field of the state considered, and Γ as its total ionization rate. Thus, a Floquet wave function describes a *mode* of ionization, i.e., a flow of population from the vicinity of the nucleus towards large distances, characterized by the law of exponential decay: $(N / N_0) = e^{-\Gamma t}$, [10].

Floquet theory has its merits, as well as its limitations. One merit is that it describes the behavior of *individual* atomic states in the radiation field, which are atomic characteristics that do not depend on the way the atom is manipulated. The theory is closely related to that of the decay of metastable states, a major problem encountered in scattering theory (although the situation is considerably more complicated in the present case). One can rely, therefore, on the reservoir of knowledge accumulated in that area. On the other hand, Floquet theory deals with idealized states, and compromises have to be made to connect them to physical reality. These states are awkward mathematical objects (lying outside Hilbert space), that have to be handled properly in order to extract correct physical information. For a long time their interpretation has been subject to uncertainty and criticism. Moreover, the theory has been devised for constant field amplitude, whereas in practice intense fields are generated in sub-picosecond pulses. This defficiency can partly be corrected by noting that, even for such short pulses, the variation of the amplitude of the field is still slow on the atomic time scale (recall that one a.u. of time is $2.42\ 10^{-17}$ sec), i.e. the atom has time to adjust "adiabatically" to the variation of the field strength. If the adiabaticity condition and some others are met, it can be shown that the results at constant amplitude can be used for the "instantaneous" description of laser pulses [11], [12]. So far Floquet theory has used a single-state description for the interpretation of the ionization dynamics; this imposes the condition that Γ be sufficiently small. Very recently, the advantages of a *multi-state description* have been illustrated (Wells, Simbotin, and Gavrila [13]).

A variety of methods have been applied to solve for the quasienergies in intense fields: complex-coordinate rotation (Shih-I Chu and Reinhardt, 1977, [14] ; Moiseyev and collaborators since 1990, e.g., see [15]), Sturmian basis diagonalization (Potvliege, Shakeshaft, and collaborators since 1988 , e.g., see [16]), close-coupling (Dimou, Faisal, since 1987, e.g., see [17]; Marte and Zoller [18]), R-matrix (Burke, Dörr, Francken, Joachain, and collaborators, since 1990, e.g., see [19]). A version, specially devised for high-frequencies, is the high-frequency Floquet theory (HFFT), developed by Gavrila and Kaminski (since 1984; e.g., see [20], [21]), which will be described in some detail in Sec.2 because of its relevance to stabilization. The HFFT solves the Floquet system of differential equations by iterations, yielding approximations of successively higher order in ω^{-1} for the quasienergies and ionization rates. Precursors of the HFFT are the works by Henneberger (1968; [9c]), and Gersten and Mittleman (1976; [22a]). More recently, the emphasis of Floquet methods has shifted to two-electron atoms, with electron correlation included (see [2c]).

Wave-packet dynamics (WPD) integrates numerically the time-dependent Schrödinger equation for the interaction of the atom with the field, in one of its forms mentioned earlier, starting from a given initial condition. A variety of numerical techniques have been devised to increase the stability of the integration over large regions of space and long intervals of time. The desired physical information is extracted by applying the quantum-mechanical rules. WPD has merits and limitations of its own. The power of the method lies in the fact that it can taylor any experimental circumstances, and its predictions express directly the laws of quantum mechanics. On the other hand, the information obtained is purely numerical, with no intrinsic structuring guidelines. New phenomena have to be recognized by sifting through a mass of numerical information.

Realistic 3D computations for the study of ionization of one-electron atoms have been initiated by Kulander (since 1987; e.g., see [23]), continued by many others (e.g., DeVries [24], LaGattuta [25]; Pindzola, Bottrell, and Bottcher [26]; Pont, Proulx, and Shakeshaft, [27]; Tang, Rudolph, and Lambropoulos [28]; Gajda, Piraux, and Rzazewski, [29]; Horbatsch, [30]; etc.). As with Floquet methods, the emphasis of WPD lies at present with two-electron atoms, a formidable numerical problem. A special place among WPD calculations is held by the "numerical experiments" made on 1D models with various sorts of potentials (e.g., Eberly, Grobe, Javanainen, Law, and Su, since 1988, [31]; Burnett, Knight, and Reed [32]). As oposed to the unwieldy 3D calculations, numerical experiments are simple enough to be carried out for a large variety of conditions, allowing to check the role played by the laser parameters, initial conditions, etc. They have been quite useful in identifying new phenomena.

The two approaches, Floquet and WPD, are manifestly complementary in nature, one emphasizing the energy domain, the other the time domain. For very short laser pulses of a few cycles, the concept of energy is not useful (there is a large spread in the energy content of the wave packet, extending far into the continuum), so that the WPD approach is the sole one effective. For realistic pulses (in the fs range, containing many cycles) the Floquet approach becomes valuable. In fact, we believe that only a *dual approach*, based on both, can give definitive answers to the subtle problems which abound in strong-field laser-atom interactions. Floquet theory can be used as a guidance for identifying unusual phenomena (such as stabilization, light-induced states, etc.), hard to accept if not substantiated otherwise, while WPD is needed to place them in a realistic physical context (for this dual approach, see Wells, Simbotin and Gavrila [13]).

The development of Floquet theory and WPD has lead to the parallel discovery of the ultra-strong field stabilization, albeit in somewhat different forms. The essence of the phenomenon is that, contrary to the predictions of LOPT , beyond a certain intensity, the ionization probability of a one-electron atom starts *decreasing* with the intensity, i.e. the atom becomes more stable. The conditions for this to happen differ between the two theories. Pont and Gavrila (1989; [33a]) have shown that stabilization is a direct consequence of the HFFT, and have given an approximate analytic formula for its description, displayng the dependence on I and ω. This form of the phenomenon was later termed *adiabatic stablization*, because it refers to the ionization of individual atomic states, while adjusting adiabatically to the variation of the field. It is a high-frequency phenomenon: the frequency has to be high with respect to the average atomic excitation energy in the field. On the other hand, the results showed that ground-state stabilization was not accessible to the prevalent laser technology (too high frequencies and intensities were needed). Nevertheless, it was pointed out by Vos and Gavrila (1992; [33b]) that by considering Rydberg atomic states, the phenomenon should be observable. Meanwhile, adiabatic stabilization has been substantiated by all methods of Floquet theory, with good agreement among themselves ([34]; see the discussion in Sec.3).

The 1D numerical experiments of Su, Eberly, and Javanainen (1990; [31b]) were first to draw attention to the alternative form of the phenomenon, termed *dynamical stabilization*. This form of stabilization refers to the global survival of the atom (as a neutral system) in the field, no matter what happens to it during the laser pulse. It was shown that this process can occur even at relatively low frequencies (e.g., when several photons are needed for ionization). Further 1D studies were done by Reed, Burnett and Knight [32b]. The results of Kulander, Schafer, and Krause (1991; [35]), followed by others (see the discussion Sec.4), have documented the phenomenon also in the realistic 3D case.

Although *stabilization experiments* are not possible at present for the ground state of Hydrogen, the calculations of Vos and Gavrila [33b] have paved the way for experiments on Rydberg states. These were carried out soon afterwards by Muller and a large group of collaborators at FOM - Amsterdam. In the first one, the breakdown of LOPT and the onset of the stabilization regime was demonstrated in a two-stage experiment (deBoer, Hoogenraad, Vrijen, Constantinescu, Noordam, and Muller, 1993; [36a]). In a subsequent, independent experiment, a third stage was added to monitor more stringently the end products (van Druten, Constantinescu, Schins, Nieuwenhuize, and Muller, 1996; [36b]). The latter experiment confirmed the first one.

After having presented in this introduction the general background of intense-field ionization, we proceed in Sec.2 to give an account of the fundamentals of the HFFT, which is the natural approach to understand the physics of adiabatic stabilization. The results obtained with the HFFT and other Floquet theories are presented in Sec.3. Stabilization for the ground state of atomic H, for Rydberg states, and for two-electron atoms, are considered. In Sec.4 we present results from WPD. We start with dynamic stabilization, and compare it with the adiabatic case. We then show that adiabatic stabilization also follows from WPD, if the initial condition is adequately chosen. Sec.5 contains a description of the stabilization experiments. Finally, in Sec.6 we consider some open problems and future directions.

2. Fundamentals of high-frequeny Floquet theory (HFFT)

In this Section we shall recall some pertinent results of the HFFT, [21]. We start from the one-electron Schrödinger equation in the lab frame and momentum gauge (with $\hbar = 1$):

$$\left[\frac{1}{2m}\left(\mathbf{P}-\frac{e}{c}\mathbf{A}\right)^2+V(\mathbf{r})\right]\psi_L=i\frac{\partial\psi_L}{\partial t} \quad . \tag{1}$$

By assuming the validity of the dipole approximation (i.e., **A** depends only on t, and not on **r**) it is possible to eliminate the $\mathbf{A}^2$ term from the equation, by extracting a time-dependent phase factor from ψ_L. At the same time one can perform a time-dependent translation of the electron coordinates from **r** to $\mathbf{r} - \alpha(t)$, where α is defined by:

$$\alpha(t) = -\frac{e}{mc}\int^t \mathbf{A}(t')\,dt' \ . \tag{2}$$

This amounts to passing to a new frame of reference, the "oscillating frame" (also called the Kramers-Henneberger frame). The new, "space-translated" form of the Schrödinger equation is [9]:

$$\left[\frac{1}{2m}\mathbf{P}^2+V(\mathbf{r}+\alpha(t))\right]\psi=i\frac{\partial\psi}{\partial t} \quad . \tag{3}$$

The physical significance of the transformation is the following. The vector $\alpha(t)$ describes the classical quiver motion of an electron in the field $\mathbf{A}(t)$, and hence the oscillating frame is that in which the classical electron is at rest. By the same token, the nucleus, while fixed in the lab frame, will have the trajectory $-\alpha(t)$ in the oscillating frame.

For a monochromatic field of arbitrary polarization $\mathbf{E}(t)$ [with amplitude E_o and frequency ω], $\alpha(t)$ is given by $\alpha(t) = (\alpha_o / E_o)\,\mathbf{E}(t)$, where $\alpha_o = -(eE_o/m\omega^2)$.

We make the Floquet ansatz, to seek solutions describing a regime of steady multiphoton ionization:

$$\psi(\mathbf{r},t) = e^{-iEt}\sum_n \phi_n(\mathbf{r})\, e^{-in\omega t} \quad . \tag{4}$$

By inserting this in Eq.(3) we obtain a system of coupled differential equations for the Floquet components $\phi_n(\mathbf{r})$:

$$\left[\frac{1}{2m}\mathbf{P}^2+V_o-(E+n\omega)\right]\phi_n = -\sum_m^{(m\neq n)} V_{n-m}\,\phi_m \ . \tag{5}$$

where $V_n(\alpha_o, \mathbf{r})$ are the Fourier components of the oscillating potential $V(\mathbf{r} + \alpha(t))$. The system Eq.(5) has to be solved with specified boundary conditions, representing the physics of the problem; this will also insure the unicity of the solution. For ionization, these have been chosen mostly of the "resonance state" (Gamow - Siegert) type. As a consequence, the solution of Eq.(5) becomes an eigenvalue problem. For the case in which one-photon ionization is possible (high frequencies), we label the channels such, that those with $n \geq 1$ are open, and those with $n \leq 0$ are closed. The complex eigenvalues obtained in this way are the "quasi-energies" of the atomic states, $E = W - i\,(\Gamma/2)]$, where, as already noted, W is interpreted as the average energy of a state in the field, and Γ as its total ionization rate. The lifetime of the state is given by: $\tau = 1 / \Gamma$. All these quantities depend on ω and α_o.

The HFFT solves Eq.(5) at large ω by successive iterations. A pragmatic convergence criterion for the iteration scheme is that ω be large with respect to some average excitation energy of the atom in the field, $\mathcal{W}_e$:

$$\omega \;>>\; \mathcal{W}_e \quad . \tag{6}$$

In general, $\mathcal{W}_e$ is of the order of magnitude of the largest binding energy of the symmetry manifold to which the initial state of the atom belongs. Eq.(6) necessarily requires that one photon ionization be possible from that manifold; it expresses the high-frequency character of the theory [37].

To lowest order in ω^{-1}, the HFFT yields the structure equation:

$$[\frac{1}{2m}\,\mathbf{P}^2 + V_o(\alpha_o; \mathbf{r})]\; u = W\, u \quad , \tag{7}$$

containing the "dressed potential" $V_o\,(\alpha_o; \mathbf{r})$, which is the time average of the oscillating potential $V(\mathbf{r} + \alpha(t))$. The occurrence of V_o appears classically quite natural if one notes that, in the oscillating frame and at large ω, the electron will not react to the rapid variation of the potential generated by the oscillating nucleus, but rather to its time average (see [21], Sec.IV C). $V_o(\alpha_o; \mathbf{r})$ can also be also interpreted as the electrostatic potential obtained by smearing out the nuclear charge along its trajectory $\{-\,\alpha(\chi)\}$, with variable charge density. This "line of charge", is a segment of length $2\alpha_o$ for linear polarization, a circle of radius α_o for circular polarization, and an ellipse in general case of elliptic polarization. The (real) eigenvalues $W(\alpha_o)$ of Eq.(7) represent the limiting form of the quasienergies $E(\alpha_o\,,\,\omega)$ at large ω. The structure equation Eq.(7), was first obtained heuristically by Henneberger [9c], but it was Gersten and Mittleman [22a] who recognized its high-frequency character (no reference to stabilization was made; see however [22b]).

The imaginary parts Γ of the quasienergies can be obtained from the next order approximation within the HFFT. This yields for the differential n-photon ionization rates the expression [20b]:

$$\frac{d\Gamma_n}{d\Omega} = \frac{k_n}{m}|\, f_n(\hat{\mathbf{k}})\,|^2 \quad , \qquad f_n(\hat{\mathbf{k}}) \;=\; -\;\frac{m}{2\pi}< u^{(-)}_{\mathbf{k}_n}\;|\,V_n\,|\,u_o > \; . \tag{8}$$

Here $\mathbf{k}_n$ is the final electron momentum ($W + n\,\omega \;=\; (k_n^2/2m)$; $\mathbf{k}_n \;=\; k_n\hat{\mathbf{k}}$) ; u_o and $u^{(-)}_{\mathbf{k}_n}$ are the initial, repectively final wave functions of the atom, both solutions of Eq.(7), adequately normalized. Angular integration in Eq.(8) gives the n-photon ionization rate Γ_n, the total rate being $\Gamma = \sum \Gamma_n$.

In the case two-electrons, by applying a space translation of vector $\alpha(t)$ given by Eq.(2) to both electrons, the Schrödinger equation becomes:

$$[\frac{1}{2m}(\mathbf{P}_1^2 + \mathbf{P}_2^2) + V(\mathbf{r}_1 + \alpha(t)) + V(\mathbf{r}_2 + \alpha(t)) + \frac{e^2}{r_{12}}]\,\psi = i\frac{\partial\psi}{\partial t}, \qquad (9)$$

where $V = -Ze^2/r$. Further, by making the Floquet ansatz Eq.(4), a system of differential equations similar to Eq.(5) can be derived. In the high-frequency limit one now extracts the structure equation (see Gavrila and Shertzer [38] for details):

$$[\frac{1}{2m}(\mathbf{P}_1^2 + \mathbf{P}_2^2) + V_0(\alpha_0, \mathbf{r}_1) + V_0(\alpha_0, \mathbf{r}_2) + \frac{e^2}{r_{12}}]\,u = W\,u\ . \qquad (10)$$

Here, $V_0(\alpha_0, \mathbf{r})$ is the dressed (Coulomb) potential, defined in the same way as for the one-electron problem. From the next order approximation of the HFFT one finds for the n-photon, single-electron ionization rate, the expression:

$$\frac{d\Gamma_\sigma^n}{d\Omega} = \frac{k_{n\sigma}}{m}|F_{fi}(\hat{\mathbf{k}})|^2, \qquad F_{nf,i}(\hat{\mathbf{k}}) = -\frac{m}{2\pi} \ll u_f^{(-)}| V_n(\mathbf{r}_1) + V_n(\mathbf{r}_2) | u_i \gg . (11)$$

Here $u_i(\mathbf{r}_1, \mathbf{r}_2)$ is the initial bound state eigenfunction (energy W), and $u_f^{(-)}(\mathbf{r}_1, \mathbf{r}_2)$, the adequately normalized final state eigenfunction corresponding to the residual atom / ion being left in state σ (energy W_σ) and the ionized electron having asymptotic momentum $\mathbf{k}_{n\sigma} = k_{n\sigma}\hat{\mathbf{k}}$ $[W + n\omega = W_\sigma + (k_{n\sigma}^2/2m)]$. Both u_i, $u_f^{(-)}$, are symmetrized solutions of Eq.(10). V_n has the same meaning as earlier. The validity condition of the theory is still given by Eq.(6), except that now $\mathcal{W}_e$ stands for the average excitation energy of the two-electron system as a whole.

3. Adiabatic Stabilization

A. *Ground state of H*

We beginn by briefly describing the *structure of the atom* in the high-frequency regime. A variety of numerical methods have been applied to solve the (nonseparable) structure equation Eq.(7) for H [39], to the effect that the eigenvalues and eigenfunctions are known with a high degree of accuracy. We show in Fig.1 the result for the eigenvalues of the ground state in linearly and circularly polarized fields. The characteristic feature is the rapid decrease with α_0 of the binding energy. This strongly contrasts the case of lower frequencies (i.e., several photons are needed for ionization), when the ground-state binding energy starts by increasing with α_0.

The ground-state eigenfunction is given for linear polarization at various α_0 in Fig.2. A remarkable feature at *large* α_o is the *dichotomy* of the electronic cloud into two non - overlapping lobes localized around the end points of the line of charge extending between $\pm \alpha_0$ **e**, where **e** is the polarization vector. Indeed, it has been shown (Pont, Walet, and Gavrila, [40]; see also [21], Sec.V.A 3), that the electronic wave function can be expressed at large α_0 as follows:

$$u_0^P(\mathbf{r}) = 2^{-1/2}\left[\, 1 + (-1)^P \mathcal{P} \,\right] v_0^{(-)}(\mathbf{r}) \; . \tag{12}$$

Here P = 0, 1, is the parity of the state, $\mathcal{P}$ is the parity operator. We are introducing the functions $v_0^{(\pm)}(\mathbf{r})$, localized around $\pm \alpha_0$ **e**, respectively, having the form:

$$v_0^{(\pm)}(\mathbf{r}) \equiv \mathcal{V}_0(\mp \;\; \alpha_0^{-1/3} \mathbf{r}_{\pm}) = \mathcal{P}\, v_0^{(\mp)}(\mathbf{r}) \quad , \tag{13}$$

where $\mathcal{V}_j$ is the solution of a certain differential equation (Eq.(147) of [21]), and $\mathbf{r}_{\pm} = \mathbf{r} \pm \alpha_0$ **e** . The functions $v_0^{(\pm)}(\mathbf{r})$ dilate with α_0 according to a $\alpha_0^{1/3}$ power law. (Note that the dilation cannot compensate the recession of the end points, whence the non-overlap of the lobes). The two wave functions P = 0 , 1, in Eq.(12) are degenerate; it was shown that their energy scales with α_0 as $W_0 = \alpha_0^{-2/3}\, Q_0$, where Q_0 is a constant.

For circular polarization, the characteristic large - α_0 feature is the *toroidal shaping* of the electronic cloud, i.e., the concentration of its wave function in a torus lying around the circle of charge, e.g. see [21], Sec.V.A, Fig.6. Scaling laws similar to those for linear polarization have been derived.

We now return to the case of atbitrary α_0. In order to be able to calculate the ionization rates according to Eq.(8) one needs also the continuum wave function $u_{\mathbf{k}_n}^{(-)}$. In previous work [33] we have used for the latter the Born approximation (replacement of $u_{\mathbf{k}_n}^{(-)}$ by a plane wave). This greatly simplifies the computation of the amplitude f_n , and allows the derivation of analytical formulas which illustrate the physical trends. On the other hand, it limits the validity of the results to larger values of ω than required by the HFFT (see [41]).

Thus, in the case of *circular polarization* one finds for the differential and angle integrated n-photon rates the simple formulas [33a], [21], [42] :

$$\frac{d\Gamma_n}{d\Omega} \;\cong\; 4\, k_n^{-3} \;\; J_n^2(\alpha_0 k_n \sin\theta)\;\; [u_0^c]^2 \;\; ; \tag{14}$$

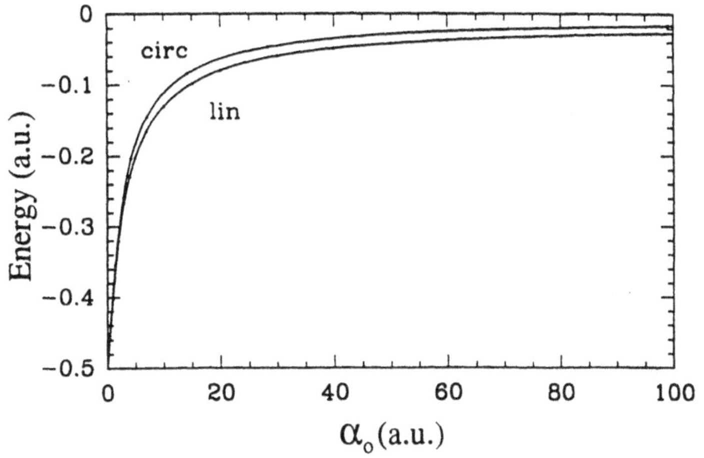

Figure 1. Ground-state energy of the H atom in high-frequency laser fields, of linear and circular polarizations, versus $\alpha_0 = E_0\,\omega^{-2}$, in a.u. From Pont, Walet and Gavrila [40], and Vos and Gavrila (unpublished).

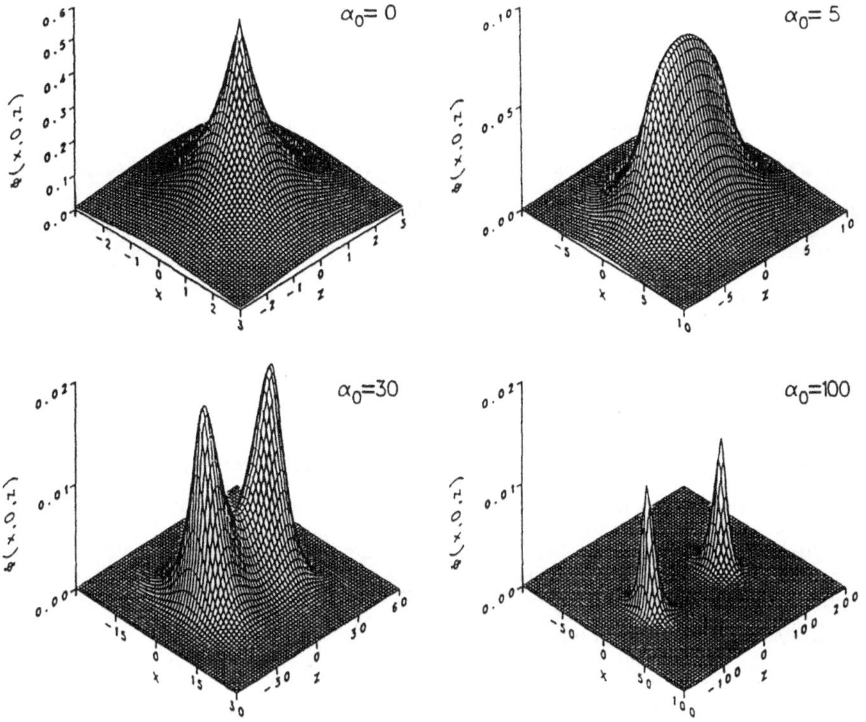

Figure 2. Ground-state energy eigenfunction of the H atom, in the oscillating frame of reference, at high frequencies and linear polarization. $\phi(x,0,z)$ is the normalized wave function in the xz plane, with the z axis along the polarization direction (in a.u.); the axis is arbitrary due to cylindrical symmetry. From Pont, Walet, and Gavrila, [40].

$$\Gamma_n \cong \frac{8\pi}{\alpha_o k_n^4} [u_o^c]^2 \int_0^{2\alpha_o k_n} J_{2n}(\xi)\, d\xi \quad . \tag{15}$$

Here, J_n are Bessel functions, θ is the angle of $\mathbf{k}_n$ and the propagation direction, and u_o^c is the constant value (due to cylindrical symmetry) of the ground state wave function u_o on the "circle of charge".

At low intensities (small α_o) Eq.(15) gives a Γ_1 proportional to I, with a ω dependence of $\omega^{-9/2}$; this is a well known result from LOPT at high frequencies. Moreover, as $\Gamma_n \sim I^n$, to lowest order in the intensity only Γ_1 contributes to Γ : $\Gamma \cong \Gamma_1$.

For intense, nonpertutbative fields ($\alpha_o k_n >> 1$) , the Γ_n of Eq.(15) can be summed over n. If also $\alpha_o >> 1$, we have $u_o^c \cong (0.147 / \alpha_o)$, and the result is:

$$\Gamma \cong \frac{0.223}{\alpha_o^3 \, \omega^2} = 0.631 \, \frac{\omega^4}{I^{3/2}} \quad . \tag{16}$$

This equation exhibits the *adiabatic stabilization* result*: at fixed* ω *the total ionization rate decreases with* I ; besides, at fixed I, Γ increases with ω . These results were termed "counter-intuitive", meaning in fact, that they were contrary to what was known from LOPT. Similar results have been obtained for linear polarization (e.g., see [21], Sec.VI. B2 ; [22b]).

The intensity dependence of the lifetimes for circular polarization, as computed from Eq.(15), is shown in Fig.3 at various ω . For the larger ω values considered ($1 \leq \omega \leq 8$), the condition Eq.(6) is (approximately) satisfied, so that the lifetime curve τ can be followed from small intensities to superintensities. In a log-log plot the curve has a linearly descending LOPT branch, a linearly ascending stabilization branch, and an intermediate branch with a minimum, marking the breakdown of LOPT and the onset of the stabilization regime at some critical intensity $I^{(c)}$. At the values of ω considered, $I^{(c)}$ lies between 1 and 1000 a.u., [43].

Also included in Fig.1 are three lower values of ω (ω = 0.125 , 0.25, 0.5) for which, at small I (or α_o), the high-frequency condition Eq.(6) is violated. Nevertheless, even for such ω , if I (or α_o) is allowed to increase sufficiently, Eq.(6) will end up by being satisfied (see Fig.1), so that the HFFT predicts the existence of states, whose lifetimes are given by the curves shown. These states are, however, of a very special nature ("light-induced states"), and their physical significance is now under scrutiny.

All other Floquet methods have subsequently confirmed the existence of adiabatic stabilization: Sturmian basis diagonalization (Dörr, Potvliege, Proulx, and Shakeshaft, [34a], Fig.1); close-coupling (Marte and P. Zoller [18], Fig.7 ; Dimou and Faisal [34b], Fig.1); R-matrix (Dörr, Burke, Joachain, Noble, Purvis, and Terao-Dunseath [34c], Fig.1). As an illustration, we reproduce here in Fig.4 the result of the R matrix Floquet theory for the total ionization rate $\Gamma = (1 / \tau)$, in a field of linear polarization; Γ exhibits a maximum, and thereafter a decrease corresponding to stabilization. The results obtained with the various theories for τ or Γ of the ground state of H, agree rather well with each other, which is quite gratifying in view of the diversity of the calculations. These theories can claim higher accuracy than the HFFT results presented here, because they integrate the exact form of the Floquet equations, and hence are equivalent to including higher order corrections in ω^{-1} within the HFFT. Besides, no approximation is made for the description of the ionized electron. On the other hand, all these theories have used so far one-center partial-wave expansions, which have convergence difficulties at larger intensities. Fortunately, the onset of stabilization occurs at intensities at which they still converge. We also mention a special method for extracting ionization rates in the stabilization regime, by manipulation of the standard perturbation theory expansion to high order, combined with analytic continuation to nonperturbative field strengths (Moon-Gu Baik, Pont, and Shakeshaft, [44]).

The frequencies and intensities required for the adiabatic stabilization of the ground state of H are not available, compare Eq.(6) with Fig.3. (This handicap may be alleviated to some extent in the years to come, see Sec.8.) Moreover, the minimum lifetime of the ground state (which lies below 1 fs in Fig.3) is much shorter than the duration of the shortest available superintense laser pulses (of some 50 fs at present). A neutral atom subjected (adiabatically) to such a pulse will not survive to enter the stabilization regime: the bottom of the valley-shaped lifetime curve in Fig.3 acts like a *Death Valley* for it. This does not mean, however, that the beyond of Death Valley for the ground state could not be accessed by other means, see Sec.8. Meanwhile, however, it turned out that adiabatic stabilization could be achieved for Rydberg states.

B. Rydberg states

The advantage of considering adiabatic stabilization for Rydberg states, instead of the ground state, hinges on the high-frequency condition Eq.(6). The field should be chosen to have *linear polarization*, and the Rydberg state should have a high magnetic quantum number m referred to the field axis, for the following reasons. High atomic m states are essentially hydrogenic, so that a H-atom calculation will apply. For linear polarization, the initial state can couple only to states of the same m , whatever the intensity. In this case, the $\mathcal{W}_e$ appearing in Eq.(6) should refer to the m manifold to which the initial state belongs, and is of the order of magnitude of the binding energy of the lowest state of the manifold. In the absence of the field ($\alpha_0 = 0$), this is $\mathcal{W}_m = 2^{-1}(|m|+1)^{-2}$, for "circilar states" $|m| = \ell = n - 1$. As it turns out that $|W_m(\alpha_0)| < \mathcal{W}_m$ at all α_0 (see [33b], Fig.1, or [21], Fig.3), the condition is even better satisfied at larger α_0. For high enough m, $\mathcal{W}_m$ can be made small with respect to the frequencies of current intense lasers, so that condition Eq.(6) for the validity of the HFFT can be effectively satisfied.

To prove the feasibility of this approach Vos and Gavrila [33b] have carried out a calculation for $|m| = 5$, at $\omega = 1.17$ eV = 0.0428 a.u. (the NdYAG fundamental) and $\omega = 2.0$

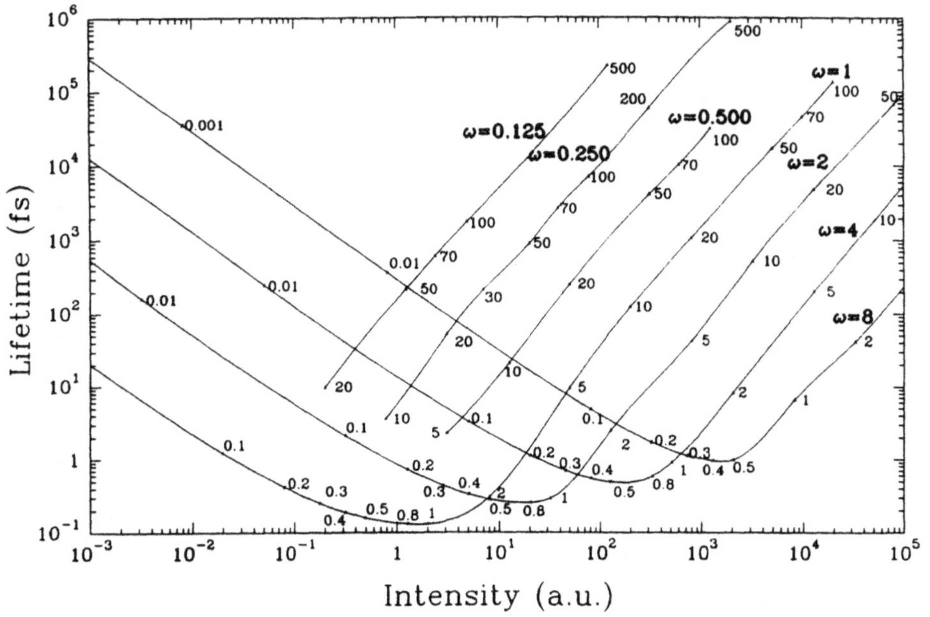

Figure 3. Lifetime of the H atom in the ground state, from the HFFT Eq. (15), versus intensity, at various ω (in a. u.); circular polarization. Numbers adjacent to points on the curves are the corresponding values of α_0. The descending branches of the curves corresponding to LOPT, the ascending ones to adiabatic stabilization. From Pont and Gavrila, [33a].

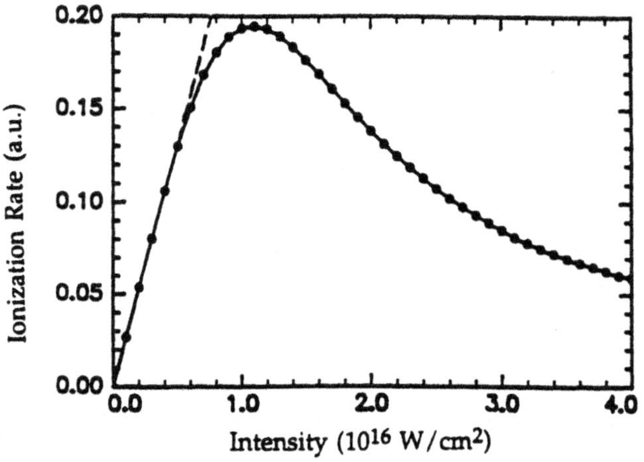

Figure 4. R-matrix Floquet theory total ionization rate Γ of the ground state of the H atom versus intensity, at ω = 0.65 a. u. and linear polarization. The broken line is the LOPT result. From Dörr, Burke, Joachain, Noble, Purvis, and Terrao-Dunseath, [34c].

eV = 0.0735 a.u (relevant to the experiments described in Sec.5). The lowest lying states associated to this m are in the field-free case the circular states of quantum numbers n = 6, ℓ = 5 (ungerade case), and n = 7, ℓ = 6 (gerade case). Their decay rates were evaluated according to Eq.(8), with the Born approximation. The intensity dependence of the lifetimes τ was given in a log-log plot in Fig.2 of [33b], see also Fig.10 of [21]. The two curves obtained are similar in shape to those of the present Fig.3 for the ground state (m=0), but there is an essential difference: *for the circular Rydberg states the minimal lifetimes exceed by far the duration of the superintense short laser pulses available* (e.g., of about 50 - 100 fs), and hence Death Valley is not longer lethal for the neutral atom. Thus, for the state evolving from n = 7, ℓ = 5, at ω = 0.0735, the minimum lifetime was calculated to be 2000 fs. Besides, if the turn-on time of the intensity is not shorter than about 100 fs, it was shown that the field variation can be considered as adiabatic, i.e., the atom will stay in its adiabatically evolving initial state. It was therefore concluded that all conditions were satisfied in this case to be able to detect stabilization with present technological means.

There are also other differences from the case of the ground state: the onset of stabilization occurs now at lower intensities (below 0.1 a.u.), and the stabilization branch increases more steeply with I . Moreover, the total lifetime for high m states is determined entirely by one-photon ionization, i.e. $\Gamma \cong \Gamma_1$, for it was shown that Γ_n varies roughly as $n^{-|m|-2}$, [33b]; this means that there is only one peak in the electron spectrum, as at low intensities.

More accurate Floquet calculations of Rydberg state lifetimes have been carried out meanwhile by Potvliege and Smith [45], based on the Sturmian Floquet theory (without the Born approximation), and by others (Scrinzi, Elander, and Piraux [46]; Buchleitner and Delande, [47]; Dimou and Faisal [48], see also [17b]; Moon-Gu Baik, Pont, and Shakeshaft, [44]). These calculations are in good agreement with each other, and have produced lifetime curves similar in shape to those of [33b], but with considerably larger values for the minimal lifetimes, and therefore, even more favorable for experiment.

C. Two-electron atoms

There are only few studies of two-electron atoms in the superintense regime (Mittleman, [49]; Muller and Gavrila, [50]; Gavrila and Shertzer, [38]). A characteristic prediction of the HFFT is that H^-, which in the absence of the field has only one bound state below the n = 1 level of hydrogen, acquires infinitely many light-induced excited states at values of α_0 larger than 3.5 (Muller and Gavrila, [50]). Besides, in the large-α_0 regime, two-electron atoms undergo dichotomy, similarly to their one-electron counterparts (Mittleman, [49]; Gavrila and Shertzer, [38]). While in the dichotomy regime, they are also stabilized; an outline of the proof [38] runs as follows.

We assume that the atom is initially in its ground state. This is degenerate because of the dichotomy, as the electrons can settle around opposite end points in a symmetric or antisymmetric state, with total spin S = 0, 1. The two eigenfunctions, can be constructed from the functions $v_o^{(\pm)}(\mathbf{r})$, Eq.(13), for the ground-state of the one-electron problem as follows:

$$u_i(\mathbf{r}_1, \mathbf{r}_2) = 2^{-1/2}\,[1 + (-1)^S\, \mathcal{Q}_{12}]\, v_o^{(+)}(\mathbf{r}_1)\, v_o^{(-)}(\mathbf{r}_2) \quad , \tag{17}$$

where $\mathcal{Q}_{12}$ is the operator interchanging the electron coordinates (see [38]). Note that $P_{12} = S$, where P_{12} is the total parity quantum number ($P_{12} = 0, 1$). As we are dealing in fact with two independent electrons, the total energy is twice that for one electron: $E_o = 2W_o = 2\alpha_o^{-2/3}\, Q_o$, see Sec.3.A.

Concerning the final continuum state, we shall assume that the photon and ejected electron energies are sufficiently large, so that we can use the Born approximation, and that the residual one-electron atom / ion is left in its ground state of dichotomous wave function $u_o^P(\mathbf{r})$ given by Eq.(12). Thus, for the final state we take:

$$u_f^{(B)}(\mathbf{r}_1,\mathbf{r}_2) = 2^{-1/2}\,[1 + (-1)^{S'}\, \mathcal{Q}_{12}\,]\; u_o^P(\mathbf{r}_1)\, \exp(\, i\, \mathbf{k}\cdot\mathbf{r}_2\,) \quad , \tag{18}$$

where S' is the total spin quantum number of the state. As the total spin is conserved in the transition, $S' = S = P_{12}$.

The Born approximation for the n-photon ionization amplitude will be denoted by $F^{(B)}_{Pn,P_{12}}$, where we have indicated the relevant quantum numbers of the initial and final states. By inserting Eqs.(17) and (18), into Eq.(11), we find to dominant order in ω^{-1}:

$$F^{(B)}_{Pn,P_{12}} \cong (-1)^P\, f^{(B)}_{n,\Pi} \quad , \qquad f^{(B)}_{n,\Pi} \equiv -\,(2\pi)^{-1} < \mathbf{k}_n | V_n | u_o^{\Pi} > \quad , \tag{19}$$

with $| \mathbf{k}_n > \equiv \exp(i\, \mathbf{k}_n \cdot \mathbf{r})$. $f^{(B)}_{n,\Pi}$ is precisely the Born approximation amplitude for one-electron n-photon ionization Eq.(8), from the initial state with parity $\Pi \equiv P + P_{12}$ (*modulo* 2). Eq.(19) leads to the equality of rates:

$$(d\Gamma^{(B)}_{Pn,P_{12}} / d\Omega) \cong (d\gamma^{(B)}_{n,\Pi} \,/\, d\Omega). \tag{20}$$

Thus, in the dichotomy regime (and in the Born approximation), the n-photon ionization rate for the two-electron atom is reduced to that for a one-electron atom. Eq.(20) assumes that the states of different parity considered can be distinguished. Otherwise, rates $\overline{\Gamma}^{(B)}_n$ and $\overline{\gamma}^{(B)}_n$, summed up over final states and averaged over the initial ones, should be defined. Eq.(20) then yields the connection:

$$\bar{\Gamma}_n^{(B)} \cong 2\, \bar{\gamma}_n^{(B)} \; . \tag{21}$$

The factor 2 represents the number of electrons of the two-electron atom.

For the one-electron problem the dichotomy regime sets in for the ground state at about $\alpha_0 \cong 40$, see Fig.2, whereas stabilization sets in around $\alpha_0 \cong 1$ at $\omega \cong 1$. Consequently, at the values of α_0 for which the Eqs.(20), (21), apply, we are well in the stabilization regime of the one-electron problem. We conclude that *stabilization occurs for two-electron atoms* as well. In fact, just as for H, stabilization probably sets in at much lower α_0-values, see Fig.3 .

Finally, we want to mention the case of *multiply charged negative ions of H*. It has been proven theoretically that a (field-free) proton cannot bind more than two electrons, i.e., the doubly-charged negative ion of hydrogen does not have bound states. Nevertheless, it has been predicted recently on the basis of the HFFT , that in an intense high-frequency field (of linear or circular polarization), the proton can indeed bind more than two electrons (van Duijn, Gavrila, and Muller, [51]). The values of α_0 and ω at which the negative threefold and fourfold charged ions appear, are within physical reach. Moreover, at the α_0 at which they are formed, the ions are already in the stabilization regime.

4. Wave-packet dynamics (WPD)

A. Dynamic stabilization

In the WPD approach, the time dependent Schrödinger equation, e.g., in its lab-frame form Eq.(1), is integrated numerically, starting from a specific initial condition. This is usually taken to be a field-free energy eigenstate of the atom. The laser pulse is applied, and the wave packet $|\psi_L(t)>$ is propagated until the end of the pulse τ. Ionization probability densities $P_E(\tau)$ are then calculated according to the quantum-mechanical rules, by projecting the wave packet, at the end of the pulse, onto field-free continuum states of the atom $|u_E>$, normalized in the energy scale: $P_E(\tau) = |<u_E|\psi_L(\tau)>|^2$, [52]. The total ionization probability is:

$$P_{ion}(\tau) = \int P_E(\tau)\, dE = 1 - \sum_n P_n(\tau) \quad . \tag{22}$$

where the $P_n(\tau)$ represent excitation probabilities to discrete states. Various forms of turn-on have been used, but quite often the calculations were done without a turn-off. The total ionization probability of the atom will depend, obviously, on the peak field-strength E_0 of the pulse, and on its frequency ω, but it turns out to depend critically also on its shape (turn-on, turn-off, duration). Thus, in general, $P_{ion}(\tau)$ will not be a characteristic of the initial atomic state.

The studies of the dependence of $P_{ion}(\tau)$ on E_o and ω that have lead to stabilization, were initiated by Eberly and collaborators on a 1D model [31]. This model, widely used since, contains the "soft Coulomb" potential:

$$V_{sc}(x) = -(1 + x^2)^{-1/2} \quad , \tag{23}$$

that simulates the Coulomb potential at large distance, eliminating however its singularity at the origin (untractable in 1D). The model has an infinite sequence of Rydberg levels; the field-free ground state energy is $W_o = -0.67$ a.u. The calculations have shown that, the ionization probability $P_{ion}(\tau)$ is an increasing function of the peak field strength E_o up to a certain critical value, and thereafter it starts to decrease. We show in Fig.5 results of Law, Su and Eberly [31c] illustrating this feature. In this case ω was taken 2 [which satisfies moderately well the high-frequency condition Eq.(6)], and the turn-on and turn-off of the pulse was over 4.25 cycles. As apparent, $P_{ion}(\tau)$ is less for $E_o = 2.0$ than for $E_o = 0.5$. This is *dynamic stabilization*, thus called to distinguish it from the adiabatic version. Its existence at high frequencies appears natural in view of what was said for adiabatic stabilization. Surprisingly, however, it was discovered by Su, Eberly and Javanainen [31b], see also [31c], that dynamic stabilization can occur also at *low frequencies*, when, e.g., two or three photons are needed to ionize the unperturbed atom. In the following we shall try to understand what mechanism causes stabilization in this case.

A turn-on of the field over few cycles, as applied in dynamic stabilization studies, represents a hard shock for the atom, which projects some of the initial population to excited states. The analysis of the populations in *dressed states* (i.e., solutions of Eq.(7) in the oscillating frame) has been initiated by Reed, Knight, and Burnett [32b], see also [2b]. It was shown that for very short turn-ons, over a few cycles, much of the population can be projected directly into the continuum: this kind of ionization is of the "shake-up" type (see also [35]). It is a consequence of the sudden turn-on of the perturbation, and is independent of the subsequent oscillatory nature of the field, and in particular of its frequency. For some of the bound dressed states thus populated, the frequency will be high enough compared with their binding energy, so that they will ionize in a high frequency regime, of the kind encountered in adiabatic stabilization (i.e. with small rates). If the frequency is high also with respect to the ground state (as is the case in our Fig.5), its population will also ionize with a small rate, and part of it may survive until the end of the pulse. If, however, for some of the lower populated states (as well as for the ground state), the frequency should be low (as was the case in [31b], [31c]), these will ionize rapidly, and will not survive until the end of the pulse. When, at the end, the pulse is turned off suddenly, the atom will endure a new shock, giving possibly rise to another burst of ionization. In spite of this, much of the population parked in excited dressed states, will remain in the vicinity of the atom, and will eventually fall down to the (field free) ground state. The fraction of the initial population that has survived the pulse is represented by the probability: $1 - P_{ion}(\tau)$. The picture presented fosters the representation of dynamic stabilization as a shake-up excitation to bound dressed states, followed by high-frequency stabilization. These ideas can be substantiated by an analysis of the wave packet in terms of Floquet states (Wells an Gavrila, [53]).

In other 1D models, short-range potentials were considered (finite-range in [54], zero-range in [55], [56]). In contrast to the soft Coulomb potential, these were chosen with only one field-free bound state. A different situation emerges now, due to the lack of excited dressed states in which to park the population during the pulse: the shake-up following the sudden turn-on projects most of the population into the continuum. This was illustrated by Grobe and Fedorov [54], who have shown that after a 2 cycle turn-on, the wave packet is driven for some time $t < t^*$ by the field, like a free particle Volkov wave packet, with no heed to the existance of the center of force (see Fig.1 of [54b]). During this time much of the wave packet spreads out towads infinity (ionization due to shake-up to the continuum). Nevertheless, for $t > t^*$ the spreading of a part of the wave packet is stopped, and this part will oscillate around the center of force; besides, it has a dichotomous structure of the type described in Sec.3A (see Figs.2 and 3 of [54b]). The authors have shown that the ionization probability $P_{ion}(\tau)$ can exhibit stabilization. Somewhat similar conclusions were reached by Sonnenmoser [55] in a more formal way, for a $-B\delta(x)$ potential, by using an intense-field approximation which neglects the atomic potential during the ionization process. Su, Irving, Johnson and Eberly [56] have considered the same $- B\delta(x)$ potential, and again found stabilization. They emphasized the need for a smooth (instead of a step-wise) turn-on in order to achieve it, for the case that the system has only one bound state.

1D model calculations are quite convenient for exploring the physical features of phenomena, when 3D calculations are unwieldy. On the other hand, one should be aware that they might introduce features not shared by their realistic 3D counterparts. For example, it was pointed out by Ménis, Taïeb, Véniard, and Maquet [57] that, in comparison to the 3D Coulomb potential, the potential Eq.(22) can lead to an overestimation of stabilization, because of the softening of the Coulomb singularity.

Kulander, Schafer, and Krause [35] were first to document by 3D calculations the dynamic stabilization of the H atom. The signature of ionization was considered to be the decrease of the confinement probability in a finite volume around the nucleus (taken as the computational grid for the finite difference representation of the wave function). After the turn-on of the field, an approximately exponential decay of the confinement probability was observed, and a global atomic decay rate could be extracted. By analyzing the peak-field dependence of the rate, it was shown that beyond a critical intensity, it starts to decrease (see Fig.16 of [35b]), which is the signature of dynamic stabilization [58]. Similar calculations for the stabilization of the ground state of H were done in the oscillating KH frame by Horbatsch [30a]. Studies by others have concentrated on the behavior of the excited states: Pont, Proulx and Shakeshaft [27],[59] have considered the field-free states n = 3 - 5, whereas Huens and Piraux, [60], and Gajda, Piraux, and Rzazewski, [29] have concentrated on n = 2. In all cases dynamic stabilization could be demonstrated.

B. Adiabatic stabilization

Natural questions arrise: can also adiabatic stabilization be obtained from WPD ? If this is the case, what are the conditions to be fulfilled ? We present here answers to these questions, following Wells, Simbotin, and Gavrila, [13]. We shall consider only the case high frequencies in the sense of Eq.(6), which is a requirement for adiabatic stabilization. A crucial issue to be decided first is what kind of initial condition should be used for WPD. We recall that, whereas WPD works with realistic initial conditions, Floquet theory, as well as the HFFT, operate with the idealized (non-normalizable) states Eq.(4), assumed to follow adiabatically the variation of the field amplitude. This adiabatic variation of cannot be duplicated numerically at present, as it would require excessively long computation times. Hence, the problem: what alternative initial condition can one use to establish contact with the HFFT ? The answer is to take, instead of field-free atomic states, states that describe the atom already in the stationary regime imposed by a field at its constant peak amplitude. These are approximately the dressed states $u_\mu(x)$ considered in Sec.2 [eigenfunctions of the lowest order HFFT structure equation Eq.(7)], taken at the value of α_0 corresponding to the peak field. The wave function $\psi_\mu(x,t)$, solution of Eq.(3) with the initial condition $\psi_\mu(x, 0) = u_\mu(x)$, will represent the evolution of $u_\mu(x)$ according to WPD, and the probability $P_\mu(t) = |< u_\mu(x) | \psi(x,t) >|^2$ will describe its decay in time.

The system taken as test case was the 1D atomic model with the "soft Coulomb" potential, Eq.(23). The initial condition (in the oscillating frame), is that the atom be in its dressed ground state at the α_0 considered. The frequencies chosen were $\omega = 2$ and 4 , and α_0 was allowed to vary from 0 to 20 . The dressed ground-state binding energy decreases monotonically with α_0 , from the field-free value of 0.67 to about one third at $\alpha_0 = 20$, e.g., see [2a], p.326, Fig.20. The frequencies chosen are about three, respectively, six times larger than the binding energy at $\alpha_0 = 0$, and about 9, respectively 18, times larger at $\alpha_0 = 20$. The frequencies are, therefore, well suited for comparison with the HFFT, according to the criterion Eq.(6).

The first goal of the calculation, was to ascertain that the evoulution of the population in the ground dressed state was given by an exponential decay law $\exp(- \Gamma t)$, as predicted by the HFFT (and by general Floquet theory). This was found to be indeed the case, for although the populations show in general a jitter of a few percent, their *average* is could be fitted quite well to decreasing exponentials. To improve the accuracy on Γ , the evolution of the population was followed as long as possible, over thousands of cycles in some cases. The values obtained for Γ at the two frequencies considered are shown in the Fig.6 by dots. Next, we have carried out a HFFT calculation of the rates based on the analogue of Eq.(8) for the 1D case (for a very large number of α_0 values). The results are represented in Fig.6, by continuous curves. As apparent, the WPD and HFFT results are in good agreement, particularly at the higher frequency $\omega = 4$. The agreement extends over a large α_0 range ($\alpha_0 \leq 20$; not shown), and four orders in the magnitude of Γ. Fig.6 obviously displays adiabatic stabilization. This proves that *adiabatic stabilization also results from WPD.*

5. Experimental results

The experiments carried out were based on the theoretical ideas discussed in Sec.3B. It was decided that the most favorable candidate under the prevalent experimental circumstances was the Ne atom, and the circular 5g (n=5, ℓ = m = 4) Rydberg state was selected as the test case. Two independent, extremely elaborate, experiments were carried out. The first one, by de Boer, Hoogenraad, Vrijen, Constantinescu, Noordam, and Muller [36a], involved two stages: the preparation of the initial Rydberg state, and the detection of its ionization by an intense laser field of (relatively) high frequency. In the second one, by van Druten, Constantinescu, Schins, Nieuwenhuize, and Muller [36b], a third stage was added to monitor the population left in the Rydberg state. The second experiment confirmed the first one; we shall present it in the following.

The three stages of the experiment used different laser pulses. These were shone successively into the ionization chamber of the "magnetic bottle" electron spectrometer containing the Ne atoms. [This device uses an inhomogeneous magnetic field to collect the electrons ionized inside the chamber, and send them in a beam parallel to its direction, towards a multi-channel plate detector, which energy analyzes them according to their time of flight. The laser beam is shone into the instrument perpendicular to its axis, which coincides with the direction of the magnetic field.] The first pulse, the "preparation pulse", consisted of intense circularly polarized UV light (ω = 4.34 eV, pulse duration τ = 1 ps, I $\approx$ 200 TW / cm^2). Its role was to populate the 5g state. The second, "main pulse", was the one driving the ionization to be stabilized. It consisted of red light (ω = 2.0 eV = 0.073 a.u.), was ultrashort (τ = 90 fs) and intense (I $\approx$ 100 TW / cm^2), and was linearly polarized. It could be delayed by a variable time τ_d with respect to the preparation pulse. The high-frequency condition Eq.(6) is relatively well satisfied for these photons and for the m = 4 manifold, whose largest binding energy is $\mathcal{W}_4$ = 0.02 a.u. Finally, the "probe pulse", consisted of green light (ω = 2.33 eV) of low-intensity (I $\leq$ 10 GW / cm^2). Its pulse duration was long (τ = 5 ns), it was

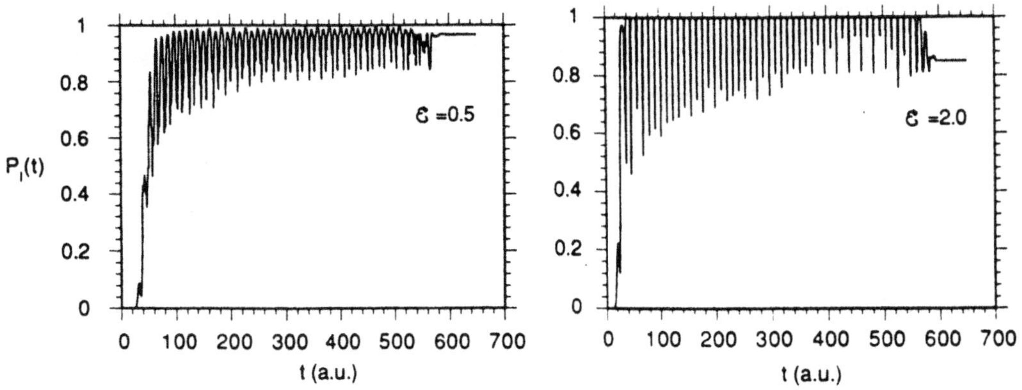

Figure 5. Electron ionization probability as a function of time for a 1D model with the "soft Coulomb potential" Eq. (23), and for two laser pulses of peak field strengths $E_0 \cong$ 0.5 and 2 a. u., $\omega = 2$ a. u. The pulses are turned on and off over 4.25 cycles = 93 a. u. From Law, Su, and Eberly [31c]. The physical ionization probability $P_{ion}(\tau)$, Eq. (22), should be read at the end of the pulse.

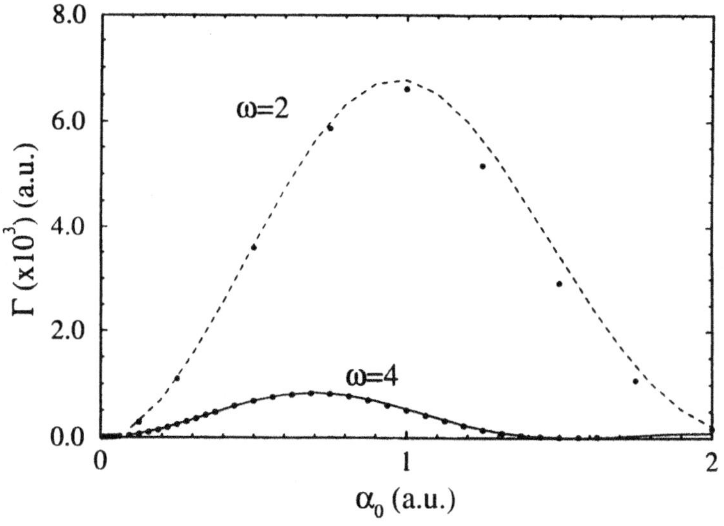

Figure 6. Total ionization rate Γ versus a0 (in a. u.) for a ID model with the "soft Coulomb potential" Eq. (23), at two high frequencies $\omega = 2$ and 4 a. u. Wave-packet dynamics results: circles; HFFT results: continuous curves. The ascending branches correspond to LOPT, the descending ones to adiabatic stabilization. From Wells, Simbotin, and Gavrila, [13].

weakly focused, and had sufficiently high fluence to ionize all the population left in the 5g and neighboring states. It was delayed by 14 ns with respect to the preparation pulse.

The excited $(2p^5)5g$ state was produced from the ground state $2p^6$, by absorbing 5 photons from the preparation pulse. In a one-electron approximation, this corresponds to a transition from the (2p, m_ℓ = -1) orbital to the (5g, m_ℓ = 4) one. The magnetic quantum numbers are referred to the axis of propagation of the preparation beam. At zero field, 5 photons barely suffice to access the continuum. However, there is a large increase in the ionization potential of the ground state during the turn-on of the field, which means that the ground state level is shifted downwards with respect to the continuum, whereas the Rydberg states remain close to it in enrgy. Consequently, the 5-photon resonance from the ground state is shifted through the (n,g)-Rydberg series during the turn-on. Population is transferred to these states in the process, and a large fraction of it survives the preparation pulse. This means that along with the 5g states, also the 6g, 7g, etc. states are populated (but to a lesser extent). It was exceedingly difficult to produce large 5g populations, so that the optimization of this stage was essential for the success of the experiment.

The main pulse was applied to the excited atoms with a delay of τ_d with respect to the preparation pulse, and the electron energy distribution was recorded with the magnetic bottle spectrometer. When performing high-intensity experiments, there is always a spatial range of intensities in the focal region, which substantially complicates the interpretation of the results. This difficulty was circumvented in the present case by allowing the focal region of the main pulse to be much larger than that of the preparation pulse, containing the excited atoms. Thus, the atoms were subject to only the central part of the main puse, containing the peak intensities. On the other hand, the fact that the two beams are parallell creates a problem: the axis of quantization of the magnetic states prepared by the first pulse (parallel to the beams) is perpendicular to the direction of the linear polarization of the main beam, contrary to the requirements of the theory. This was solved elegantly, by making use of the very magnetic field of the electron spectrometer, which is perpendicular to the beams. In this field, the m = 4 excited atomic states undergo Larmor precession, which will rotate their axis of quantization, allowing it to become perpendicular to the beam direction. This requires obviously that the delay time τ_d be adequately chosen; its optimal value was measured to be 20 ps (quarter Larmor period).

A typical electron spectrum, for the case that all three pulses were applied in succession, is shown in Fig.7 (full lines), for τ_d = 20 ps. Two series of photoelectron peaks are displayed: one is due to the main laser pulse, the other to the probe pulse (as indicated). The peaks represent one-photon absorption originating in each case from the different (n,g) states populated by the preparation pulse (as indicated). The energy difference of corresponding peaks equals the energy difference of the photons involved, 2.33 - 2.00 = 0.33 eV. The figure contains for reference also a dashed curve, representing the spectrum for τ_d = - 5 ps *without* the probe laser. (Negative τ_d means that the main pulse precedes the preparation pulse.) Thus, the dashed curve represents the electron yield from the ground state due to the main pulse, i.e., gives the background signal under the main-photon absorption peaks in the three pulse experiment. Fig.7 contains also a dotted curve, giving the spectrum for $\tau_d = -5$ ps *with* the probe laser on. This gives the electron yield in the absence of the main pulse, i.e., the total population created in the 5g state. By subtracting the dashed curve from the full one under the main-photon peaks, and the full one from the dotted one under the probe photon peaks, one could determine independently the fraction of the 5g population that has ionized, respectively, survived.

These fractions are shown in Fig.8. The error bar on the results is indicated. Note first that, within the errors, the two fractions add to 1, indicating that there are no significant decay channels other than one-photon ionization. This agrees with the theoretical prediction made in Sec.3.B. Next, above a critical peak intensity of the main pulse approximatively equal to 60 TW / cm^2 , hardly any ionization is observed. The population stabilizes at about 25% of its initial value, and conversely, some 70% of it is left in the initial state. The results contradict LOPT, represented in Fig.8 by the continuous curves. This is a clear indication of adiabatic stabilization.

6. Open problems

Stabilization is a striking pheomenon, revealing the existence of a new regime of atomic behavior in superintense laser fields. It has been documented theoretically in a large number of cases, by a variety of methods, in both the adiabatic and dynamic forms. It has been confirmed by two extremely elaborate experiments, although, until now, under limited circumstances. This represents, however, only a beginning stage in its study, and much more remains to be done in order get the complete physical picture.

On the experimental side, one has just explored the onset of the stabilization regime for Rydberg states. Beyond this onset, at higher intensities, lies the ascending branch of the lifetime curve predicted by the HFFT (see Fig.3), that needs to be investigated. For this, more powerful lasers are needed. The importance of the linear polarization requirement for Rydberg states needs to be verified. Lower bound states need to be explored, to check the role of the frequency in adiabatic stabilization. And, of course, there is the persistent issue of the ground state.

The practical barrier in the study of ground-state stabilization, and of many other of the predicted superintense-field phenomena, is the occurrence of Death Valley. It excludes the possibility that a light atom will survive an adiabatic turn-on of a superintense laser pulse, meant to bring it into the stabilization regime. Ground states are, however, essential for any possible application of the phenomenon. Are there any prospects of exploring the physics of the Beyond Death Valley? Important steps in this direction are the current efforts to built and

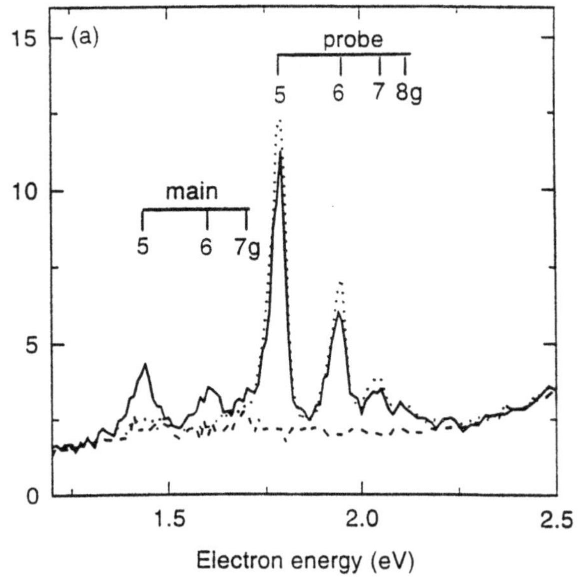

Figure 7. Photoelectron spectra from the stabilization experiment of van Druten, Constantinescu, Schins, Nieuwenhuize, and Muller [36b], after all three laser pulses (preparation, main and probe) have been applied (see text). Delay time between preparation and main pulses: $\tau_d = 20$ ps. The expected electron energies for one-photon ionization from the $(2p^5)$ ng states of Ne are indicated for the main and probe pulses.

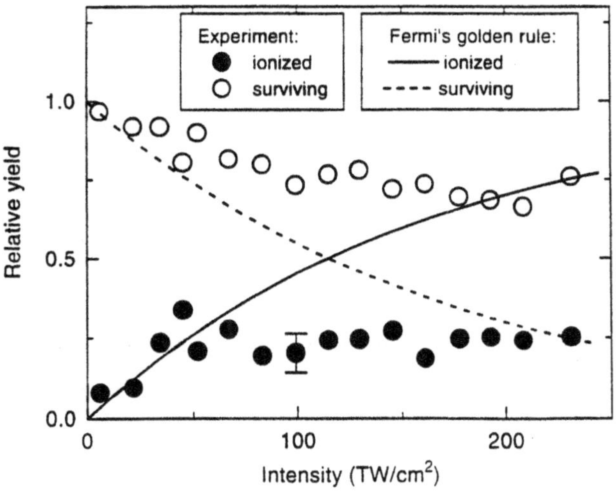

Figure 8. Measure fraction of ionized (open circles), and surviving (solid circles) populations in the 5g state, following exposure to the main pulse, versus main-pulse intensity. Continuous curves are LOPT ("Fermi golden rule") results: ionization fraction (full line), surviving fraction(dashed line). Error bars as indicated. The data are a clear indication of adiabatic stabilization.

operate superintense free-electron lasers in the high frequency regime (i.e., with photon energies larger than one rydberg), and (sub)femtosecond pulse durations [61]. There is no doubt that such efforts will eventually succeed. With the radiation parameters thus becoming available, a new generation of experiments will be possible, concerning not only atomic hydrogen, but also hydrogenic atoms such as H^- and higher charged negative ions of H, as well as hydrogenic molecules, such as H_2, H_2^+.

A parallel avenue of research emerges from the realization that schemes can be devised to access the Beyond Death Valley, *without having to actually cross the Valley*, by using the existing technology in an artful way. For example, this could be achieved by a combination of excitation-deexcitation steps involving excited states, with three or four lasers in the experiment. The experimental state of the art has reached a point where such schemes have become possible.

Much remains to be done also regarding the development of the theory. For example, in the case of dynamic stabilization, even for 1D models there are open questions left. The results for the 3D case, on the other hand, which should provide the direct lead for experiments, are considerably more sketchy for obvious computational reasons. There is no systematic study to clarify the limits of observability of the phenomenon, particularly at low frequencies with respect to the atomic binding energy, which are the physically interesting ones. Possible schemes of accessing the Beyond Death Valley regime should be explored, to guide experiments. Another open problem is the connection of Floquet calculations to their WPD counterparts, in other words, the issue of the extent to which the unusual Floquet predictions are validated by WPD. This emhasizes the necessity of a concerted dual approach to the solution of physical problems, based on both Floquet theory and WPD.

Finally, we mention the issue of relativistic calculations. Ultra-strong fields of 10^{18} W / cm^2 or higher, impose a large quiver motion on weakly bound electrons, which, at high frequencies, implies velocities nearing the speed of light. This requires an approach different from that used so far, based on the nonrelativistic Schrödinger equation and the dipole approximation. Firstly, one needs to start from the Dirac equation. Secondly, as the size of the atom in the field (of order $2\alpha_0$) becomes comparable to the wavelength of the light, the dipole approximation breaks down, and the effects of retardation in wave propagation have to be taken into account. It is an open question how will these changes will affect stabilization as known so far. This, and related problems, have just begun to be investigated.

Acknowledgment

This work was partially supported by the EC Network "Atoms in Superintense Laser Fields".

References

1. For linear polarization, this corresponds to the atomic unit of intensity, 3.51 10^{16} W / cm^2.

2. For an overview of intense-field laser-atom interactions, and in particular ionization, see: (a) the volume *Atoms in Intense Laser Fields*, Ed. M. Gavrila (Academic Press, 1992), and the review articles: (b) *Atoms in Ultra-intense Laser Fields*, by K. Burnett, V.C. Reed, and P.L. Knight, J.Phys. B **26**, 561 (1993); (c) *Ionization Dynamics in Strong Laser Fields*, by L.F. DiMauro and P. Agostini, Adv.At.Mol. Opt.Phys. **35**, 79 (1995).

3. Other types of stabilization have been considered in the literature, e.g see the discussion by H.G. Muller in *Superintense Laser-Atom Physics IV*, Eds. H.G. Muller and M. Fedorov (Kluwer Acad.Publ., 1996), p.1. We shall limit ourselves here to ultra-strong field stabilization. No reference will be made to the classical form of stabilization.

4. An overview of superintense-field stabilization, at a more general level, was given by J. H. Eberly and K.C. Kulander, Science **262**, 1229 (1993).

5. N.B. Delone and G.S. Voronov, JETP Letters **1**, 66 (1965); J.L. Hall, E.J. Robinson, and L.M. Branscomb, Phys.Rev.Lett. **14**, 1013 (1965).

6. P. Agostini, F. Fabre, G. Mainfray, G. Petite, and N.K. Rahman, Phys.Rev. Lett. **42**, 1127 (1979).

7. M. Göppert-Mayer, Ann.Phys.(Leipzig) **9**, 273 (1931).

8. P. Kruit, J. Kimman, H.G, Muller, and M.J. van der Wiel, Phys.Rev.A **28**, 248 (1983).

9. (a) W. Pauli and M. Fierz, Nuovo Cimento **15**, 167 (1938); (b) H. A. Kramers, see *Collected Scientific Papers* (North Holland, Amsterdam, 1956), p.866; (c) W.C. Henneberger, Phys.Rev.Lett. **21**, 838 (1968); (d) F. H. Faisal, J.Phys.B **6**, L89 (1973).

10. Historically, the method of complex eigenvalues (in the absence of the external periodic field) was invented by J. J. Thomson [Proc.London. Math.Soc.**15** (1),197 (1884)]; in the context of quantum mechanics it was introduced by G. Gamow [Zeits.f. Phys. **51**, 204 (1928)]. The Floquet theory for differential equations originates with G. Floquet, Ann.Ec.Norm.Suppl. **12**, 47 (1883).

11. In the adiabatic case, the exponential decay law is expressed as : $(N / N_0) = \exp(-\int_0^t \Gamma(t')\, dt')$; see M. H. Mittleman and A. Tip, J.Phys.A **17**, 571 (1984).

12. M. Dörr, O. Latinne, and C.J. Joachain [Phys.Rev.A **52**, 4289 (1995)], have shown, by comparing with wave-packet calculations, that the Floquet description can give rather accurate values for the ionization probabilities.

13. J.C. Wells, I. Simbotin, and M. Gavrila, submitted for publication.

14. Shih-I Chu and W. P. Reinhardt, Phys.Rev.Lett **39**, 1195 (1977), see also Shih-I Chu, Adv.At.Mol.Phys. **21**, 197 (1985); Adv. Chem. Phys. **73**, 739 (1989).

15. N. Moiseyev and H.J. Korsch, Phys.Rev.A **41**, 498 (1990); N. Ben-Tal, N. Moiseyev, and R. Kosloff, Phys.Rev.A **48**, 2437 (1993).

16. R. M. Potvliege and R. Shakeshaft, Phys.Rev.A **40**, 3061 (1989); **41**, 1609 (1990), see also [2a], p.373.

17. (a) L. Dimou and F. H. M. Faisal, Phys.Rev.Lett. **59**, 872 (1987). (b) F.H.M. Faisal, L. Dimou, H.J. Stiemke, and M. Nurhuda, J. Nonlin.Opt.Phys.Mat. **4**, 701 (1995).

18. P. Marte and P. Zoller, Phys.Rev.A **43**, 1512 (1991).

19. P. G. Burke, P. Francken, and C. J. Joachain, Europhys.Lett. **13**, 617 (1990); J.Phys.B **24**, 761 (1991); M. Dörr, M. Terao-Dunseath, J. Purvis, C.J. Noble, P.G. Burke, and C.J. Joachain, J.Phys.B **25**, 2809 (1992).

20. (a) HFFT for scattering: M. Gavrila and J. Z. Kaminski, Phys. Rev. Lett. **52**, 614 (1984). (b) HFFT for ionization: M. Gavrila, in *Fundamentals of Laser Interactions*, Editor F. Ehlotzky (Lecture Notes in Physics, vol. **229**; Springer, Berlin, 1985), p. 3.

21. A systematic account of the HFFT for the ionization of one-electron atoms was given in by the author in *Atomic Structure and Decay in Hygh-frequency Fields*, see [2a] p.435.

22.(a) J. I. Gersten and M. H. Mittleman, J.Phys.B **9**, 2561 (1976). (b) The papers by J. Gersten and M. Mittleman [Phys.Rev.A **10**, 74 (1974); **11**, 1103 (1975)], are not related to adiabatic stabilization, as they are specifically dealing with low frequencies. They lead to a different dependence of Γ on I and ω than our equations [e.g., Eq.(16) for circular polarization].

23. K. Kulander Phys.Rev.A **35**, 445 (1987); **36**, 2726 (1987); **38**, 778 (1988); see also [1], p.247.

24. P. DeVries, J.Opt.Soc.Am.B **7**, 517 (1990).

25. K.J. LaGattuta, J.Opt.Soc.Am.B **7**, 639 (1990); Phys.Rev.A **43**, 5157 (1991).

26. M.S. Pindzola, G.J. Bottrell, and C. Bottcher, J.Opt.Soc.B **7**, 659 (1990).

27. M. Pont, D. Proulx, and R. Shakeshaft, Phys.Rev.A **44**, 4486 (1991).

28. X. Tang, H. Rudolph, and P. Lambropoulos, Phys.Rev.Lett. **65**, 3269 (1990).

29. M. Gajda, B. Piraux, and K. Rzazewski, Phys.Rev.A **50**, 2528 (1994).

30. M. Horbatsch, (a) Phys.Rev.A **44**, R5346 (1991); (b) J.Phys.B **24**, 4919 (1991); (c) J.Phys.B **25**, 1745 (1992).

31. (a) J. Javanainen, J.H. Eberly, and Q.Su, Phys.Rev.A **38**, 3430 (1988). (b) Q. Su, J. H. Eberly, and J. Javanainen, Phys.Rev. Lett. **64**, 862 (1990); Q. Su and J.H. Eberly, J. Opt.Soc.Am.B **7**, 564 (1990). (c) C.K. Law, Q. Su, and J.H. Eberly, Phys.

Rev.A **44**, 7844 (1991). See also J.H. Eberly, R. Grobe, C. K. Law, and Q. Su, in [2a], p.301, and [4].

32. (a) V. Reed and K. Burnett, Phys.Rev.A **42**, 3152 (1990). (b) V. C. Reed, P. L. Knight, and K. Burnett, Phys.Rev.Lett. **6**, 1415 (1991). (c) R.M.A. Vivirito and P.L. Knight, J.Phys.B **28**, 4357 (1995).

33. Adiabatic stabilization: (a) Ground state of H: M. Pont and M. Gavrila, Phys.Rev.Lett. **65**, 2362 (1990); presented at the SILAP I conference, Rochester, NY, June 1989. (b) Rydberg states: R. J. Vos and M. Gavrila, Phys.Rev.Lett. **68**, 170 (1992).

34. Adiabatic stabilization calculations for the ground state of H, according to method: (a) Sturmian: M. Dörr, R. M. Potvliege, D. Proulx, and R. Shakeshaft, Phys. Rev.A **43**, 3729 (1991). (b) Close-coupling: P. Marte and P. Zoller, see [17] ; L. Dimou and F.H. Faisal, Phys.Rev.A **46**, 4442 (1992); see also [16b]. (c) R-matrix: M. Dörr, P.G. Burke, C. J. Joachain, C. J. Noble, J. Purvis, and M. Terao-Dunseath, J.Phys.B **26**, L275 (1993).

35. (a) K.C. Kulander, K.J. Schafer, and J.L. Krause, Phys.Rev.Lett. **66**, 2601 (1991); (b) see also [2a], p.247, and [4].

36. (a) M.P. de Boer, J.H. Hoogenraad, R.B. Vrijen, R.C. Constantinescu, L. D. Noordam, and H.G. Muller, Phys.Rev.Lett. **71**, 3263 (1993); Phys.Rev.A **50**, 4085 (1994); (b) N. J. van Druten, R. Constantinescu, J. M. Schins, H. Nieuwenhuize, and H.G. Muller, Phys.Rev.A **55**, 622 (1997).

37. The condition $\alpha_o^2\,\omega >> 1$, mentioned in [20a] and reproduced by others (e.g., in [2b]), has proven to be superfluous.

38. M. Gavrila and J. Shertzer, Phys. Rev.A **53**, 3431 (1996).

39. The methods applied were: diagonalization in a multicenter Gaussian basis set [40]; two different finite element programs[33b], [38]; diagonalization in a Slater-type basis set in spheroidal coordinates [50].

40. M. Pont, N. Walet, and M. Gavrila, Phys.Rev.A **41**, 477 (1990).

41. M. Marinescu and M. Gavrila, Phys.Rev.A **53**, 2513 (1996).

42. From now on we shall be using atomic units, unless otherwise specified.

43. Note that the stabilization branches of the curves have barely perceptible undulations. However, stabilization calculations done for 1D atomic models with potentials of short-range [G. Yao and Shih-I Chu, Phys.Rev.A **45**, 6735 (1992)] , or long-range [M. Marinescu and M. Gavrila, Phys.Rev.A **53**, 2513 (1996)], display a quite prominent superimposed oscillatory behavior. Similar oscillations appear also for the ionization probabilities of 1D models calculated from WPD, see Su et al.[56]. It was concluded, however, by Marinescu and Gavrila that this is a specific feature of 1D models.

44. Moon-Gu Baik, M. Pont, and R. Shakeshaft, PRA **51**, 3117 (1995).

45. R.M. Potvliege and P.H.G. Smith, Phys.Rev.A.**48**, R46 (1993).

46. A. Scrinzi, N. Elander, and B. Piraux, Phys.Rev.A **48**, R2527 (1993).

47. A. Buchleitner and D. Delande, Phys.Rev.Lett. **71**, 3633 (1993).

48. L. Dimou and F.H. Faisal, Phys.Rev.A **49**, 4564 (1994); see also Faisal et al. [17b].

49. M. H. Mittleman, Phys.Rev.A **42**, 5645 (1990).

50. H.G. Muller and M. Gavrila, Phys.Rev.Lett. **71**, 1693 (1993).

51. E.van Duijn, M. Gavrila, and H.G. Muller, Phys.Rev.Lett. **77**, 3759 (1996).

52. The correct way of deriving the ionization probability $P_E(\tau)$ from the wave function ψ in the oscillating frame [solution of Eq(3)], was given by Vivirito and Knight, [32c].

53. J.C. Wells and M. Gavrila, in preparation.

54. Grobe and M.V. Fedorov, Phys.Rev.Lett. **68**, 2592 (1992); J.Phys.B **26**, 1181 (1993).

55. K. Sonnenmoser, J.Phys.B **26**, 457 (1993).

56. Q.Su, B.P. Irving, C.W. Johnson, and J.H. Eberly, J.Phys.B **29**, 5755 (1996).

57. T. Ménis, R. Taïeb, V. Véniard, and A. Maquet, J.Phys.B **25**, L263 (1992).

58. Although the global rate has the shape of an adiabatic rate curve (e.g., our Fig.4), it should be remembered, however, that it represents the decay of a coherent superposition of dressed states.

59. M. Pont and R. Shakeshaft, Phys.Rev.A **44**, R4110 (1991).

60. E. Huens and B. Piraux, Phys.Rev.A **47**, 1568 (1993).

61. We mention in this respect the free-electron laser, now under construction at Brookhaven National Laboratory, with projected photon energy in the range 10 - 25 eV, intensity in the range 1 - 100 a.u., and pulse duration ~ 5 fs.

LASER ASSISTED AND LASER PRODUCED SCATTERING

J H Sanderson and W R Newell

Department of Physics and Astronomy
University College London
Gower Street
London WC1E 6BT
United Kingdom

INTRODUCTION

The influence of radiation fields on collision processes and interactions is quite universal since all scattering processes take place in radiation fields even if it is only the thermal background. In the present article we will be concerned with radiation field intensities varying between 10^5 and 10^{15} W/cm^2 and how they either assist or produce scattering processes.

Electron-atom scattering in the presence of an external electromagnetic radiation field was initially studied by Goppert-Mayer (1931) in association with her theoretical work on two-photon excitation. However, it is only by combining the more recent developments of laser technology and the expertise of high-resolution electron scattering that an experimental study of the simultaneous interactions of photons and electrons with a discrete atom or molecule can be investigated. Such interactions are of fundamental interest in the understanding of three-body interactions and the relative coupling between radiation and particles.

In Free-Free (FF) scattering and Simultaneous Electron Photon Excitation (SEPE) electrons and photons are incident on the atomic and molecular target; these scattering processes are termed laser assisted. In laser assisted electron scattering we can in general distinguish between two types of reaction; those in which the laser is resonant or is non-resonant with a stationary state in the target atom or molecule. In the laser resonant case the projectile electron scatters, either elastically, super elastically or inelastically from a stationary excited state whereas in the non-resonant case the projectile electron and atomic target are dressed by the laser field through virtual absorption of radiation quanta.

In the resonant case laser-assisted electron scattering has, in the experimental domain, been restricted to super-elastic electron scattering of sodium and barium which have been pumped to their first excited levels using low-power turnable dye lasers (see Standage, 1995). This approach has been extended in recent work by Johnstone et al (1993) to carbon dioxide in which super elastic electron scattering from thermally excited vibrational levels are studied. Hammond (1996) has used an electron beam to excite discrete states in rare gases followed by laser excitation to Rydberg levels.

In the non-resonant case we can measure the transfer of photons to and from electrons while they are undergoing an elastic collision, described by the equation

$$e(E_i) + mh\nu + ATOM \rightarrow e(E_i \pm nh\nu) + (m \mp n)h\nu + ATOM \tag{1}$$

In this case the atom is not excited by the projectile nor by the photons but the electron does absorb or emit n photons. In this type of scattering the atom acts only as a third body to conserve the energy and momentum in the reaction. These Free-Free or Inverse and Stimulated Bremsstrahlung transitions have been reported by several authors (e.g. Weingartshofer et al 1979, 1983) and require high-power laser intensities of $10^8 \rightarrow 10^{12}$ W/cm^2 in order to achieve a multi-photon ($n > 1$) transfer to the scattered electrons. Single-photon transfer to the scattered electrons is achieved with the lower laser intensity of $10^4 \rightarrow 10^5$ W/cm^2. Andrick and Langhans, (1976), in a pioneering experiment demonstrated this process for low energy (11 eV) electron-argon scattering in the field of a continuous wave (cw) 50W CO_2 laser. The relatively large laser wavelength used in this work only permitted the dressing of the projectile electron and not the target atom.

In free-free scattering, equation (1), the atom plays only a passive role but there exists an excitation process in which the incident electron (E_i) combines with a photon ($h\nu$) to cause an excitation of the atom called Simultaneous Electron Photon Excitation (SEPE). This particular reaction in helium ($1^1S \rightarrow 2^3S$) is described by the equation

$$e(E_i) + mh\nu + He(1^1S) \rightarrow e(E_j = E_i - \Delta E + h\nu) + (m - 1)h\nu + He(2^2S) \tag{2}$$

in which the excitation of the 2^3S stationary state in helium, which lies at ΔE above the 1^1S ground state, is accomplished by the absorption of one quantum of radiation $h\nu$ from a laser field combined with a simultaneous inelastic electron scattering in which the electron provides the energy decrement (ΔE - $h\nu$) required to excite the 2^3S state. Since no stationary state exists between the 1^1S and 2^3S levels this excitation must proceed by a virtual interaction.

The effect of high intensity (~10^{14} W/cm^2) laser radiation on atoms and molecules has been well demonstrated in the area of multiphoton ionisation (see for example Agostini et al, 1989) with the uncovering of such phenomena as above threshold ionisation (ATI) and the coulomb explosion (Codling et al, 1989). In molecular dissociation novel phenomena such as bond softening and molecular alignment have been recently reported; these effects are laser produced.

In the molecule, H_2, several different path ways are possible to produce H^+ and H_2^+. These can be described by the equation

$$mh\nu + H_2 \rightarrow \underset{\substack{\downarrow \\ H+H^+}}{H_2^+} + e + (m-n)h\nu \tag{3}$$

in which the molecule is first ionized into a vibrational state of H_2^+ with the absorption of n photons followed by dissociation of the H_2^+ either by bond softening or the absorption of further photons. The production of the (H, H^+) pair by a direct process appears unlikely. In molecules a process of above threshold dissociation (ATD) in which groups of higher energy protons are produced is analogous to ATI in atoms.

In the present article we will confine our discussion to an up-date and review of the experimental progress of free-free scattering, simultaneous electron photon excitation and multiphoton dissociative ionization. In the case of MPDI only the work in H_2 will be discussed.

OUTLINE OF THEORY

The general problem of the interaction of an electron, atom and laser field can be written symbolically as

$$| P_f A_f \nu_f > = S | P_i A_i \nu_i > \tag{4}$$

where P_i A_i and ν_i represent the initial states of the electron, atom and photon respectively, S is the operator for the reaction and $P_f A_f$ and ν_f are the final states of the competing particles between which photons are virtually exchanged.

The electron will have a function form, i.e. a plane wave, outside the laser beam but a different form inside the laser beam and the electron will also 'scatter' off the pondermotive potential U_p produced by the electric field in the focused laser. However, if we neglect the pondermotive potential we can easily consider how the electron state is modified when the electron is inside the laser field.

The Schrödinger equation for an electron moving in a linearly polarised laser potential $A(r,t)$ is

$$i\hbar \frac{\partial \chi_q}{\partial t} = \frac{\hbar^2}{2m}\left[P + \frac{e}{c}A(r,t)\right]^2 \chi_q(r,t) \tag{5}$$

where

$$A(r,t) = \frac{e}{w} E(r) \cos(k.r - wt) \tag{6}$$

In the situation where laser wavelength is >> electron wavelength (i.e. $k >> q$), which normally applies, then the amplitude $E(r)$ of the radiation field will change slowly over λ ($=2\pi/q$) giving (Low, 1958; Kroll and Watson, 1973)

$$\lambda \frac{dE}{dr} << E \tag{7}$$

Then the solution of the Schrödinger equation is given by

$$\chi_q = \exp i\left[q.r - \frac{\varepsilon_q t}{h} - \frac{e}{m\omega^2} q.E(r)\sin(\omega t - k.r)\right] \tag{8}$$

which is the Volkov solution and shows how the radiation field dresses the electron wave with virtual photons. This is shown schematically in figure 1 for zero, first and second order dressings. In addition the radiation field can also dress the target atom or molecule. A direct perturbative picture of the coupling between the radiation field, (considering only n = 1) electron and atom is given by the four Feynman diagrams in figure 2 where each diagram represents one particular scattering amplitude. The single line, ↑, represents the electron and the double line, ⇑, denotes the atom. Time runs vertically in the figure and -

x- denotes the electron-atom interaction with ~ denoting the photon interaction. The role of these four Feynman diagrams in free-free scattering has been studied by Dubois et al (1985) in atomic hydrogen while the SEPE cross-section in atomic hydrogen and helium have been calculated by Jetzke et al (1984, 1987), and in helium by Fainstein et al (1995).

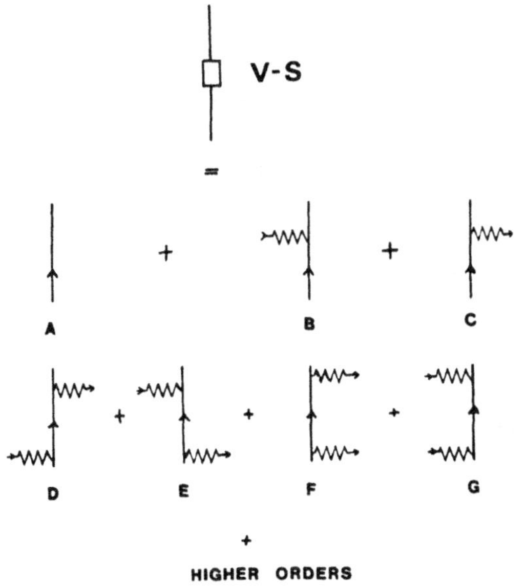

Figure 1. Volkov solution (v-s) of Schrödinger equation for the interaction of EM wave (~) and electron (↑). The solution consists of an infinite number of amplitudes given in increasing order of photon coupling, i.e. A zero order, B and C first order, DEF and G second order etc.

The interaction of electrons and photons with atomic systems has been discussed by Kroll and Watson (1973) who derived the formula

$$\frac{d\sigma}{d\Omega}(qq'n) = \frac{q'}{q} J_n^2(\Gamma) \frac{d\sigma}{d\Omega} \text{ (elastic)} \tag{9}$$

for the transfer of n photons while the electron momentum changes from q to q'. The argument of the Bessel function is given by

$$\Gamma = 1.944\ 10^{-12}\ \lambda^4\ F\ E_i\ (\varepsilon \bullet \Delta q/2q)^2 \tag{10}$$

where λ is the laser wavelength in μm, F is the laser intensity in W/cm^2 and E_i is the incident electron energy in eV. The physics of the dynamical scattering is described by the vector product of the laser polarisation direction and the momentum transfer of the scattered electron. It is clear from equation 9 that the cross section $d\sigma(qq'n)/d\Omega$ is a simple product of a radiation term represented by the Bessel function and a scattering term $d\sigma$(elastic)$/d\Omega$ within the soft photon approximation $h\nu < E$. However, if the elastic scattering cross section is determined using the first order Born approximation then the radiation term is correct to all orders in the radiation field whereas if the elastic cross section is determined to all orders the radiation term will be confined to first order. This equation can be generalised to include inelastic scattering events (Newell 1992 and references therein) and a completely non perturbative solution to these problems has been given by Burke et al (1991) in which the R-matrix method is coupled with Floquet theory.

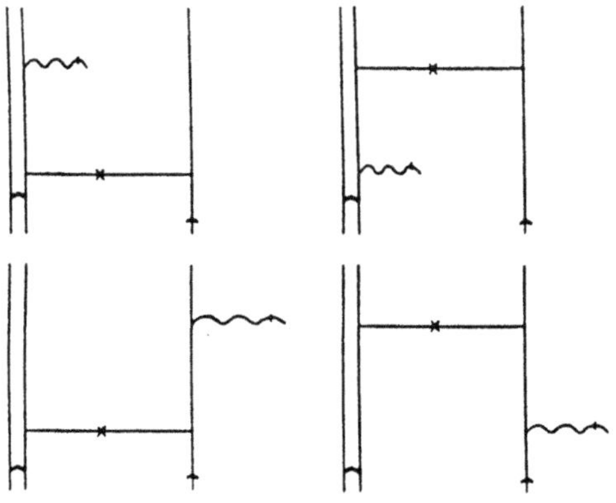

Figure 2 First order Feynman diagrams for SEPE

FREE-FREE SCATTERING

All experimental work to date on FF scattering has been done with CO_2 laser radiation at 10.6 μm (~0.117 eV). This requires the use of high resolution electron scattering techniques to enable primary electrons of energy E to be resolved from the FF scattered electrons E ± *nh*ν. The pioneering work of Andrick and Langhans (1976) employed a 50 W CW laser with field intensities of 10^4-10^5 W/cm^2 which resulted in first order (n = 1) FF scattering. Similar work was reported by Curry et al (1983). Second (n = 2) and higher order (n > 2) FF scattering requires the use of pulsed CO_2 lasers with powers of ~10^8 W/cm^2.

In all cases the atom is not excited by the projectile nor by the photons but the electron does absorb or emit *n* photons. In this type of scattering the atom acts only as a third body to conserve the energy and momentum in the reaction. These high order free-free or Inverse and Stimulated Bremmstrahlung transitions have been reported by several authors (e.g. Weingertshofer et al 1979) and require high-power laser intensities of $10^8 \rightarrow 10^{12}$ W/cm^2 in order to achieve a multi-photon (n>1) transfer to the scattered electron.

A typical multiphoton exchange free-free electron spectrum in helium is given in figure 3 for the scattering of 11 eV electrons at θ = 153°. When the laser beam is off the elastic peak only (spectrum B) is obtained whereas with the laser beam on (spectrum A) side peaks at one photon energy spacings (hν = 0.117 eV) are also obtained. These free-free peaks result from the scattered electron either absorbing or emitting photons; in this case a maximum of three photons is exchanged although multiple photon exchanges of n=11 have been reported. (Wallbank and Holmes 1994). The Kroll-Watson formula, equation 9, explains quite well the observed spectra obtained at large scattering angles and in addition a sum rule (Weingartshofer and Jung 1984) can be used to show that the total signal intensity in the free-free peaks is equal to the reduction of signal intensity in the elastic feature. In addition the dynamical dependence of the momentum transfer vector, Δq, and the laser polarisation vector, ε, has been well demonstrated experimentally; when ε is parallel to Δq the maximum free-free signal is observed whereas when ε is perpendicular to Δq the free-free signal falls to almost zero. However it should be noted that the Kroll-Watson formulae predicts an exact zero in the free-free cross section when Δq is perpendicular to ε.

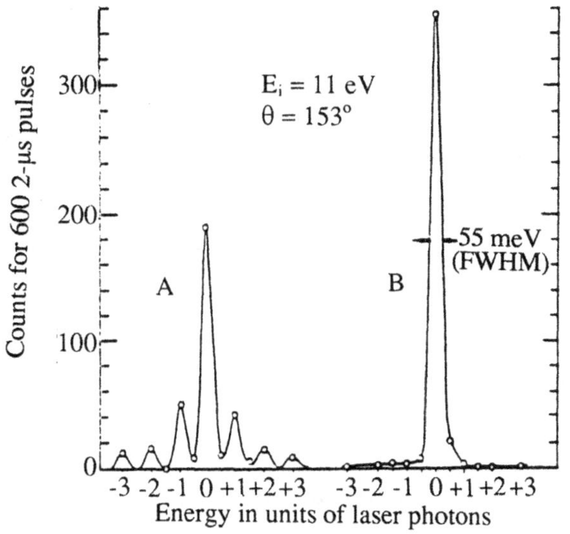

Figure 3 Multiphoton exchange in free-free electron scattering in helium (see text for discussion)

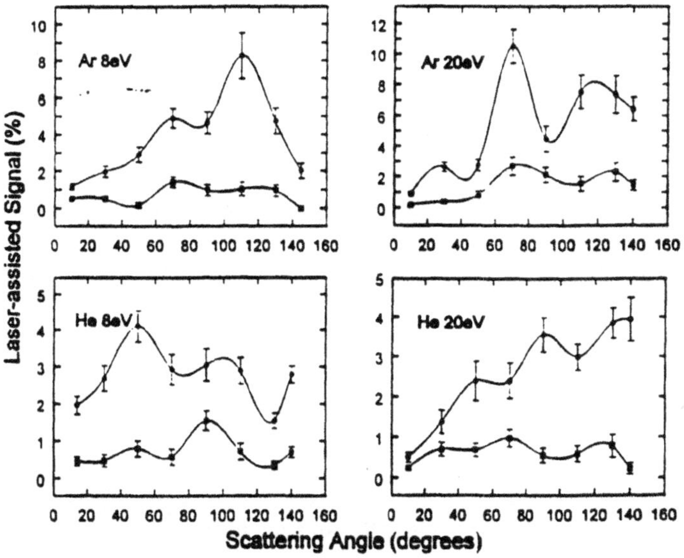

Figure 4 Free-free scattering in Argon and Helium

Recent work by Wallbank and Holmes (1994) has reported free-free cross sections in argon and helium which are in serious disagreement with the predictions of the Kroll-Watson formula. See also the disucssions by Wallbank (1995). Free-Free scattering cross sections measured at small angle scattering ($\theta = 9°$ to $30°$) with the laser polarisation aligned along the primary electron beam direction have yielded values of the free-free differential cross sections of 1% (for n=1), 0.4% (for n=2) and 0.2% (for n = 3), see figure 4. In this case the laser assisted free-free cross section is expressed as a percentage of the field free elastic cross section. Here values are at variance to that given by the Kroll-Watson formula which are 10^{-2}, 10^{-6} and 10^{-11} for the n = 1, 2 and 3 photon processes respectively. This surprising result is further compounded when the data obtained with the laser polarisation aligned perpendicular to Δq yields a laser free-free differential cross section of ~2% in argon and helium for the scattering of 8 eV electrons at $\theta = 10°$ (Wallbank and Holmes 1994). Recently Geltman (1995) has considered the role of atomic polarisation in the laser field to see what influence it has on the Kroll-Watson theory. The maximum effect that polarisation has would only increase the Kroll-Watson value by ~0.1 in the case of small angle scattering. It is however possible that some experimental effect associated with the purity of the laser polarisation or multiple scattering could be present. Consideration of the role of the laser-atom interaction in relation to the laser-electron interaction (Byron et al 1984, 1987) has been shown to be negligibly small (Rabadan et al 1994) with the present laser intensities.

SIMULTANEOUS ELECTRON PHOTON EXCITATION

In free-free scattering the atom plays only a passive role but there exists an excitation process in which the incident electron (E_i) combines with a photon ($h\nu$) to cause an excitation of the atom, Simultaneous Electron Photon Excitation (SEPE). This particular reaction in helium ($1^1S \rightarrow 2^3S$) is described by the equation

$$e(E_i) + nh\nu + He(1^1S) \rightarrow e(E_j = E_i - \Delta E + h\nu) + (n-1)h\nu + He(2^3S)$$

in which the excitation of the 2^3S stationary state in helium, which lies at ΔE above the 1^1S ground state, is accomplished by the absorption of one quantum of radiation $h\nu$ from a laser field combined with a simultaneous inelastic electron scattering in which the electron provides the energy decrement ($\Delta E - h\nu$) required to excite the 2^3S state. Since no stationary state exists between the 1^1S and 2^3S levels this excitation must proceed by a virtual interaction.

The different possible combinations of electron and photon energies are given in figure 5. In cases B and D the electron can be superelastically scattered whereas in case A it will lose energy in exciting the 2^3S state. Only in case C, in which neither the electron nor the photon has sufficient energy to separately excite the 2^3S state, does it require the joint scattering process to achieve excitation.

The main experimental requirement in this work is that the energy spread ($\Delta\varepsilon$) of the electron scattering apparatus is much less than the photon energy, $\Delta\varepsilon \ll h\nu$, in order that the photon transfer effects are resolved. The experimental procedures and requirements for different laser wavelengths and for both pulsed and c.w. laser systems have been detailed by several authors (Mason and Newell 1987a,b, 1989, Wallbank et al 1988; Luan et al 1991). In all the experimental work discussed in this paper the signature of the excitation is the production of the metastable 2^3S state. These metastable atoms have sufficient internal energy (19.8 eV) to make them easily detectable using a channel electron multiplier. The time-of-flight spectra of the metastable atoms are recorded with and without the laser beam present with the difference in the metastable flux being a monitor of the SEPE cross sections.

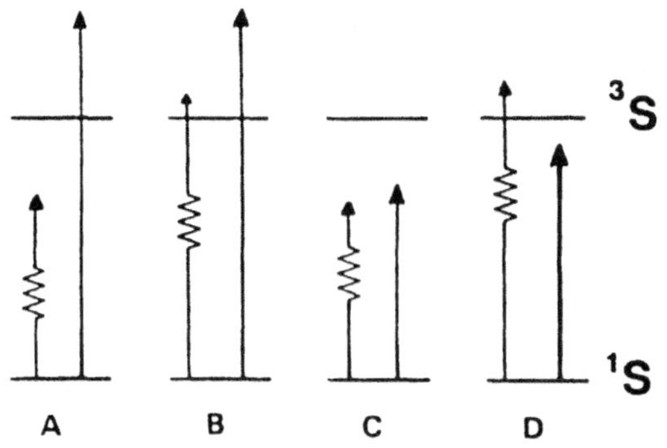

Figure 5 Different combinations of electron (→) and photon energies (~~) in exciting the (2^3S) level of helium.

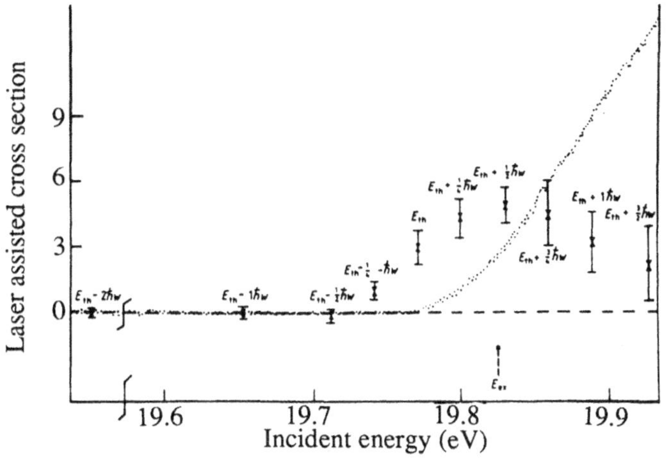

Figure 6 SEPE cross section in helium (•) in c.w. CO_2 laser field and the direct cross section (...) for the 2^3S level. E_{ex} is the excitation threshold of the 2^3S level and the E_{th} is the energy at which the direct signal is first detected.

The simultaneous electron photon excitation cross section of the triplet spin exchanged level in He (2^3S) is given in figure 6 (Mason and Newell 1989) which shows the laser assisted cross section and the direct non laser assisted cross sections of plane polarised laser radiation. In this work only the one photon transfer process is detected as a CW laser (CO_2 300 W) was employed.

In total SEPE experiments have been reported by three groups of workers; Mason and Newell, Wallbank et al and Luan et al. In essence the same technique was employed by all the groups, namely TOF detection of the excited helium atoms which recoil through an angle of ~17° from the primary gas beam direction. The main difference between the experiments is in the laser intensities and wavelengths employed, with the pulsed lasers measuring multiphoton effects.

Wallbank et al used a pulsed CO_2 laser (10^8 W/cm^2) whereas Luan et al used a pulsed YAG system (10^{10} W/cm^2). The SEPE cross section measured by Luan et al is given in figure 7. Theory predicts a λ^4 dependence of the SEPE cross section on the wavelength of the laser radiation. Consequently there should be a reduction of 10^4 when performing the experiments with YAG laser radiation as opposed to CO_2 laser radiation. The larger than expected signal in the experiment of Luan et al can be understood when it is realised that the negative ion resonance at 19.37 eV is within one photon energy (1.17 eV) of the threshold of the 2^3S state (19.8 eV); this is not the case with CO_2 laser radiation (0.117 eV). The life time of the resonance (~10^{-13}s) effectively increases the classical collision time (~10^{-17}s) by four orders of magnitude consequently enhancing the apparent SEPE cross section. The turn over in the SEPE cross section seen by Mason and Newell and Wallbank et al is also apparent in the work of Luan et al which was done with a higher photon energy (1.17 eV). In addition an unpublished experiment by Khakoo et al (1982) also detected evidence of SEPE in helium using YAG laser radiation.

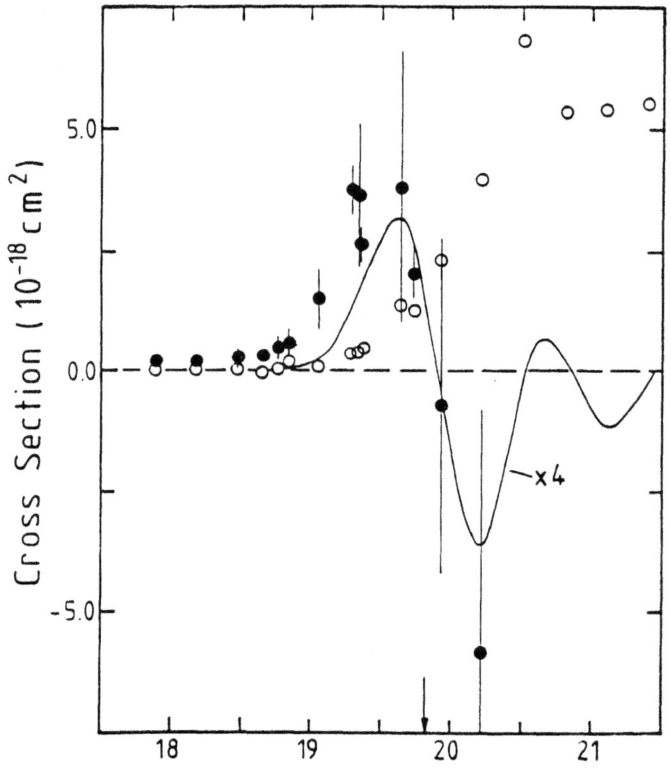

Figure 7 SEPE cross section in helium (•) in a pulsed Nd:YAG laser field and the direct cross section (•) for the 2^3S level. See Luan et al for discussion of theory curve (—).

A review of the calculations of the SEPE process has been given by Maquet (1991). Using an extension of the Kroll-Watson formulae and the field-free total electron impact excitation cross-section of the He(2^3S) given by Fon et al (1981), Geltman and Maquet (1989), have calculated the laser-assisted excitation cross-sections for the conditions prevailing in both CO_2 laser experiments. In the case of Wallbank et al the calculated SEPE cross-section is lower than the direct cross-section above $E_{th} + 1h\nu$ and this derives from the large argument in the Bessel function produced by the high-laser intensity. In contrast, the lower laser power used by Mason and Newell only invokes the use of the zero-order Bessel function and the argument is such that no zeros are accessed. Recent calculations on the differential SEPE cross sections in helium have been reported by Fainstein et al (1995). In this case a 29-state R-matrix method was used for the laser off cross-sections (Fon et al, 1981) with a modified Kroll-Watson formulae for the laser on cross sections. More detailed experiments are required to test these new calculations

MULTIPHOTON DISSOCIATION

The process of FF scattering and SEPE are laser assisted scattering mechanisms where as multiphoton dissociation (MPD) is a laser produced process. The fragmentation of polyatomic molecules under laser impact has been pioneered by Colding and Frasinski (see Frasinski et al 1987), and others (see for example Bandrauk 1994 and references therein). In this present paper we will focus on the MPD of the H_2 molecule using picosecond and femtosecond laser pulses.

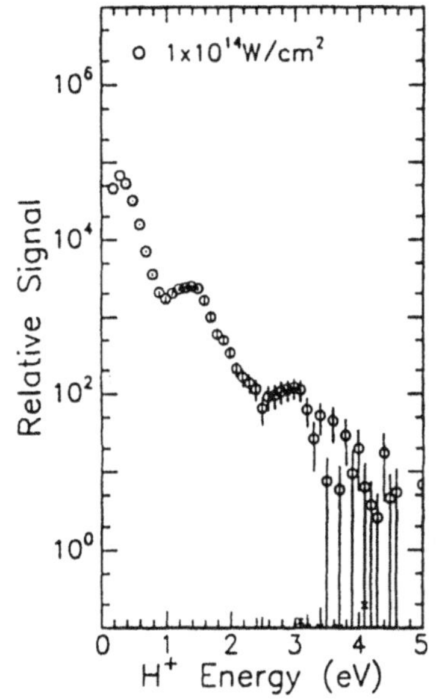

Figure 8 Proton spectrum from MPDI of H_2 using 70 ps pulses at 532 nm (Zavriyev et al)

It is generally accepted that the multiphoton dissociative ionisation (MPDI) of H_2 proceeds by the path in which H_2 is first ionised to H_2^+, in a distribution of excited vibrational states, which then absorbs further photons and dissociates. The H_2^+ (1sσg) is strongly optically coupled to the 2pσ$_\mu$ dissociative state; any odd number of photons will effect a strong coupling. Orientation of the H_2^+ ion along the laser polarisation direction during the multiphoton absorption and bond softening will also occur in the laser beam, both effects influencing the proton spectrum. Using a mode locked Nd:YAG laser (70 ps, 25 mJ) at 532 nm focussed to give an intensity of 10^{14} W/cm^2, Zavriyev et al (1990) measured the proton spectrum, figure 8. Sanderson et al (1996) using a similar laser system (35 ps, 40 mJ) at 532 nm with a focused intensity of 5.10^{14} W/cm^2, measured the proton spectrum, figure 9. The spectra are similar in that both span the same proton energy range of 0.2 eV to 4 eV. However the structure observed is quite different. Zavriyev et al observed peaks seperated by half a photon energy (1.115 eV) with the intensity of each peak decreasing by an order of magnitude. In the work of Sanderson et al there is evidence of structure but the intensity of the proton signal over the range up to 3 eV falls by only a factor of 5 compared with 100 in the work of Zavriyev et al.

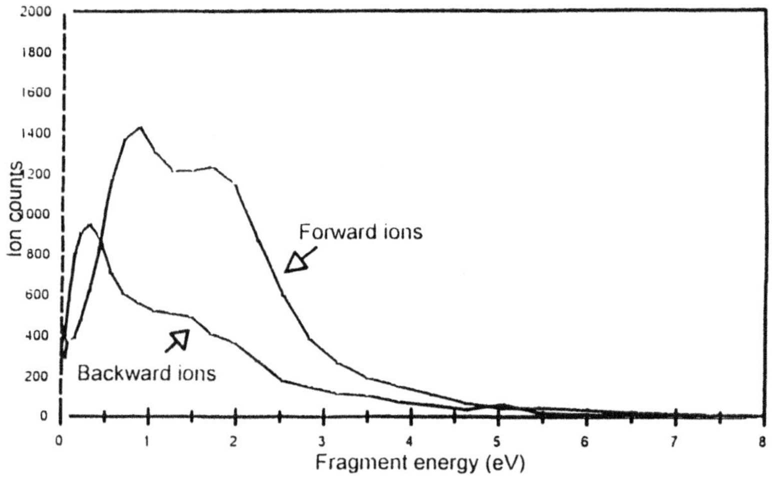

Figure 9 Proton spectrum from MPDI of H_2 using 35 ps pulses at 532 nm (Sanderson et al)

Sanderson et al repeated the same experiment using 50 fs pulses at 760 nm focused to 10^{15} W/cm^2 which gave the forward proton spectrum shown in figure 10. Clearly this is quite different from the spectra obtained with ps pulses with a proton peak centred around 0.5 eV and a broad peak extending over the region 1.0 to 5.0 eV. The low energy peak probably arises from above threshold dissociation or bond softening. It is also important to note that when the pulse duration decreases from ps to fs, a factor of 10^3, the rotational motion of H_2^+ (~500 fs) will be frozen. The vibrational motion of H_2^+(~ 25 fs) will not be completely frozen. Nevertheless in a 50 fs pulse the pulse intensity will be changing rapidly during one vibrational period and this will effect the ionisation process. Additionally in high field intensities the stabilisation of light induced bound states have been predicted by Fedorov et al (1975) and Bandrauk and Sink (1981). Some experimental evidence for a light induced boundd state of H_2^+ has been reported by Zavriyev et al (1993) using 100 fs pulses at 769 nm which shows structure similar to that detected by Sanderson et al. If this model is correct then these faster protons result from photoionisation of the light induced trapped vibrational state of H_2^+. However Posthumus et al (1996) and Codling and Fransinski (1993) have also measured the proton spectrum from H_2 using 400 fs and 55 fs pulses at 750 nm and have obtained spectra similar to Sanderson et al but with no structure on the high energy proton peak. They suggest that vibrational trapping is not responsible for the high energy proton spectrum. A fuller discussion is given in the review by Guisti-Suzor et al (1995).

OUTLOOK

The general area of laser assisted and laser produced collision processes and of fast processes in molecular physics is one of unfolding interest. New work will concentrate on polarizable targets, excited states and ionic species with vibrational and rotational control of reaction dynamics in the biological sciences.

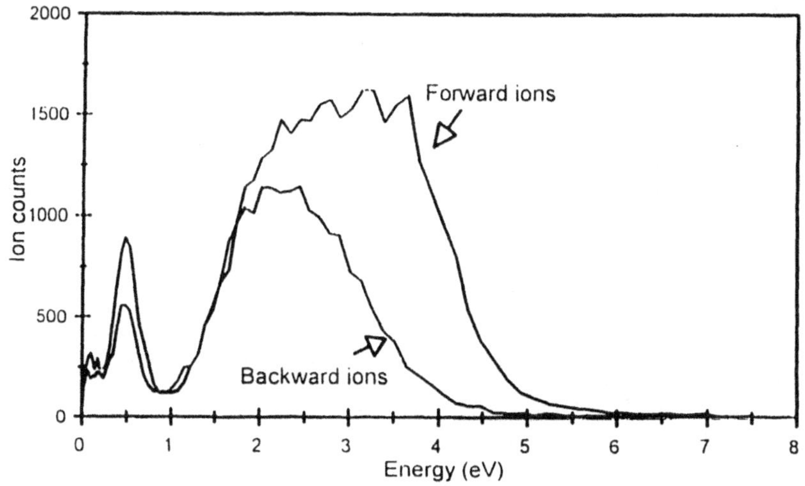

Figure 10 Proton spectrum from MPDI of H_2 using 50 fs pulses at 710 nm (Sanderson et al)

ACKNOWLEDGEMENTS

The authors would like to acknowledge the facilities in the Department of Pure and Applied Physics, QUB, which were used in the preparation of this paper.

References

Agostini, P., Antonetti, A., Breger, P., Crance, M., Migus, A., Muller, G. and Petite, G. J.Phys.B. 22 1971 (1989).

Andrick, D. and Langhans, L., J. Phys B, 9, L459 (1976).

Bandrauk, A. D., (editor). Molecules in Fields, Marcel Dekker Inc. (1994).

Bandrauk, A. D., and Sink, M. L., J. Chem. Phys., 74, 1110, (1981); Chem. Phys. Lett, 57, 569, (1978).

Burke, P. G., Francken, P., and Joachain, C. J., J. Phys. B. 24, 761, (1991).

Byron, F. W. Jr., Francken, P., and Joachain, C. J., J. Phys. B., At. Mol. Phys. 20, 5487 (1987).

Byron, F. W., Jr., and Joachain, C. J., J. Phys. B: At. Mol. Phys. 17, L295-L301, (1984).

Codling, K. and Fransinski, L. J., J. Phys. B. 783, (1993).

Codling, K., Fransinski, L. J., and Hatherly, P. A., J. Phys. B 22, L321, (1989).

Curry, P., Newell, W. R. and Smith, A., ICPEAC, XIII, Berlin, p.682, eds. J. Eichler, W. Fritsch, I. V. Hertel, N. Stolterfoht and V. Wille. (1983).

Dubois, A., Maquet, A., and Jetzke, S., Phys. Rev. A34, 1888, (1985).

Fainstein, P.D., Maquet, A., and Fon, W.C., J. Phys. B. 28, 2723, (1995).

Federov, M. V., Kudrevotova, O. V., Makarov, V. P., and Samokhin, A. A., Opt Comm, 13, 299, (1975).
Fon, W C., Berrington, K. A., Burke, P. G., and Kingston, A. E., J. Phys. B. 14, 2921, (1981).
Frasinski, L. J., Colding, K., Hatherly, P. A., Barr, J., Ross, I. N, Toner, W. T., Phys. Rev. Lett. 58, 2424 (1987).
Geltman, S. Phys.Rev.A. 51, R34, (1995).
Geltman, S. and Maquet, A., J. Phys. B 22, L419, (1989).
Goppert-Mayer, M, Ann Phys Lpg. 9:273, (1931).
Guisti-Suzor, A., Mies, F. H., Di Mavro, L. F., Charron, E. and Yang, B., J. Phys. B. 28, 309 (1995).
Hammond, P., private communication (1996).
Jetzke, S., Broad, J. and Maquet, A. J.Phys.B. 20 2887 (1987).
Jetzke, S., Faisal, F. H. M., Hippler, R. and Lutz, H. O., Z. Phys. A315, 271, (1984).
Johnstone, W. Mason, N. and Newell, W. R., J Phys., B. L147, (1993).
Khakoo, M., Newell, W. R., Toner, W. T., and Eason, R. W., Rutherford Appleton Laboratory Annual Report, Laser Facility Committee, P608, (1982).
Kroll, N. M., and Watson, K. M., Phys. Rev. A8, 804, (1973).
Low, F. E., Phys. Rev. 110, 974, (1958).
Luan, S. Hippler, R., and Lutz, H. O., J. Phys, B 24, 3241, (1991).
Maquet, A., ICPEAC, XVII, Brisbane, eds. W. R MacGillivray, I.E. McCarthy and M.C., Standage, Adam Hilger, Bristol. 327 (1991).
Mason, N. J., and Newell, W. R., J Phys, B 23, L179, (1990).
Mason, N. J., and Newell, W. R., J. Phys. B 20, 1357, (1987a).
Mason, N. J., and Newell, W. R., J. Phys. B 20, L323, (1987b).
Mason, N. J., and Newell, W. R., J. Phys. B. 22, 777, (1989).
Newell, W. R., Comm. Atm. Mol. Phys. 28 59 (1992).
Posthumus, J. H., Giles, A. J., Thompson, M. R. and Codling, K., to be published (1996).
Rabadán, I., Méndez, L., and Dickinson, A. S., J. Phys.B., 27, L535 (1994).
Sanderson, J., Thomas, R., Williams, I., Langley, A., and Shaikh, W., private communication (1996).
Standage, M., ICPEAC, XIX, Whistler,. eds L. J. Dube, J. B. A Mitchell,. J. W. McConkey,. and C. E. Brion, A.I.P., 173 (1995).
Wallbank, B. and Holmes, J. K., J. Phys. B. 27, 1221, (1994).
Wallbank, B., Holmes, J. K., LeBlanc, L., and Weingartshofer, A., Z. Phys. D10, 467, (1988).
Wallbank, B., ICPEAC XIX, Whistler. Eds L. J. Dube J. B. A. Mitchell,C. E. Brion, and J. W.McConkey, A.I.P. 189, (1995).
Weingartshofer, A., and Jung, C., "Multiphoton Free-Free Transitions,", in Multiphoton Ionization of Atoms, S. L Chin and P. Lambropoulos, Eds., Toronto: Academic Press, 155, (1984).
Weingartshofer, A., Clarke, E. M, Holmes, J. K., and Jung, C, Phys. Rev. 19, 2371, (1979).
Weingartshofer, A., Holmes, J. K., Sabbagh, J., and Chin, S. L., J. Phys. B. 16, 1805, (1983).
Zavriyev, A., Bucksbaum, P. H., Muller, H. G., and Schumacher, D. W., Phys. Rev. A 42, 5500, (1990).
Zavriyev, A., Buckshaum, P. H., Squier, J., and Saline, F. Phys. Rev. Lett., 70, 1077, (1993).

R-MATRIX-FLOQUET THEORY OF MULTIPHOTON PROCESSES

Martin Dörr

Max-Born-Institut, D-12474 Berlin, and
Université Libre de Bruxelles

INTRODUCTION

Since its first formulation [1, 2], the R-matrix-Floquet (RMF) method has now become a useful tool in a wide variety of non-perturbative calculations of atomic multiphoton processes in intense laser fields. The R-matrix approach allows an *ab initio*, accurate description of atomic structure, while the Floquet ansatz constitutes a fully non-perturbative description of multiphoton processes. These processes can be classified into the categories laser-assisted scattering, ionization and detachment, and harmonic generation in a laser field. The R-matrix-Floquet theory has been successfully applied to the calculation of all of these processes. The development of the methods and the implementation of the RMF theory has relied heavily on previous work which used the R-matrix approach for perturbative calculations of photoionization for a wide variety of atomic and ionic species [3]. Other authors have performed nonperturbative Floquet calculations, however so far mostly for single-active-electron systems [4].

In section 1, the RMF theory is presented [2, 5]. The solutions adopted in the present calculations are discussed and some of the present and future methodological and algorithmic developments are mentioned. In section 2, the broad range of current results is summarized in an attempt to list all applications of the RMF theory to-date [6]. The method is geared towards precise total and partial cross-sections in the regime where non-perturbative dynamics, including atomic resonances is important. Laser field intensities range from the perturbative (single- or multi-photon) to the atomic order of magnitude. The intensity of a linearly polarized laser field with peak electric field of $\mathcal{E}_0 = 1$ a.u. $= 5.21 \times 10^9$ V/cm is 3.51×10^{16} W/cm^2. Laser frequencies have so far been chosen of the order of magnitude of the relevant energy differences of the atomic system under study, such that in the perturbative limit multiphoton couplings of the order 1 to about 5 photons are relevant. At larger intensity, nonperturbative effects become apparent which evidently involve a large number of photons. For illustration, some results are presented.

In the third section the prospects and challenges for future work are discussed. On the one hand, these prospects result naturally from planned and imminent further applications and developments of the R-matrix-Floquet programs, while on the other hand, they are dictated by the current progress, both experimental and theoretical, in the field of multiphoton physics and chemistry.

1. THEORY

1.1 Floquet approach

The Floquet approach uses the key idea of identifying a short timescale, namely the laser field period, during which the laser field intensity does not change appreciably. Thus, for a definite intensity and frequency, the Floquet eigenstates constitute the generalization of the field-free (quasi-) stationary eigenstates of the atomic system, such as bound states or autoionizing states. The field intensity variation which is always present in a strong laser pulse must be included in a subsequent step, which will not be discussed here [8]. Atomic units are used throughout.

The time-dependent Schrödinger equation for an $N+1$-electron atomic system in a laser field reads

$$\mathrm{i}\,\frac{\partial}{\partial t}\Psi(\underline{\boldsymbol{X}},t) = [H_{\mathrm{atom}} + H_{\mathrm{int}}(t)]\,\Psi(\underline{\boldsymbol{X}},t). \tag{1}$$

Here $\underline{\boldsymbol{X}}$ denotes the spatial and spin coordinates of all $N+1$ electrons, $\{\boldsymbol{x}_1, ..., \boldsymbol{x}_{N+1}\}$, H_{atom} is the time-independent field-free atomic Hamiltonian and the electron-field interaction is

$$H_{\mathrm{int}}(t) = \frac{1}{c}\,\boldsymbol{A}(t)\cdot\boldsymbol{p} + \frac{N+1}{2c^2}\boldsymbol{A}^2(t), \tag{2}$$

where $\boldsymbol{p} = \sum_{i=1}^{N+1}\boldsymbol{p}_i$ is the sum of all electrons' momenta. The classical external laser field is assumed to be monochromatic, linearly polarized and spatially homogeneous. It is given by the vector potential

$$\boldsymbol{A}(t) = \hat{\boldsymbol{\epsilon}}A_0\cos\omega t, \tag{3}$$

which is related to the electric field by $\boldsymbol{E}(t) = -(1/c)(d\boldsymbol{A}/dt)$.

The Hamiltonians H_{atom} and H_{int} appearing in equation (1) are given explicitly in the coordinate representation by

$$H_{\mathrm{atom}} = -\sum_{i=1}^{N+1}\left[\frac{1}{2}\nabla_i^2 + \frac{Z}{r_i}\right] + \sum_{i<j}^{N+1}\frac{1}{|\mathbf{r}_i - \mathbf{r}_j|}, \tag{4}$$

$$H_{\mathrm{int}}(t) = -\sum_{i=1}^{N+1}\left[\frac{\mathrm{i}}{c}\boldsymbol{A}(t)\cdot\nabla_i - \frac{1}{2c^2}A^2(t)\right]. \tag{5}$$

The atomic Hamiltonian is written in the nonrelativistic approximation and an infinitely heavy nucleus is taken.

Since the driving field is periodic in time, in the Floquet ansatz the wavefunction is written as

$$\Psi(\underline{\boldsymbol{X}},t) = e^{-\mathrm{i}Et}\sum_{n=-\infty}^{\infty}\psi_n(\underline{\boldsymbol{X}})\,e^{-\mathrm{i}n\omega t}. \tag{6}$$

For the ionization or detachment problem the parameter E is the complex quasi-energy

$$E = E^{(0)} + \Delta - \mathrm{i}\Gamma/2, \tag{7}$$

with Γ denoting the ionization rate or width of the atomic system and Δ denoting the shift from the field-free value. For the scattering problem, E is the real scattering energy ($\Gamma = 0$). The Floquet ansatz, when inserted into the time-dependent Schrödinger equation, leads to a tri-diagonal system of time-independent coupled differential equations

$$(H_{\text{atom}} - E - n\omega)\psi_n + D_-\psi_{n+1} + D_+\psi_{n-1} = 0, \tag{8}$$

where the dipole-operator in the velocity form is

$$D_+ = \frac{A_0}{2c}\,\hat{\boldsymbol{\epsilon}} \cdot \boldsymbol{p} = D_-^\dagger. \tag{9}$$

Expanding the wavefunction Floquet components $\psi_n(\underline{\boldsymbol{X}})$ on a basis, which in our case involves a close-coupling approach and thus an expansion in angular momenta, leads to a matrix representation for the operators in equation (8). The resulting Hamiltonian matrix consists of Floquet blocks. In actual calculations, the expansion (6) must be truncated at finite n, $-n_{\text{emi}} \leq n \leq n_{\text{abs}}$. The field couples in principle all angular momenta, but this expansion (see below) must also be truncated. To ensure convergence of the results, tests must be done on their stability with an increase of the channels retained in the calculation. As the field becomes very strong, or the frequency becomes very low, many Floquet blocks are necessary for a converged calculation. This fact is currently setting a practical limit on these parameters, not allowing an exploration of for example the low-frequency ionization of the rare gases, which would require of the order of 8 or more photons to ionize.

1.2 Inner region

As the reader undoubtedly knows, the R-matrix approach [9] is based on the observation that, for a scattering process from a multielectron target, configuration space can be partitioned into an inner and and outer region. The *inner region* is defined by $r_i < a \;\forall\; i \in \{1, ..., N+1\}$, r_i being the i-th electron's radial coordinate and a the boundary of the inner region. In this region a quantum multiparticle interaction process must be solved. Outside of the sphere a there is only a single particle. The *outer region*, which will be discussed in the next subsection, is thus defined by one of the r_i being greater than a, while all the others are smaller. In a scattering context the residual N-electron system is called the target and the electron in the outer region is called the projectile.

The wavefunction in the inner region is expressed on a basis, just as for the usual R-matrix calculations, involving a set of optimized single-particle orbitals $\varphi_j(r)$ which have decayed exponentially practically to zero at $r = a$ and a set of basis functions $u_{il}^{\Lambda}(r)$ which extend up to $r = a$. The latter are sometimes called "continuum basis functions" since they represent the outgoing electron. The Floquet component n of eigenfunction k in the configuration interaction approximation is then

$$\psi_{kn}(\mathbf{X}_{N+1}) = \mathcal{A} \sum_{\Lambda il} \bar{\phi}_i^{\Lambda}(\boldsymbol{x}_1, ..., \boldsymbol{x}_N, \hat{\boldsymbol{r}}_{N+1}, \sigma_{N+1})\; r_{N+1}^{-1}\; u_{il}^{\Lambda}(r_{N+1})\; a_{ilkn}^{\Lambda}, \tag{10}$$

where $\mathcal{A}$ is the antisymmetrization operator and the a_{ilkn}^{Λ} are the expansion coefficients. The channel functions $\bar{\phi}_i^{\Lambda}(\boldsymbol{x}_1, ..., \boldsymbol{x}_N, \hat{\boldsymbol{r}}_{N+1}, \sigma_{N+1})$ are formed by coupling the configuration interaction N-electron 'target' wavefunctions $\phi_i(\boldsymbol{x}_1, ..., \boldsymbol{x}_N)$ with the spin-angle functions of the scattered or ejected electron (which depend on $\hat{\boldsymbol{r}}_{N+1}$ and σ_{N+1}) to give a state with quantum numbers $\Lambda \equiv \lambda LSM_LM_S\pi$. Here L is the total orbital angular momentum quantum number, S the total spin quantum number, M_L and M_S are the corresponding magnetic quantum numbers and π is the parity of the $(N+1)$-electron system. The quantity λ indicates the remaining quantum numbers required to completely define the channel Λ.

Therefore, before the RMF calculation proper is started, there is a first step which consists in finding the basis orbitals φ_j and in defining the combinations of these orbitals which make up the configurations to be retained in order to build the wavefunctions ϕ_i. There is unfortunately not a general algorithm available for this purpose, since for reasons of computational expense the basis should be kept as small as possible, and the basis must be able to represent all relevant physical states. Which states are important depends on the problem to be studied and thus there is no universally useful CI basis for a given atomic system. A lot of physical insight should thus go into the judicious construction of the basis [10].

In practice, quadratically integrable configurations $\chi_{i'}^{\Lambda'}(\underline{\boldsymbol{X}})$ are included in expansion (10) as an additional term $\sum_{\Lambda' i'} \chi_{i'}^{\Lambda'}(\underline{\boldsymbol{X}}) b_{i'kn}^{\Lambda'}$. These configurations are built out of the basis orbitals φ_j. It must be ensured that the other basis functions u_l are orthogonal to the orbitals φ_j employed in these purely bound $N+1$-electron configurations.

A general basis in the inner region can in principle be obtained using any complete set of basis functions. Such an approach appears feasible only for two-active electron systems outside a closed (inactive) shell and a promising basis is given by the B-splines [11]. Alternative pseudostate methods are presented in [41].

It is advantageous to solve the inner region problem in the length gauge rather than in the velocity gauge, since convergence of the length gauge wavefunction with the basis is generally much better. In particular, the convergence in the continuum basis is very slow in the velocity gauge and the usual way of improving the convergence by adding a diagonal field-free Buttle correction to the R-matrix does not work in this case [12]. Therefore equation (1) is transformed to the length gauge for all electrons [2].

The surface terms in the resulting eigenvalue problem in the inner region are removed by means of a Bloch operator and thus one obtains a hermitian eigenvalue problem defining a set of eigenvalues and eigenfunctions in the internal region. From the eigenvalues and their eigenvectors, the spectral decomposition of the R-matrix can be obtained [13], since the R-matrix $\mathbf{R}$ is just the resolvent of the Hamiltonian, with solution matrix $\mathbf{F}(r_{N+1})$:

$$\mathbf{F}(a) = \mathbf{R}(E) \left[r_{N+1} \frac{d\mathbf{F}}{dr_{N+1}} \right]_{r_{N+1}=a}, \tag{11}$$

evaluated at the boundary a. There are as many linearly independent solutions $F(r_{N+1})$, compatible with the physical boundary conditions at small r, as there are channels in the problem : the matrices $\mathbf{F}$ and $\mathbf{R}$ are therefore square. The R-matrix element ij is calculated as

$$R_{ij}(E) = \frac{1}{2a} \sum_k \frac{w_{ik}\, w_{jk}}{E_k - E}, \tag{12}$$

with the amplitudes on the surface $r_{N+1} = a$ defined as

$$w_{ik} = a \langle \bar{\phi}_i(\boldsymbol{x}_1, ..., \boldsymbol{x}_N, \hat{\boldsymbol{r}}_{N+1}, \sigma_{N+1}) | \psi_{kn} \rangle |_{r_{N+1}=a} . \tag{13}$$

The index k denotes the different solutions obtained from the diagonalization of the hermitian eigenvalue problem in the inner region. The R-matrix obtained with a finite number of basis functions u_{il}^{Λ} is improved by adding a Buttle correction to the diagonal elements. The boundary condition $u_{il}^{\Lambda}(r)^{-1}du_{il}^{\Lambda}(r)/dr = 0$ is imposed at $r = a$ for all l.

Current efforts to improve the computational methods include using matrix inversion rather than diagonalization to compute the R-matrix in a direct way [14]. The spectral form of equation (12) is then replaced by a linear equation, which must be solved to obtain the R-matrix. The advantage of the spectral form, equation (12), is that the R-matrix can be readily calculated for any E, but its big disadvantage is that an eigenvalue decomposition of a matrix must be solved for the inner region, and this matrix can become very large : since computational expense scales as the linear matrix dimension d^3, and since roundoff error grows with the number of operations, the largest basis that has been used in RMF calculations so far has been $d = 5000$ — at a huge expense [27]. The direct method by linear equation solving is less costly since there are less equations to solve (though it scales also as d^3) and it allows a parallel solution by iteration. Its disadvantage is that for every energy E the inversion problem must be solved.

In the outer region, which is the subject of the following subsection, it is important to compute the outer electron's wavefunction in the velocity gauge, since the length gauge's interaction term, $\boldsymbol{E}(t) \cdot \boldsymbol{r}$ would become too large for large r. Therefore, at $r = a$ the R-matrix must be transformed to the velocity gauge. This transformation is local, it is given by a matrix $\mathbf{A}$ for the wavefunction expressed as a vector in channel space. The corresponding transformation for the R-matrix is

$$\mathbf{R}_V = \mathbf{A}\left[\mathbf{R}_L^{-1} + \frac{1}{2}\mathbf{P}\right]^{-1}\mathbf{A}^{-1}. \tag{14}$$

The matrix $\mathbf{A}$ is obtained as the channel decomposition of the gauge transform function $\exp[i\boldsymbol{r} \cdot \boldsymbol{A}(t)]$ and thus a problem with the present approach is immediately apparent : for large $r = a$ or large fields the elements of $\mathbf{A}$ become large and the truncated matrix $\mathbf{A}$ is no longer orthogonal. Thus, in order to obtain sufficient accuracy for this transformation, many Floquet and angular channels must be included, which tends to increase the computational effort. One way to overcome this problem would be to compute the R-matrix directly in the velocity gauge in the inner region, for the continuum electron : this approach has to face the slow convergence discussed above.

1.3 Outer region

Since exchange effects between the outer electron and the N inner electrons can be neglected, in the outer region only a one-electron problem must be solved. The outer electron is subject to both the external laser field and to the multipole potential from the core $\mathbf{V}(r)$ which includes the nuclear Coulomb potential. In the following, we abbreviate the outer electron's radius r_{N+1} by r.

The coupled differential equations to be solved in the outer region can be written in matrix form as

$$\left[\frac{d^2}{dr^2} + \mathbf{P}\frac{d}{dr} - \mathbf{V}(r) + 2E\right]\mathbf{F}(r) = 0, \tag{15}$$

where $\mathbf{P}$ is an antihermitian matrix proportional to A_0 and $\mathbf{F}(r)$ is the solution matrix, all matrices being square.

These equations are solved at intermediate r, $a < r < a'$, by propagating the R-matrix using standard propagators [15]. The constant first-derivative couplings

$\mathbf{P}d/dr$ can be removed by the transformation

$$\left[\frac{d^2}{dr^2} - \frac{\mathbf{P}^2}{4} - e^{\mathbf{P}r/2}\,\mathbf{V}(r)\,e^{-\mathbf{P}r/2} + 2E\right] e^{\mathbf{P}r/2}\,\mathbf{F}(r) = 0. \tag{16}$$

The R-matrix corresponding to the solution of this equation

$$\mathbf{G}(r) = \exp(\mathbf{P}r/2)\,\mathbf{F}(r) \tag{17}$$

is given by

$$\mathbf{R}(a) = \mathbf{G}(a)\left\{ r\frac{d\mathbf{G}}{dr}\bigg|_{r=a} \right\}^{-1} \tag{18}$$

$$= e^{\mathbf{P}a/2}\left\{ r\frac{d\mathbf{F}/dr}{\mathbf{F}}\bigg|_{r=a} + \frac{\mathbf{P}}{2} \right\}^{-1} e^{-\mathbf{P}a/2}. \tag{19}$$

Within the close-coupling approach, the asymptotic channels of the outer electron are given by its angular momentum and its energy, which are coupled to the final state of the residual 'target'. Boundary conditions must be set in a frame in which the channels decouple asymptotically, which for the present case is the so-called acceleration or Kramers-Henneberger frame. Then total and partial cross-sections into the various channels can be computed. Since we are considering processes taking place fully inside the field, the asymptotic states consist on the one hand of the field-dressed free electron, which has a well-defined average direction and energy, and which goes over, as the field is turned off, to a free electron with well-defined momentum. The residual 'target', on the other hand, is also dressed by the field. At large r, $r = a'$ the equations are therefore solved by an asymptotic expansion, incorporating the appropriate physical (resonance or scattering) boundary conditions [5]. The frame transformation to the acceleration frame could be performed at any radius $r > a$. It is however nonlocal and it has been found that it is hard to perform accurately at finite r. Thus this transformation is performed formally at $r = \infty$ and it is incorporated in the asymptotic expansion. The asymptotic form of the frame transformation between the acceleration and velocity frames is

$$\mathbf{F}(r) \underset{r\to\infty}{\longrightarrow} \mathbf{A}^{(\mathrm{asy})}\,\mathbf{\Phi}(r), \tag{20}$$

where $\mathbf{A}^{(\mathrm{asy})}$ is a square unitary matrix and $\mathbf{\Phi}(r)$ is the diagonal matrix of asymptotic solutions in the acceleration frame. In the case of multiphoton ionization

$$\mathbf{\Phi}(r) = \frac{1}{\sqrt{\boldsymbol{k}}}\exp(i\boldsymbol{k}r + iZ\boldsymbol{k}^{-1}\ln 2\boldsymbol{k}r - i\boldsymbol{\ell}\pi/2 + i\boldsymbol{\sigma}_{\ell}), \tag{21}$$

where $\boldsymbol{k}$ is the set of asymptotic channel momenta, which are shifted due to truncation of the problem [5, 16]. The channel angular momenta and Coulomb phases are denoted by $\boldsymbol{\ell}$ and $\boldsymbol{\sigma}_{\ell}$. We assume an asymptotic expansion in analogy to the field-free case

$$\mathbf{F}(r) = \sum_{n=0}^{\infty} r^{-n}\,\mathbf{A}^{(n)}\,\mathbf{\Phi}(r). \tag{22}$$

Recurrence relations for the $\mathbf{A}^{(n)}$ are obtained by substituting the above expansion into the close coupling equations (15) and using a multipole expansion for the potential $\mathbf{V}(r)$. The recurrence is started by setting $\mathbf{A}^{(0)} = \mathbf{A}^{(\mathrm{asy})}$.

At $r = a'$ the solution from the inner region is matched to the solution from the asymptotic expansion to obtain either the K-matrix for a scattering calculation [2, 21] or to obtain the quasienergy E in an ionization calculation. In the latter case, the energy is found by an iterative procedure searching for a zero in the determinant of the matching matrix **B**

$$\mathbf{B} = \mathbf{R}_V(E)\left[r\frac{d\mathbf{G}(E)}{dr} + \frac{1}{2}\mathbf{P}\mathbf{G}(E)\right] - \mathbf{G}(E), \tag{23}$$

with $\mathbf{G}(E)$ the solution matrix fulfilling the asymptotic boundary conditions for the energy E and all quantities evaluated at the matching radius $r = a'$.

This direct specification of boundary conditions (as opposed to complex rotation calculations) allows the computation of virtual states [17] or of shadow poles, as long as the propagation is not performed over a too large distance.

Most RMF ionization calculations at low frequency to-date have been plagued by the fact that the process converges to the true physical solution only if the initial complex energy guess is very close to the correct value. Two reasons are responsible for this behaviour : first, there are many complex energies for which there is a zero determinant but these do not correspond to a physical solution. Second, the matching matrix (or rather, its determinant) contains poles in addition to zeros which are due to the poles of the R-matrix on the real energy axis; standard zero-search algorithms can easily be led astray if the function has poles near the zero. It is therefore important to start the search with a rather good guess of the physical complex energy.

2. RESULTS

2.1 Ionization

The first test of the RMF method for a real atomic system was the calculation of multiphoton ionization rates for atomic hydrogen. First results were presented in [5] and a more detailed study was performed in [18], where also partial rates (or branching ratios) were presented. These results were compared with Sturmian Floquet results [4] for the same processes and excellent agreement was obtained. In our calculations, a special inner region method was developed, which took advantage of the fact that there was only a single electron. The general atom code was adapted to compute results for this case also, which were in agreement. The agreement for the branching ratios at low frequency was satisfactory only if the inner region radius was not too large [33].

Results have been obtained for other atoms, namely the noble gases helium [19, 20, 30], neon [29] and argon [23] and the alkaline earth magnesium [31, 34]. The focus has been on the precise calculation of total rates for both perturbative and nonperturbative field intensities and the effects of resonances. At perturbative intensities, excellent agreement was found with other high-accuracy perturbative calculations. At nonperturbative intensities, other calculations have been performed mainly within a time-dependent approach, at much lower frequencies (*e.g.* [37, 38]).

At resonances between bound and autoionizing states, the Floquet approach exhibits complex-energy degeneracies, which lead to interesting structure in the total and partial rates and shifts [23, 27, 30]. These have been termed "laser-induced degenerate states" or LIDS, in analogy with "laser-induced continuum structure" (LICS), to which they are closely related, for two-color experiments [7].

The generalization of the RMF approach to two frequencies, either incommensurable [30] or commensurable [32] has allowed the experimentally important study

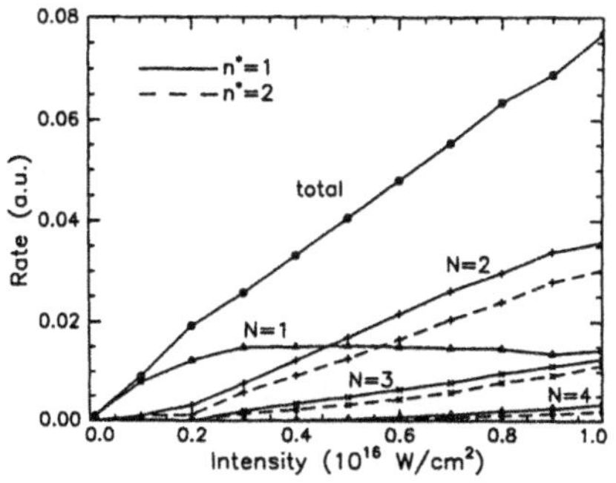

Fig. 1: Ionization rate for Helium in a laser field of frequency 1.6 a.u. The partial rates into the N photon channels are also shown, differentiated into ground state ($n^* = 1$) and excited ($n^* = 2$) residual ion. The target wavefunction is restricted to the 1s, 2s and 2p orbitals. Convergence is reached at the highest intensity for roughly 150 channels. Note the indications of 'stabilization' in the partial rates, but a linear (although not perturbative) behaviour in the total rate.

of double-resonances and of coherent interaction between the fundamental frequency and one of its harmonics, for the commensurable case [35].

At low frequencies, it is generally recognized that Single-Active-Electron model potential calculations can give a good approximation to the evolution of an atom in a laser field : at higher frequencies, however, when for example doubly excited resonances are important, these model potentials will no longer be appropriate.

Multiphoton ionization of positively charged ions using the RMF theory has so far been computed only in [33] for He^+ : this is an important process, leading to multiply charged ionic species in an intense laser pulse by sequential ionization.

2.2 Detachment

Negatively charged ions have usually a rather weakly bound 'outer' electron and are notoriously difficult to describe precisely. The RMF theory has been applied extensively to the negative hydrogen ion H^- [19, 22, 23, 24, 25], exploring total and partial rates. At low frequency, excellent agreement was found with other perturbative approaches. It was found to be very useful to semi-empirically adjust the binding energy to the exact binding energy in order to obtain precise rates.

The perturbative regime for the field intensity depends quite sensitively on the frequency. Generally, at larger frequency, larger intensity can be reached within the

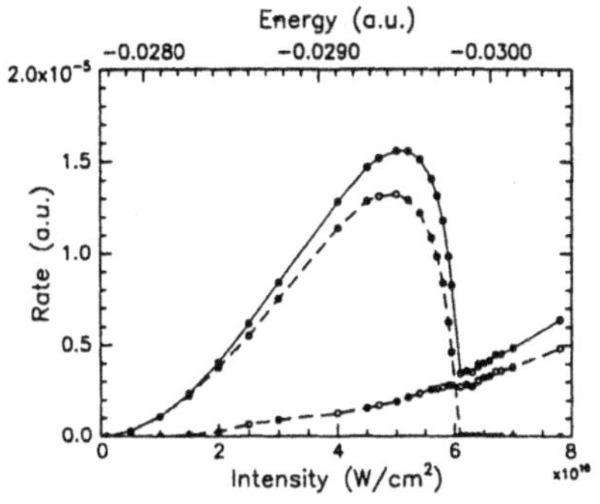

Fig. 2: Ionization rate of the negative hydrogen ion in a laser field of frequency $\omega = 0.0149$ a.u. The total rate and the partial rates into the $N = 2$ and 3 photon channels are given. The two-photon channel closes at an intensity of 6×10^{10} W/cm^2. From [22].

perturbative regime. Near resonances, however, also the ones involving autoionizing states, perturbation theory breaks down at much smaller intensities [24].

The RMF codes have been adapted to variable noninteger nuclear charge which allows the (purely theoretical) connection between negatively charged ions and the corresponding neutral atoms [25]. The question of the spectrum of H^- at high frequency and high intensity has also been addressed.

The negative halide ions F^- and Cl^- have been systematically studied at low frequencies and perturbative and non-perturbative intensities [26]. The one-photon ionization of the negative lithium ion has been computed in [28].

2.3 Scattering

Just as the multiphoton ionization of hydrogen was the first nontrivial test case of our RMF method for ionization, thus the electron-proton scattering was the first case to be considered for laser-assisted scattering. The R-matrix approach being in fact primarily used for scattering calculations, "the electron-proton problem" is the general name for the two processes, just as "the electron-hydrogen problem" denotes also the H^- detachment problem. Although "the electron-proton" field-free scattering problem is too trivial to have warranted its inclusion in the options of the original general atom R-matrix codes, it is far from trivial when a laser field is present [4]. We have obtained excellent agreement with other approaches for the widths and positions of laser-induced resonances and also for cross-sections. The latter depend on the proper coherent superposition and the definition of asymptotic momenta and it is much harder to reach convergence for such differential information. Not all calculations mentioned in [4] agree in this case.

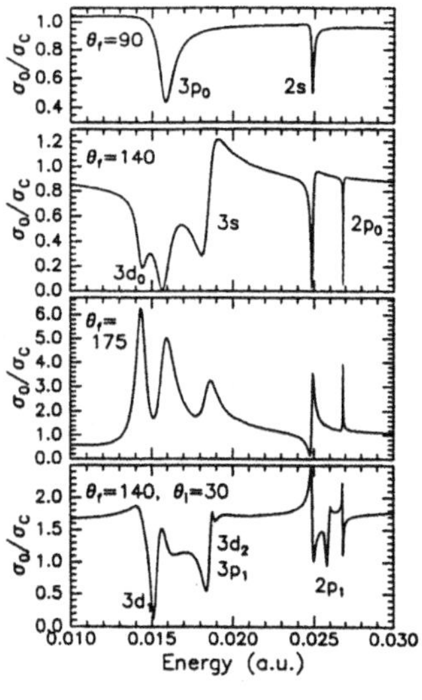

Fig. 3: Laser-assisted electron-proton scattering in a field of intensity 10^{12} W/cm^2 and frequency $\omega = 0.074$ a.u. The ratio for elastic scattering (no net photons absorbed or emitted) to the field-free differential scattering cross-section is presented. The upper three plots have incoming angle θ_i zero (along the laser polarization axis) and various outgoing angles θ_f. Note that the one- and two-photon resonance profiles are strongly dependent on geometry. From [21].

The next step involves evidently more complex targets. It is crucial to define the dressed target states in the field properly in order to compute processes involving excitation of the target. Experimental results are available *e.g.* for electron-helium scattering. New theoretical RMF results for this problem will be presented by M. Terao-Dunseath *et al* at this conference.

2.4 Harmonic generation

The process of high-order harmonic generation in atomic gases constitutes an important source of short wavelength, high-brilliance radiation. Harmonic generation has been calculated with the RMF approach [33, 34].

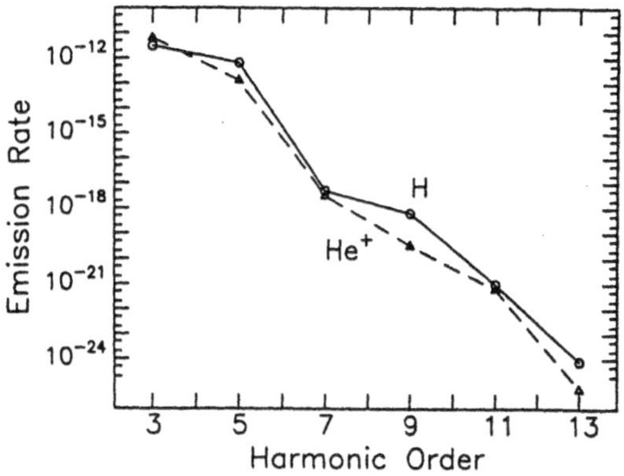

Fig. 4: Harmonic generation total rates for hydrogen (dashed line) at 5.2×10^{14} W/cm^2 and He$^+$ (solid line) at 10.3×10^{14} W/cm^2, both at frequency $\omega = 0.367$ a.u. Note that the absolute harmonic rates are similar (on this log-scale), although the binding energies and ionization rates are very different. After [33].

In [33] harmonic generation up to order 13 to 19 has been calculated for hydrogenic systems. Excellent agreement with previous Sturmian Floquet results for hydrogen have been obtained. The RMF results have been obtained with the inner region method specific for hydrogen. In [34] the general atom program has been used to compute harmonic generation in hydrogen and helium. A detailed study of third-harmonic generation in magnesium has been performed, investigating the role of an autoionizing state resonance.

3. FUTURE

In the field of multiphoton physics, theoretical developments have always been closely associated with technical developments regarding new powerful laser sources. The concomitant experimental developments are continuing at a very rapid pace. Evidently the range of intensities and available wavelengths is continuously expanding and also high-repetition-rate lasers enable measurements of effects over many orders of magnitude in signal yield [36].

In terms of experimentally available wavelengths and intensities several light sources will in the near future provide radiation at the order of the atomic unit of both frequency and intensity, such as synchrotron-based radiation sources, harmonic conversion devices and other XUV radiation sources. There are several RMF results at rather high frequency and intensity, which therefore should be within experimental reach in the near future.

In the search for ever more intense light pulses, the laser pulses are necessarily simultaneously becoming ever shorter, the shortest pulses now being of the order of 10 fs at a wavelength of 800 nm : this means that the pulse comprises only a few cycles of the field and evidently has a very large bandwidth. For these ultrashort intense laser pulses eventually the Floquet approach will no longer be applicable. Time-dependent calculations for complex atoms modeled within a CI-'target' approach, thus within a basis similar to our R-matrix basis and with a single outgoing electron have been performed in [37]. It is desirable to produce a fully time-dependent R-matrix code. For this case, the inner region basis can be retained in its present form, however with time-dependent coefficients, while in the outer region the boundary conditions must be reformulated into fully absorbing or transparent boundary conditions, which again must be set in the accelerated frame.

Even when the Floquet approach is useful, it is important to improve the present methods in order to be able to compute high-intensity or low frequency cases where many channels are required with sufficient accuracy. The codes have been ported to parallel computing [14] and some of the problems mentioned in section 1 have been addressed.

The question of correlated double multiphoton ionization is another topic of crucial interest. This it is known as the "three (or more) body Coulomb problem" which even in the absence of a laser field is far from trivial. The question of nonsequential *versus* sequential multiple ionization is since long a key discussion on the way towards understanding correlated electron dynamics [39]. Our present RMF approach is restricted to a single outgoing electron and can thus only describe sequential multiple ionization. There has been much progress recently on the "intermediate energy R-matrix" [40] or "R-matrix with pseudostates" [41] approaches and double electron ejection [42] in the context of field-free scattering. These new techniques, once fully operational, should be readily though non-trivially adaptable to the R-matrix-Floquet approach.

A large area of current research in multiphoton physics which is within the scope of the present RMF method concerns multiphoton processes in molecules. The outer region solution should require no significant modification from the atomic case so that the main part of the work consists in providing the required matrix elements for the Floquet Hamiltonian in the inner region and for the field-dressing of the 'target' states.

Acknowledgments

The work presented here has been the result of a team effort and the author is grateful for a long and fruitful and pleasant collaboration, principally with the "hard core" involved in the first few RMF publications and subsequently with the new collaborators on the team. The author was supported by the Deutsche Forschungs-Gemeinschaft.

References

[1] P. G. Burke, P. Francken and C. J. Joachain, Europhys. Lett. **13**, 617 (1990)

[2] P. G. Burke, P. Francken and C. J. Joachain, J. Phys. B **24**, 761 (1991) "R-matrix-Floquet theory of multiphoton processes" (RMF 1)

[3] Some of these ongoing studies are presented in the contributions of M. LeDourneuf and of K.A. Berrington to this conference.

[4] A. Giusti-Suzor and P. Zoller, *Phys. Rev.* A **36**, 5178 (1987); L. Dimou and F.H.M. Faisal, *Phys. Rev. Lett.* **59**, 872 (1987); A. Franz, H. Klar, J.T. Broad

and J.S. Briggs, *J. Opt. Soc. Am.* B **7**, 545 (1990); L.A. Collins and G. Csanak, *Phys. Rev.* A **44**, R5343 (1991); G. Csanak and L.A. Collins, *Phys. Rev.* A **47**, 3240 (1993); C.T. Chen and F. Robicheaux, *J. Phys.* B (1996).
The Sturmian Floquet method by R.M. Potvliege and R. Shakeshaft is reviewed in : M. Gavrila ed., Adv. At. Mol. Phys., Suppl. **1**, 453 (1992).
A "multi-electron multi-photon" method is being pursued by the group of T. Mercouris and C.A. Nicolaides, *e.g.* J. Phys. B **23**, 2037 (1990).

[5] M. Dörr, M. Terao-Dunseath, J. Purvis, C. J. Noble, P. G. Burke and C. J. Joachain, J. Phys. B **25**, 2809 (1992) "RMF 2 : solution of the asymptotic equations in the velocity gauge"

[6] The author is grateful for any comments, additions and corrections to the list of current and future applications of RMF theory.

[7] Resonances are discussed in much more detail by the contribution of N. Kylstra to this book.

[8] See *e.g.* H. Rottke, B. Wolff-Rottke, D. Feldmann, K.H. Welge, Martin Dörr, R.M. Potvliege and Robin Shakeshaft, *Phys. Rev.* A **49**, 4837 (1994) "Atomic hydrogen in a strong optical radiation field"

[9] E. P. Wigner and L. Eisenbud, *Phys. Rev.* **72**, 29 (1947)

[10] Luckily, there are tools to assist in this artful exercise, foremost the program CIV3 by A. Hibbert and the program SuperStructure by W. Eissner.

[11] See the contribution of H.W. van der Hart to this conference concerning "B-Spline basis R-matrix calculations of e-H scattering".

[12] M. Dörr, M. Terao-Dunseath and C.J. Noble, unpublished (1994) "Extrapolating the R-matrix continuum basis". An alternative to extrapolation would be a variable boundary condition continuum (u_l) basis – for example B-splines [11].

[13] P.G. Burke, A. Hibbert and W.D. Robb *J. Phys.* B **4**, 153 (1971) "Electron scattering by complex atoms"

[14] D. Glass, C.J. Noble, P.G. Burke and H.W. van der Hart, unpublished (1996) "RMF 9 : a direct R-matrix method (by linear equation solution) with an application to He"

[15] J.C. Light and R.B. Walker, *J. Chem. Phys.* **65**, 4272 (1976);
K.L. Baluja, P.G. Burke and L.A. Morgan *Comp. Phys. Comm.* **27**, 299 (1982)

[16] H. Day, A.S. Fearnside and R.M. Potvliege, unpublished (1995) "Shifted momenta in truncated Floquet calculations"

[17] A. Cyr and M. Terao-Dunseath, unpublished (1994) "A virtual state in the O^- ion"

[18] M. Dörr, P. G. Burke, C. J. Joachain, C. J. Noble, J. Purvis and M. Terao-Dunseath, J. Phys. B **26**, L275 (1993) "RMF 3 : multiphoton ionization of atomic hydrogen"

[19] J. Purvis, M. Dörr, M. Terao-Dunseath, C. J. Joachain, P. G. Burke and C. J. Noble, Phys. Rev. Lett. **71**, 3943 (1993) "RMF theory of multiphoton ionization of two-electron atomic systems"

[20] M. Dörr, M. Terao-Dunseath, P. G. Burke, C. J. Joachain, C. J. Noble, and J. Purvis, unpublished, "RMF : multiphoton ionization of He"

[21] M. Dörr, M. Terao-Dunseath, P. G. Burke, C. J. Joachain, C. J. Noble, and J. Purvis, J. Phys. B **28**, 3545 (1995) "RMF 4 : electron-proton scattering"

[22] M. Dörr, J. Purvis, M. Terao-Dunseath, P. G. Burke, C. J. Joachain, and C. J. Noble, J. Phys. B **28**, 4481 (1995) "RMF 5 : multiphoton detachment of H^-"

[23] O. Latinne, N. J. Kylstra, M. Dörr, J. Purvis, M. Terao-Dunseath, C. J. Joachain, P. G. Burke and C. J. Noble, Phys. Rev. Lett. **74**, 46 (1995) "Laser-induced degeneracies involving autoionizing states in complex atoms"

[24] A. S. Fearnside and H.W. van der Hart, unpublished (1995) "Total and partial detachment rates for two-photon resonances between the H^- ground state and autoionizing states"

[25] A. S. Fearnside, unpublished (1996) "Appearance of light-induced resonances in H^- in a strong high-frequency laser field"

[26] H.W. van der Hart, J. Phys. B **29**, 3059 (1996) "RMF 6 : the negative halide ions"; and unpublished results at non-perturbative intensities.

[27] O. Latinne and A. Cyr, unpublished (1996) "A three-photon degenracy in Ar"

[28] G.B. Wöste *et al*, unpublished (1995) "RMF results for photoionization of Li^-"

[29] N.J. Kylstra, M. Dörr, C.J. Joachain, and P.G. Burke, J. Phys. B **28**, L685 (1995) "Resonant two-photon ionization of atoms via two coupled Rydberg series"

[30] H.W. van der Hart, "RMF 7 : a two-color approach" (for incommensurable frequencies) J. Phys. B **29**, 2217 (1996)

[31] H.W. van der Hart and N.J. Kylstra, unpublished (1996) "Two-color resonances involving autoionizing states in Mg"

[32] B. Borca *et al*, unpublished (1996) "RMF theory for two commensurable frequencies"

[33] M. Bensaid, thesis, Université Libre de Bruxelles (1996) "Etude théorique de la géneration d'harmoniques par la méthode R-matrix-Floquet"

[34] R. Gebarowski, M. Bensaid *et al*, unpublished (1996) "RMF theory of harmonic generation"

[35] See also the contribution of V. Véniard to this book.

[36] B. Walker, B. Sheehy, L.F. DiMauro, P. Agostini, K.J. Schafer and K.C. Kulander, *Phys. Rev. Lett.* **73**, 1227 (1994)

[37] J. Zhang and P. Lambropoulos *J. Phys.* B **28**, L101 (1995) "Non-perturbative time-dependent theory and ATI in two-electron atoms"

[38] J. Parker, K.T. Taylor, C.W. Clark and S. Blodgett-Ford, *J. Phys.* B **29**, L33 (1996) "Intense-field multiphoton ionization of a two-electron atom"

[39] On single-photon double-ionization, see the contribution of R. Shakeshaft to this conference.

[40] P.G. Burke, C.J. Noble and M.P. Scott, *Proc. Roy. Soc.* A **410**, 289 (1987) "R-matrix theory of electron scattering at intermediate energies"

[41] K. Bartschat, E.T. Hudson, M.P. Scott, P.G. Burke and V.M. Burke, *J. Phys.* B **29**, 115 (1996) "Electron-atom scattering at low and intermediate energies using an R-matrix basis with pseudo-states"

[42] K.M. Dunseath, M. LeDourneuf, M. Terao-Dunseath and J.M. Launay, *Phys. Rev.* A **54**, 561 (1996) "Two-dimensional R-matrix propagator : application to e-H scattering"

RESONANCE EFFECTS IN MULTIPHOTON IONIZATION

N J Kylstra*

Physique Théorique CP 227,
Université Libre de Bruxelles,
B-1050 Bruxelles, Belgium

INTRODUCTION

The advent of table-top laser systems which are capable of producing electric field strengths of the order of the atomic unit has allowed the study of laser-atom interactions in a regime where the radiation field can no longer be considered as simply a perturbation to the atom. In this nonperturbative regime interesting new phenomena, which are collectively referred to as multiphoton processes, have been observed. Three notable examples include: above threshold ionization (ATI), in which electrons absorb more photons than are required to ionize the atom [1]; resonantly enhanced multiphoton ionization (REMPI), the shifting into resonance of Rydberg states due to the strong ac-Stark shifts induced by the intense laser field [2]; and harmonic generation, the emission by the atom of high-order, odd harmonics of the laser field frequency [3].

From the theoretical point of view, the study of intense laser-atom interactions is challenging: at high intensities (typically of the order of 10^{13} W/cm^2), the inherent difficulties of the required nonperturbative treatment have restricted the bulk of the theoretical investigations to one-electron systems: either one dimensional models [4], hydrogen [5] or single electron approximations [6]. One exception is the *ab initio* R-matrix-Floquet theory of multiphoton processes [7, 8] which can be applied to atoms of arbitrary complexity. In this approach the time dependent problem is reduced to a time independent, eigenvalue system, which is then solved within the framework of the R-matrix theory [9]. The R-matrix-Floquet theory has already been successfully applied to the calculation of multiphoton rates in the two electron systems H^- and helium [10] as well as in other complex atoms [11–14]. In this contribution we consider some resonant processes occurring in multiphoton ionization processes involving complex atoms in strong laser fields. In particular we will discuss recent results obtained using the R-matrix-Floquet approach and the use of simple model atoms to understand the qualitative behavior of the resonances (see also [15]). In the next two sections some of the basic concepts of, respectively, the theory of strong laser-atom interactions and the R-matrix-Floquet approach will be introduced. We then discuss laser intensity effects on the one-photon resonant, two-photon ionization of atoms via two coupled Rydberg series. Next, some aspects of autoionizing resonances in strong fields will be examined. In particular, the existence of laser induced degeneracies (LIDS) and the effects of LIDS on some double resonant processes will be considered. Finally, we comment on double poles which occur in the laser-assisted electron-atom scattering S-matrix. Atomic units are used throughout, unless otherwise indicated.

*email: nkylstra@ulb.ac.be

GENERAL CONSIDERATIONS

The laser field is taken to be a time dependent, monochromatic, spatially homogeneous electric field

$$\mathbf{E}(t) = \boldsymbol{\epsilon}\, f(t) E_0 \cos\omega t\,, \tag{1}$$

where $f(t)$ describes the shape and duration of the pulse, E_0 is the maximum field strength of the pulse, ω is the angular frequency of the laser and $\boldsymbol{\epsilon}$ is a unit vector in the direction of the laser polarization (we consider only linear polarization). We have used the dipole approximation and we assume that the laser field can be described classically. The cycle averaged intensity of the laser field is related to the electric field strength of the laser field by $I = (c/8\pi)E_0^2$.

The most appropriate theoretical framework for studying the interaction of an atom with a laser field depends on the characteristics of the laser pulse. For our purposes, the relevant laser parameters are the frequency, maximum intensity and duration of the laser pulse. When considering the laser pulse duration, we can, though in a somewhat artificial manner, distinguish between three regimes: the i) long pulse, ii) short pulse and iii) ultra-short pulse regime. In the long pulse regime, which is typically characterized by nanosecond (10^{-9} s; we note that the atomic unit of time is 2.4×10^{-17} s) or longer laser pulses, the decay, or ionization, of the atom can be characterized by a single ionization rate $\Gamma(I, \omega)$ which depends only on the angular frequency and maximum intensity of the laser. The exact details concerning the turn-on and turn-off of the pulse are not relevant, except for the requirement that they be adiabatic, so that the atom evolves adiabatically in the laser field. In this regime, the ionization rate of the atom in the laser field can be calculated using time-dependent perturbation theory when the field is weak, while for strong fields the Floquet approach [16], which will be outlined below, can be utilized. In the short pulse regime (pulse durations on the order of picoseconds, or 10^{-12} s) the evolution of the atom in the laser field can still be assumed to be adiabatic. However now the shape of the laser pulse is important and therefore, if a comparison is to be made with experiment, the relevant quantity to be calculated is the ionization probability which is determined from the instantaneous intensity dependent ionization rate of the atom in the laser field,

$$P(t) \;=\; 1 - \exp[-\int_{-\infty}^{t} \Gamma(I(t'), \omega) dt']\,. \tag{2}$$

We note that in general, when comparing with experiment the spatial profile of the laser in the interaction region will also have to be considered [17]. Once the ionization rates have been calculated for the range of intensities of the pulse, the ionization probability can be determined for some laser pulse shape $f(t)$. Finally, in the ultra short pulse regime the evolution of the atom cannot be considered to be adiabatic so that the only recourse is to obtain information about the ionization processes from a direct integration of the time-dependent Schrödinger equation [6] (the reader is also referred to the contributions of V. Véniard and K. Taylor in this volume). Obviously, the boundaries which define the three regimes cannot be precisely defined. Moreover they depend on the atom being considered. In all our further considerations we will assume that we remain in a regime in which the adiabatic assumption is valid, i.e. either the long pulse or the short pulse regimes.

Within this adiabatic regime, our goal is to study, using the R-matrix-Floquet approach, the multiphoton ionization of atoms for laser intensities at which a lowest order perturbative description is no longer valid. With respect to the frequency of the laser, this approach is most suitable for laser frequencies which are not "too low" or not "too high". This "intermediate" frequency regime is sometimes referred to as the multiphoton regime since the ionization process can be readily viewed as being composed of many laser-induced transitions between the initial state of the atom and the continua. In order to characterize the frequency of the laser in a general way, we can compare the energy of the photons with the ionization potential U_0 of the atom. When $\hbar\omega \ll U_0$ (the low frequency or tunneling ionization regime) the valence electrons of the atom can be viewed as moving in a slowly varying, time dependent effective potential composed of the superposition of the potential due to the Coulomb interactions and the potential due to the time dependent electric field of the laser. Ionization probabilities can then be determined from static field tunneling ionization rates. At the other extreme, for frequencies so that $\hbar\omega \gg U_0$, a high-frequency Floquet approach, which is formulated in the Kramers reference frame, has been developed [18] (see the contribution of M. Gavrila in this volume). In both of these limiting cases, resonant processes play essentially no role. This is in contrast with the intermediate regime, where the ionization can occur via intermediate bound states or autoionizing states of the atom. We stress the three regimes described as such are only indicative: a more rigorous categorization must depend on the intensity of the field.

THE R-MATRIX FLOQUET APPROACH

In this section we recall some basic elements of the R-matrix-Floquet theory. More details can be found in [7, 8], as well as the contribution by M. Dörr in this volume. The goal is to solve the time-dependent Schrödinger equation for the $N+1$ electron atom in a laser field:

$$\begin{aligned} \mathrm{i}\frac{\partial}{\partial t}\Psi(\mathbf{X},t) &= [H_{\text{Atomic}} + H_{\text{Int}}(t)]\Psi(\mathbf{X},t) \\ &= \left[-\sum_{i=1}^{N+1}\left[\frac{1}{2}\nabla_i^2 + \frac{\mathrm{i}}{c}\mathbf{A}(t)\cdot\boldsymbol{\nabla}_i - \frac{1}{2c^2}A^2(t) + \frac{Z}{r_i}\right] + \sum_{i<j}^{N+1}\frac{1}{|\mathbf{r}_i - \mathbf{r}_j|}\right]\Psi(\mathbf{X},t), \end{aligned} \tag{3}$$

where all the spatial and spin coordinates of the electrons have been collected in the symbol $\mathbf{X}$ and the electromagnetic vector potential $\mathbf{A}(t)$ is related to the time-dependent electric field by $\mathbf{E}(t) = -(1/c)(d\mathbf{A}/dt)$. This is a formidable task and obviously we can only hope to obtain approximate solutions. We proceed as follows. First, as discussed above, we assume that the variations in the intensity envelope function $f(t)$ occur slowly (i.e. $df/dt \ll \omega_0/2\pi$, where $\omega_0/2\pi$ corresponds to the typical orbital frequencies of the atom) so that the atom evolves adiabatically in the laser field. Thus within our adiabatic picture, we have an atom in some initial bound state which is subjected to an adiabatically applied laser pulse. This allows solutions of the time-dependent Schrödinger equation to be sought for constant laser intensities. The time-dependent Hamiltonian, and therefore the solutions of the time dependent Schrödinger equation, can be labeled by the parameters I and ω

$$\mathrm{i}\frac{\partial}{\partial t}\Psi_{(I,\omega)}(\mathbf{X},t) = H_{(I,\omega)}(t)\Psi_{(I,\omega)}(\mathbf{X},t). \tag{4}$$

Since the laser intensity is constant in time, the Hamiltonian $H_{(I,\omega)}(t)$ is periodic in time and it follows from the theory of ordinary differential equations with periodic coefficients that solutions of the time-dependent Schrödinger equation exist and can be written in the form (Floquet's theorem [19])

$$\Psi_{(I,\omega)}(\mathbf{X},t) = \exp[-\mathrm{i}Et]P_{(I,\omega)}(\mathbf{X},t), \tag{5}$$

where $P_{(I,\omega)}(\mathbf{X},t)$ is a periodic function in ω. The quantity E, which depends on I and ω, is the time independent quasi-energy of the system. The periodicity of $P_{(I,\omega)}(\mathbf{X},t)$ is exploited by making a Fourier expansion,

$$P_{(I,\omega)}(\mathbf{X},t) = \sum_{n=-\infty}^{\infty}\exp[-\mathrm{i}n\omega t]F_n^{(I,\omega)}(\mathbf{X}). \tag{6}$$

The $F_n^{(I,\omega)}$ are called the Floquet-Fourier components of the wavefunction. Inserting the Floquet-Fourier form of the wavefunction into the time dependent Schrödinger equation leads to the time independent Floquet eigenvalue problem

$$H_{\mathrm{F}}^{(I,\omega)}|F^{(I,\omega)}\rangle = E|F^{(I,\omega)}\rangle, \tag{7}$$

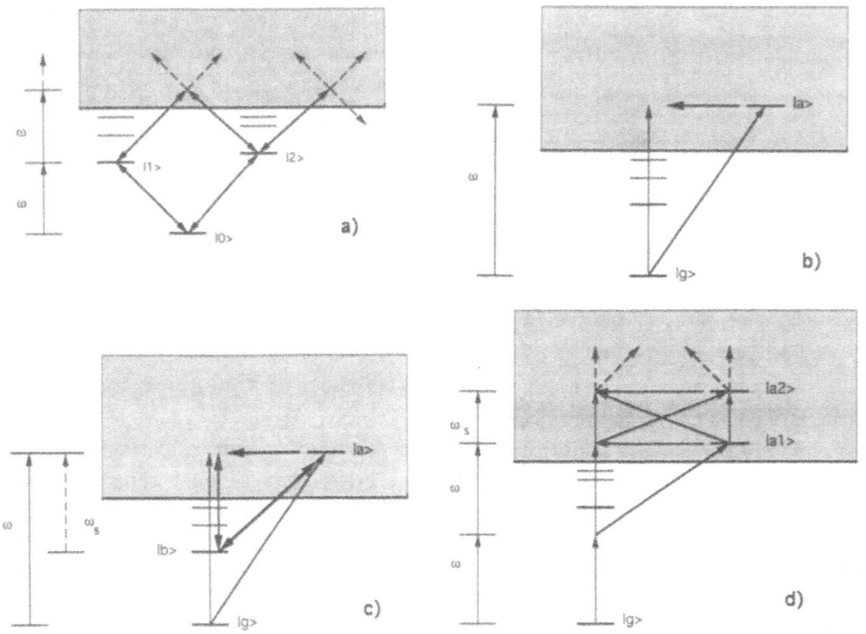

Figure 1: Resonant processes discussed in this contribution.

which is the fundamental equation which must be solved. Here $F^{(I,\omega)}$ is a vector with the components $F_n^{(I,\omega)}$ and the Floquet Hamiltonian is given by

$$[H_{\mathrm{F}}^{(I,\omega)}]_{nn'} = [H_{\mathrm{Atomic}} - n\omega]\delta_{n,n'} + \tfrac{1}{2}[D^{\dagger}\delta_{n,n'+1} + D\delta_{n,n'-1}] \,. \tag{8}$$

The dipole interaction operator is given by D and is proportional to the electric field strength.

The Floquet equations are now solved within the framework of the R-matrix method. In this approach, which has proven an effective method for treating many-body problems in a wide variety of atomic processes [9], the configuration space is divided into an inner region in which the charge distribution of initial state of the atom and the remaining ion are confined and an outer region where only the ejected electron is present. For the R-matrix inner region, a modified version of the existing inner region R-matrix codes is used to generate a field-free atomic basis. Using this basis, the resulting matrix representation of the Floquet Hamiltonian (which has been modified by a Bloch operator in order to remove the surface terms at the inner region boundary) is diagonalized and the R-matrix calculated on the inner region boundary. This R-matrix is then propagated in the outer region using a modified Light-Walker propagator to the asymptotic region where the correct boundary conditions must be imposed on the solution. Since the laser field couples the bound states of the atom to the continua, these states are no longer true bound states due to the fact that they now ionize (or decay) in the field. Because we are seeking a solution of the time-dependent Schrödinger equation for a constant intensity, the atom will ionize at a constant rate (i.e. the probability that the atom will ionize at any given instant is independent of the ionization probability at previous times). Therefore these states can be viewed as being resonance states or quasi-bound states in complete analogy with autoionizing states. For this reason, a solution of a modified, non-Hermitian Floquet Hamiltonian is sought. One way of achieving this is to use the complex rotation, or dilation method [20]. This results is a non-Hermitian problem

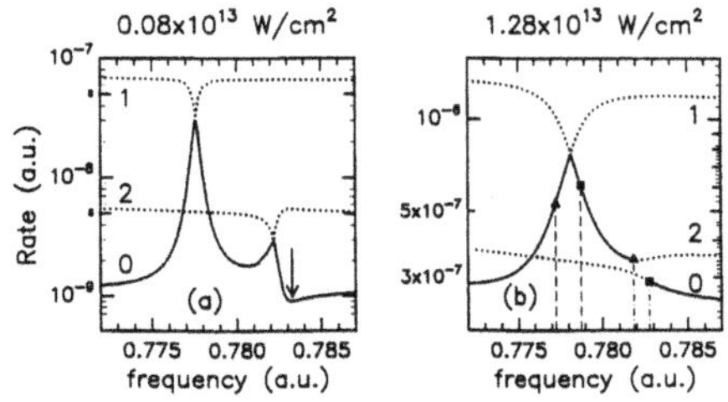

Figure 2: The rates, as a function of frequency, of the ground state of neon (denoted by 0), the 5s ^{1}P Rydberg state (denoted by 1) and the 4d ^{1}P Rydberg state (denoted by 2). The solid lines indicate the rates associated with the ground state. (Figure taken from [12].)

which can be solved in terms of a basis of L^2 functions. Examples of such an application of the Floquet theory to the study of one electron systems include calculations by Maquet *et al* [21] and by Potvliege and Shakeshaft [5]. In the R-matrix-Floquet approach, the close-coupling equations in the outer region are solved with Siegert boundary conditions imposed, i.e. the asymptotic outgoing boundary conditions are such that the momenta (and thus the quasienergy) in the ionization channels are complex. Writing the complex quasienergy in the form $E(I,\omega) = E_0 + \Delta(I,\omega) - \frac{i}{2}\Gamma(I,\omega)$, where E_0 is the energy of the field free state and $\Delta(I,\omega)$ is the so-called dynamic Stark shift (the field induced energy shift) and using the Floquet-Fourier form of the wavefunction, one finds that

$$\langle \Psi_{(I,\omega)}(t) \,|\, \Psi_{(I,\omega)}(t) \rangle \quad \propto \quad \exp[-\Gamma(I,\omega)t]\,. \tag{9}$$

Hence, for some given intensity I and frequency ω, the Floquet state decays in time at the constant rate $\Gamma(I,\omega)$.

RESONANT TWO-PHOTON IONIZATION

The first type of resonant process which we consider involves intermediate Rydberg states which lie below the first ionization threshold. When the frequency of the laser is such that the ground state and a Rydberg state are resonantly coupled, the degenerate two-level subsystem must be treated non-perturbatively. For adiabatic pulses, the result is a crossing of the widths, or rates, of the two resonant states as a function of the laser frequency. Since it is usually the case that the ionization rate of the Rydberg state just off of resonance is much larger than that of the ground state (because fewer photons are required to couple the Rydberg state to the continuum) a resonance between the ground state and the Rydberg state gives rise to a large peak in the multiphoton ionization rate of the ground state of the atom (however pronounced interfences can occur between neighboring resonances [22]). Very close to the crossing, i.e. when the rates are nearly equal, the smaller of the two rates should be associated with the ground state since it is this state which is adiabatically connected to the field-free ground state in the limit of small intensities.

For the case of two-photon ionization, due to the dipole selection rules, resonances can in general occur between the ground state and the members of two different Rydberg series. Now, when two Rydberg states, each belonging to a different Rydberg series and lying close in energy are resonantly coupled to the ground state by a single photon transition, interference between the two resonant ionization pathways will occur (see figure 1a). This interference, which depends on the laser intensity, will in turn modify the ionization rate of the ground state. This effect has been demonstrated in Ne [12]. The laser frequency was chosen to be at a one photon resonance with the 5s and the 4d Rydberg states. The results of the R-matrix-Floquet calculations are shown in figure 2. At the relatively weak intensity of 8.0×10^{11} W/cm^2 (figure 2a), the ionization rate as a function of the laser frequency, the solid line, displays the characteristic crossings with the two Rydberg states, the dotted lines. The effect of interference between the ionization pathways is evident at a frequency of 0.783 where a pronounced minimum occurs in the ionization rate. At the much higher intensity of 1.28×10^{13} W/cm^2 the behavior at the resonance is very different (figure 2b). The crossing of the rates of the ground state and the 4d Rydberg state is no longer present. This implies that in order to determine the ionization rate as a function of the laser frequency in the resonance region we must identify the solution which is adiabatically connected to the ground state in the weak field limit. This is shown in figure 2b, where again the solid line indicates the rate of the ground state. In the long pulse regime, interestingly enough, this rate is not a continuous function of the frequency. Another example of this phenomenon is given in the next section. We also point out that the minimum in the ionization rate of the ground state is no longer present. This is due to the importance of higher order processes at this high intensity.

AUTOIONIZING RESONANCES IN STRONG FIELDS: LASER INDUCED DEGENERACIES

Photoionization cross sections of atoms demonstrate characteristic, asymmetric peaks. These structures are due to Rydberg series of doubly excited, or autoionizing, states that converge to a threshold of an excited state of the remaining ion. The shape of these peaks can be well reproduced using the well known Fano formula for the ratio $W(\omega)$ of the photoionization cross section to the nonresonant contribution to the cross section [23],

$$W(\omega) = \frac{(\omega - E_a + E_g + q\Gamma_a/2)^2}{(\omega - E_a + E_g)^2 + (\Gamma_a/2)^2}. \quad (10)$$

The energies of the ground state and autoionizing state are given respectively by E_g and E_a. The width of the autoionizing state is Γ_a while q is the Fano asymmetry parameter which describes the interference between the direct and indirect (i.e. via the autoionizing state) ionization pathways. Clearly the resonance profile is not symmetric about $E_a - E_g$ and goes to zero when $\omega = E_a - E_g - q\Gamma_a/2$. In deriving this expression it is assumed that E_a, E_g, Γ_a and q are are constant functions of the frequency in the neighborhood of the resonance, which is usually a good approximation. Most importantly, it is also assumed that the atom-laser interaction can be treated perturbatively. We now consider what happens in the nonperturbative regime. As discussed in the previous section, an atom in a laser field can be viewed as being a resonant state. With this in mind, it is evident that a perturbative description of the ground state-autoionizing state coupling will fail when the intensity is large enough such that the laser-induced width of the ground state close to, but not at, resonance becomes comparable to the width Γ_a of the autoionizing state.

An *ab initio* study of the influence of the strong laser induced coupling of the ground state of Ar and the 3s3p^{6}4p ^{1}P autoionizing Rydberg state has been recently carried out using the R-matrix-Floquet approach [11]. The results of the calculation are shown in figure 3. The trajectories of the complex quasienergies $E(I,\omega)$ of the ground state and the autoionizing state are plotted in the complex energy plane for a fixed values of the frequency, for intensities ranging from 0 to 5×10^{13} W/cm^2. The zero-field position of the ground state on the real energy axis is $E_g = -0.57816$, while the energy of the autoionizing state is shifted by $-\omega$. Thus the zero field position of the autoionizing state (denoted by the big circles) changes with ω, and is at the complex energy $(-\omega + 0.40936 - i\,0.00119) = -\omega + E_a - i\Gamma_a/2$, where Γ_a is the field-free width of the autoionizing state. For a fixed frequency there are two curves: one which, in the limit of small intensity, is connected to the weak field position of the ground

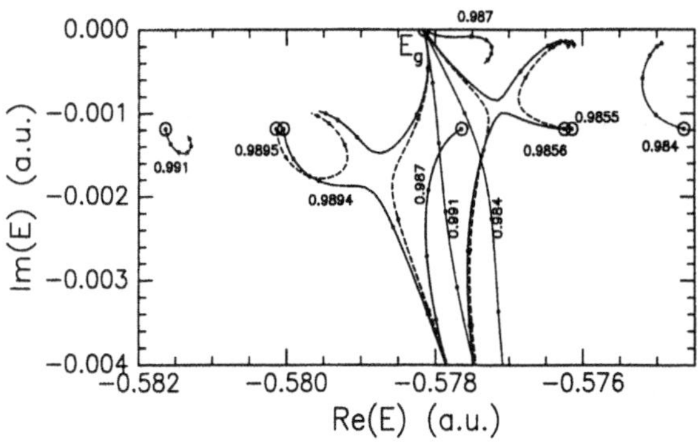

Figure 3: Trajectories of the complex Floquet quasienergies of the ground state and the $3s3p^64p\ ^1P$ autoionizing state for intensities up to 5×10^{13} W/cm^2. The laser frequencies are indicated next to the trajectories. (Figure taken from [11].)

state and the other which is connected to the weak field position of the autoionizing state. The small dots give the intensity in steps of 9×10^{12} W/cm^2. The detuning from resonance is defined as $\delta = E_a - E_g - \omega$. At large (positive or negative) detunings (e.g. $\omega = 0.984, 0.991$) the width of the ground state increases with intensity. At very small detunings ($\omega = 0.987$), just the opposite happens: the curve connected to the autoionizing state increases in width with intensity, while the ground state is "trapped" close to the real axis. We observe that there are two critical points, for each of which there corresponds a critical frequency and intensity of the laser where the two complex quasienergies are exactly degenerate, i.e. where laser induced degenerate states (LIDS) occur. Thus, owing to the existence of LIDS, for a fixed frequency lying between the two critical frequencies the rate of ionization of the ground state will first increase with intensity and then exhibit a typical "stabilization", namely a decrease of the ionization rate with increasing intensity. This is evident for the trajectory corresponding to the frequency 0.9894. This effect is even more pronounced for the frequency 0.9855. For frequencies far enough outside this region, the ground state's ionization probability increases monotonically with intensity. The existence of the degeneracies can be exploited to transfer population from the ground state to an autoionizing state. For Ar (figure 3) this could be achieved by (i) choosing $\omega = 0.9856$ and increasing I from 0 to 2×10^{13} W/cm^2, (ii) changing ω slightly, from 0.9856 to 0.9855, and (iii) decreasing I back to 0. The trajectories must be timed carefully due on one side to the constraint of adiabaticity (the changes in frequency and intensity must not be too fast) and on the other side by the fact that the atomic system is decaying in the field. Instead of step (iii) above, further adjustments in the frequency and intensity allow a circuit to be completed around the degeneracy. This has been discussed for degeneracies which occur due to resonances in the presence of a two-colour field [24].

The occurrence of such degeneracies is a quite general phenomenon in strong field autoionization. This can be seen by considering a model atom consisting of a bound level, an autoionizing level and a continuum (see figure 1b). To solve the Schrödinger equation for this system, the wavefunction is first written in the Floquet-Fourier form, (5) and (6), which results in a time independent eigensystem. The Feshbach projection operator formalism is then applied to the Floquet Hamiltonian, from which a non-Hermitian effective Floquet Hamiltonian can be derived in the discrete subspace of the resonant states. Keeping only the resonantly coupled sub-matrix (rotating wave approximation, RWA) results in a two by two

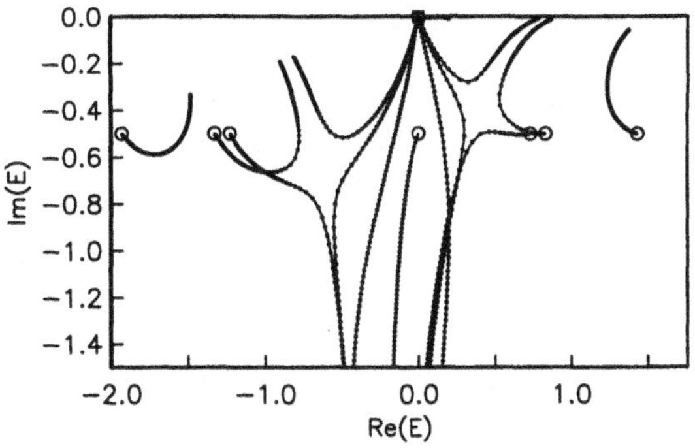

Figure 4: Complex quasienergy trajectories given by (11). The energy is measured in units of Γ_a.

matrix whose complex quasienergies are given by

$$\begin{aligned} E_{1,2}(I,\omega) &= \frac{1}{2}\left[E_a + E_g - \omega - \frac{\mathrm{i}}{2}[\Gamma_a + \Gamma_g] \pm \Omega\right], \\ \Omega &= \left[\left(\delta - \frac{\mathrm{i}}{2}[\Gamma_a - \Gamma_g]\right)^2 + \Gamma_a\Gamma_g[q - \mathrm{i}]^2\right]^{\frac{1}{2}}. \end{aligned} \tag{11}$$

The parameters are the same as those in equation (10) and $\Gamma_g = I\gamma_g$ is the width acquired by the ground state in a bare continuum, with I being the intensity of the laser. Again, all the atomic parameters are taken to be constants in the neighborhood of the resonance. From the expression (11) it is clear that degeneracies in the complex energy spectrum, or LIDS, occur when $\Omega = 0$. This condition is fulfilled by the critical detunings and intensities given by

$$\begin{aligned} \delta_{\mathrm{c}}^{\pm} &= \Gamma_a[q \pm \sqrt{q^2+1}], \\ I_{\mathrm{c}}^{\pm} &= \frac{[\delta_{\mathrm{c}}^{\pm}]^2}{\gamma_g\Gamma_a} \end{aligned} \tag{12}$$

The critical frequencies are thus given by $\omega_{\mathrm{c}}^{\pm} = E_a - E_g - \delta_{\mathrm{c}}^{\pm}$. The validity of this model can be verified by plotting the quasienergy trajectories and comparing with the results for Ar. For $q = -0.25$, which corresponds to the Ar example, the quasienergies (11) are plotted in figure 4 as a function of intensity for fixed frequencies. Comparing with figure 3, we see that the qualitative agreement between the trajectories obtained from the model and the quasienergy trajectories given by the R-matrix-Floquet calculations is remarkably good. This demonstrates clearly that even at the high intensities considered here, the resonant ground state-autoionizing state coupling dominates other higher-order multiphoton processes.

Important differences do exist, though. The model predicts that one of the quasienergies becomes purely real, for any fixed laser intensity, when the frequency is such that

$$\omega_{\mathrm{T}}(I) = E_a - E_g + \frac{q}{2}[\Gamma_a - \Gamma_g]. \tag{13}$$

This is a consequence of the perfect destructive interference between the "direct" and "indirect" ionization pathways. In the limit of small intensities, this frequency coincides with the zero in the photoionization line shape in (10). In reality, as is demonstrated in the R-matrix-Floquet calculations, due to the presence of additional multiphoton ionization channels, this quasienergy never becomes real. Moreover, the model considered here does not take into account the intensity dependent AC Stark shifts, i.e. E_g and E_a do not depend on the laser intensity.

While bearing in mind its shortcomings, we now calculate using this model the ionization rate of the ground state, as a function of the laser frequency, for different laser intensities. Again taking $q = -0.25$, the ratio of the resonant ionization rate to the nonresonant ionization rate, $-2\mathrm{Im}(E)/\Gamma_g$, is plotted in figure 5a. This expression is the strong field analogue of the Fano profile given by (10), where E is chosen from $E_{1,2}$ using the criterion that for some fixed frequency $E_{1,2}$ is adiabatically connected to E_g in the limit of zero intensity. Three different intensities are considered: $I/\Gamma_a = 0.05 < I_c^+/\Gamma_a$, solid line; $I_c^+/\Gamma_a < I/\Gamma_a = 0.9 < I_c^-/\Gamma_a$, dotted line; and $I/\Gamma_a = 3.0 > I_c^-/\Gamma_a$, dashed line. The vertical lines indicate the critical detunings $\delta_c^\pm/\Gamma_a$. At the lowest intensity the resonance line profile is essentially the same as the one given by the usual Fano profile. At the intermediate intensity, between the two critical intensities, the profile is distinguished by a discontinuity at δ_c^+/Γ_a, while at the highest intensity there exist discontinuities at both critical detunings. Thus, at high intensities the usual Fano profile is markedly altered: we clearly see that for intensities greater than I_c^- and I_c^+, the rate is not continuous at respectively ω_c^+ and ω_c^- and that there is strong suppression of the ionization rate between these two frequencies.

The line shapes plotted in figure 5a can be directly related to experiment only in the long pulse regime. However, as is the case with the Ar example, high intensities, and thus short pulses, will be required to observe the non-perturbative effects. In view of calculating the ionization probability for some given laser pulse shape, we show in figure 5b the ionization rate (in units of Γ_a) as a function of intensity (in dimensionless units of Γ_g/Γ_a) for seven equally spaced laser frequencies. The solid lines are for two frequencies $\omega < \omega_c^+$, the dotted lines are for three frequencies $\omega_c^+ < \omega < \omega_c^-$ and the dashed lines correspond to two frequencies $\omega > \omega_c^-$. It is evident that at low intensities all the rates increase linearly with the intensity, while at sufficiently high intensities the rates stabilize for frequencies which lie between the critical frequencies. The arrows indicate the critical intensities $I_c^\pm$. Next we calculate the ionization probability (2) for a laser pulse whose shape is given by $f(t) = \sin^2 \pi t/\tau$, where τ is the pulse duration. Since the two degeneracies are separated along the real energy axis by $\Delta E = 2\Gamma_g\sqrt{q^2+1}$, the bandwidth of the laser must be much smaller than ΔE. This means that, in addition to the adiabaticity requirement, we must have $\tau \gg 1/(2\Gamma_g\sqrt{q^2+1})$. Evidently, the larger the q, the shorter the required pulse length. Taking $\Gamma_a\tau = 30/(2\sqrt{q^2+1})$, the ionization probability is plotted in figure 5c as a function of the laser detuning from resonance. The maximum pulse intensity for each curve is the same as the intensity of the corresponding curve type in figure 5a. At the low intensity, the ionization probability as a function of frequency is reminiscent of the Fano profile. However, as the maximum intensity is increased to above I_c^+, a discontinuity appears, just as in the rates in figure 5a, at ω_c^+. Finally, for the highest intensity pulse, the ionization is practically complete, except for frequencies lying between the two critical intensities where, for this realistic pulse, the suppression of ionization between the critical frequencies is dramatic. Note that due to the intensity dependence of the trapping frequency ω_T, the ionization probability is never exactly zero.

The model can give an indication of which atomic systems would be amenable for the experimental investigation of LIDS. In order to obtain a degeneracy at low intensity, it is desirable to take a system for which Γ_a/γ_g is small. This can be achieved by using a higher-n member of the autoionizing Rydberg series, since Γ_a scales with n as $(n-\mu)^{-3}$ (where μ is the quantum defect) while γ_g varies much more slowly. A degeneracy also occurs at a low intensity when q is large. For $q = 0$ the two degeneracy points are at the same intensity, but for large q one of the intensities $I_c^\pm$ becomes small. However, for very large q, the effects of the degeneracy will probably be less dramatic since the trapping frequency is proportional to q, indicating that other nearby resonances would probably destroy the pronounced minimum in the ionization rate and ionization probability. Another issue which must be kept in mind is the fact that presently all high intensity lasers produce radiation in the infrared and visible [3]. Therefore, in general, the resonant coupling between the ground state and the autoionizing

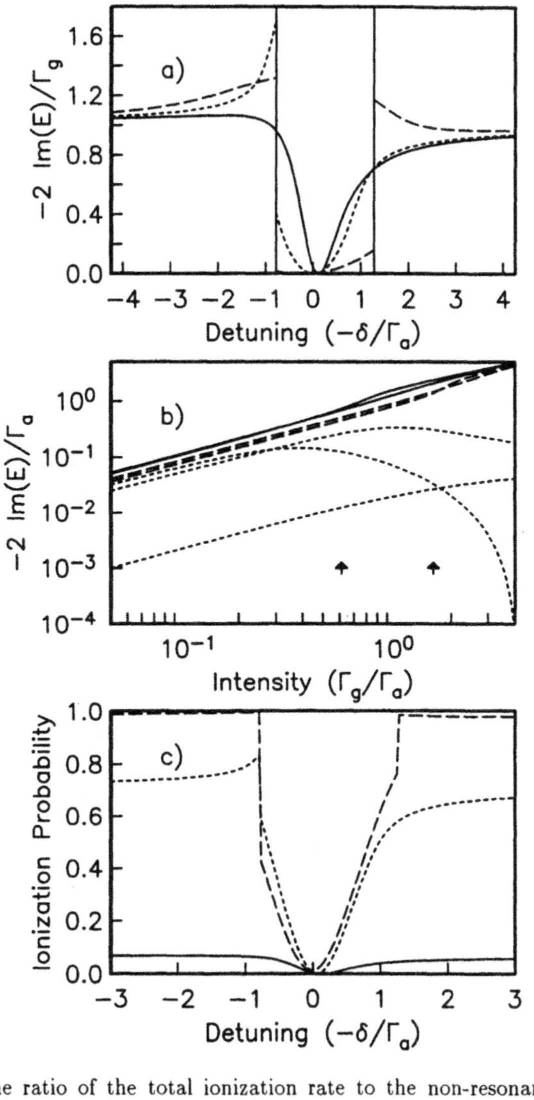

Figure 5: Plot a: the ratio of the total ionization rate to the non-resonant contribution to the ionization rate as a function of the laser frequency for three different intensities (see text). Plot b: the ionization rate in units of Γ_a as a function of the laser intensity for seven different frequencies. Plot c: the ionization probability as a function of the laser frequency for the three laser pulses described in the text.

state must occur via a multiphoton transition. The existence of LIDS in this context was demonstrated in H^- for a two-photon coupling between the ground state and lowest lying ^1S resonance [11]. Additionally, a degeneracy between the ground state of Ar and the same $3s3p^6 4p\ ^1P$ autoionizing Rydberg state has been demonstrated using a resonant three photon coupling [25]. However, the price to be paid when considering a multiphoton transition is the existence of ionization channels representing photon processes which are of the same, or even lower, order than the resonant coupling. For the examples in H^- and Ar, one photon and two photon absorption, respectively, are sufficient for ionization. Therefore, the total width will consist of a large background contribution plus the width due to the resonant process. This implies that for short laser pulses, the effects of the degeneracies will be more difficult to deduce from the ionization probability for different frequencies and maximum intensities. Another possibility is to take not the ground state, but an excited state of the atom to be the initial state. Clearly, the natural lifetime of the state must be long compared with the duration of the laser pulse. Starting from the first excited ^1P level in Mg, a degeneracy between this state and the $3p^2\ ^1S$ autoionizing level occurs at a very modest intensity [26].

A number of studies have examined various aspects of the effects of autoionizing resonances in strong field photoionization. We end this section by briefly contrasting our results with two investigations in which this resonance process was first discussed. Rzazewski and Eberly [27] determined an expression for the electron energy distribution in the long-time limit under the assumption of an instantaneous application of the laser field. Using Laplace transform methods the energy spectrum of the electrons can be determined in this long-time limit when the ionization of the atom is complete. Though it should be kept in mind that when one of the two complex quasienergies associated with the two level system becomes purely real, i.e. at the trapping frequency (13), population remains trapped in the ground state of the atom. The obtained electron energy spectrum displays a complicated behavior as a function of the laser intensity. In this context complete destructive interference between the two ionization pathways manifests itself as a zero in the electron energy spectrum. Lambropoulos and Zoller [28] also assumed an instantaneous turn-on of the field, however they considered the ionization probability after some finite interaction time. These approaches differ from the adiabatic picture which we maintain in which, in the long pulse regime, the electron energy spectrum is given by a simple Breit-Wigner profile whose width is given by the ionization rate, while for some given pulse shape an electron energy distribution can only be reconstructed from the shifts and rates of the Floquet state in the laser pulse. In addition, any population "trapping" which occurs will depend not only on the laser frequency, but also on the pulse duration and maximum intensity.

LASER INDUCED DEGENERACIES IN TWO-COLOUR, DOUBLE RESONANT PROCESSES

In this section we consider an indirect approach, which was suggested in [11], for creating LIDS. In a typical double resonant process two lasers are used to resonantly couple three discrete states of the atom. Within the adiabatic approach, two-colour laser fields can also be treated within the Floquet framework for commensurate frequencies [29] and noncommensurate frequencies [30]. We will discuss two double resonant processes in which it is possible to study the effects of the strong laser coupling of an autoionizing state with a high-lying bound Rydberg state or with another autoionizing state. In this setup, a second laser, which functions as a weak probe, is used to couple the strong laser-dressed states to the ground state of the atom. In this way, the effect of the strong laser will manifest itself in the resonant photoionization rate of the ground state. With such a scenario LIDS can occur among high lying states which are more readily accessible with existing laser frequencies.

The first situation which we investigate is shown in figure 1c. Here the strong coupling is between the autoionizing state $|a\rangle$ and a bound Rydberg state $|b\rangle$ of the atom. We note that the intensity independent dipole coupling between the ground state and the autoionizing state γ_g is in general larger then the dipole coupling γ_b between the Rydberg state and the

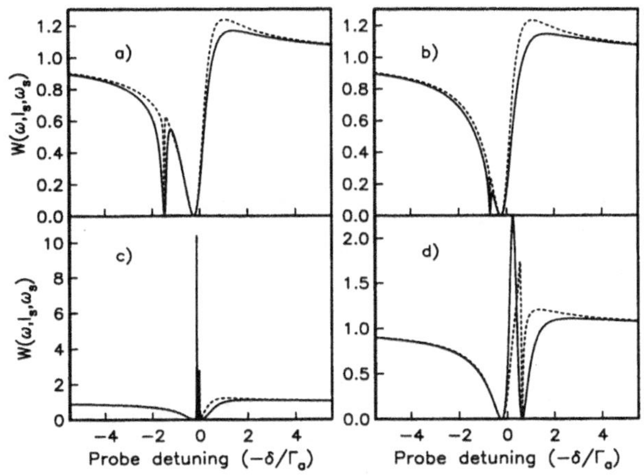

Figure 6: Expression (14) is plotted as a function of the probe laser frequency for four different frequencies ω_s (plots a–d) and two different intensities I_s (respectively the dotted and solid lines).

autoionizing state. This implies that while the frequencies necessary to produce LIDS between the Rydberg state and the autoionizing state will be smaller than the frequencies required to produce LIDS between the ground state and the autoionizing state, the required intensity will be larger.

We wish to consider this process in a general way and our goal is the following: by treating the strong laser couplings exactly and the probe laser couplings perturbatively, an expression for the ionization rate of the ground state must be determined. The first step, as usual, is to write the total wavefunction of the atom in the Floquet-Fourier form. This leads to the time-independent Floquet eigenvalue problem. We then appeal to a S-matrix approach put forth by Shore [31] for calculating the weak field resonance profile (10), which is based on the observation that photoionization can be viewed as a photon scattering process, thus allowing the standard S-matrix theory to be applied with the field-free Hamiltonian being treated exactly and the laser-atom interaction treated perturbatively. This approach can be applied directly if one now utilizes, in place of the field-free Hamiltonian, the full time-independent Floquet Hamiltonian describing the atom and its resonant interaction with the strong laser. Following this prescription and treating the resonant coupling in the RWA, the generalization of the Fano result (10), which is now also a function of the angular frequency ω_s and intensity I_s of the strong laser, is found to be

$$W(\omega, I_s, \omega_s) = \frac{\Big([\omega + E_g - E_a + \Gamma_a q/2][\omega + E_g - E_b - \omega_s] + q_b\Gamma_a\Gamma_b[q - q_b]/4\Big)^2}{|\omega + E_g - E_1(I_s,\omega_s)|^2|\omega + E_g - E_2(I_s,\omega_s)|^2}, \quad (14)$$

with, as in (11) above,

$$E_{1,2}(I_s,\omega_s) = \frac{1}{2}\left[E_a + E_b + \omega_s - \frac{\mathrm{i}}{2}[\Gamma_a + \Gamma_b] \pm \Omega\right],$$

$$\Omega = \left[\Big(\delta_s - \frac{\mathrm{i}}{2}[\Gamma_a - \Gamma_b]\Big)^2 + \Gamma_a\Gamma_b[q_b - \mathrm{i}]^2\right]^{\frac{1}{2}}. \quad (15)$$

In these expressions, E_b is the position of the bound level $|b\rangle$, Γ_b is its laser induced width,

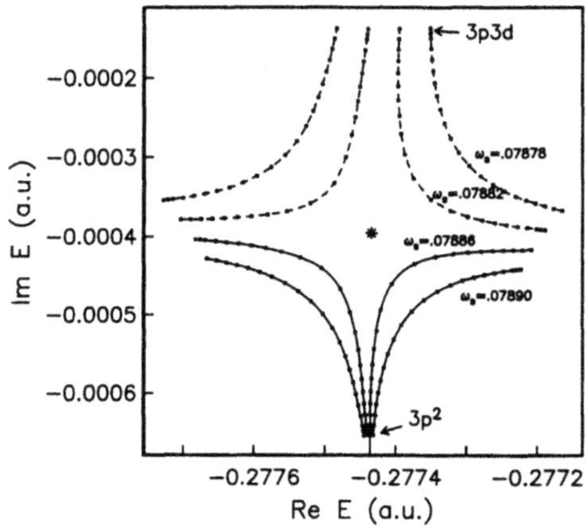

Figure 7: Trajectories of the complex Floquet quasienergies of the $3p^2\ ^1S$ (solid lines) and $3p3d\ ^1P$ (dashed lines) autoionizing states of Mg as a function of the laser intensity. The asterisk indicates the position of the degeneracy between the two states.

which depends on I_s, in the structureless continuum and q_b is the Fano asymmetry parameter associated with the this state and the autoionizing level. The detuning is given by $\delta_s = E_a - E_b - \omega_s$. Evidently, in the limit $I_s \to 0$ the expression (14) reduces to the Fano profile (10). A remarkable property of (14) is that a zero in the absorption cross section occurs at two different probe laser frequencies. We note that within the context of an instantaneous laser turn-on, Deng and Eberly [32] calculated the electron energy in the long-time limit and have shown that two zeros occur in the electron energy distribution. When the frequency of the strong laser is tuned to the trapping frequency $\omega_T(I_s)$ (where one the complex quasienergies $E_{1,2}(I_s, \omega_s)$ becomes purely real) the line shape reduces to the simple Fano expression, however with a modified effective q parameter. When the two quasienergies (15) are degenerate, double poles of $W(\omega, I_s, \omega_s)$ occur.

In figure 6 the resonance profile, given by (14) with $q = q_b = -0.5$, is shown as a function of the probe laser detuning for four equally spaced frequencies ω_s, plots a–d, and for two intensities I_s. The intensities were chosen to be $I_s = I_c^+/5$ (dotted lines) and $I_s = I_c^+$ (solid lines). In plot c the frequency ω_s is close to the trapping frequency, while in plot d one has $\omega_s = \omega_c^+$. Evident in each plot are the two zeros in the line shapes. Furthermore, by varying the frequency and intensity of the strong laser, the resonance profile can be dramatically altered. However, at the degeneracy, given by the solid curve in plot d, there is no evident signature in the resonant line profile which would indicate the presence of the degeneracy.

Instead of coupling the autoionizing state to a bound Rydberg state, the autoionizing state can be resonantly coupled by the strong laser to another autoionizing state (see figure 1d). The first observation of this process has recently been reported [33]. In this experiment the ground state of Mg was coupled by a two-photon transition to the $3p^2\ ^1S$ autoionizing state while another laser coupled this autoionizing state to the $3p3d\ ^1P$ state. Using a two-colour version of the R-matrix-Floquet code for noncommensurate frequencies [34], calculations for this process have been carried out. A description of the CI wavefunctions employed can be found in [14]. In figure 7, the complex quasienergies of the $3p^2$ and 3p3d autoionizing states are plotted as a function of the coupling laser intensity for different frequencies ω_s. The intensity of the laser is increased from 1.0×10^9 W/cm^2 to 5.0×10^9 W/cm^2, with the arrows indicating the positions of the autoionizing states at the weakest intensity considered.

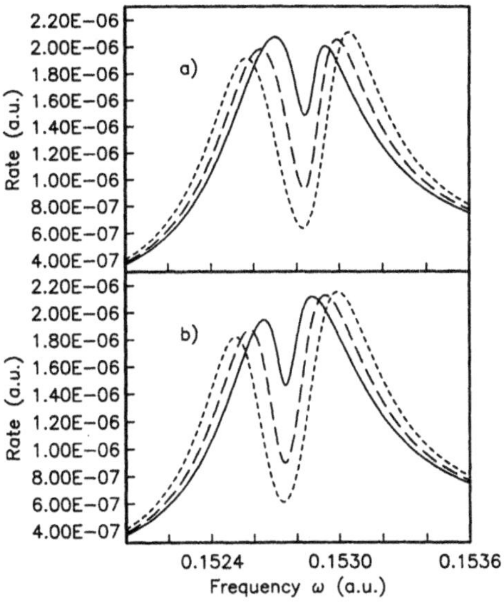

Figure 8: The total ionization rate of the ground state of Mg as a function of the probe laser frequency. In plot a and plot b we have, respectively, $\omega_s = 0.07864$ and $\omega_s = 0.07884$. The three curves correspond to three different intensities I_s (see text).

A degeneracy between the two autoionizing states appears at the intensity $I_c = 3.1 \times 10^9$ W/cm^2 and frequency $\omega_c = 0.07884$. In figure 8 the ionization rate of the ground state of Mg is shown for the frequency $\omega_s = 0.07864$, plot a, and $\omega_s = \omega_c$, plot b. The solid curves are for $I_s = 1.0 \times 10^9$ W/cm^2, the long-dashed curves for $I_s = I_c = 3.1 \times 10^9$ W/cm^2 and the short-dashed curves for $I_s = 6.2 \times 10^9$ W/cm^2. Once again, there is no apparent signature of the degeneracies on the ionization rate of ground state. Indeed, despite the simple double peaked form of the resonant profile in figure 8, in this intensity and frequency range of the strong laser there is no obvious relation between these peaks and the quasienergy trajectories in figure 7. This can be seen from the following formula for the resonant profile,

$$W(\omega, I_s, \omega_s) = \Big[\Big(\omega + E_g - E_{a_1} + \Gamma_{a_1} q/2\Big)^2 \Big([\omega + E_g - E_{a_2} + w_s]^2 + [\Gamma_{a_2}/2]^2\Big)$$
$$-D_{12}^2\Big(2[\omega + E_g - E_{a_1} + \Gamma_{a_1} q/2][\omega + E_g - E_{a_2} + w_s] - [1+q^2]\Gamma_{a_1}\Gamma_{a_2}/4 - D_{12}^2\Big)\Big]$$
$$\times \frac{1}{|\omega + E_g - E_1(I_s,\omega_s)|^2 |\omega + E_g - E_2(I_s,\omega_s)|^2} . \tag{16}$$

In deriving this expression, we have assumed that the coupling between the two autoionizing states is dominated by the direct dipole matrix element, i.e. that the couplings via the continuum are small. This is a very good approximation for the Mg example, however see [28, 35] for more detailed discussions. The complex quasienergies are now given by

$$E_{1,2}(I_s, \omega_s) = \frac{1}{2}\Big[E_{a_1} + E_{a_2} - \omega - \frac{i}{2}[\Gamma_{a_1} + \Gamma_{a_2}] \pm \Omega\Big] ,$$

$$\Omega = \left[\left(\delta_s - \frac{i}{2}[\Gamma_{a_1} - \Gamma_{a_2}]\right)^2 + 4D^2_{a_2a_1}\right]^{\frac{1}{2}} . \qquad (17)$$

In this expression $\delta_s = E_{a_1} - E_{a_2} + \omega_s$ and the laser induced coupling, which depends on I_s, between the two autoionizing states is $D_{a_2a_1}$. As $I_s \to 0$ we see that (16) reduces to (10). It is also clear that degeneracies occur when

$$\delta_c = 0 ,$$
$$I_c = \frac{[\Gamma_{a_1} - \Gamma_{a_2}]^2}{[4d_{a_2a_1}]^2} , \qquad (18)$$

where $D_{a_2a_1} = I^{\frac{1}{2}} d_{a_2a_1}$. At the degeneracy, for large values of q, the numerator in (16) has a minimum at $\omega = E_{a_1} - E_g$ while at the same frequency the denominator has a sharp maximum. For the Mg example $q \simeq 27$, and therefore this gives an explanation for the general shape of the resonance profile in the neighborhood of the degeneracies. Overall, this model was found to reproduce well the R-matrix-Floquet calculations.

While the two double resonant processes described here are interesting and can be used to study different aspects of the resonant coupling between the ground state of an atom and discrete states in the continuum, it is not immediately clear how such processes can be exploited to investigate LIDS. The basic problem is that it would seem that it is not possible to separate in a simple way the effects due to the strong resonant coupling between the high lying states and the weak resonant coupling with the ground state. To study the effects of LIDS, we must have degeneracies between the initial ground (or excited) state and the autoionizing state. This does not mean that double resonant processes are not useful. Indeed, it has been found in R-matrix-Floquet two-colour calculations that degeneracies occur between the ground state of Mg and the 3p3d ^{1}P autoionizing resonance at rather modest intensities of the high frequency laser which couples the ground state and the 3p^2 ^{1}S autoionizing state [14].

DOUBLE POLES OF THE S-MATRIX IN LASER-ASSISTED ELECTRON-ATOM SCATTERING

The complex Floquet quasienergies which are obtained by imposing Siegert boundary conditions on the Floquet states correspond to poles of the laser-assisted electron-atom scattering matrix, or S-matrix, when scattering boundary conditions are imposed. Therefore LIDS, in a scattering context, correspond to double poles of the S-matrix. A multiple pole of the S-matrix implies that the time evolution of the resonant, or quasi-stationary, state does not follow the usual exponential decay law, but is modified by a polynomial in time [36]. Atomic and molecular physics could provide a physically realizable setting for studying this phenomenon since the resonant states created by the application of external fields can be manipulated by adjusting the field parameters. For example, it has been pointed out that Stark mixing in an atom [37] and the decay of Rabi oscillations in a two-level system [38] can display double pole decay. For both of these cases, the realization of the double pole decay requires the instantaneous preparation of the resonant state so that the external fields are applied suddenly, i.e. as a step pulse. This liability is avoided in a purely scattering context. In this section we discuss in very general terms the effect of the double poles on the resonant contribution to the cross section. In analogy with the double resonance examples above, we consider the case in which the energy of the projectile electron is such that there is a resonance due to an autoionizing state of the composite electron-atom system and that the frequency of the laser field is tuned to a resonance between this autoionizing state and a bound state, or another autoionizing state, of the composite system.

When describing the scattering of an electron by the atom in the presence of a laser, it is assumed that the electron enters the adiabatically applied laser pulse far away from the target atom, is scattered by the atom in the field of the laser and subsequently the pulse vanishes when the scattered electron is far from the atom. Therefore, when theoretically studying the modifications to field-free electron-atom scattering, the laser-assisted electron-atom scattering quantities can be calculated under the assumption that the field remains on for all times [39]. Since the time dependence of the atom-field interaction is periodic, the general solution of the time dependent Schrödinger equation can be written in the Floquet-Fourier form, leading to the time independent Floquet eigenvalue equation. The usual methods of the time independent theory of scattering, and in particular the R-matrix Floquet theory, can then be applied to solve the Floquet equation.

The first resonant process we consider is the laser-assisted scattering of an atom in the neighborhood of an autoionizing resonance with the laser frequency chosen such that the autoionizing state is in resonance with a quasi-bound state, i.e. a state that would be bound in the absence of the laser. The laser will dress the projectile and the target and we assume, as usual, that at resonance this dressing is dominated by the resonantly coupled states. Clearly our discussion is valid for not too high intensities, which is always the case for laser-assisted electron-atom scattering. The cross section due to this resonant process [40], which depends on the angular frequency of the laser ω, its intensity I and the energy of the electron E, is given by

$$\sigma(E,I,\omega) \propto \frac{[\Gamma_a+\Gamma_g]^2[E-E_{\mathrm{W}}(I,\omega)]^2}{|E-E_1(I,\omega)|^2|E-E_2(I,\omega)|^2}. \tag{19}$$

The resonant laser coupling has been treated in the RWA. The complex quasienergies, $E_{1,2}$, are given by (15). At the window energy,

$$E_{\mathrm{W}}(I,\omega) = \frac{\Gamma_a[E_g+\omega]+\Gamma_g E_a-\Gamma_a\Gamma_g q}{\Gamma_a+\Gamma_g}, \tag{20}$$

the two resonant processes interfere fully destructively and the resonant contribution to the total cross section vanishes. Degeneracies in the complex energy spectrum of the dressed states, or LIDS, occur when the quasienergies become degenerate. At the degeneracy, or double pole, the window energy is given by

$$E_{\mathrm{W}}(I_{\mathrm{c}}^{\pm},\omega_{\mathrm{c}}^{\pm}) = \mathrm{Re}(E_{\mathrm{c}}^{\pm}), \tag{21}$$

where $E_{\mathrm{c}}^{\pm}=E_1(I_{\mathrm{c}}^{\pm},\omega_{\mathrm{c}}^{\pm})=E_2(I_{\mathrm{c}}^{\pm},\omega_{\mathrm{c}}^{\pm})$. It can thus be concluded that the cross section is a symmetric function about the window energy at the degeneracy and that the window energy is located at the position of the degeneracy along the real energy axis. The symmetry of the cross section about the window energy is a unique signature of the double pole.

We now consider the case in which the laser frequency is chosen such that the laser resonantly couples two autoionizing states. Again, we neglect the direct laser-induced coupling between the autoionizing states and the continua and the laser-induced continuum-continuum coupling. For the case in which the energy of the incident electron is resonant with the lower lying autoionizing state, the resonant elastic and inelastic cross sections are [40], respectively,

$$\begin{aligned}\sigma_{a_1a_1}(E,I,\omega) &\propto \frac{\Gamma_{a_1}^2[4(E-E_{a_2}+\omega)^2+\Gamma_{a_2}^2]}{4|E-E_1(I,\omega)|^2|E-E_2(I,\omega)|^2},\\ \sigma_{a_2a_1}(E,I,\omega) &\propto \frac{\Gamma_{a_2}D_{a_1a_2}^2\Gamma_{a_1}}{|E-E_1(I,\omega)|^2|E-E_2(I,\omega)|^2}.\end{aligned} \tag{22}$$

Expressions for $\sigma_{a_2a_2}$ and $\sigma_{a_1a_2}$ are found by interchanging the labels a_1 and a_2 and by making the appropriate change in the energy scale. The complex quasienergies $E_{1,2}$ are given by (17). The resonant laser-induced coupling has again been treated within the RWA. A degeneracy occurs at the critical detuning and intensity given by (18). In this example, the existence of a degeneracy manifests itself in a different manner. The elastic and inelastic cross sections are symmetric about the position of the double pole along the real axis for all intensities when the frequency is such that $\delta=\delta_{\mathrm{c}}$ since Ω is purely imaginary for $I<I_{\mathrm{c}}$ and purely real for $I>I_{\mathrm{c}}$. From this and the fact that the elastic cross sections $\sigma_{a_1a_1}$ and $\sigma_{a_2a_2}$ are equal when respectively $E = E_{a_2} - \omega$ and $E = E_{a_1} + \omega$ we can conclude that when, for example, $\Gamma_{a_1} > \Gamma_{a_2}$ the cross sections $\sigma_{a_2 a_2}$ and $\sigma_{a_2 a_1}$ are singly peaked functions with a maximum at respectively $E = E_{a_2}$ and $E = E_{a_1}$ for $\delta = \delta_c$ and $I \leq I_c$ (however they are much narrower than the usual Breit-Wigner profile), while for $I > I_c$, all three cross sections are doubly peaked, with the separation of the peaks increasing as the square root of the intensity and the local minimum being at either $E = E_{a_1}$ or $E = E_{a_2}$.

ACKNOWLEDGEMENTS

It is a pleasure to acknowledge my collaborators who have contributed to the work described above: P.G. Burke, M. Dörr, C.J. Joachain, O. Latinne and H.W. van der Hart, as well as C.J. Noble, J. Purvis and M. Terao-Dunseath. The author has been supported by the European Community HCM Program as well as the Belgian Institut Interuniversitaire des Sciences Nucléaires (IISN).

REFERENCES

[1] P Agostini, F Fabre, G Mainfray, G Petite and N Rahman 1979. *Phys. Rev. Lett.* **42** 1127.

[2] R R Freeman, P H Bucksbaum, H Milchberg, S Darack, D Schumacher and M E Geusic 1987. *Phys. Rev. Lett.* **59** 1092.

[3] A L'Huillier, L-A Lompré, G Mainfray and C Manus 1992. *Atoms in Intense Laser Fields*, M Gavrila, ed., Academic Press, San Diego.

[4] J H Eberly, R Grobe, C K Law and Q Su 1992. *Atoms in Intense Laser Fields*, M Gavrila, ed., Academic Press, San Diego.

[5] R M Potvliege and R Shakeshaft 1988. *Phys. Rev. A* **38** 6190; R M Potvliege and R Shakeshaft 1992. *Atoms in Intense Laser Fields*, M Gavrila, ed., Academic Press, San Diego.

[6] K C Kulander 1987. *Phys. Rev. A* **36** 2726; 1988. *Phys. Rev. A* **38** 778; K C Kulander, K J Schafer and J L Krause 1992. *Atoms in Intense Laser Fields*, M Gavrila, ed., Academic Press, San Diego.

[7] P G Burke, P Francken and C J Joachain 1990. *Europhys. Lett.* **13** 617; P G Burke, P Francken and C J Joachain 1991. *J. Phys. B: At. Mol. Opt. Phys.* **24** 761.

[8] M Dörr, M Terao-Dunseath, J Purvis, C J Noble, P G Burke and C J Joachain 1992. *J. Phys. B: At. Mol. Opt. Phys.* **25** 2809.

[9] P G Burke and K A Berrington, ed., 1992. *Atomic and Molecular Processes: An R–matrix Approach*, Institute of Physics Publishing, Bristol.

[10] J Purvis, M Dörr, M Terao–Dunseath, C J Joachain, P G Burke and C J Noble. *Phys. Rev. Lett.* **71** (1993) 3943; M Dörr, J Purvis, M Terao–Dunseath, C J Joachain, P G Burke and C J Noble 1995. *J. Phys. B: At. Mol. Opt. Phys.* **28** 4481.

[11] O Latinne, N J Kylstra, M Dörr, J Purvis, M Terao–Dunseath, C J Joachain, P G Burke and C J Noble 1995. *Phys. Rev. Lett.* **74** 46.

[12] N J Kylstra, M Dörr, C J Joachain and P G Burke 1995. *J. Phys. B: At. Mol. Opt. Phys.* **28** L685.

[13] O Latinne 1995. *Thèse de doctorat.* (Université Libre de Bruxelles)

[14] N J Kylstra 1996. *Thèse de doctorat.* (Université Libre de Bruxelles)

[15] P Lambropoulos and X Tang 1992. *Atoms in Intense Laser Fields*, M Gavrila, ed., Academic Press, San Diego.

[16] S-I Chu 1985. *Adv. At. Mol. Phys.* **21** 197.

[17] H Rottke, B Wolff-Rottke, D Feldmann, K H Welge, M Dörr, R M Potvliege and R Shakeshaft 1994. *Phys. Rev. A* **49** 4837.

[18] M Pont, N R Walet, M Gavrila and C.W. McCurdy 1988. *Phys. Rev. Lett.* **61** 939; M Gavrila 1992. *Atoms in Intense Laser Fields*, M Gavrila, ed., Academic Press, San Diego.

[19] G Floquet 1883. *Ann. Ecole Norm. Sup.* **12** 47; M S P Eastham 1973. *The Spectral Theory of Periodic Differential Equations.* (Scottish Academic Press: Edinburgh)

[20] B R Junker 1982. *Ad. At. Mol. Phys.* **18** 207.

[21] A Maquet, S-I Chu and W P Reinhardt 1983. *Phys. Rev. A* **27** 2946.

[22] J Morellec, D Normand, G Mainfray and C Manus 1980. *Phys. Rev. Lett.* **44** 1394.

[23] U Fano 1961. *Phys. Rev.* **124** 1866.

[24] M Pont, R M Potvliege, R Shakeshaft, and P H G Smith 1992. *Phys. Rev. A* **46** 555.

[25] A Cyr and O Latinne 1996. Private communication.

[26] H W van der Hart 1996. Private communication.

[27] K Rzazewski and J H Eberly 1981. *Phys. Rev. Lett.* **47** 408.

[28] P Lambropoulos and P Zoller 1981. *Phys. Rev. A* **24** 379.

[29] R M Potvliege and P Smith 1992. *J. Phys. B: At. Mol. Opt. Phys.* **25** 2501.

[30] T-S Ho, S-I Chu and J V Tietz 1983. *Chem. Phys. Lett.* **96** 464; M Dörr, R M Potvliege, D Proulx and R Shakeshaft 1991. *Phys. Rev. A* **43** 3729.

[31] B W Shore 1967. *Rev. Mod. Phys.* **39** 439.

[32] Z Deng and J H Eberly 1984. *J. Opt. Soc. Am.* B **1** 102.

[33] N E Karapanagioti, O Faucher, Y L Shao, D Charalambidis, H Bachau, E Sanchez and E Cormier 1996. *Phys. Rev. A* **53** 2587.

[34] H W van der Hart 1996. *J. Phys. B: At. Mol. Opt. Phys.* **29** 2217.

[35] H Bachau, P Lambropoulos and R Shakeshaft 1986. *Phys. Rev.* A **34** 4785.

[36] M L Goldberger and K M Watson 1964. *Phys. Rev.* **136** B1472; J S Bell and C J Goebel 1965. *Phys. Rev.* **138** B1198.

[37] K E Lassila and V Ruuskanen 1966. *Phys. Rev. Lett.* **17** 490.

[38] P L Knight 1979. *Phys. Lett.* **72**A 309.

[39] M Gavrila 1989. *Collision Theory for Atoms and Molecules*, F A Gianturco, ed., Plenum, New York.

[40] N J Kylstra and C J Joachain 1996. To be published.

MULTIPHOTON PROCESSES IN A TWO-ELECTRON ATOM USING A CRAY T3D

K T Taylor, J S Parker, D Dundas, E Smyth and S Vivirito

Department of Applied Mathematics and Theoretical Physics
The Queen's University of Belfast
Belfast BT7 1NN, U K

INTRODUCTION

Many-electron atoms interacting with intense coherent radiation presents many new challenges for the theorist. Theory was stimulated enormously in recent years by the discovery of high harmonic generation in closed-shell atoms[1,2] , the discovery of above-threshold ionization[3], and the demonstration of double and multiple-electron ionization in intense fields[4]. An ab initio theoretical description of such processes is greatly hampered by the non-perturbative nature of the Schrödinger equation in the intense field limit where the effects are most prominent, as well as by the extreme difficulty of the quantum mechanical few-body problem. Nevertheless, especially as laser pulses get ever shorter, there is an overwhelming to directly integrate the full time-dependent Schrödinger equation (TDSE) for the laser-driven atom. The direct numerical integration of the TDSE for a laser-driven one-electron atom (which in linearly polarized light reduces to a PDE in two spatial-dimensions and the time), is straightforward on modern workstations. By contrast, the TDSE for two-electron atoms in linearly polarized light involves five spatial-dimensions. Until recently this was beyond the power of supercomputers to solve without severe approximation.

Such approximations have taken various forms. They range over single active electron SAE models and extensions[5]; time-dependent Hartree-Fock models[6,7]; lower-dimensionality models[8,9,10]; time-dependent Thomas-Fermi approximation[11] ; treatments via classical mechanics[12,13]; density-functional formalism[14]; R-matrix-Floquet formalism[15,16,17]; and time-integration of the TDSE over a field-free basis[18,19].

In this article we describe calculations on multiphoton absorption by helium, carried out by direct numerical integration of the full TDSE. The remainder of the article is arranged as follows. The next section gives brief details of the Cray T3D supercomputer we use and this is followed by a short description of the algorithm devised for this type of massively-parallel machine. Some early results are then presented, followed by examples of visualisations and by indications of possible extensions to our work. Finally some preliminary conclusions are drawn.

THE CRAY T3D AT EDINBURGH PARALLEL COMPUTER CENTRE

First installed in Edinburgh in the summer of 1994 the machine is now a Cray T3D MCN512-8 (512 150MHz 21064 Alpha processors known as "Processor Elements", each with 64MB local memory). The Cray T3D is physically constructed of "nodes", each containing 2 Processor Elements. Nodes are arranged in a 3-D torus with six links from each node, 2 each in X, Y and Z dimensions. Each link has a bandwidth of 300 MB/s and latency time (i.e. the time taken to initiate message passing) of the order of 1-2 microsecs. Two IO gateways supporting MPP to host communication down 200 MB/s HISP links, connect the Cray T3D to several front-end machines including a Cray J916/10-1024 (10 processors and 1 GB shared memory) with 304GB disc in 32 x Cray DDS-30 units. The aspect of the Cray T3D architecture crucial to our algorithm is the high-bandwidth of 300 MB/s between nodes.

THE ALGORITHM

To perform the integration of the TDSE we use a mixed basis set, finite-difference method, in which the two radial coordinates r_1 and r_2 are modelled on a finite-difference grid, and the four angular coordinates $\theta_1, \theta_2, \phi_1, \phi_2$ are handled by writing the wavefunction on a basis set of coupled spherical harmonics ("partial waves"), $|l_1 l_2 LM>$. In linearly polarized light, the z component of the total angular momentum, M, is constant resulting in only 5 independent spatial-dimensions. Since the initial state has $M = 0$, we write the wavefunction $\Psi(r_1, r_2, \theta_1, \theta_2, \phi_1, \phi_2)$ as,

$$\Psi = A \sum_{L,l_1,l_2} f_{L,l_1,l_2}(r_1, r_2, t)|l_1 l_2 L>, \tag{1}$$

where A is the Pauli symmetrization operator. Electron spin S is conserved: in the present case of a singlet initial state, $S = 0$ at all times, so that A enforces even symmetry of the wavefunction under exchange of electron spatial coordinates.

The radiation field has the freedom to exchange angular momentum with the electrons, and the number of partial waves coming into play can become quite large. The largest number of partial waves we have used to date is 256. With this setting, the Schrödinger equation becomes a set of 256 coupled two-dimensional time-dependent, partial differenential equations, each of which is solved by finite-difference techniques. The kinetic energy operators of the atomic Hamiltonian are approximated as 5 point finite-difference operators. The momentum operators of the (velocity gauge) interaction Hamiltonian are approximated as 2 or 4 point finite-difference operators. The electron-electron coulomb potential is expanded in the well-known series,

$$\frac{1}{|\mathbf{r}_1 - \mathbf{r}_2|} = \sum_l \frac{{r_<}^l}{{r_>}^{l+1}} P_l(cos(\theta_{12})). \tag{2}$$

In our calculations up to now, the RHS of Eq. (2) has been truncated at the $l = 1$ dipole term; and 2 point finite-difference operators used to represent the interaction Hamiltonian. Retaining 256 partial waves permits a partial wave expansion as far as

$l_{1\,max} = l_{2\,max} = 9$; $L_{max} = 18$; alternatively the retention of 128 partial waves allows $l_{1\,max} = l_{2\,max} = 6$; $L_{max} = 12$. With 248 nm light incident on ground state He, the 128 partial wave expansion appears to be a good approximation up to 4×10^{14} W/cm^2.

The ground state of the finite-difference grid is found by running the Schrödinger equation as a diffusion equation, rather than as a wave equation. States of higher energy than the ground state dissipate rapidly. With a grid spacing of $\delta r = 0.5$ Bohr radii, and enforcing the correct functional form near the nucleus, the two-electron ground state is found to have an energy of -2.79 Hartrees, significantly higher than the true value of -2.9 Hartrees. Convergence to -2.9 Hartrees occurs only at spacings below $\delta r = 0.1$ Bohr radii, a setting far beyond that which our present allocation of computing resources can accommodate for the subsequent time propagation of the Schrödinger equation. However, the smooth convergence to -2.9 Hartrees verifies that the finite-difference technique is capable of faithfully modelling the electron-electron electrostatic repulsion which, largely through the screening of one electron by the other, determines the helium ground state energy.

The basis state decomposition is particularly well suited to parallelization. The present version of the code runs well on a Cray T3D. At present, each of the 256 partial waves is mapped to one of the 256 processors of the Cray T3D. On the T3D a typical 15 field-period run using 256 processors now takes about 5 wall-clock hours, assuming 2000 time steps per field period. The results discussed here were obtained in a $50a_0$ box in r_1 and r_2. We do not use an absorbing potential at the outer boundary but rather split the wavefunction into a coherent superposition of an outer part and an inner part. The outer part is zero at $r_1, r_2 \leq 30a_0$ but grows so that it contains the total wavefunction at the 50 a_0 boundary, with the consequence that the inner part smoothly approaches zero at the boundary. At present we discard this outer part and so the net effect is equivalent to that of an absorbing boundary.

This numerical method is thus capable of faithfully modelling double ionization, autoionization, or any other effect associated with highly correlated states of the two electrons. The degree to which the numerical integration approximates the Schrödinger equation is governed by four parameters in the code: the spacing, δr, of the finite-difference grid points representing the two radial variables; the maximum angular momentum L_{max} present in the partial-wave expansion; the number of terms retained in a series expansion of the electron-electron interaction; and the size of the integration volume. By changing these parameters, the model can be made arbitrarily close to the Schrödinger equation from which it is derived. In the data presented here, the dominant source of error is the rough radial grid. The grid spacing δr is set to 0.5 Bohr radii, a setting chosen to minimize CPU usage in exploratory work.

RESULTS

All results generated up to now have been for 248 nm radiation ramping up linearly over two cycles to peak intensity which is maintained for another 5-10 cycles. Convergence studies have already been presented elsewhere[20] and so here we present and discuss some new results for single-atom harmonic generation rates. But first we display in Figure 1 a plot (full line) of the decay of the He electron probability density within 4.5 a_0 of the nucleus for a low peak intensity of 2×10^{14} W/cm^2. The other full line, remaining at unity, is the decay (negligible) of He^+ that occurs for this intensity. The dashed line, which follows the He result closely, comes from a SAE treatment of He over our spatial grid where the inner effective charge used (Z = 1.33) comes from

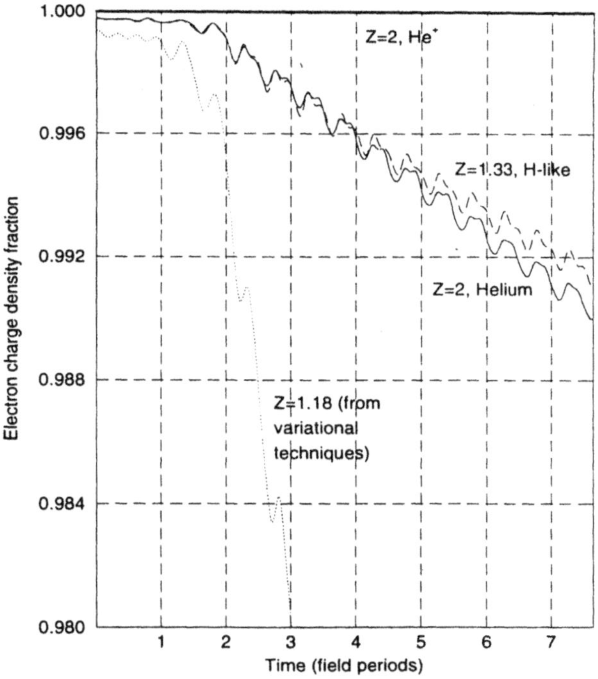

Figure 1: Fraction of electron charge density within 5 Bohr of the nucleus in presence of a 248 nm laser pulse ramping linearly over 2 cycles to a peak intensity of 2×10^{14}Wcm^{-2}. The full-line remaining at unity is for He^+ and the falling full-line is for He over the grid. The dashed line is for a SAE approximation for He chosen to have an ionization potential identical to the first ionization potential of our He model over the grid, while the dotted line is obtained also using a SAE approximation but now with an ionization potential appropriate to the outer electron in a simple two-parameter variational form.

the first ionisation potential $[(2.79 - 1.9) = 0.89$ Hartrees$]$ over our grid. The remaining dotted line is produced from a SAE description, but this time using $Z = 1.18$ (which comes from the exponential form for the outer electron in the simplest two-parameter variational wavefunction for He[21]. We note that at this low intensity only single electron ionization occurs, which can be calculated accurately using a SAE model provided the correct value is chosen for the residual charge seen by the escaping electron.

In figure 2 we present results for the single-atom harmonic response of He at a variety of peak intensities, together with, in the last frame, the response of He^+ at the maximum peak intensity considered. 128 partial waves were retained for the lowest two intensities, and 256 for the others. Not only does the harmonic train from He lengthen under increasing intensity, as expected, but clear indications of the emergence of a secondary plateau can be seen. Moreover the comparison with He^+ indicates that the He atom is more efficient at generating high-order harmonics than the He^+ ion.

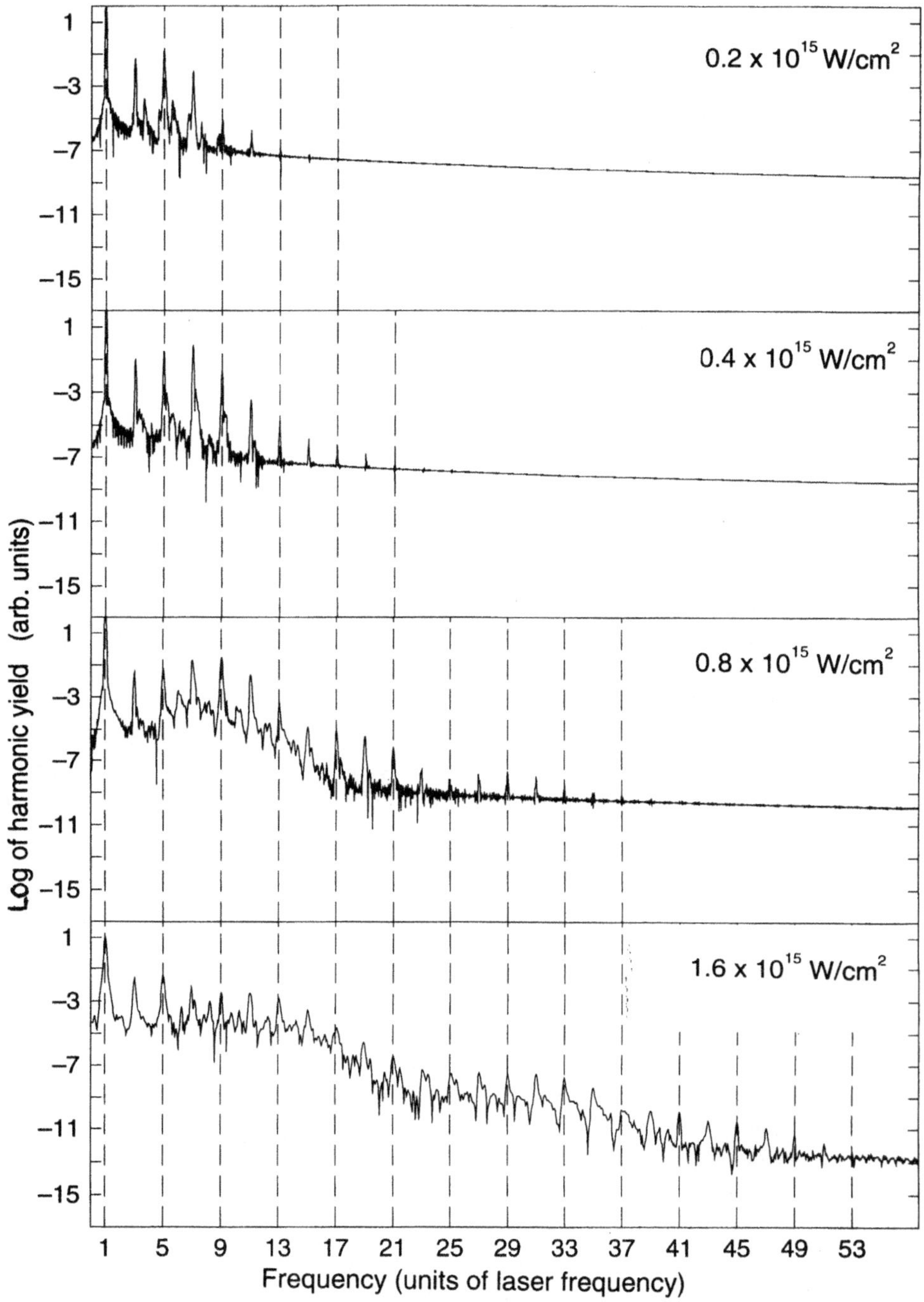

Figure 2: Single-atom harmonic response of He and He^+ for a selection of increasing peak intensities of 248 nm laser radiation. In each case the pulse rises linearly to peak intensity over 2 cycles.

VISUALIZATIONS

In order to gain more insight into the dynamics underlying our results, some of which have been reported above, we have developed a number of visualizations that allow us to follow the time-evolution of the 5 dimensional wavefunction generated for laser-driven helium. In practice this amounts to taking a 2 dimensional cut, with the other dimensions essentially averaged over. Such post-processing and subsequent visualization display is carried out on local graphical workstations to which Cray T3D generated data (typically several Gb per pulse) is downloaded from Edinburgh.

The first visualization we consider is the construction of radial plots of the probability density. These plots are produced by integrating the probability density over all angular variables:

$$P(r_1, r_2, t) = \int_{\Omega_1} d\Omega_1 \int_{\Omega_2} d\Omega_2 \Psi^*(\mathbf{r}_1, \mathbf{r}_2, t)\Psi(\mathbf{r}_1, \mathbf{r}_2, t)) = \sum_{l_1 l_2 L} |f_{l_1 l_2 L}(r_1, r_2, t)|^2 , \qquad (3)$$

noting that the basis functions are orthonormal. Plots of this quantity over a sequence of time-instant provides us with useful information on the ionization dynamics. In such plots three different regions can be identified as indicated schematically in Figure 3. Migration of electron flux between these regions identifies distinct ionization pathways. Region 1, extending to about 5 a_0 from the nucleus in both radial co-ordinates encompasses the innermost part of the two-electron wavefunction. Ground state helium lies almost entirely within this region. Region 2, where one radial co-ordinate is large and the other small, is where singly-excited and singly-ionized states of He are mostly to be found, while region 3, where both radial co-ordinates are large, contains most wavefunction of doubly-excited and doubly-ionized states. Thus a movement of electron flux from region 1 to region 2 corresponds to single-electron excitation and ionization of the atom; from region 1 directly to region 3 corresponds to simultaneous-double-electron excitation and ionization. Finally, movement from region 2 to region 3 indicates both second electron excitation and second electron ionization of already singly-excited or singly-ionized He.

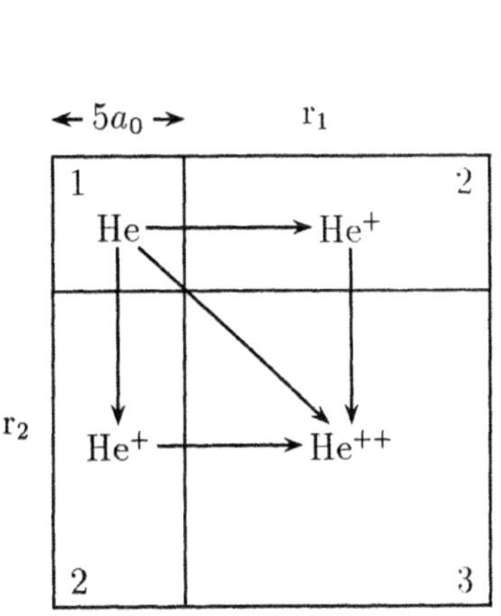

$1 \rightarrow 2$: $He + n_p\hbar\omega \longrightarrow He^+ + e^-$

$2 \rightarrow 3$: $He^+ + n_q\hbar\omega \longrightarrow He^{++} + e^-$

$1 \rightarrow 3$: $He + n_r\hbar\omega \longrightarrow He^{++} + 2e^-$

Figure 3: Schematic diagram of the radial plots showing the main ionization mechanisms.

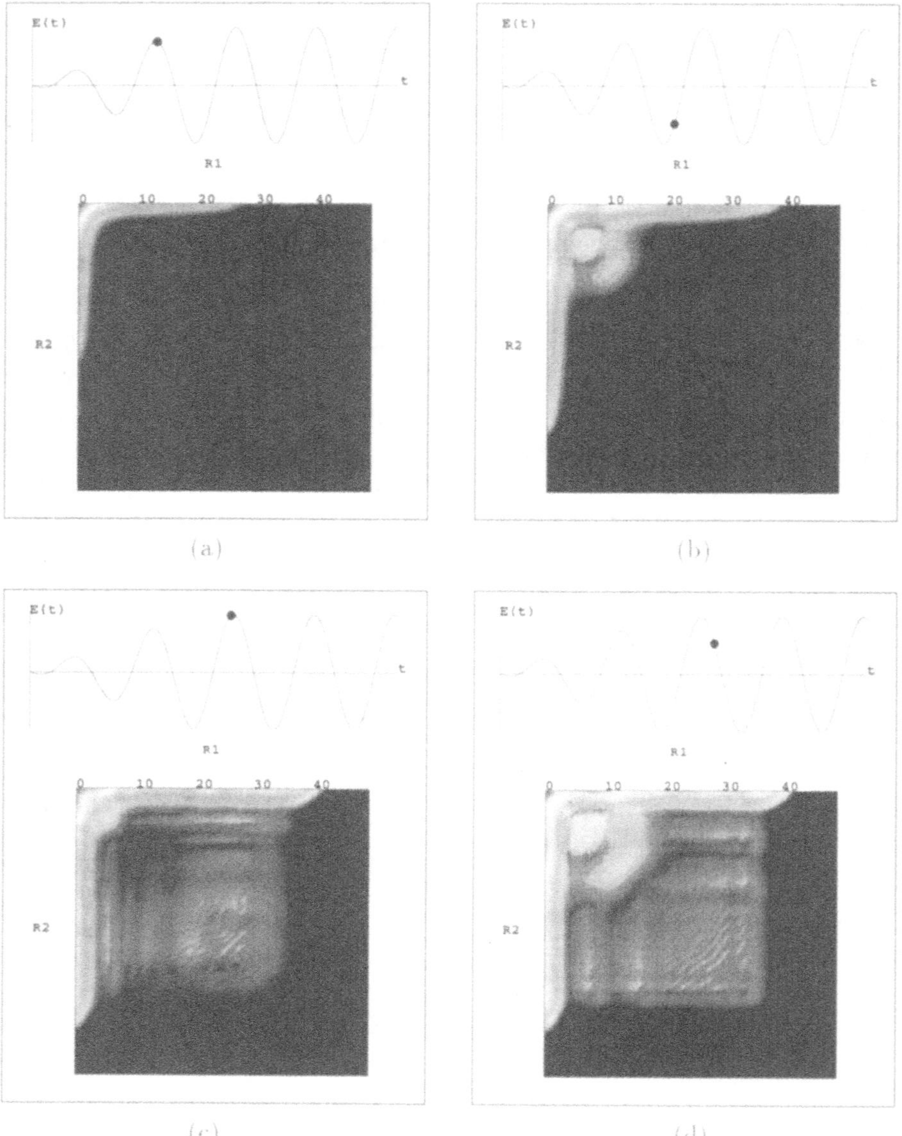

Figure 4: Radial plots at various instants in a 248nm pulse which linearly ramps up over two optical cycles to an intensity of $2 \times 10^{15}\mathrm{Wcm}^{-2}$; red and blue areas corresponding to high and negligible probability respectively. Each plot includes a profile of the pulse, the red dot indicating the instant in the pulse. In (a) during the initial ramp on, single-electron ionization occurs. In (b) where the field has ramped on fully the onset of simultaneous-double ionization is apparent. In (c) at a later time where the pulse is passing through a maximum, sequential double ionization occurs in addition to simultaneous double ionization . In (d), shortly after the field has passed through another maximum, another burst of simultaneous double ionization is seen to occur.

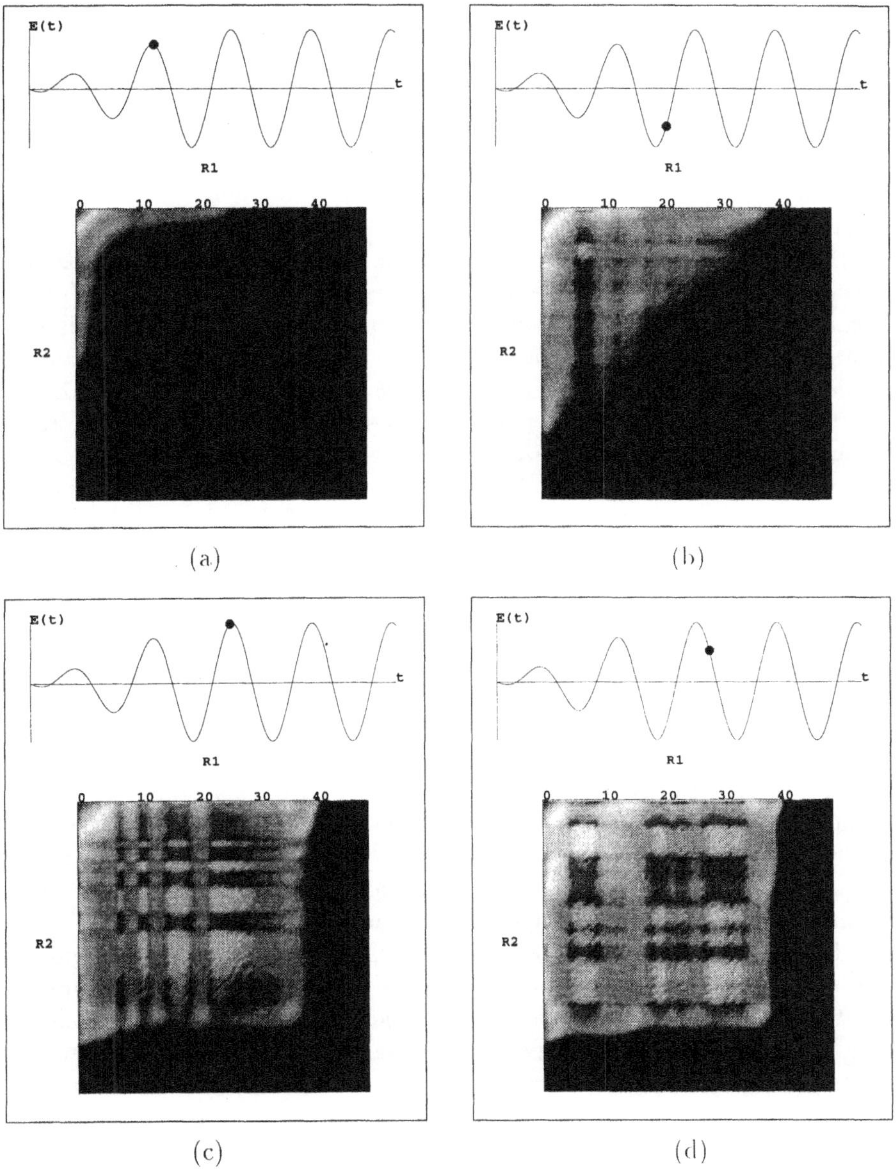

Figure 5: Radial plots from a calculation dropping $1/r_{12}$ at the same instants as the corresponding plots in Figure 4 for a 248nm pulse which linearly ramps up over two optical cycles to an intensity of $2 \times 10^{15}\text{Wcm}^{-2}$; red and blue areas corresponding to high and negligible probability respectively. No indication of simultaneous-double-ionization is present.

Figure 4 displays radial plots at four time-instants during a pulse rising to a maximum intensity of 2×10^{15}Wcm^{-2}. In (a), at an instant during the initial ramp on, single-electron excitation and ionization is indicated by a population in region 2. In (b) where the field has ramped on fully a packet entering region 3 from region 1 indicates the onset of simultaneous-double-electron ionization and excitation. In (c) where the field is passing through a maximum, we see that in addition to simultaneous-double -electron ionization and excitation, sequential-double-electron ionization and ionization is indicated by wavefronts passing from region 2 to region 3. In (d), shortly after the field has passed through a maximum, another burst of population moving from region 1 to region 3 indicates further simultaneous-double -electron ionization and excitation. A subsequent burst occurs every half-cycle and sequential-double-electron ionization and ionization continues to occur.

In order to unambiguously identify the role of the $1/r_{12}$ term in these excitation and ionization pathways the calculation was repeated dropping this term from the Hamiltonian. Figure 5 displays radial plots from this calculation at precisely the same instants as in Figure 4. Although single-electron excitation and ionization and sequential-double-electron excitation and ionization are clearly occurring the absence of any burst of population migrating from region 1 to region 3 indicates the entire absence of simultaneous-double -electron ionization and excitation in this case.

Other visualizations yielding angular information on electron dynamics have either been produced or are in course of development. Details will be reported elsewhere.

FURTHER WORK

An important extension to the algorithm would be integration of the electron wavefunctions to larger distances from the nucleus than the 50 a_0 used at present. We believe this can be done (possibly with some rearrangement of our calculation over processors) by dropping the explicit point-to-point inclusion of the $1/r_{12}$ term for larger radial distances. The thinking is illustrated in Figure 6. In this figure the boundary of the inner kernel region (that treated at present) is at $r_1 = r_2 = R_2$. Integration of the inner region solution would proceed, alone, and without the present damping, until this boundary was first reached. Integration in the outer region ($r_1 > 0, r_2 > R_2$ i.e. the bottom of Figure 6(a)) would then be started up. The schematic points indicated by x's in Figure 6(b) and (c) represent values of $r_1, \theta_1, r_2, \theta_2$ in the inner region that would be involved in the outer region propagation. The double arrow in Figure 6 (a) indicates that allowance must be made in the inner and outer region integrations for wavefunction flow back and forth between these regions. The extent of this flow about the boundary $r_2 = R_2$ would be governed by the quiver amplitude for the laser frequency and intensity under consideration. This clearly requires the inner and outer regions be calculated in step, which could be done on the Cray T3D by handing over the outer region integration to a limited number of processors, retaining the bulk of them for the inner region work. Such an extension would allow the present calculations of harmonic generation to be extended to longer wavelengths and higher peak intensities (both of which increase the quiver amplitude) and would also make possible the calculation of other quantities such as rates for ATI.

A general point to note is that our method is not restricted to any particular pulse-shape and so we welcome details of experimental forms, especially in the 248 - 500 nm wavelength range which might be produced in the 20 fs laser systems under construction.

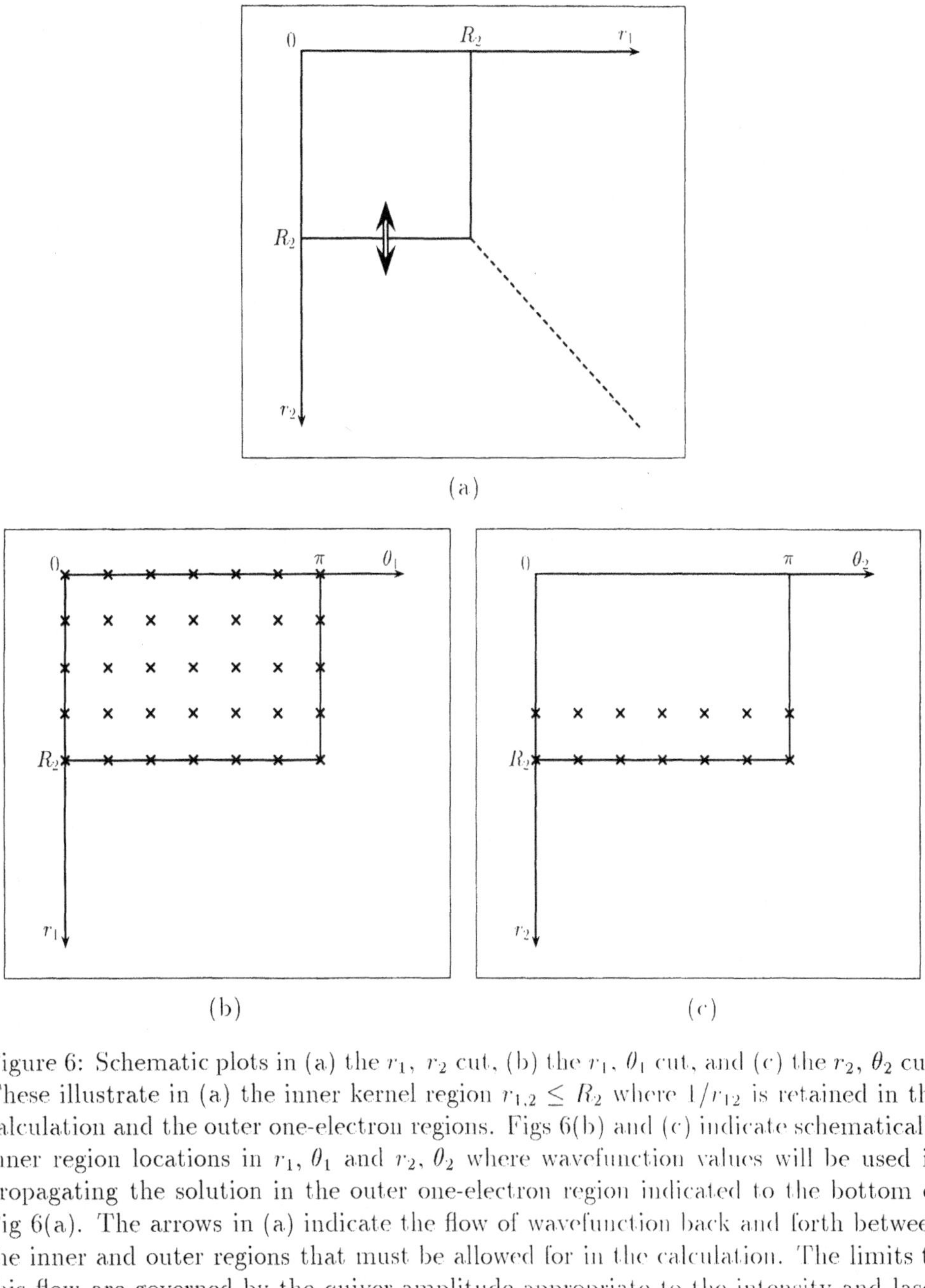

Figure 6: Schematic plots in (a) the r_1, r_2 cut, (b) the r_1, θ_1 cut, and (c) the r_2, θ_2 cut. These illustrate in (a) the inner kernel region $r_{1,2} \leq R_2$ where $1/r_{12}$ is retained in the calculation and the outer one-electron regions. Figs 6(b) and (c) indicate schematically inner region locations in r_1, θ_1 and r_2, θ_2 where wavefunction values will be used in propagating the solution in the outer one-electron region indicated to the bottom of Fig 6(a). The arrows in (a) indicate the flow of wavefunction back and forth between the inner and outer regions that must be allowed for in the calculation. The limits to this flow are governed by the quiver amplitude appropriate to the intensity and laser frequency under consideration.

CONCLUSIONS

We believe our *ab initio* approach is especially valuable for short laser pulses (20 fs) where a significant amount of the outer electron is still present when the inner electron begins to respond significantly. We expect in these circumstances that the role of the inter-electronic interaction will be increasingly important as one moves to higher stages of ionization along an isoelectronic sequence.

Acknowledgements

This work has been supported in part by Cray Research Inc and the UK Engineering and Physical Sciences Research Council through the award of a Cray Research Inc fellowship to Dr Jonathan Parker and resources on the Cray T3D at Edinburgh. Daniel Dundas and Edward Smyth acknowledge receipt of Department of Education Northern Ireland studentships and Dr Sara Vivirito acknowledges support from the EU through a post-doctoral fellowship.

REFERENCES

1. X.F. Li, A. L'Huillier, M. Ferray, L.A. Lompre', G. Mainfray and C. Manus 1989 *Phys. Rev. A* **39** 5751
2. G. Gibson, H. Jara, U. Johann, T.S. Luk, I. McIntyre, K. Boyer, and C.K. Rhodes 1987 *J. Opt. Soc. Am B* **4** 595
3. P. Agostini, F. Fabre, G. Mainfray, G. Petite and N.K. Rahman 1979 *Phys. Rev. Lett.* **42** 1127
4. A. L'Huillier, L.A. Lompre', G. Mainfray and C. Manus 1983 *Phys. Rev. A* **27** 2503
5. K.C. Kulander 1988 *Phys. Rev. A* **38** 778
6. K.C. Kulander 1987 *Phys. Rev. A* **36** 2726
7. M. Horbatsch, H.J. Ludde and R.M. Dreizler 1992 *J. Phys. B: At. Mol. Opt. Phys.* **25** 3315
8. R. Grobe and J.H. Eberly 1992 *Phys. Rev. Lett* **68** 2905
9. A. Artemyev, R. Grobe and J.H. Eberly 1995 *Phys. Rev. A* **51** 155
10. D.G. Lappa, A. Sanperra, J.B. Watson, K. Burnett, P.L. Knight, R. Grobe and J.H.Eberly 1996 *J. Phys. B: At. Mol. Opt. Phys.* **29** L619
11. M. Brewczyk, K. Rzążewski and C.W. Clark 1995 *Phys. Rev. A* **52** 1468.
12. D.W. Wasson and S.E. Koonin 1989 *Phys. Rev. A* **39** 5676
13. P.B. Lerner, K. LaGattuta and J.S. Cohen 1993 *Laser Physics* **3** 331
14. E. Runge E and E.K.U. Gross 1984 *Phys. Rev. Lett* **52** 997
15. P.G. Burke, P. Francken and C.J. Joachain 1991 *J. Phys. B: At. Mol. Opt. Phys.* **24** 761.
16. J. Purvis, M. Dörr, M. Terao-Dunseath, C.J. Joachain, P.G. Burke and C.J. Noble 1993 *Phys. Rev. Lett* **71** 3943
17. M. Dörr, J. Purvis, M. Terao-Dunseath, P.G. Burke, C.J. Joachain and C.J. Noble 1995 *J. Phys. B: At. Mol. Opt. Phys.* **28** 4481.
18. X. Tang, H. Rudolph and P. Lambropolous 1990 *Phys. Rev. Lett.* **65** 3269
19. J. Zhang and P.Lambropoulos 1995 *J. Phys. B: At. Mol. Opt. Phys.* **28** L101
20. J. Parker, K.T. Taylor, C.W. Clark and S. Blodgett-Ford 1996 *J. Phys. B: At. Mol. Opt. Phys.* **29** L33
21. B.L. Moiseiwitsch 1961 *The variational method* in *Quantum Theory* Ed by D R Bates vol 1 (New York and London: Academic Press)

APPLICATION OF PARALLEL COMPUTERS TO ELECTRON-MOLECULE SCATTERING CALCULATIONS

N. Sanna[1] and F. A. Gianturco[2]

[1] Center for Supercomputing Applications to Research (CASPUR), The University of Rome, Cittá Universitaria, 00185 Rome, ITALY
[2] Department of Chemistry, The University of Rome, Cittá Universitaria, 00185 Rome, ITALY

1. INTRODUCTION

The treatment of the scattering processes which occur in the collision of electrons with polyatomic targets has been object of intensive theoretical studies in the last ten years or so, and several methods to solve the final scattering equations have been proposed[1, 2, 3, 4] by different groups each with specific features which treat at various levels of sophistication the physics of the electron-molecule scattering process. The intrinsic complexity of the models adopted to understand the behaviour of the scattering process requires the use of algorithms which exploit at best the programming features of the available top-class hardware and software technologies. While a full comprensive description of the different models has been given elsewhere,[5] some general and common requirements that fit the needs of parallel computing have still to be exploited, in particular when huge cpu-demanding power is necessary such that required for the solution of the electron-molecule scattering equations. We outline below how we implemented our set of codes on parallel machines within the Single-Center-Expansion (SCE) framework description, showing specific numerical results on a typical, parallel MIMD architecture.

2. THE SINGLE CENTRE EXPANSION (SCE) METHOD

2.1. The SCE bound state wavefunction

A general discussion on the computational aspects of the present model has been given before[1] and specific results have also been analysed elsewhere.[6, 9] We will therefore only outline here the procedure for obtaining the bound state wavefunction to be included later into the scattering equations.[10] In the present computational approach one needs to interface an initial, general-purpose quantum chemistry code that is employed to generate the Single Determinant description, (near to the Hartree-Fock (HF) limit) of the target electronic wavefunction, with a numerical procedure that can give us all the necessary quantities as being referred to the molecular C.O.M.[1]

In most of the numerical methods employed to solve the scattering equations one then converts the CC equations into a set of coupled radial equations by making first a single centre expansion (SCE) of the bound and continuum functions and then by integrating over all the angular coordinates. One therefore writes down the bound and continuum orbitals as expansions around the centre of mass, from which the body-fixed (BF) frame of reference originates:

$$\phi_i(\mathbf{r}) = r^{-1}\sum_{hl} u^i_{hl}(r)\, X^{i,p\mu}_{hl}(\theta,\phi) \quad , \tag{1a}$$

$$F_{p\mu}(\mathbf{r}) = r^{-1}\sum_{hl} f^{p\mu}_{hl}(r)\, X^{p\mu}_{hl}(\theta,\phi) \quad . \tag{1b}$$

Here i labels a specific, multicentre molecular orbital (MO), which contributes to the density of the bound electrons in the nonlinear target, while the indices $|p\mu>$ for the continuum functions label one of the relevant IRs and one of its components, respectively. The index h labels a specific basis, at a given angular momentum l, for the p-th IR that one is considering.[12]

One important point is the construction of the symmetry-adapted, generalized harmonics X as linear combinations of spherical harmonics $Y_l^m(\theta,\phi)$ which, for a given l, form a basis of the $(2l+1)$ dimensional IR of the full rotation group. If such generalized harmonics satisfy the usual orthonormality relations, then the coefficients over the Y's correspond to a unitary transformation between the X's and the spherical harmonics:

$$X^{p\mu}_{hl}(\theta,\phi) = \sum_m b^{p\mu}_{hlm}\, Y_l^m(\theta,\phi) \tag{2}$$

and the coefficients b's can be obtained from a knowledge of the character tables for each of the IR appearing in the relevant molecular point group.[13]

In order to perform expansions (1), one needs, therefore, to start from the multicentre wavefunction which describes the target molecule and then generate by quadrature each of the $u^k_{hl}(r)$ coefficients; they were obtained for the first time by numerical quadrature of the multicenter GTO's given as Cartesian Gaussian functions:[6]

$$g^{kj}_v(\alpha_j,\overline{r}_k) = N_j(a,b,c;\alpha)\; x^a\cdot y^b\cdot z^c\; e^{-\alpha\overline{r}_k^{\,2}} \tag{3}$$

labelled by the k-th atomic center, of which $g^{kj}_v(\alpha_j,\overline{r}_k)$ is the j-th function, and by the global index v of the GTO $(a,b,c;\alpha)$ parameters included in the normalisation constant N:

$$N_j(a,b,c;\alpha) = \left[\left(\frac{2\alpha}{\pi}\right)^{3/2}\frac{(4\alpha)^{a+b+c}}{(2a-1)!!(2b-1)!!(2c-1)!!}\right]^{1/2} \tag{4}$$

The corresponding radial coefficients of eq. (1a) are thus given by the angular quadrature:

$$u^i_{hl}(r;R) = \sum_k\sum_j\sum_v\sum_m \int_0^{\pi}\sin(\theta)\,d\theta\int_0^{2\pi} b^i_{lm}Y_l^m(\theta,\phi)C^i_{kj}(R)\cdot d^{kj}_v\cdot\, g^{kj}_v(\alpha_j,\overline{r}_k)\,d\phi \tag{5}$$

where the first two terms on the r.h.s. describe explicitly the generalized real, symmetry-adapted harmonics X_{hl} with $|p\mu> = |i>$, C^i_{kj} the GTO coefficient of the v-th GTO at a given molecular geometry R and the d^{kj}_v the contraction coefficients of the primitive

Gaussians $g_v^{kj}(\alpha_j, \overline{r}_k)$. The quadrature were carried out using Gauss-Legendre abscissas and weights for θ and Gauss-Chebyschev abscissas and weights for ϕ, on a discrete variable radial grid.

2.2. The SCE one-electron density

After one obtains the radial coefficients from eq. (5), each bound one-electron wavefunction is also expanded about the C.O.M. in terms of the X functions

$$u_i(\mathbf{r}) = r^{-1} \sum_{hl} u_{hl}^i(r) X_{hl}^{p\mu}(\theta, \phi) \tag{6}$$

then, the one-electron density function can be written as usual

$$\rho(\mathbf{r}) = \int |\psi(x_1, x_2, ..., x_N)|^2 \, dx_2...dx_N = 2 \times \sum_i |u_i(\mathbf{r})|^2 \tag{7}$$

where the factor 2 is due to the sum over spin, and the i sum is over each doubly occupied orbital. Once the quantity $\rho(\mathbf{r})$ is obtained from the bound-state wavefunction $u_i(\mathbf{r})$ of eq. (6), it can be expanded in terms of symmetry-adapted functions belonging to the A_1 irreducible representation as

$$\rho(\mathbf{r}) = r^{-1} \sum_{hlm} \rho_{hlm}(r) X_{hlm}^{A_1}(\theta, \phi) \tag{8}$$

where

$$\rho_{hlm}(r) = 2 \times \sum_i \int_0^{\pi} \sin(\theta) \, d\theta \int_0^{2\pi} u_i(\mathbf{r}) \cdot u_i(\mathbf{r}) \, d\phi \tag{9}$$

2.3. The SCE Static Potential

From the one-electron density of eq. 8 it is straightforward to perform the calculation of the static potential exerted by the electrons and nuclei on the surrounding the molecular volume as

$$V^{ST}(\mathbf{r}) = \int \rho(s) \frac{1}{|r - s|} ds + \sum_i^N \frac{Z_i}{|r - R_i|} \tag{10}$$

where the first term of the right-hand-side refers to the electron contribution and the second, with the the sum over the number of atoms N, to the shielding due to the positive nuclar charges. The integral of the one-electron density expressed in the short hand notation of eq. 10 means actually a double integration outward from the C.O.M. coordinates to r_i, being r_i the current r integration point, and inward from the asymptotic r value to r_i. This integration is carried out in our set of codes using a very simple extended Simpson rule [7] that have been proved to be reliable enough since its first formulation due to Gianturco and Thompson.[8] The V_{ST} is then, as usual, expanded over symmetry adapted spherical harmonics (SASH) as

$$V^{ST}(\mathbf{r}) = \sum_{hlm} V_{hlm}^{ST}(r) X_{hlm}^{A_1}(\theta, \phi) \tag{11}$$

where $V_{hlm}^{ST}(r)$ is the radial part of the electron and nuclear contributions to the total static potential for a given l, m couple of values of the h representation under the A_1 i.r. one is considering.

2.4. The Free Electron Gas Exchange (FEGE) Model Potential

One way to simplify the solution of integro-differential radial equation due to the calculation of the exact exchange potential[14], is the use of a model - energy dependent - exchange potential which could reproduce the relevant features of the interactions among the inner electrons with the incoming charged projectile. One of the most widely used for diatomic and polyatomic targets has been the Free-Electron-Gas (FEG) introduced by Hara [15] as an extension to scattering problems of the older Slater's average exchange potential for bound states. In our procedure the FEGE potential represents the starting point of the Exact Static Exchange (ESE) solution of the scattering equations, where it will be iteratively substituted by the exact exchange potential. Nonetheless, the FEGE potential represents the best initial guess to the exchange interaction between the incoming electron and the bound state electrons, hence permitting to solve the iterative exchange scheme in relatively few iterations. The actual form of the FEGE exchange potential derives from two aproximations: (i) the molecular electrons are treated as a free-electron gas, with a given charge density $\rho(\mathbf{r})$ determined from the ground electronic state; (ii) the distortion due to interaction with the continuum function is neglected and the impinging projectile is treated as a plane wave. The resulting FEGE potential is then given by the following form:

$$V^{FEGE}(\mathbf{r}) = -\frac{2}{\pi}K_F(\mathbf{r})\left[\frac{1}{2} + \frac{1-\eta^2}{4\eta}ln\left|\frac{1+\eta}{1-\eta}\right|\right] \tag{12}$$

where the wavevector up to the top of the Fermi surface is given by the usual free-electron relation

$$K_F(\mathbf{r}) = \left[3\pi^2\rho(\mathbf{r})\right]^{1/3} \tag{13}$$

and η holds the ratio between the actual wavevector for the scattering electron and the one at the top of the Fermi surface for the elctron gas, $\eta = k/K_F$, where the local momentum is given as

$$k(\mathbf{r}) = \left[2(E_{coll} + I_P) + K_F^2(\mathbf{r})\right]^{1/2} \tag{14}$$

the collision energy E_{coll} is in turn given by the initial-channel energy of $1/2k_0^2$. I_P now is the ionisation potential for the neutral target molecule. The final FEGE exchange potential is then obtained with the integration scheme outlined so far for the bound state wavefunction and the electron density, as

$$V^{FEGE}(\mathbf{r}) = \sum_{hlm} V_{hlm}^{FEGE}(r)X_{hlm}^{A_1}(\theta,\phi) \tag{15}$$

2.5. The VCP Polarisation Potential

Another important type of interaction between the impinging electron and the target molecule comes from the description of the response function of the target to the incoming projectile. The modeling of such interaction is rather difficult because its nature depends on the distance of the incoming electron, varying from a purely polarisation effects at long R up to including exchange-correlation interaction with the bound electrons within the molecular volume. The model correlation-polarisation potential, V^{cp}, that have been implemented within a SCE treatment of polyatomic molecules is therefore distinguishing between a long-range region of interaction (where perturbative polarisation effects are dominant, the potential acquires correctly a local analytic form and is both adiabatic on the projectile's velocity and independent of the sign of its charge) and a short-range region where non-adiabatic, non-local effects play an important role and differences appear between the behaviour of electrons and positrons as projectiles. One therefore writes the overall Correlation-Polarisation interaction as given by

$$V^{cp}(\mathbf{r}) = V_{corr}(\mathbf{r}) \qquad for \qquad \mathbf{r} \leq \mathbf{r_c} \tag{16a}$$

$$= V_{pol}(\mathbf{r}) \qquad for \qquad \mathbf{r} > \mathbf{r_c} \tag{16b}$$

where the connecting spatial factors $\mathbf{r_c}$ are usually obtained from the crossing radii of the lower coefficients in the expansion

$$V^{cp}(\mathbf{r}) = \sum_{hl} V_{hl}^{cp}(r) X_{hl}^{A_1}(\theta, \phi) \tag{17}$$

In early implementation of the V^{cp} potential, the short-range correlation interaction was simulated via a FEG modelling in order to evaluate the correlation between the bound electrons, given as the SCF density for the target ground electronic state, and the impinging electron. The corresponding E_{corr} turns out to be an analytic function of such densities within the Kohn-Sham description of the single-particle orbitals[16] and therefore one obtains an analytic expression for both V_{corr} and V_{pol} in eq.s (16), thereby producing the corresponding crossing region to obtain the required r_c value.

The above formulation of the V^{cp} polarisation potential has been modified during the last ten years or so and at present a new implementation of the V^{cp} is available where the V_{corr} is calculated as a modified FEG model which has a weaker short-range correlation potential with the functional form

$$V_{corr}^{FEG}(\mathbf{r}) = 0.0311 ln r_s - 0.0584 + 0.00133 r_s ln r_s - 0.0084 r_s \qquad for \qquad \mathbf{r_s} < \mathbf{1.0} \tag{18a}$$

$$= \frac{\gamma(1 + \frac{7}{6}\beta_1 r_s^{1/2} + \frac{4}{3}\beta_2 r_s)}{1 + \beta_1 r_s^{1/2} + \beta_2 r_s)^2} \qquad for \qquad \mathbf{r_s} \geq \mathbf{1.0} \tag{18b}$$

where $\gamma = 0.1423$, $\beta_1 = 1.0529$ and $\beta_2 = 0.334$. The asymptotic correction which depends on the polarizability, has been included in a more general fashion as described in the next section 2.7.

2.6. The VDFT Polarisation Potential

A full description of the V_{corr}^{DFT} correlation polarisation potential has been given elsewhere by Gianturco and Rodriguez [17, 18, 19] and we only outline here its leading features. As for the relevant quantities reported so far the V_{corr}^{DFT} potential must be expanded over SASH

$$V_{corr}^{DFT}(\mathbf{r}) = \sum_{hlm} V_{hlm}^{DFT}(r) X_{hlm}^{A_1}(\theta, \phi) = \frac{\delta}{\delta\rho} E_c[\rho(\mathbf{r})] \tag{19}$$

where E_c is the correlation energy which depends on the one-electron density $\rho(\mathbf{r})$. Apart from the one-electron density, the $V_{corr}^{DFT}(\mathbf{r})$ depends on the first and second functional derivatives of $\rho(\mathbf{r})$ as reported below:

$$V_{corr}^{DFT}(\mathbf{r}) = -a(F_1'\rho + F_1) - abC_F\rho^{5/3}\left[G_1'\rho + \frac{8}{3}G_1\right] -$$

$$\frac{ab}{4}\left[G_1''\rho\,|\nabla\rho|^2 + G_1'(3\,|\nabla\rho|^2 + 2\rho\nabla^2\rho) + 4G_1\nabla^2\rho\right]$$

$$-\frac{ab}{72}\left[3G_1''\rho\left|\nabla\rho\right|^2 + G_1'(5\left|\nabla\rho\right|^2 + 6\rho\nabla^2\rho) + 4G_1\nabla^2\rho\right] \tag{20}$$

where

$$F_1\left[\rho(\mathbf{r})\right] = \left[1 + d\rho^{1/3}\right] \tag{21a}$$

$$G_1\left[\rho(\mathbf{r})\right] = F_1(\rho)^{-5/3}\exp(c\rho^{-1/3}) \tag{21b}$$

$$C_F = \frac{3}{10}\left(3\pi^2\right)^{2/3}. \tag{21c}$$

Unlike the other potentials discussed before, the V_{corr}^{DFT} requires a further Single Centre Expansion of the first and second functional derivatives of eq. 20 in addition to the SCE of the bound state wavefunction. For this reason the calculation of the V_{corr}^{DFT} represents the most time-consuming task in the preprocessing section of our set of codes as the data reported in section 5 will clearly demonstrate.

2.7. The Asymptotic Correction to the Polarisation Potential

In the outermost aymptotic region the local polarisation potentials defined so far (V_{corr}^{FEG} and V_{corr}^{DFT}) can not correctly take into account all the relevant features of the interaction between the molecular system and the incoming electron. For this reason, and since the earliest attempts to describe electron-molecule interactions, a long range potential has been used which depends mainly on the static electrical properties of the molecules (i.e. dipole and higher moments and associated polarizabilities) with the general form, for a given molecular geometry R,

$$V_{pol}^{A1}(\mathbf{r}) = \lim_{r\to\infty} -\sum_{l=1} \frac{\alpha_l}{2r^{2l+2}} \tag{22}$$

Let us limit our attention, for now, to the case of the permanent dipole moment and of its associated polarizability α. When only the symmetrical spherical polarisability is known (often referred to as the average dipole polarizability $\bar{\alpha}_{av} = \frac{1}{3}(\alpha_{xx} + \alpha_{yy} + \alpha_{zz})$ it is customary to assume that $\bar{\alpha}_{av} = \alpha_{xx} = \alpha_{yy} = \alpha_{zz}$ and $\alpha_{xy} = \alpha_{xz} = \alpha_{yz} = 0$. In this case the polarisability tensor is written as

$$\begin{matrix} \bar{\alpha}_{av} & 0 & 0 \\ 0 & \bar{\alpha}_{av} & 0 \\ 0 & 0 & \bar{\alpha}_{av} \end{matrix}$$

On the other hand, if all of the six terms of α are known, then the polarisability tensor becomes

$$\begin{matrix} \alpha_{xx} & \frac{1}{2}\alpha_{xy} & \frac{1}{2}\alpha_{xz} \\ \frac{1}{2}\alpha_{xy} & \alpha_{yy} & \frac{1}{2}\alpha_{yz} \\ \frac{1}{2}\alpha_{xz} & \frac{1}{2}\alpha_{yz} & \alpha_{zz} \end{matrix}$$

A common case is when only the α_0 and $\alpha 2$ polarizabilities are known; in this case the polarisation potential, for a linear molecule, has the simple form

$$V_{pol}(r) = -\frac{\alpha_0}{2r^4} - \frac{\alpha_2}{2r^4}P_2[cos(\theta)] \tag{23}$$

where the Legendre polynomial $P_2[cos(\theta)] = \frac{1}{2}[3cos^2(\theta) - 1]$. Then we have: $\alpha_{xx} = \alpha_{yy} = (\alpha_0 - \frac{1}{2}\alpha_2)$ and $\alpha_{zz} = (\alpha_0 + \alpha_2)$, with the other off-diagonal terms of the polarisability tensor being zero.

Besides this special case, one has to take into account all the components of the dipole static polarizability so that the long range polarisation potential becomes, in atomic units

$$V_{pol}(x,y,z) = -\frac{1}{2r^6}\sum_{i=1}^{3}\sum_{j=1}^{3} q_i q_j \alpha_{ij} \qquad (24)$$

where $q_{i,j} = x, y, z$ for $i, j = 1, 2, 3$ respectively. In our asymptotic potential code is this last equation implemented with the options to select all the different polarizabilities reported above, as a function of the available data.

Once the $V_{pol}(x, y, z)$ is calculated and expanded over the SASH of the A_1 i.r. as the other potentials

$$V_{pol}(\mathbf{r}) = \sum_{hlm} V_{hlm}^{pol}(r) X_{hlm}^{A_1}(\theta, \phi) \qquad (25)$$

one has to choose a matching function $f(r; r_c)$ between it and the local V_{corr}^{DFT} or V_{corr}^{FEG} potentials in order to find the correct matching point r_c. In our code this can be accomplished in various ways by selecting the matching direction along the line connecting the C.O.M. and a given atom, or a generic x, y, z point provided by the user and the C.O.M. and by selecting different l values for which both the above procedures can be followed.

3. THE PARALLEL IMPLEMENTATION OF THE SCE EXPANSION

The implementation of our SCE approach to the solution of the scattering equations on parallel architectures has been given in details before and we will not repeat it here.[20] We worked on both parallel SIMD and MIMD architecture machines and we exploited the corresponding leading features of each architecture in order to obtain the best performances in terms of numerical production. Starting from those experiences we choose to have as a reference parallel architecture the MIMD one, due to the fact it represent a *de-facto* standard in parallel computing although we kept the SIMD release for future works as the architecture becomes more mature.

Between the models we exploited, i.e. *embarassing parallelism* and *domain decomposition*, we preferred to follow the latter because we found it more amenable to be improved in terms of scalability as the computational needs will increase. In order to obtain the best performances from the parallel hardware architecture we decided to completely rewrite the preproccessing part of our set of codes, i.e. the calculation of the SCE wavefunction and potentials, so that now a new code is available to perform this task in parallel.

The new SCELib codes can be run either in parallel or in serial fashions and in the latter mode a direct comparison can be carried out with the full FORTRAN set of codes, POLY2.0, already used for production runs by our group during the last three years or so. A brief overview of our set of programs is given in Fig.1 where a flow chart is reported. The doubly dashed box includes the codes currently available on both SCELib and POLY2.0 and which are the subject of the test runs reported here.

The SCELib code, is written in C and is particularly tailored to get the maximum peformance from a MIMD architecture (low message-passing load, negligible node I/O, etc.) using PVM[21] as well as MPI[22] message-passing parallel programming tools. The data reported in this paper refer to the PVM test only, because the SCELib package is

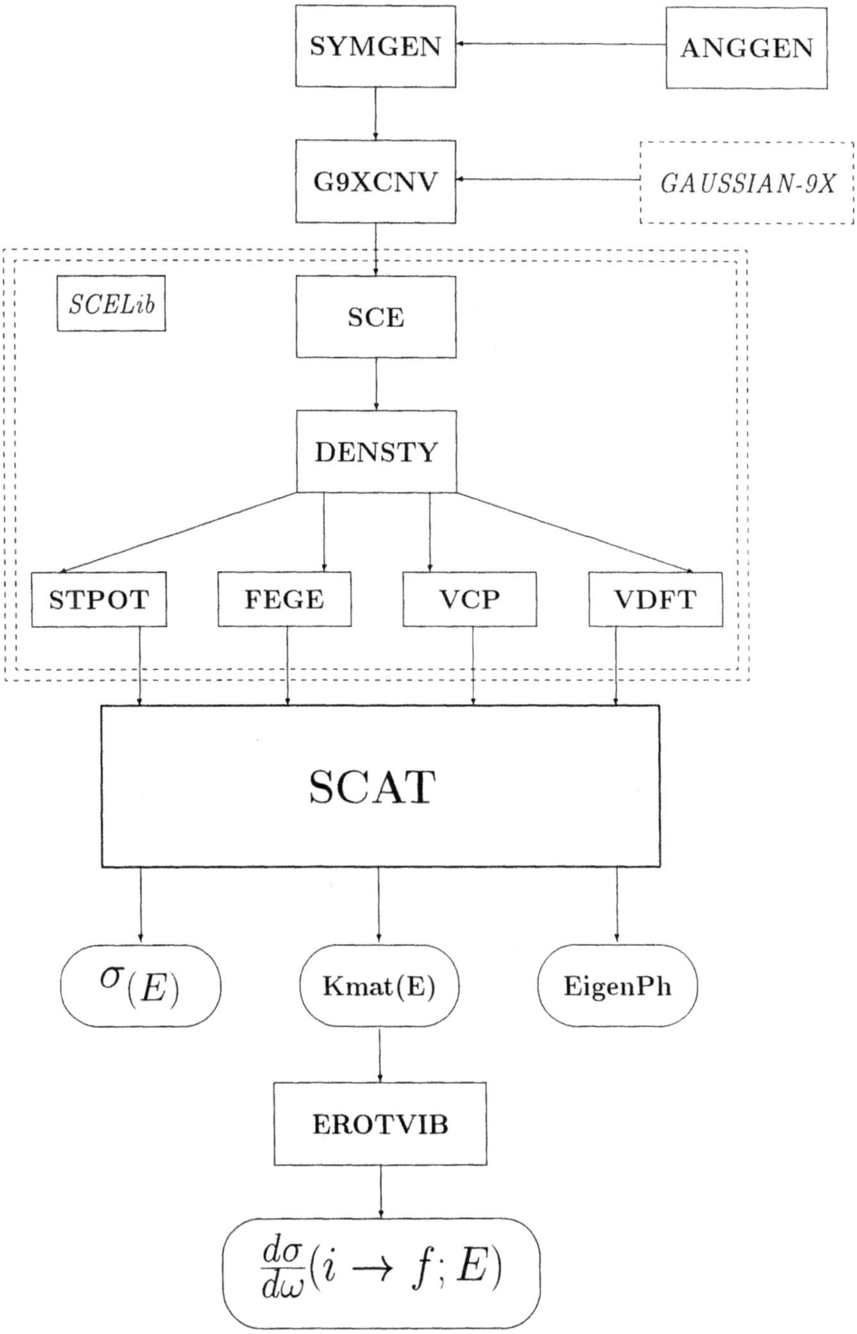

Figure 1. Flow-chart of SCELib and POLY2.0 set of codes for electron-molecule scattering calculations

currently under development and we used a prototype to carry out all the parallel tests. For example, of the initial SCE code we ported in parallel the calculation of the GTO wavefunction only (*sphins*) while the final code will contain the *rotgto* and *sphinc* part too. With respect to the original work on the SCE code, also the *vdft* calculation has been ported in parallel and its release can be considered as definitive. The choice to port in parallel only the *sce* and *vdft* programs, is due to the fact that they account for more than 90 % of the total serial run time and then any attempt to parallelize other sections of the package would not give any significant improvement of the overall performances.

The main functional difference between SCELib and POLY2.0 is the I/O. In fact, the C codes have a main (the master) which launch a sequence of calculations while the FORTRAN programs are executed by a c-shell script and communication among them is accomplished via external files. Of course, the comparison of the performance of each C/FORTRAN equivalent program is carried out taking into account this difference of functionality. Besides efficiently porting in parallel the *sphins* and *vdft* programs, in the prototype version of SCELib we took care to optimize as best as possible the serial code with respect to the original FORTRAN version; the results in the next section 5 will show this very clearly and we expect to obtain even a further improvement in the final release.

As in the original parallel version of the SCE code [20] the model adopted is MASTER/SLAVE as implemented under the PVM parallel tool. With respect to the original parallel prototype, in the present parallel version we implemented two types of *domain decomposition*: the first along the θ component and the second along the R component of the whole (R, θ, ϕ) domain. In the case of the decomposition along θ, the master has to dispatch to the slave nodes the computation of a chunk of the domain, $(R_i, \delta\theta, \phi)$ and then get, accumulating along θ, the results for each R_i. On the other hand, decomposing along R means that the master dispatches and receives one at the time a few fixed chunks $(\delta R, \theta, \phi)$ from the slaves.

As an example, consider the CH_4 molecule with 5 bound state orbitals and suppose a (R, θ, ϕ) domain of $150 \times 48 \times 21$ points for R, θ and ϕ respectively with 6 slave nodes. In the case of the θ decomposition the master receive 150 times a chunk of $5 \times 8 \times 21 \times 8$ bytes $= 6670$ bytes from each slave; on the contrary, in the case of the R decomposition the master collect once 6 chunks of $5 \times 25 \times 48 \times 21 \times 8$ bytes $= 1008000$ bytes.

This makes a great difference in the communication behaviour: in the first case we have many small chunks of data to be received by the master while in the second one few large chunks must be collected by the master itself. This different communication pattern would show the performances of the low layer send/receive drivers and subsystems: while in the case of the θ decomposition one stresses the latencies and interprocess syncronism, in the case of the R decomposition the key factor is the bandwidth of the communication subsystem. The tests performed in parallel and reported below have the aim to show how these different decomposition approaches can affect the final computational performances of the parallel architecture and of the programs under test.

4. COMPUTATIONAL DETAILS

4.1. Hardware

For the serial runs we used the following hardware:

- IBM workstation model 390, POWER2 RISC processor at 67 MHz, 256 KB of L2 cache, 128 MB of RAM memory and 256 MB of swap space.[23]

- DEC workstation Alpha AXP 3000/500, ALPHA ev4 RISC processor at 150 Mhz, 256 KB of L2 cache, 128 MB of RAM memory and 256 MB of swap space.[24]
- PC class workstation (no brand), INTEL Pentium-Pro processor at 150 Mhz, 256 KB of integrated L2 cache, 128 MB of RAM and 128 MB of swap space.[25]
- PC class workstation (no brand), INTEL Pentium processor at 100 Mhz, 128 KB of L2 cache, 48 MB of RAM and 64 MB of swap space.[25]

The parallel tests were carried out using the IBM 9076 SP2 equipped with 16 nodes, each configured as the IBM 390 workstation above but connected with a High Performance Switch (HPS) capable of 80 MB/s burst communication bandwidth.[26]

4.2. OSes, Languages and Tools

For either the serial and parallel tests we used the C and FORTRAN compilers provided with the machine and those released by the FSF (Free Software Foundation) under the GNU public license for the tested architectures.[27] In all cases we used the standard compiler options to get the maximum performaces from a given architecture without any particular tuning of the compiler itself and we made no explicit use of any external preprocessor (KAP or VAST for example). In this way we would reproduce the standard end-user developing environment on a typical workstation with the most used compiler's options under a given Operating Systems as reported in Table 1.

Table 1. OSes and Compiler Options used in Serial and Parallel Tests

CPU	OS Level	Compiler	Options
Pentium (Pro)	Linux 1.3.97	gcc 2.7.2	-O2
		g77 0.5.18	-O2 -fno-f2c
POWER2	AIX 4.1.4	xlc 3.1.2.4	-O3 -qarch=pwr2 -qtune=pwr2
		xlf 3.2.3.2	-O3 -qhot
			-qcache=assoc=4:line=128: size=256:type=d
			-qarch=pwr2 -qtune=pwr2
Alpha ev4	DEC Unix 3.2	DEC cc 3.11	-O4
		DEC f77 3.8	-O5 -tune ev4 -math_library fast

In one serial test case we used gcc V2.7.2 with the options reported above on all the machines we tested.

For the parallel test cases on the SP2 machine we adopted the compiler and options of the IBM-POWER2 architecture with PVM V3.3.11 and PVMe 2.1 (to which our results reported here refer) and we used PSSP and POE V2.1 under AIX 4.1.4.

5. CASE STUDIES AND NUMERICAL RESULTS

As described in full detail thus far, we have obtained two set of codes: the POLY2.0 FORTRAN package and the SCELib C prototype preprocessor. The tests we carried out on the various architectures (serial and parallel) had to reflect this duality in order to correctly compare performances between different codes (in C and FORTRAN), implementations (serial and parallel), and machine architectures. For this reason we started to analyze the results of the serial C programs SCELib and, where applicable, we compared these data with the equivalent codes of the FORTRAN POLY2.0 package. Then we carried out the tests on the parallel architecture described above and in this case we used the SCElib codes only. We took care to use either the same compiler or compiler release within a set of tests trying to avoid misleading comparisons with the aim to reproduce the machine configurations one usually has on a standard workstation. In short: we are not doing here any kind of *benchmarketing* as one sometimes sees on vendor's datasheets but our aim is instead to look at the real expected perfomances of a given architecture on a selected computational task. It is for this reason that we performed our tests with the machines under standard operative conditions (configuration, services, etc) thus avoiding any misleading configuration which can outperform others on a particular task.

5.1. Test Case Descriptions

As test cases we chose the CH_4 and the CF_4 molecular systems. For both systems the maximum angular l value (l_{max}) was 10, with an angular grid of 48 points in θ and 21 points in ϕ for the *sphins* and *vdft* codes. The number of R points were 150 and 208 for the CH_4 and CF_4 molecules, respectively. In the case of the CF_4 molecular system, it is worth noting that we extended the molecular description by a factor of 15 by artificially increasing the weight of the wavefunction calculation. This means that we can think of these results as referred to a molecular system with a 15 times larger GTO wavefunction or any combination of $N_A \times N_{GTO} \times N_{CG}$ being N_A the no. of atoms, N_{GTO} the no. of bound state orbitals and N_{CG} the no. of contracted gaussians for each GTO. The real calculation on the CF_4 molecule would require an increased R, θ, ϕ grid (besides of a larger basis set) but the timing required to perform an SCE or DFT calculation is so huge that one can miss the detection of the expected performances of the parallel code with respect to its corresponding serial counterpart. For these reasons we preferred to use the CH_4 angular grid for the CF_4 molecule and to adopt the same GTO basis sets (DZP) augmented by a factor of 15; in this way the computational times increased only one order of magnitude thus permitting a direct useful comparison between the serial and parallel programs. The comparison would not be so straightforward if we were using the actual operating parameters for the CF_4 where a SCE calculation can require, for example, dozen of CPU hours in a serial code, depending on the l_{max} and the architecture being chosen.

5.2. Conventional Tests

Let us begin this discussion by looking at some very basic results: the performances of the serial C and FORTRAN equivalent codes from SCELib and POLY2.0 on the methane molecule. The test runs were carried out using the GNU C compiler (gcc v2.7.2) and a beta version of the GNU FORTRAN compiler (g77 v0.5.18). These two compilers use the same *back-end* while, of course, the *front-end* is different in order to analyze the syntax of C and FORTRAN source codes. The results are reported in Table 2 for two INTEL machine: a Pentium @ 100 MHz and a Pentium Pro @ 150 MHz configured as indicated in section 4.1.

The first comment is on the overall performance of the two processor: the Pentium Pro (PPro) is almost twice as fast as the Pentium. This result is not completely in line with the INTEL declared performances for the PPro but we have to say that neither the gcc nor the g77 have any particular option for the tuning of the executable for the PPro processor. In fact, the resulting executables are actually compiled as runnable on any x86 processor and is obvious that this is done at the expenses of the performance. On the other hand, the SCELib executables are two or more times faster than the equivalent POLY2.0 ones, with the exception of the unexpected results for *vfege* and *vcp* on the Pentium.

Table 2. CH_4 SCELib and POLY2.0 serial runs on INTEL architectures. CPU Time in seconds per program element.

	g77 P100	gcc P100	g77 P150	gcc P150
sphins	21.5	16.7	12.2	13.3
rotgto+sphinc	14.1	9.2	5.7	3.2
densty	11.2	1.4	4.1	0.5
vst	10.5	0.2	3.8	0.1
vfege	11.1	17.9	5.1	3.4
vcp	10.5	17.8	4.7	3.4
vdft	114.6	52.5	90.1	45.0
vasymp	9.4	1.7	4.5	0.9
Total	202.9	117.4	130.2	69.8

In Table 3 the results of the serial SCELib runs on different architectures, using the same compiler and options (gcc v2.7.2), are reported. In this way one can have a rough estimate of the brute CPU power of each architecture even if we are mixing here different processor technologies (RISC and CISC). The data show that the PPro has twice the floating point CPU power of the Pentium and surpasses the old Alpha ev4 technology as well, while its performances are still well below those of the POWER2.

Table 3. CH_4 SCELib serial runs on all architectures with gcc v2.7.2. CPU Time in seconds per program element.

	P100	P150	ev4	pwr2
sphins	16.7	13.3	16.0	10.1
rotgto+sphinc	9.2	3.2	7.5	3.9
densty	1.4	0.5	1.2	0.7
vst	0.2	0.1	0.2	0.1
vfege	17.9	3.4	11.2	2.9
vcp	17.8	3.4	11.1	2.7
vdft	52.5	45.0	50.5	20.7
vasymp	1.7	0.9	1.3	0.8
total	117.4	69.8	99.0	41.9

These trends become more evident if one uses the compilers provided with the vendor machines (DEC, IBM) as reported in Table 4. In this run one not only test the processor hardware as it stands, but also the software technology inside the C compiler. In fact, the performances of the Alpha machine are enhanced by its compiler and even if the overall CPU times are not as good as the PPro, the latter is overtaken on the *vdft* test. As a final comment, the POWER2 processor has shown to have top performances on all tests and to outperform all the other processors. However, it must be said that this depends on the hardware and to a large extent on the xlc C compiler. The extensive experience of the IBM into the Unix arena plays here a crucial role and relatively new technologies like the PPro and Alpha ones stand behind the POWER2 performance due to the absence or to lack of dedicated compiler for its processors.

Table 4. CH_4 SCELib serial runs on all architectures with the best available compiler. CPU Time in seconds per program element.

	gcc P100	gcc P150	cc ev4	xlc pwr2
sphins	16.7	13.3	15.3	8.1
rotgto+sphinc	9.2	3.2	6.0	2.3
densty	1.4	0.5	0.9	0.5
vst	0.2	0.1	0.2	0.1
vfege	17.9	3.4	9.4	2.1
vcp	17.8	3.4	9.4	1.9
vdft	52.5	45.0	38.4	15.0
vasymp	1.7	0.9	1.0	0.7
total	117.4	69.8	80.6	30.7

5.3. Parallel Tests

On the SP2 architecture we carried out two series of calculations: (i) on the methane molecule with the parameters as in the corresponding serial jobs and (ii) on the fluoromethane molecule with the increased GTO expansion as reported in section 5.1. The tests were performed while keeping in mind to imitate the production environment one could experience on a typical parallel machine.

The first test case was the methane molecule with the parameter setup reported so far, and the results are shown in Table 5 and Table 6. In this case the same serial job was ran in parallel with the aim of offering the best comparison between the two execution environments. Unfortunately, as the number of nodes increase, the final results are affected by errors due to the difficulty of measuring the total job elapsed time. For this reason, the tests were limited to 8 nodes which is equivalent to no less than 5-6 seconds of elapsed time. This limit has been experimentally calculated as half of the slaves spawning time which is about 1.5 secs per slave over a single-segment ethernet network. Below this limit the uncertainty of the measured elapsed time is unknown and in this way the reported results must be evaluated. Nonetheless, the observed speedups are almost linear with the exception of the *sphins* test for the R decomposition where a lower value of 6.3 on 8 nodes is obtained. We must say that we are getting closer to the low limit for the measure of the elapsed time and, considering that the *sphins* code is still under development, the obtained results are extremely good.

Table 5. CH_4 SCELib Parallel Runs on SP2 with PVMe v2.1. θ Decomposition. Elapsed Time (E.T.) in seconds and Speedup vs. the number of nodes.

	SPHINS		*VDFT*	
Nodes	E.T. (sec)	Speedup	E.T. (sec)	Speedup
1	32.7	1.0	60.3	1.0
3	10.9	3.0	20.6	2.9
6	5.6	5.8	10.3	5.8
8	4.2	7.8	7.8	7.7

Table 6. CH_4 SCELib Parallel Runs on SP2 with PVMe v2.1. R Decomposition. Elapsed Time (E.T.) in seconds and Speedup vs. the number of nodes.

	SPHINS		*VDFT*	
Nodes	E.T. (sec)	Speedup	E.T. (sec)	Speedup
1	32.7	1.0	60.3	1.0
3	13.0	2.5	22.3	2.7
6	6.0	5.4	10.8	5.6
8	5.2	6.3	8.2	7.3

At this point becomes more evident the reason why we chose to extend our tests on a molecular system one order of magnitude larger than the CH_4, the fluoromethane molecule extended as reported before. In Table 7 and Table 8 the results are reported, like for the CH_4, with the elapsed time vs. the number of nodes for the two program elements under test (*sphins* and *vdft*) along the two different decompositions (over θ and R).

The results are extremely good with a linear speedup up to 13 nodes with the only exception of the last point of the *sphins* parallel code for the θ decomposition which confirm once more its prototype nature. If one considers the CF_4 results, it becomes evident how the choosen decomposition affects the overall performance. In fact, the decomposition along R seems to be more suitable to the problem under test, i.e. the angular integration of polycentric GTO function over a fixed grid domain. Nonetheless, when the code will be finally released, also the decomposition along θ could perform well enough if the communication pattern between the master and the slaves is simplified.

Table 7. CF_4 SCELib Parallel Runs on SP2 with PVMe v2.1. θ Decomposition. Elapsed Time (E.T.) in seconds and Speedup vs. the number of nodes.

	SPHINS		*VDFT*	
Nodes	E.T. (sec)	Speedup	E.T. (sec)	Speedup
1	751.0	1.0	578.0	1.0
3	259.6	2.9	191.0	3.0
6	129.0	5.8	97.2	5.9
8	98.2	7.7	72.9	7.9
12	72.6	10.3	49.2	11.8

6. CONCLUSIONS

In relation to the serial tests which we carried out on the reported architectures, it is worth noting the fact that these kind of calculations were confined, until few years ago, to the high-end workstation domain or above and now, thanks to new silicon technologies, they become also feasible on very low-cost machines as those of the class of high-end PCs.

Furthermore, we exploited here one of the most diffused parallel architectures, the IBM SP2, where it is surprising to see how one could get with a minimal effort, a high performance ratio using standard parallelization techniques and tools. This fact is even

Table 8. CF_4 SCELib Parallel Runs on SP2 with PVMe v2.1. R Decomposition. Elapsed Time (E.T.) in seconds and Speedup vs. the number of nodes.

	SPHINS		*VDFT*	
Nodes	E.T. (sec)	Speedup	E.T. (sec)	Speedup
1	751.0	1.0	578.0	1.0
2	388.5	1.9	291.3	2.0
4	194.2	3.9	145.3	4.0
8	99.3	7.6	73.3	7.9
13	61.5	12.2	45.7	12.7

more important if one considers that it becomes comparatively more feasible to obtain one order of magnitude (and more) on the affordable physical problem with parallel machines of such a class where quasi-conventional programming styles are combined with high performance technologies.

As an example we are approaching in our group the calculation of the scattering of electrons from C_{60} and preliminary results show that with a fairly small GTO basis set (3-21G*) the preprocessing calculation is more than two orders of magnitude larger than for the case of the CF_4 reported before. Taking into account that for a real calculation on this molecule we have to increase the angular parameters and use a larger GTO basis set, we estimate the CPU time needed to complete this task to be three-four orders of magnitude larger than that for the CF_4 system. This means we need several millions of seconds of a POWER2 CPU time, which is too demanding for a serial conventional processor. With the use of the parallel SCE version, on the other hand, we can reduce the computational effort within the order of days instead of weeks, thereby permitting us to solve this challenging physical problem which at this time is still unsolved.

7. ACKNOWLEDGEMENTS

The financial support of the Italian National Research Council (CNR) and of the Italian Ministery for University and Research (MURST) is gratefully acknowledged. We are also grateful to the staff at CASPUR center for their helpful support on hardware usage. We finally thank L. Valcamonici, G. Chillemi and E. Buonomo for their generous help in the various stages of the calculations.

References

[1] F.A. Gianturco, V. Di Martino and A. Jain, Nuovo Cim. D, **14**, 411 (1992).

[2] P. G. Burke, A. Hibbert and W. D. Robb, J. Phys. B: At. Mol. Phys., **4**, 153 (1971).

[3] R. R. Lucchese, K. Takatsuka and V. McKoy, Phys. Rep., **131** 147 (1986)

[4] T. N. Rescigno, C. W. McCurdy, A. E. Orel, and B. H. Lengsfield III, "The Complex Kohn Variational Method", in: "Computational methods for Electron-Molecule Collision", W. M. Huo and F. A. Gianturco Eds, Plenum Publ. Co., New York (1995).

[5] "Computational methods for Electron-Molecule Collision", W. M. Huo and F. A. Gianturco Eds, Plenum Publ. Co., New York (1995).

[6] F.A. Gianturco, R.R Lucchese and N. Sanna, J. Chem. Phys., **100 (9)**, 1 (1994).

[7] W. H. Press et al., Numerical Recipes, Second Edition, Cambridge University Press (1992).

[8] F. A. Gianturco and D. G. Thompson, Chem. Phys., **14**, 111 (1976).
[9] F.A. Gianturco, R.R. Lucchese, N. Sanna and A. Talamo, A Generalized Single Centre Approach For Treating Electron Scattering From Polyatomics Molecules, in: "Electron Scattering Processes in Molecules Clusters and Surfaces", H. Erhardt and L. Morgan Eds, Plenum Publ. Co., New York (1994).
[10] F.A. Gianturco, R.R. Lucchese and N. Sanna, J. Chem. Phys., **102**, 5743 (1995).
[11] F.A. Gianturco, J.A. Rodriguez-Ruiz and N. Sanna, Phys. Rev. A, **52 (2)**, 1257 (1995).
[12] S.L. Altmann, The Cubic group, Proc. Camb. Phil. Soc., **53**, 343 (1957)
[13] Tinkham "Group Theory and Quantum Mechanics", McGraw-Hill, New York (1964).
[14] F.A. Gianturco, D.G. Thompson and A.K. Jain, Electron-Scattering from Polyatomic Molecules using a Single-Center-Expansion Formulation, in "Computational methods for Electron-Molecule Collision", W. M. Huo and F. A. Gianturco Eds, Plenum Publ. Co., New York (1995).
[15] S. Hara, J. Phys. Soc. Japan, **22**, 710 (1967).
[16] J. O'Connell and N.F. Lane, Phys. Rev. A, **27**, 1893 (1983).
[17] F. A. Gianturco and J. A. Rodriguez-Ruiz, J. of Mol. Struct. (Theochem), **260**, 99 (1992).
[18] F. A. Gianturco A. K. Jain and J. A. Rodriguez-Ruiz, Phys. Rev. A, **48**, 4321 (1993).
[19] F. A. Gianturco and J. A. Rodriguez-Ruiz, Nuovo Cim. D, **3**, 267 (1994).
[20] F. A. Gianturco, N. Sanna, and R. Sarno, "A Study of the Porting on SIMD and MIMD Machines of the Single Centre Expansion Code to Treat Electron Scattering from Polyatomics Molecules", in "Computational methods for Electron-Molecule Collision", W. M. Huo and F. A. Gianturco Eds, Plenum Publ. Co., New York (1995).
[21] A. Beguelin, J. Dongarra, A. Geist, R. Manchek and V. Sunderam, Oak Ridge Natl. Laboratory Tech Report **TM-11826:1** (1991).
[22] W. Gropp, E. Lusk and A. Skjellum, "Using MPI: Portable Parallel Programming with the Message Passing", MIT press (1994).
[23] PowerPC and POWER2. Technical Aspects of the new IBM RISC Systems/6000, IBM Press (1994), (Ref. SA23-2737-00).
[24] DECchip 21064/21064A Microprocessor Hardware Reference Manual, DEC Press (1995).
[25] Visit the INTEL microprocessor's WEB site (http://pentium.intel.com) for a full description of the Pentium and Pentium Pro architectures and systems.
[26] T. Argewala et al., SP2 System Architecture, IBM System Journal, Vol. 34 N.2 of Scalable Parallel Computing, Reprint Order No. G321-5563.
[27] Free Software Foundation Inc., 675 Mass. Ave., Cambridge, MA 02139, USA.

CHAOS IN ATOMS IN STRONG FIELDS: PERIODIC ORBITS AND DIFFRACTION

P.A. Dando and T.S. Monteiro

Department of Physics and Astronomy
University College London
Gower Street
London WC1E 6BT
United Kingdom

ABSTRACT

The problem of reconciling the quantum spectra of non-hydrogenic atoms in strong fields (either calculated using a combination of R-matrix and quantum defect techniques or measured experimentally) with the results of the standard theory of quantum chaology—Periodic Orbit theory—has been the subject now of several studies by ourselves and others. However, previous work has considered the subject *not* within the framework of Periodic Orbit theory *per se* (the Gutzwiller trace formula), but rather using a related theory—Closed-Orbit theory—due to Delos. We show here that the effect of the multi-electron core of a Rydberg atom manifests itself differently in the Delos (photoabsorption spectrum) and Gutzwiller (energy level spectrum) theories. For the former, the leading terms represent new modulations; for the latter they represent 'shadow' terms which reduce the amplitude of modulations associated with the primitive orbits. We consider the Gutzwiller formula explicitly and demonstrate, for the main quasi-Landau modulation, that the core-effects are analogous to diffractive corrections to Periodic Orbit theory developed recently and applied to quantum billiard problems.

INTRODUCTION: PERIODIC-ORBIT THEORY, CLOSED-ORBIT THEORY AND DIFFRACTION

For time-independent atomic and molecular spectroscopy involving highly excited states in the chaotic regime, the typical quantum spectrum looks extremely irregular, see Fig. 1(a). Eigenstates generally defy classification in terms of quantum numbers or

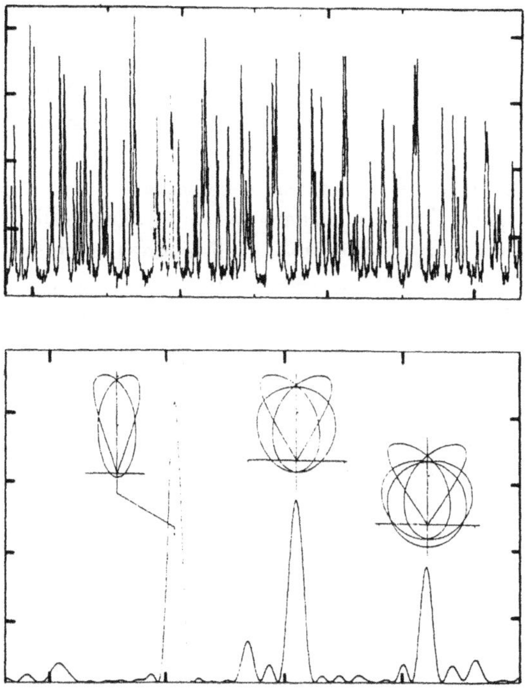

Figure 1. (a) Scaled Spectrum of hydrogen in the range of excitation energy -77.7cm^{-1} to -54.4cm^{-1} and magnetic fields from 5.19T to 3.03T. (b) Corresponding Fourier transform showing that the very irregular looking spectrum is in fact very well modulated by oscillations corresponding to closed orbits (from Holle et al.[3]).

channels. Instead, Periodic Orbit theory[1] has provided atomic spectroscopists with an extremely powerful framework for the analysis of quantum spectra in the classically chaotic limit as it has been shown that the quantum spectra show periodicities associated with isolated periodic orbits, Fig. 1(b). In the semiclassical Periodic Orbit theory the amplitudes of these *quantum* modulations are related to the elements of a *classical* entity—the stability matrix M—obtained by linearizing the classical motion in the vicinity of the periodic orbit. The amplitude of the quantum modulation due to a given orbit is proportional to $(\text{Tr}(M) - 2)^{-1/2}$. For unstable orbits, the eigenvalues (Liapunov exponents) of the stability matrix quantify the rate of exponential divergence of neighbouring trajectories, a well-known property of chaotic dynamics. In the context of atomic and molecular spectroscopy it is the elements of the stability matrix and periods of the short periodic orbits which provide a quantitative description of the spectral features of the finite resolution experimental spectrum. Hydrogen atoms in external static fields have become paradigms in the study of these effects.[2]

However, Periodic Orbit theory—in the shape of the well-known Gutzwiller trace formula—in fact relates classical periodic orbits to the *energy level* spectrum in the semiclassical regime. In order to investigate experimental *photoabsorption* spectra (which involve a quantum spectrum weighted by the appropriate absorption probability) a separate semiclassical theory for the recurrence spectrum was developed.[4–6] Rather than taking the trace of an energy domain Green's function as required for the Gutzwiller formula, a matrix element of an outgoing Green's function evaluated over initial states acted on by dipole operators is obtained in terms of classical orbits. Since the full quantum Green's function is needed near the nucleus the orbits need not be periodic, though in effect, for atoms in magnetic fields, the relevant orbits are periodic orbits of the diamagnetic Kepler problem.

There are important differences between the Gutzwiller and Delos theories. Though the periodicity of the spectral modulations is the same, the amplitudes are different. For example, the Delos theory is weighted by an off-diagonal element of the stability matrix; the Gutzwiller theory by a factor related to its trace. The Delos/photoabsorption formula uses only those orbits which deliver amplitude back to overlap with the localized initial state: only orbits which pass through the nucleus contribute. For the Gutzwiller/eigenvalue formula the important short periodic orbits may sample any part of phase-space.

The Delos/photoabsorption theory has proved extremely successful for the interpretation of experimental spectra of hydrogenic atoms in static fields. But subsequent studies showed[21,20] that the theory did not give a quantitative description for non-hydrogenic spectra. The amplitude of many modulations was incorrect and new modulations were found in non-hydrogenic spectra which were not associated with classical trajectories of hydrogen but coincided with combinations of two or more orbits ('combination recurrences'). In 1995, we were able to show[14,15] that, by putting in quantum corrections which were higher order in $\hbar$, excellent agreement with fully quantal results, calculated from a combination of R-matrix and quantum defect-type approach, could be recovered for the Delos theory. An alternative approach using a model potential attempted to describe non-hydrogenic atoms within the standard theory.[16,17] However, the implications for the Gutzwiller trace formula have not been examined to date.

In parallel with the work in atomic physics, there has been much recent interest in $\hbar$ corrections to Periodic Orbit theory. The Gutzwiller formula and other related semiclassical theories (*e.g.* scar theory) is only correct to leading order in $\hbar$. Much important work has been carried out on systematic $\hbar$ corrections[7] and also more specific $\hbar$ corrections *e.g.* due to divergences caused by bifurcations or due to complex orbits and tunnelling.[19]

In fact, one of the most significant recent developments has been in the development of $\hbar$ corrections for diffraction. Periodic Orbit theory's reliance on stationary phase conditions in effect means that it represents a quantum analogue of the geometric-optics limit of wave optics. Hence, much progress has been made in incorporating, as a further step forward, the geometric theory of diffraction[22,23] into Periodic Orbit theory, especially important where there are dynamical structures in the potential smaller than the de Broglie wavelength of the quantum particle. This work has been applied in the main to billiard problems. Here, by considering the Gutzwiller formula, we argue below that the effect of a non-hydrogenic core may be quantitatively understood as a manifestation of diffraction in a 'real' physical system.

In our previous work, we showed that corrections to the basic semiclassical theory of the photoabsorption spectrum appeared as a series in powers of $\hbar^{1/2}$. The leading terms (which are actually of order $\hbar^{1/2}$) represented combinations of two different orbits, or reductions to the amplitudes of harmonics of the periodic orbits; the primitive periodic orbits themselves are unaffected. This has implications for the experiments, where the well-known quasi-Landau modulations will appear with similar amplitudes for different Rydberg atoms (and the basic modulation depends on the primitive orbit). However the harmonics—and hence the shape of the modulation— varies from one Rydberg atom to another.

However, here we find that for the Gutzwiller formula (*i.e.* the density of states) the leading corrections appear as a reduction of the amplitudes of the primitive periodic orbits. The additional modulations seen only in non-hydrogenic atoms, resulting from combinations of periodic orbits, appear to higher order relative to the photoabsorption theory (they are proportional to $\hbar$ at lowest order in the energy level spectrum, rather than $\hbar^{1/2}$ in the photoabsorption spectrum).

So why the qualitative difference in the core effect in these two such closely related theories ? This is best understood by considering how one obtains the basic two-orbit combination. In Closed Orbit theory, semiclassical amplitude propagating along a core-intersecting trajectory is partially redistributed at the core into some other trajectory. The fraction of scattered amplitude (proportional to $\hbar^{1/2}$) propagates along the second orbit until it returns to overlap with a localized initial state. A stationary phase integral is not carried out all the way back to the initial point. Hence the combination of two trajectories need not be periodic. A single collision with the core is sufficient to connect two orbits. For the Gutzwiller trace formula the requirement that the initial and final momentum should be equal—the condition for a periodic orbit—is enforced, so a second core scattering is needed to make the combination periodic. This implies a further reduction of the scattered amplitude by a factor of order $\hbar^{1/2}$ so in the Gutzwiller formula a two-orbit combination is weaker, being only of order $\hbar$.

Diffractive effects in atomic systems represent corrections to the amplitudes of short periodic orbits of order 10–50% or so. However the cumulative effect of multiple scattering on higher harmonics and the proliferation of small combination recurrences has a profound effect on the detailed diamagnetic Rydberg spectrum. For example where the Stark spectrum of Hydrogen looks quite regular, the equivalent Lithium spectrum appears extremely irregular. One particular 'symptom' of a classically chaotic limit is in the energy level statistics, in particular the Nearest Neighbour Spacings (NNS). For a regular system the NNS have a characteristic Poisson distribution, associated with large numbers of degeneracies between eigenvalues. In the chaotic limit, a Wigner (GOE) distribution is found. Diffraction/core-scattering has the effect of producing a distribution close to the GOE limit in a lithium-like atom, even at low energies where hydrogen is near the Poisson limit.[18] Hence, although these diffractive corrections are not extremely large in the low resolution spectrum, the substantial differences in the high resolution spectra originate from the same physical mechanism.

It should be recalled that, for diffraction, the obstacle should be smaller than the wavelength of the incident wave. In the atomic case both the core and the de Broglie wavelength are of order 1 au. This means that, in some cases, the core scattering may be approximated by an effective non-Coulombic potential within the framework of the standard theory. Although, as shown below, this can fail substantially, it gives much physical insight and so we proceed with a brief discussion of this approach.

MODEL-POTENTIAL DYNAMICS OF NON-HYDROGENIC ATOMS

In contrast to the corresponding hydrogenic system, there is no straightforward classical limit for the non-hydrogenic atom. One approach[8,16,17] is to approximate the interaction of the outer Rydberg electron with the ionic core by using a one-electron model potential. The usual parametric form for the potential involves a screened nuclear charge:

$$V_{\rm core}(r) = \frac{(Z-1)}{r}\left(1+\frac{r}{a}\right)e^{-r/a} \qquad \text{(atomic units)} \tag{1}$$

where Z is the nuclear charge. The parameter, a, which determines the range of the core, is optimized numerically so as to reproduce the field-free energy levels, and hence the quantum defects, for the alkali atoms. Classically, the presence of a non-Coulombic core-potential causes the Runge-Lenz vector, which for a pure Coulomb potential lies along the semi-major axis of the Kepler ellipse and is a constant of the motion, to precess. Semiclassically, an interpretation for the effect of an electronic core in the field-free case has long been available:[10] a classical trajectory, incident with angular momentum, $L = (l + 1/2)\hbar$, is deflected through an angle which can be directly related to the quantum defects, μ_l, describing the core:

$$\Theta(l) = 2\pi \frac{d\mu_l}{dl}. \tag{2}$$

Thus, we can expect model potentials which yield the same quantum defects to provide a reasonable description of the classical motion provided the dependence of μ_l on l is reasonably smooth. However, for smaller atoms such as lithium or helium, only the s-wave quantum defect differs appreciably from zero.

By solving Hamilton's equations numerically with the core interaction described by Eq. (1) we can begin to investigate some aspects of the classical motion of a non-hydrogenic Rydberg atom in the presence of an external field. In Fig. 2 we show the Poincaré surface-of-section (SOS) for the model non-hydrogenic atom at scaled energies (a) $\varepsilon = -0.5$ and (b) $\varepsilon = -0.316$ with $\gamma = 10^{-5}$ ($B = 2.35\,\mathrm{T}$) and with the parameter, a, chosen to correspond to the spectrum of lithium: for comparison we also show the SOS for the diamagnetic hydrogen problem at the same scaled energies.

For the hydrogen atom at $\varepsilon = -0.5$, Fig. 2(c), the system is still close to its integrable limit, $\varepsilon \to -\infty$: almost all trajectories are regular and confined to tori. Phase space is dominated by the islands associated with the fixed points corresponding to the stable straight-line orbits parallel and perpendicular to the field, conventionally denoted V_1 and R_1 respectively.[3] Between these islands lies a separatrix and the hyperbolic fixed points associated with the almost circular orbit, C.

The SOS for the model lithium atom shown in Fig. 2(a) is very different. In particular, the large islands of stability seen in Fig. 2(c) are no longer visible. The points on this SOS were obtained by following a *single* trajectory for a long time in contrast to the corresponding hydrogenic SOS where each different torus was obtained from the crossings of a *different* trajectory. Note that a qualitatively similar SOS to that shown in Fig. 2(a) would be obtained if we were to consider *any* other trajectory (*i.e.* a trajectory starting from an arbitrary point on the SOS). Thus, at this scaled energy, a trajectory of the non-hydrogenic system explores most of phase-space via a process of 'torus-hopping': regular motion associated with tori of the hydrogenic system interspersed with scattering from the ionic core.

The SOS at $\varepsilon = -0.316$ are shown in Fig. 2(b) for lithium and Fig. 2(d) for hydrogen. Both SOS's are qualitatively similar with the chain of four elliptic islands being particularly prominent. Detailed studies of the classical motion of the hydrogenic system in the vicinity of $\varepsilon = -0.316$ have shown that, close to this scaled energy, there is a sequence of two period–4 bifurcations.[9,11] Most importantly a chain of four large islands, associated with a period–4 orbit which does not approach the nucleus, is created. This bifurcation has a dramatic effect on the energy level statistics of hydrogen. While away from this energy the NNS move monotonically from the regular (Poisson) to the chaotic (Wigner) limit, close to $\varepsilon = -0.316$ they revert abruptly to a more 'regular' distribution.

The results of these classical calculations have shed light on some of the features observed in the NNS of $\varepsilon = -0.316$. Firstly, we note that NNS reflect behaviour on

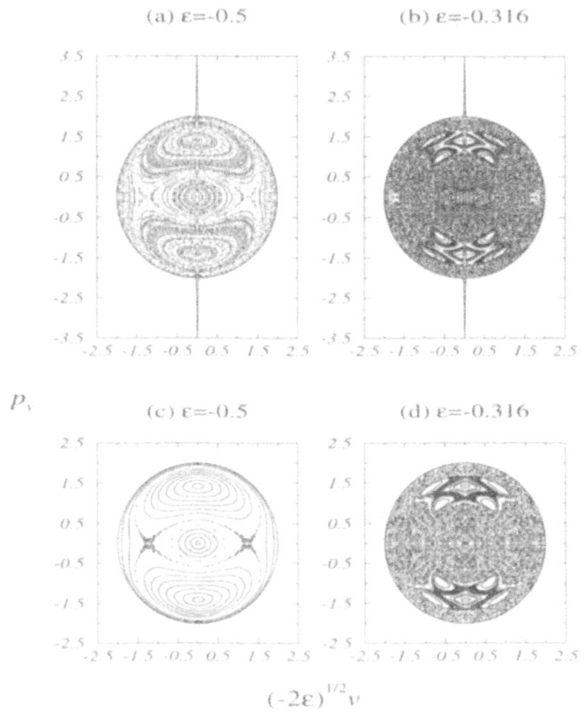

Figure 2. Classical Poincaré surface-of-section for a model lithium atom in a magnetic field with $\gamma = 10^{-5}$ at (a) $\varepsilon = -0.5$ and (b) $\varepsilon = -0.316$. Also shown are the corresponding hydrogenic surfaces in (c) and (d) respectively. The sections are taken through $u = 0$ and represent the v, p_v plane where u and v are semi-parabolic coordinates. Note that the position coordinate has been scaled by a factor of $\sqrt{-2\varepsilon}$ so that the energy shell for $u = 0$ maps into an area bounded by a circle of radius 2.

the finest energy scales and hence on the longest time scales. The classical calculations show that, even at low scaled energies, typical trajectories of the non-hydrogenic system, while remaining on hydrogenic tori in the short term (between collisions with the core), are ergodic in the long-time limit and that a single trajectory will eventually fill the whole of the energy surface. Hence, the NNS for non-hydrogenic atoms are at the chaotic (Wigner) limit, even at low scaled energies where the hydrogenic problem is nearly regular.

However, the quantum calculations show that the abrupt change to a more regular NNS distribution at $\varepsilon = -0.316$ is equally evident in the non-hydrogenic case,[18] so one concludes that this is associated with the existence of the chain of stable islands present in the SOS at and near this energy. The stable periodic orbit associated with these islands does not pass through the core region and hence the stable region persists undisturbed. For moderate values of $\hbar$ these stable islands occupy a sufficient area in phase-space to support quantum states which are nearly degenerate with other states associated with the central island. These near-degeneracies account for the more regular NNS distribution observed near to $\varepsilon = -0.316$.

A DIFFRACTIVE APPROACH TO CORE-SCATTERING

While the use of a model potential to describe the ionic core has been helpful in explaining some features of the behaviour of non-hydrogenic Rydberg atoms, on theoretical grounds it is difficult to justify a classical description of the core-scattering process because the core might only affect one or two partial waves. The most general approach, valid for any regime of atomic core scattering in Rydberg atoms must allow for the breakdown of classical path methods at the core. In the next section we review briefly our 1995 work on core-scattering corrections to the closed orbit theory of Delos and co-workers[4,5,12,13] which describes semiclassically the photoabsorption spectrum. In the subsequent section we consider Periodic Orbit theory (*i.e.* the energy level spectrum) and the diffractive corrections to the Gutzwiller trace formula.

Core-scattered closed orbits and photoabsorption spectra

By starting with the physical picture of the photoabsorption process suggested by Gao and Delos,[12,13] but including the core scattered waves consistently, we have been able to extend closed-orbit theory to reproduce successfully the core-induced phenomena. The theory underlying the closed-orbit formula has been presented previously in great detail[4,5,12,13] so we give only an outline here. Briefly:

- when an atom absorbs a photon, the electron propagates outwards in a near zero-energy Coulomb wave.
- At sufficiently large distances from the nucleus the wave propagates semiclassically along classical trajectories with wave fronts perpendicular to those trajectories.
- Eventually, the trajectories and their associated waves are turned back by the action of the external field. Some of the trajectories return to the vicinity of the nucleus and the waves associated with them, which are (cylindrically modified) incoming Coulomb waves $\psi^j_{\mathrm{Coul}}(\theta^j_f)$ associated with the closed orbit indexed by j, interfere with the outgoing waves generating oscillations in the absorption spectrum.
- The incoming waves produce Coulomb scattered waves and, in the case of non-hydrogenic atoms, core-scattered waves, $\psi^{j,k}_{\mathrm{core}}(\theta^j_f, \theta^k_i)$ which re-distribute amplitude from an incoming closed orbit, j, into the closed orbit, k.

The Coulomb scattered waves are strongly backwardly focussed and back-trace the orbit resulting in repeated traversals of the trajectory. In this respect, quantal and classical Coulomb scattering are equivalent. The core-scattered waves, on the other hand, redistribute amplitude into other closed orbits, isotropically for an s-wave ($l = 0$) quantum defect or as $\cos\theta$ for p-wave scattering.

The method for incorporating the core-scattered contributions has been presented previously[14,15] so here we reproduce only the main result. For constant scaled energy spectra, the oscillatory part of the density of oscillator strengths averaged over the experimental resolution of the laser is given by

$$f_{\mathrm{osc}}(\hbar_{\mathrm{eff}}, \varepsilon) = -\frac{4\sqrt{2}}{\hbar_{\mathrm{eff}}^{1/2}} \mathrm{Im} \sum_j \mathcal{N}_j \mathcal{Y}^*(\theta^j_f) \tag{3}$$

where the sum is taken over the closed orbits, j, which depart from the nucleus at an angle θ^i_j and return at an angle θ^f_j. The angular function, $\mathcal{Y}^*(\theta)$, depends on the dipole operator, the initial state wavefunction and the quantum defects, μ_l, of the ionic core. For the situation considered here, namely photoabsorption from s states using π polarization, only the final states with $m = 0$ and odd l are excited and $\mathcal{Y}^*(\theta)$ takes the simple form

$$\mathcal{Y}^*(\theta) = -B_{10}e^{i\pi\mu_1}\sqrt{\frac{3}{4\pi}}\cos\theta \tag{4}$$

where B_{10} represents only a simple multiplicative factor determined by the photoabsorption process. The coefficients, $\mathcal{N}_j$, are found by an iterative procedure. To first order they are given by,

$$\mathcal{N}_j = \hbar_{\text{eff}}^{\nu_j/2}\mathcal{D}_j\mathcal{A}_j^n e^{i\Delta_j^n} + \hbar_{\text{eff}}^{(\nu_j+\nu_k)/2}\sum_k \mathcal{D}_k\mathcal{A}_k^p\mathcal{A}_j^n\mathcal{B}_k^p e^{i(\Delta_k^p+\Delta_j^n)} \tag{5}$$

where $\nu_j = 2$ for $\theta_i = 0$ or $\theta_i = \pi$ and $\nu_j = 1$ otherwise; a summation over repetitions n and p is assumed. The $\mathcal{A}_j^n$ are semiclassical amplitudes involving elements of the stability matrix and the Δ_j^n are phases which include the actions (and hence the modulation frequency) of the jth periodic orbit. All symbols were defined in Dando et al.:[15] the important feature here though is that the expression for $\mathcal{N}_j$ brings in combinations of two hydrogenic actions, $\Delta_j^n + \Delta_k^p$—so the spectral modulation frequency will also be a sum of actions—and that the modulations of the photoabsorption probability induced by the core-scattering are at least order $\hbar_{\text{eff}}^{1/2}$ smaller than the modulations induced by the hydrogenic closed orbits. Hence, far in the semiclassical limit, the contribution of an individual core-scattered closed orbit gets progressively weaker, but remains significant throughout the experimental range. In fact, when going to larger and larger actions, the core-scattered orbits proliferate exponentially and may give contributions to the photoabsorption cross-section.

We have used our semiclassical method, which combines closed orbit theory with core-scattering, to calculate absorption spectra for diamagnetic non-hydrogenic Rydberg atoms, the Fourier transform (the 'recurrence' spectrum) of which shows two effects:

1. there are core-scattered contributions (or 'combination recurrences') due to the electron travelling along one closed orbit and then being scattered by the ionic core on to another closed orbit;

2. the core casts a 'shadow' in the backwards direction thus depleting the recurrence strength on subsequent returns of each orbit to the nucleus.

In Fig. 3 we compare the results for $m = 0$, odd-l and $\gamma^{-1/3}$ in the range 60 to 120 in a magnetic field with constant scaled energy $\varepsilon = -0.3$. The photoabsorption spectrum is calculated from an initial s state using π polarization. Both the 'quantum' (using the numerical method of Delande et al.[20]) and the 'semiclassical' (using the method of Dando et al.[14,15]) spectra are computed and further Fourier transformed with respect to $\gamma^{-1/3}$ to obtain recurrence spectra.

Fig. 3(a) shows the Fourier-transformed spectrum of hydrogen. Fig. 3(b) shows the quantum spectrum for an atom with $\mu_1 = 0.5$ (comparable to caesium which has $\mu_1 = 0.57$ and $\mu_{l\geq 3}$ negligible.). Fig. 3(c) shows the standard closed-orbit result for hydrogen while Fig. 3(d) shows our semiclassical calculation with quantum defect $\mu_1 = 0.5$. One can see that, even for quite small values of $\hbar_{\text{eff}}$ ($n \simeq$ 77–154), there is a substantial difference between hydrogen and the core-scattered case. Each peak in the Fourier transform is associated with a closed orbit. It appears at the scaled action $\tilde{S}_j$ of the orbit with an amplitude proportional to A_j. The core-scattered contributions are the peaks located at the sums $\tilde{S}_j + \tilde{S}_k$. We can see that the combination of closed-orbit theory

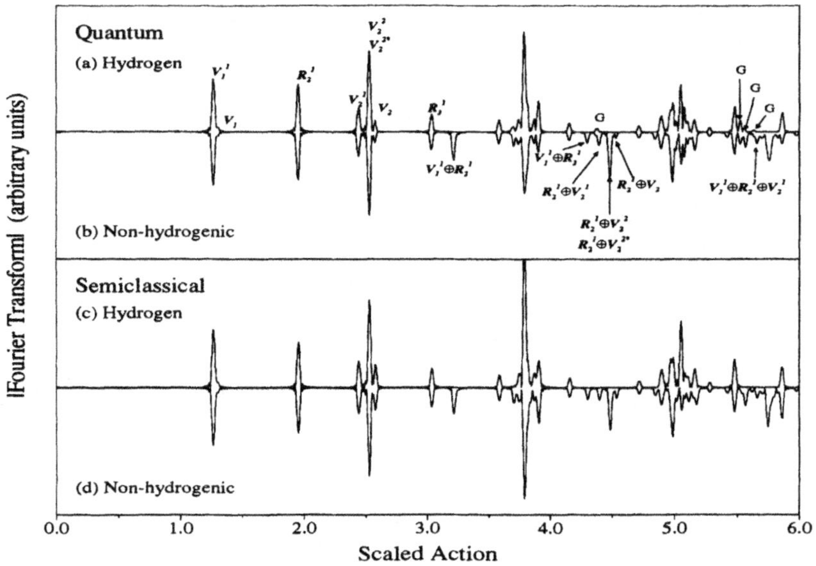

Figure 3. Fourier transforms of the photoexcitation spectra (from the ground state) of Rydberg atoms in a magnetic field at constant scaled energy $\varepsilon = -0.3$ ($m = 0$, odd parity states, $\gamma^{-1/3}$ in the range 60 to 120). (a) Quantum result for hydrogen. The peaks appear at the actions of the closed orbits which we have identified using the convention of Holle et al.[3] Also visible are 'ghost' peaks (marked 'G'). (b) Quantum results for a non-hydrogenic atom with $\mu_1 = 0.5$; core-scattered peaks appear at the sum of actions of periodic orbits. (c) Standard semiclassical calculation for the hydrogen atom, in good agreement with (a), but not with (b). (d) Our semiclassical calculation, which goes beyond the standard closed orbit theory by including core-scattering effects and successfully reproduces the quantum non-hydrogenic spectrum.

with core-scattering gives excellent agreement with the fully-quantal non-hydrogenic case.

In Fig. 3(a) we have indicated the primary orbits using the convention of Holle et al.[3] The combinations of orbits shown in Fig. 3(b), such as that denoted $V_1^1 \oplus R_2^1$, indicate the core-scattered peaks. A few peaks present in Fig. 3(a) (indicated with a 'G') are not found in our semiclassical results. They correspond to 'ghost' peaks:[19] in other words orbits which appear in the quantum spectrum below the energy of the bifurcation at which they are born. Also visible is a peak due to double scattering: an action corresponding to a combination of three orbits. In this case, scattering beyond double-scattering was found to be insignificant.

The major disagreement between our semiclassical and the quantum results is in the peak for the 3rd harmonic of V_1^1. However there is already a comparatively similar discrepancy for this peak between Figs 3(a) and 3(c), the hydrogenic case. Such discrepancies occur when the semiclassical amplitude is near a singularity, for example near a bifurcation. In this case the semiclassical result over-estimates the quantum amplitude with the discrepancy increasing for larger $\hbar_{\text{eff}}$. Nevertheless the semiclassical results shown in Fig. 3(c) agree with the quantum in predicting a reduction of about 30% in the amplitude of this peak due to the back-scattered shadow of the core. Very strong reductions in amplitudes of higher harmonics were observed by Monteiro and Wunner[21] who looked at very low lying-levels ($n \simeq 10$–30) where comparison with semiclassical results is more difficult.

In a recent alternative approach, Hüpper et al.[16,17] have used the model potential of Eq. (1) to describe the interaction of the outer electron with the core. Such a classical description of the core-scattering process certainly produces combination recurrences and 'shadows'. However, because of the sensitive dependence of the final scattering angle on the initial angular momentum, there are a great number of orbits which must be found and hence the method is computationally extremely intensive. For example, Hüpper et al.[16,17] found about 2600 closed orbits for the diamagnetic lithium problem where there are only 35 hydrogenic orbits. By using these orbits within the standard closed-orbit framework, Hüpper et al.[16,17] found good agreement for the strengths of the combination recurrences. However, we have found that the model potential approach does not reproduce the quantum results for smaller atoms such as helium in the presence of a static electric field. In Fig. 4 we show the Fourier transforms of the fixed scaled energy Stark spectra for helium at $\varepsilon = -3.0$ with $F^{-1/4}$ in the range 30 to 200. Fig. 4(a) shows the quantum result: the spectrum is dominated by contributions from the two straight line orbits parallel to the electric field, commonly referred to as the 'uphill' ($\theta_i = 0$) and 'downhill' ($\theta_i = \pi$) orbits, and their repetitions. Also visible are the small core-scattered peaks produced by the core-scattering process between the two parallel orbits. The Fourier transform of the spectrum resulting from using the orbits from the model potential calculation is shown in Fig. 4(b). As can be seen, the model potential calculation completely fails to reproduce the core-scattered peaks seen in the quantum calculation of Fig. 4(a). The reason for this is quite simple: the core potential for helium does not allow for a classical trajectory to be deflected through an angle of π and hence scattering between the uphill and downhill orbits is not possible. However, our diffractive approach shown in Fig. 4(c) does reproduce the quantum spectrum satisfactorily.

Core-scattered corrections to the Gutzwiller trace formula

In order to obtain a semiclassical description of the actual energy level, as opposed to the photoabsorption, spectrum of a non-hydrogenic Rydberg atom in an external field we need to make some corrections to the Gutzwiller trace formula. Recently, there has been considerable interest[22–24] in diffractive corrections to periodic orbit theory since the theory gives poor results when the dynamical system contains structures smaller than the characteristic de Broglie wavelength. The basic approach is to decompose a given periodic orbit into 'geometric' segments where the ordinary stationary phase approach is valid. In between these segments are 'vertices' (here each indexed by γ) which originate diffracted waves. The result is an effective 'diffractive' Green's function:

$$\mathcal{G}_{\text{diff}}(x, x'; E) = \left\{ \prod_{\gamma=1}^{n} d_\gamma \right\} G_f(x, \xi_n; E) G_f(\xi_n, \xi_{n-1}; E) \cdots G_f(\xi_1, x'; E) \tag{6}$$

which is a product of Green's functions from one vertex to another weighted by a diffractive constant d_γ for the vertex at ξ_γ. The important result[22–24] was to show that the trace integral taken over $\mathcal{G}_{\text{diff}}(x, x; E)$ is just proportional to the individual Green's functions:

$$\operatorname{Tr} \mathcal{G}_{\text{diff}}(E) = \frac{T}{i\hbar} \left\{ \prod_{\gamma=1}^{n} d_\gamma G(\xi_{\gamma+1}, \xi_\gamma; E) \right\}. \tag{7}$$

In billiard problems,[22–24] the diffractive amplitude d_γ is obtained from wave optics. Of course paths which do not intersect vertices are treated with the ordinary Gutzwiller

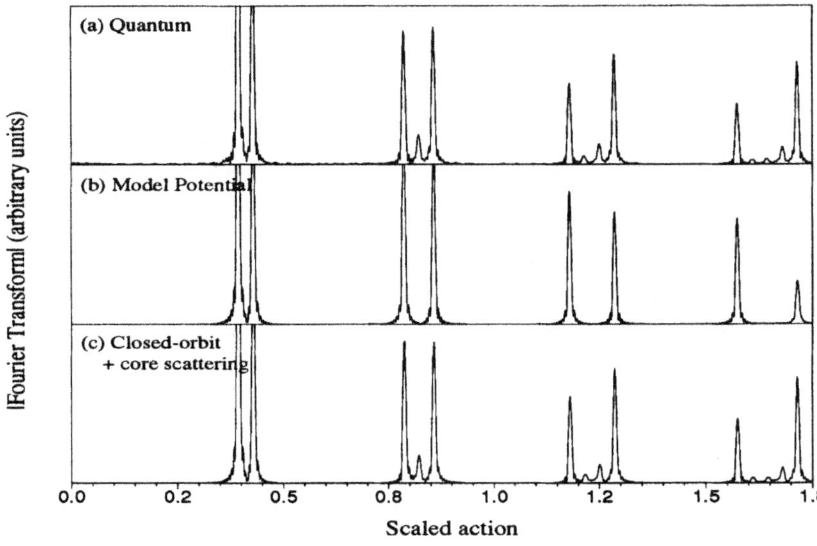

Figure 4. Fourier transforms of the photoexcitation spectra (from the ground state) of helium Rydberg atoms in an electric field at constant scaled energy $\varepsilon = -3.0$ with $m = 0$ and $F^{-1/4}$ in the range 30 to 200. (a) Quantum result for helium. (b) Standard closed-orbit theory with the parameters for the classical orbits obtained using the model potential of Eq. (1) with $Z = 2$ and $a = 0.33$. (c) Our semiclassical calculation which successfully reproduces the quantum spectrum.

formula.

For the atomic case we once again use the Delos approach which decomposes the scattered wave into a Coulomb component and, for non-zero quantum defect, a core-scattered part. We then equate Coulomb scattering with the geometric paths. The trace then results in a zeroth order (Gutzwiller) term plus diffracted contributions obtained by choosing

$$d_\gamma(\theta_f^j, \theta_i^k) = \frac{\psi_{\text{core}}^{j,k}(\theta_f^j, \theta_i^k)}{\psi_{\text{Coul}}^j(\theta_f^j)}. \tag{8}$$

The Green's function's for the diffractive terms can be written in semiclassical form:

$$G_f(r_0, \theta_f^j; r_0, \theta_i^j; E) = \mathcal{A}_j \exp\left[i\left(\frac{S_j}{\hbar} - \mu_j \frac{\pi}{2}\right)\right]. \tag{9}$$

In Fig. 5 we show Fourier transforms obtained from a fully quantal R-matrix type calculation at $\varepsilon = -0.45$ both for hydrogen and a non-hydrogenic atom in a magnetic field. A comparison is also shown for the results obtained from a semiclassical calculation for the Gutzwiller formula (c) and including the diffractive terms (d). The agreement for the basic quasi-Landau modulation (*i.e.* the periodic orbit which moves perpendicular to the magnetic field) is excellent. In fact the fractional reduction in amplitude for such the quasi-Landau periodic orbit has a simple analytic form

$$\frac{A_H - A_{NH}}{A_H} = \hbar^{1/2} 4 \sqrt{\frac{\pi}{m_{12}}} \sin \pi\nu \tag{10}$$

where m_{12} is an off-diagonal element of the stability matrix M and A_H and A_{NH} are the amplitudes of the hydrogenic and non-hydrogenic peaks respectively found from

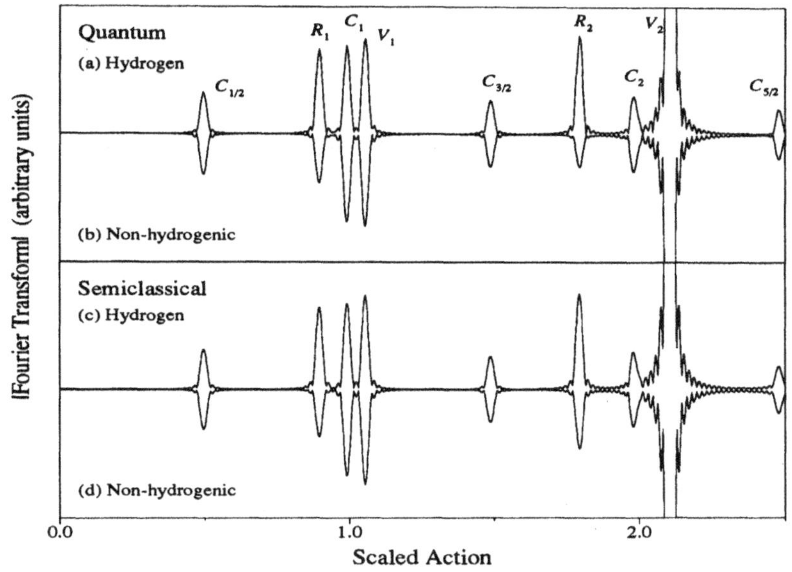

Figure 5. Fourier transforms of the energy level spectra of Rydberg atoms in a magnetic field at constant scaled energy $\varepsilon = -0.45$ ($m = 0$, even parity states, $\gamma^{-1/3}$ in the range 60 to 120). (a) Quantum result for hydrogen. The quasi-Landau modulation is the peak which we have denoted R_1. (b) Quantum results for a non-hydrogenic atom with $\mu_0 = 0.5$. Note the reduction in the amplitude of the modulation associated with R_1. The modulations associated with the orbits denoted C and V are relatively unchanged. (c) Standard semiclassical calculation for the hydrogen atom, in good agreement with (a). (d) Our semiclassical calculation, which includes the diffractive effects of the core in the standard Gutzwiller formula, and reproduces the reduction in amplitude of the modulation associated with R_1 observed in (b).

the quantum calculation. The winding number ν is obtained from $2\cos\pi\nu = \mathrm{Tr}(M)$. The second harmonic of the perpendicular orbit, labelled R_2 in Fig. 5, has an additional higher-order term dependent on $\hbar$. A full detailed comparison will be published elsewhere.

In general, obtaining agreement for peaks of the longer orbits presents more of a challenge than in the case of Closed-orbit theory. All periodic orbits contribute. Hence there are many more instances, at a given energy, of periodic orbits close to bifurcations. In that case the amplitudes are very strongly $\hbar$ dependent. As we have not corrected for bifurcations—a laborious process since each bifurcation must be considered individually—there are significant discrepancies between the fully quantal and semiclassical results. However for orbits unaffected by bifurcations, agreement is obtained with the diffractive formula.

ACKNOWLEDGEMENTS

The authors thank Dominique Delande, Ken Taylor and Niall Whelan for helpful discussions. We thank the Institut Henri Poincaré, where part of this work was undertaken, for its hospitality. This work was supported by the EPSRC and EU Network Electron and Photon Interactions with Atoms, Ions and Molecules.

REFERENCES

1. M.C. Gutzwiller, *Chaos in Classical and Quantum Mechanics*, Springer-Verlag, New York (1990).
2. H. Friedrich and D. Wintgen, The hydrogen atom in a magnetic field—an example of chaos, Phys. Rep. **183**: 37 (1989).
3. A. Holle, J. Main, G. Wiebusch, H. Rottke, and K.H. Welge, Quasi-Landau spectrum of the chaotic diamagnetic hydrogen atom, Phys. Rev. Lett. **61**: 161 (1988).
4. M.L. Du and J.B. Delos, Effects of closed classical orbits on quantum spectra: ionization of atoms in a magnetic field. I. Physical picture and calculations, Phys. Rev. A **38**: 1896 (1988).
5. M.L. Du and J.B. Delos, Effects of closed classical orbits on quantum spectra: ionization of atoms in a magnetic field. II. Derivation of formulas, Phys. Rev. A **38**: 1913 (1988).
6. E.B. Bogomolny, Light absorption by atoms in external fields near the ionization threshold, Zh. Éksp. Teor. Phys. **96**: 487 (1989) [Sov. Phys. JETP **69**: 275 (1989)].
7. P. Gaspard, Quantization of chaos—$\hbar$-expansion theory Prog. Theor. Phys. Suppl. **116**: 59 (1994).
8. P.A. Dando,T.S. Monteiro, W. Jans, and W. Schweizer, Non-hydrogenic Rydberg atoms in magnetic fields, Prog. Theor. Phys. Suppl. **116**: 403 (1994).
9. J.-M. Mao and J.B. Delos, Hamiltonian bifurcation theory of closed orbits in the diamagnetic Kepler problem, Phys. Rev. A **45**: 1746 (1992).
10. M. Born, *The Mechanics of the Atom*, Bell, London (1960).
11. W. Schweizer, R. Niemeier, G. Wunner, and H. Ruder, The diamagnetic Kepler problem—bifurcations and catastrophes, Z. Phys. D **25**: 95 (1993).
12. J. Gao, J.B. Delos, and M. Baruch, Closed-orbit theory of oscillations in atomic photoabsorption cross sections in a strong electric field. I. Comparison between theory and experiments on hydrogen and sodium above threshold, Phys. Rev. A **46**: 1449 (1992).
13. J. Gao and J.B. Delos, Closed-orbit theory of oscillations in atomic photoabsorption cross sections in a strong electric field. II. Derivation of formulas, Phys. Rev. A **46**: 1455 (1992).
14. P.A. Dando, T.S. Monteiro, D. Delande, and K.T. Taylor, Beyond periodic orbits: an example in nonhydrogenic atoms, Phys. Rev. Lett. **74**: 1099 (1995).
15. P.A. Dando, T.S. Monteiro, D. Delande, and K.T. Taylor, Role of core-scattered closed orbits in nonhydrogenic atoms, Phys. Rev. A **54**: 127 (1996).
16. B. Hüpper, J. Main, and G. Wunner, Photoabsorption of non-hydrogenic Rydberg atoms in a magnetic field: effects of core-scattered classical orbits, Phys. Rev. Lett. **74**: 2650 (1995).
17. B. Hüpper, J. Main, and G. Wunner, Nonhydrogenic Rydberg atoms in a magnetic field: a rigorous semiclassical approach, Phys. Rev. A **53**: 744 (1996).
18. W. Jans, T.S. Monteiro, W. Schweizer, and P.A. Dando, Phase-space distributions and spectral properties for non-hydrogenic atoms in magnetic fields, J. Phys. A: Math. Gen. **26**, 3187 (1993)
19. M. Kuś, F. Haake, and D. Delande, Prebifurcation periodic ghost orbits in semiclassical quantization, Phys. Rev. Lett. **71**: 2167 (1993).
20. D. Delande, K.T. Taylor, M.H. Halley, T. van der Veldt, W. Vassen, and W. Hogervorst, Scaled energy spectra of nonhydrogenic Rydberg atoms in a magnetic field, J. Phys. B: At. Mol. Opt. Phys. **27**: 2771 (1994).
21. T.S. Monteiro and G. Wunner, Analysis of quantum manifestations of chaos in general Rydberg atoms in strong magnetic fields, Phys. Rev. Lett. **65**: 1100 (1990).
22. G. Vattay, A. Wirzba, and P.E. Rosenqvist, Periodic orbit theory of diffraction, Phys. Rev. Lett. **73**: 2304 (1994).
23. N.D. Whelan, Geometric and diffractive orbits in the scattering from confocal hyperbolas, Phys. Rev. E **51**: 3778 (1995).
24. H. Bruus and N.D. Whelan, Edge diffraction, trace formulae and the cardioid billiard, Nonlinearity **9**: 1023 (1996).

THE OPACITY AND IRON PROJECTS - AN OVERVIEW

Keith A. Berrington

Department of Applied Mathematics and Theoretical Physics
Queen's University, Belfast BT7 1NN, UK

INTRODUCTION

Science profits from a two-objective aim.
One thing is to push back frontiers, understand
better the universe, its contents and its frame-
work of law. Objective two is to expand
quantitatively the base of science: to claim
new predictions, accurate data standards and
better models. These bring high rewards these days;
and in many applications, data pays.

This theme – of the link between data needs and scientific progress – describes well the development of computational atomic physics, which is profiting from a systematic programme of research to provide the data required by applications as diverse as astrophysics and laboratory plasma technology. In particular, outstanding progress has been made by two international projects: the recently-completed Opacity Project (OP)[1–27] and the current IRON Project (IP)[28–48].

These projects will be described more fully in their respective sections, but as the names imply, the OP was to calculate radiative opacity data for atoms and atomic ions involved in stellar opacity; the IP is to calculate electron collisional data mainly for the iron group of elements. The two projects have much in common on the atomic physics front, using *ab initio* theoretical and state-of-the-art computational methods based on R-matrix techniques[49,50], as summarised in the next section, and both projects involve the calculation of large amounts of good atomic data.

But of equal importance, the execution of these projects has increased our understanding of collisional and radiative physics and led to many new developments. It is these aspects which will be discussed here; in particular, we shall examine the impact on the theory of photoionization.

Recent developments in atomic physics are straining the traditional divisions set by the collisional energy regime: low energy (R-matrix and other close-coupling methods), high energy (Born, distorted wave), negative energy (atomic structure). Two of these

active introduction of gaseous recycling impurities may take place to increase the radiated power loss from the divertor plasma. Divertor studies rely on visible, vuv and xuv spectroscopic observations in conjunction with detailed models of the interplay of plasma dynamics and reaction kinetics to assess performance, impurity sources, recycling and migration. These studies have led to a need and use for very high quality atomic data of relatively complex neutral and near neutral light and medium weight elements.

In general terms, the JET plasma in both the beam penetrated and divertor zones can be described as in transient states of ionisation. Also the plasma is of medium density from the point of view of collisionality in the atomic modelling of the populations of ions in these zones. Therefore dynamic collisional radiative modelling is the relevant picture for quantitative studies. At JET a highly structured approach has been taken to ensure that there is a comprehensive modelling capability of this type and that derived atomic data is properly tuned to the actual experimental data reduction and analysis. An integrated and centralised system, called the 'Atomic Data and Analysis Structure' (ADAS) has been set up covering the fundamental data, collisional-radiative modelling and derived data aspects

TOKAMAK OVERVIEW

The geometric shape, confining magnetic field surfaces and heating systems of modern tokamaks such as JET result in a plasma usefully viewed as made up of a number of distinct regions. There is the hot core plasma region, confined within closed magnetic flux surfaces which is to be contrasted with the unconfined scrape-off-layer (SOL) region lying outside the last closed flux surface. In the SOL, plasma particles travel freely along magnetic field lines connected to material surfaces.

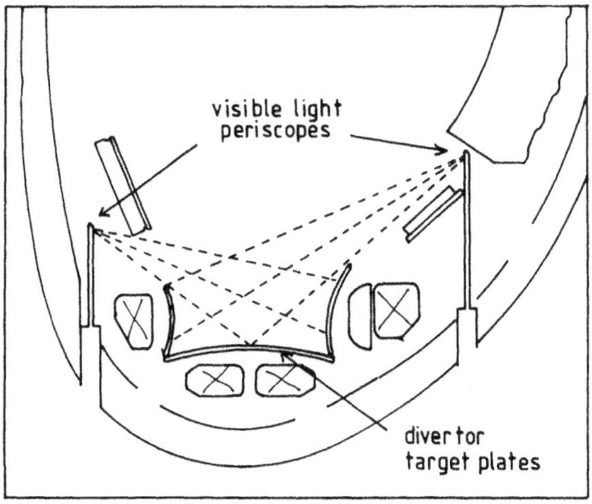

Figure 1. Divertor region of JET showing target plates and views afforded by visible light periscopes

Between the SOL and core lies the transport region, so called because diffusive transport of particles and energy across this region is pronounced but sensitive to operating regimes. Transport across this region reveals the effectiveness of the magnetic confinement and of course strongly influences attainment of thermo-nuclear fusion conditions in the core. From a tokamak design point of view, it is found that the entry of impurity species through the scrape-off-layer into the core plasma, a serious problem for fusion, may be inhibited by the construct known as a divertor. JET has a form of this called the 'axisymmetric divertor'. Such an arrangement leads unconfined plasma along field lines to a point remote from the bulk plasma where the contacts with material surfaces occur. These are at so called 'strike zones'. A divertor plasma is much cooler than the bulk plasma and relatively dense in ideal conditions forming another distinctive region. Turning to the heating of the plasma, fast neutral deuterium beams are generally present, supplementing the ohmic and ion cyclotron resonance heating (ICRH). Beam-penetrated plasma has quite characteristic radiating features forming another distinct region for the purposes of this overview. Both beam heating and ICRH produce distorted ion distribution functions. All these regions have their influences on ultimate fusion performance and so are subject to intensive study.

From the point of view of atomic processes, spectrometric observations provide the principal diagnostic insight and all modern large fusion experiment are well endowed in this respect. Soft X-ray observations of the plasma core, XUV and VUV scans of the transport region and scrape-off-layer and UV and visible observations of the divertor are generally available. At the JET experiment, the divertor and beam penetrated plasma are of particular interest - the latter because of its impact on diagnosis of the core plasma. Figures 1 and 2 show some of the spectroscopic views available of these two regions at JET .

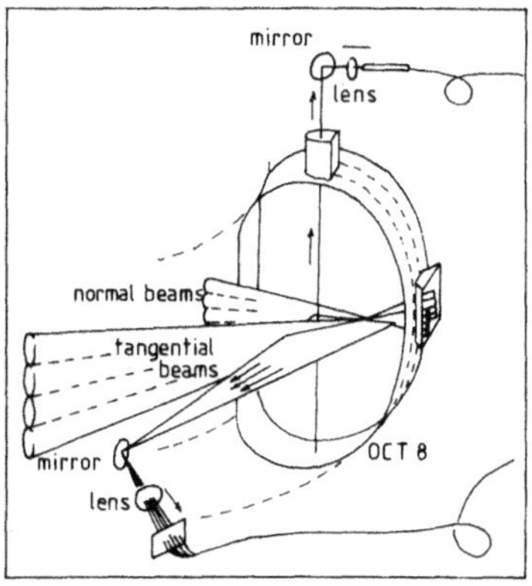

Figure 2. Paths of the four 'normal' and four 'tangential' beams from an injector showing the visible spectroscopy along the beam-line.

In this paper, we wish to summarise some current studies on atomic reaction processes and populations in these special regions. The objective of such studies is of course the development of diagnostic measurements. However, the tokamak plasma regions are not in general simple systems from an atomic point of view, but inhomogeneous with active processes which distort particle distribution functions and modify primary observational effects. Present work shows increasing awareness of these secondary effects and seeks to incorporate them in the diagnostic interpretation. Figure 3 shows some of the primary and secondary populations contributing to beam driven charge exchange spectroscopy and beam emission spectroscopy. Figure 4 shows birth and successive atom and ion generations as atoms enter the divertor plasma region from a contact surface. Ion/atom collision processes are influential in both cases and in this respect the fusion plasma is different from the solar coronal plasma with which it is often compared [1].

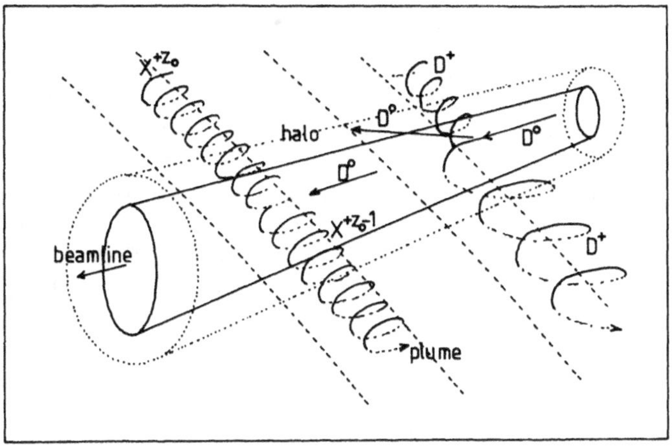

Figure 3. Schematic of first and second generation deuterium and impurity populations in D^0 beam penetrated plasma. Note the D^0 halo and the X^{+z_0-1} plume.

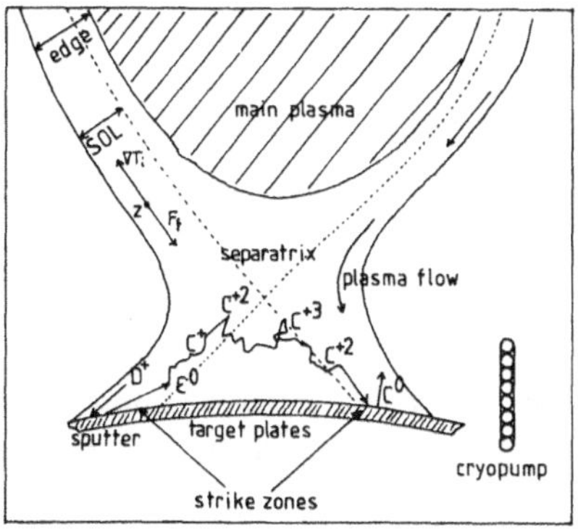

Figure 4. Schematic of the divertor region showing sputtering of C^0 from target plates by D^+ and carbon ion impact and successive ionisations. Ion movement in the SOL is indicated.

STUDIES OF THE DIVERTOR PLASMA AND SOL

The divertor region is very rich from the spectroscopic point of view since there impurity species are concentrated and partially ionised. Divertor ion temperatures, certainly at the strike zones, must be low to inhibit sputtering. Likewise divertor electron temperatures, electron densities neutral deuterium and impurity densities must be suitable for dispersing, either radiatively or by charge exchange processes, the heat load flowing into it from the bulk plasma. Finally velocity fields must be such as to inhibit impurity movement up the SOL to the bulk plasma[2]. Figure 5 shows a VUV spectrum along a line of sight directed at the JET divertor. Prominent lines from nitrogen ions are evident. This experiment series probed the effectiveness of nitrogen gas puffing in creating a strongly radiating divertor and it was shown that the primary radiating zone could in fact broaden and detach from the target - a useful effect [3]. Because of the geometrical complexity of the divertor, it has become usual to interpret divertor spectra through the intermediary of Monte Carlo simulation codes such as DIVIMP [4]. Individual atoms are tracked from birth on a surface until exit from the plasma to establish the ion distributions. Expected emission along actual spectrometer lines of sight is computed for comparison with the observations. Atomic modelling must provide the effective recombination, ionisation and radiated power coefficients to be used in these models of the divertor plasma. The low electron temperature and the relatively high density in the divertor means that true finite density 'collisional-radiative coefficients are required, that is with the modifying effects of stepwise ionisation, suppression of dielectronic recombination etc.

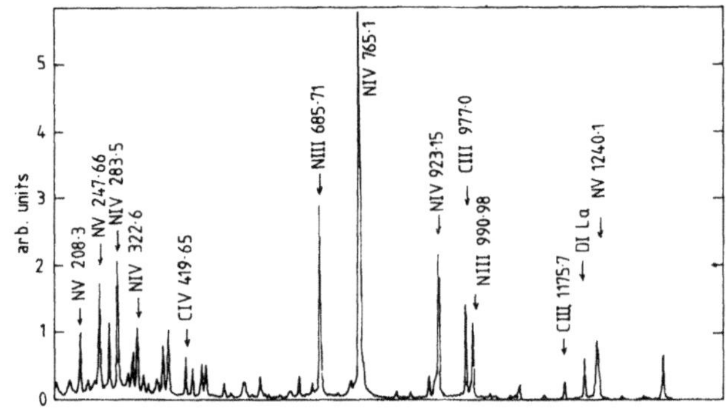

Figure 5. VUV spectrum of the JET divertor region. Prominent nitrogen lines are present due to nitrogen gas puffing during the observation. This created a detached strongly radiating divertor plasma.

Note also that in divertors, low metastable states of impurity ions may have populations comparable with those of ground states and not be in temperature equilibrium with the ground state populations. This situation is often ignored in obtaining the ionisation state of dynamic plasmas. However observations of metastable ion fluxes show the dis-equilibrium. Thus it is the metastable resolved (generalised) coefficients which are required (cf. figure 6) [5]. Charge exchange with neutral deuterium is an additional recombination mechanism for impurity ions in the deuterium rich divertor. Low energy charge transfer being a strongly state selective process can also directly modify excited state populations in a competition with electron impact excitation in this region. Simulations of emission along lines of sight scanning the two strike zones of the divertor in the CIII (λ=538.2Å) spectrum line show such effects [6]. Gaseous species such as nitrogen or neon, popularly introduced at the present time to enhance radiating efficiency in the divertor, also need such atomic coefficients to support their spectral study. There is generally a deficiency of information for the electron impact cross-sections involving the neutral and near neutral stages, especially since transitions involving the n=3 shell are required for visible spectroscopy. State selective charge exchange cross-section data of good accuracy are very incomplete and are a main source of weakness in spectral interpretation

Choice of material for the divertor strike zones for a fusion reactor is a key issue, balancing resistance to erosion, conductivity etc. against the deleterious effects of heavy species entering the core plasma. Observations of spectral emission from inflowing ions have long been used to infer impurity fluxes. A theoretical atomic physics quantity, the 'ionisation per photon' is used for the reduction [8]. For species of current and future interest such as molybdenum or tungsten, the calculation of these quantities is challenging. Figure 7 shows ionisations per photon coefficients for the MoI (λ=3845.7Å) multiplet spectrum line. This line is on the ground septet side. For the quintet metastable side, interpretation of the equivalent λ=5529.5Å line in terms of flux needs the coefficients driven from the septet ground as well as the quintet metastable.

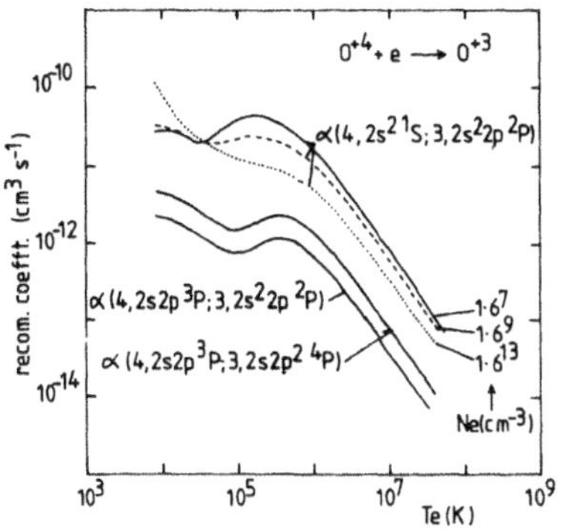

Figure 6. Density and temperature dependence of effective, metastable resolved 'recombination coefficients for $O^{+4} + e \rightarrow O^{+3}$

STUDIES OF THE BEAM AND BEAM DRIVEN PLASMA

The remarkable diagnostic impact of observations of spectral emission from the beam/plasma interaction region has been much explored in the last few years. The types of spectra involved have become familiar being divided into the emission from the beam atoms themselves ' beam emission spectroscopy' and the emission from ions in the plasma following charge transfer from beam atoms ' charge exchange spectroscopy'. Illustrations are given in figures 8 and 9 respectively. For the core plasma in the JET tokamak, the beams are the heating beams and may be composed of ^{2}D, ^{3}T, ^{3}He or ^{4}He atoms. For deuterium, the motional electric field, from the beam atom crossing the magnetic field, produces a large linear Stark resolution of the Balmer α transition array which is the primary observational phenomenon [9]. This is complicated by the presence of three velocity fractions in the beam, arising from molecular species in the source. The Stark multiplet wavelength separations are diagnostic of the internal tokamak magnetic fields and so of the current distribution.

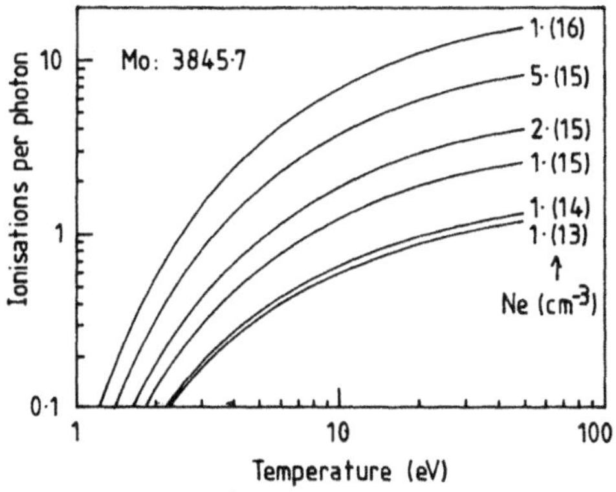

Figure 7. Ionisations per photon for MoI (a ^{7}S - z ^{7}P) (λ=3845.7Å).

The absolute intensities and excited state populations of the D^0 atoms are driven by collisions with plasma ions principally. The populations and beam stopping coefficient (the effective ionisation rate coefficient of beam atoms) determination requires full collisional radiative modelling. Non-isotropic modelling, polarisation and proper use of differential ion/atom cross-sections are development areas. Measurement of beam attenuation from the beam emission spectrum is a crucial support to absolute charge exchange spectroscopy. For He^0 beams at energies ~30-50 keV/amu, the motional Stark field is insufficient to mix the n=3 term populations but does influence the n=4 shell where it interferes with the spin system breakdown. The primary observed visible beam emission lines on the singlet (λ6678.2Å) and triplet ((λ5875.7Å) sides are critical to supporting charge exchange spectroscopy with helium beams. The very low efficiency of ground state helium beam donors compared with metastable beam donors means that quite low metastable fractions in the beam have an influence. The initial metastable beam content on entry into the plasma and the regenerated metastable content of the beam in the plasma are important points of enquiry [10]. Figure 9 shows the He^{+1}(n=4-3) (λ4685.2Å) emission region with the broad core charge exchange feature driven by the beams contrasted with narrow (composite) emission in the same line but from the thermal edge of the plasma.

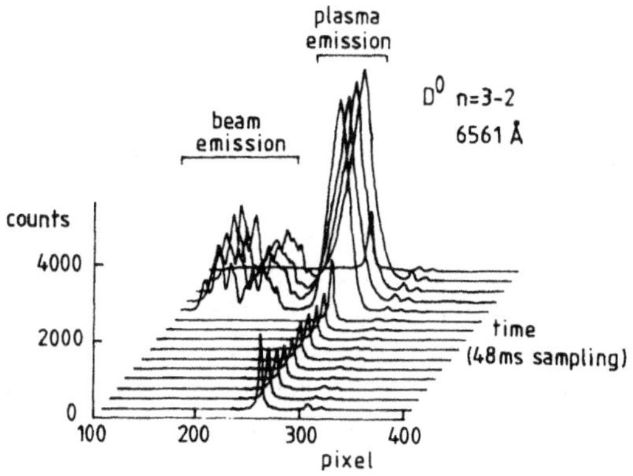

Figure 8. D^0(n=2-3) beam emission spectrum. Doppler shifted to the blue from the beam/viewing line angle of inclination. The primary beam particle energy is 40keV/amu.

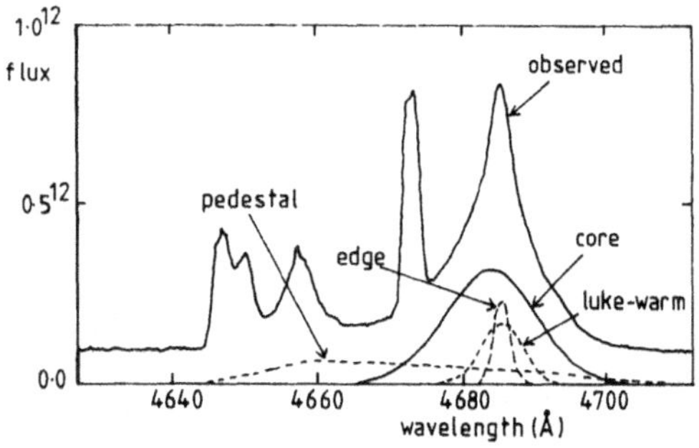

Figure 9. CX spectroscopy measurement of the vicinity of the HeII (n=4-3) (λ=4685.2Å) transition. Note the separation of the broad CX feature into core, edge and pedestal parts.

Measurement of alpha particles, the 'ash' of the D-T fusion reaction, is a key task with detection both of the slowing population from their birth energy of 3MeV and of the thermalised population required. Evidence of a slowing distribution is given by the pedestal in figure 9, but the source of the fast alpha particles is the beam itself. It is clear however from the discussion of figure 3 earlier that separability of the various portions of the He^{+1}(n=4-3) emission line is not feasible without examination of the secondary populations. The injected distribution of He^{+2} particles by ionisation of helium beam atoms is narrow in pitch angle/ velocity space. It evolution to a thermal isotropic distribution may be followed via a Fokker-Planck solution. The distribution is detected as usual by the HeII emission following charge transfer as it passes through the beams. The predictions are illustrated in figure 10. It appears that measured signals are ~factor 2 larger than predicted. There are however further contributions to the HeII envelope. He^{+1} ground state ions formed by cascade after thermal He^{+2} pass through the beams may be re-excited by electron collisions and emit He^{+1}(n=4-3) 'downstream' in the optical viewing line of sight. Survival of the He^{+1} against collisional ionisation to emit in this manner depends on speeds parallel and perpendicular to the field lines and distance from the viewing line. Consequently the profile of this 'slow plume' feature is asymmetric (Figure 11). He^{+1} is also created from first ionisation of the beam atoms extending and emitting away from the beams as a 'fast plume' He^{+2}/He^{+1} ion/ion charge exchange influences the formation and fast plume population distribution. To this already complex picture must be added the presence of a He^{+0} and H^{+0} haloes surrounding helium beams formed respectively by double and single charge transfer and an H^{+0} halo surrounding a hydrogen beam [11].

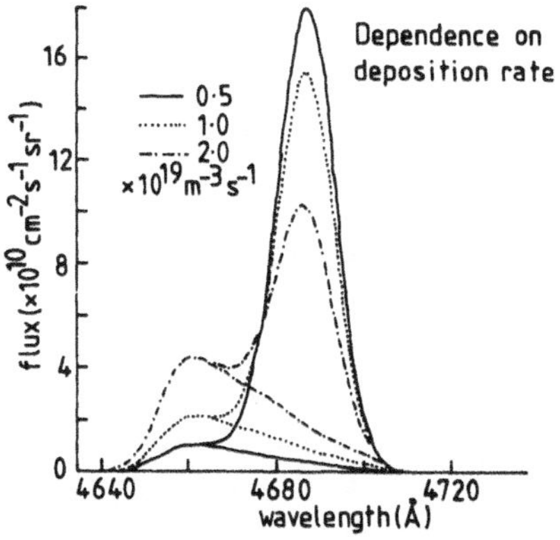

Figure 10. Fokker-Planck solution for the He^{+2} slowing population detected by CX spectroscopy. Note the slowing and thermal parts. Curves are foe various deposition rates of initial fast particles by the beams.

The complementary measurement of fast ion (D^{+}, He^{+2} etc.) populations is by detecting neutral particles exiting the plasma. Neutralisation may take place in the beams. It now appears that the picture is less simple. Fast ion populations with energies in the 1-2 MeV regime may be created by ICRH tuned to the appropriate minority ion (H^{+}, $^{3}He^{+2}$, $^{4}He^{+2}$ etc.). Results at JET for the neutral H flux have been surprising in that, even without neutral beams present for neutralisation, a large 'passive' neutral flux is observed (cf. Figure 11). With the neutral beams present, an enhanced 'active' neutral flux is observed, but its energy dependence does not match that calculated from beam neutralisation and attenuation to the plasma edge. Instead it has the energy dependence of the passive signal. Present interpretation is that there is a reservoir of thermal D^{0} in the plasma core of very low number density, but large compared with expectations from ionisation balance. The H^{0} is itself not an efficient donor to MeV H^{+} but rather thermal D^{0}/C^{+6} transfer occurs, after cascade disturbing the carbon ground state ionisation balance. C^{+5}(1s) is an effective donor to fast H^{+}. It is believed that the 'active' signal forms by the same mechanism but with and enhanced thermal D^0 density in the core sustained by the thermalising beam source. Since the actual deposition by the beam may be estimated, the difference between the passive and active signals allows absolute estimate of the ambient thermal D^0 presence [12]. This is an area of continuing study both with respect to the high ambient D^0 density and to the situation with helium beams. These theoretical studies require quite different charge exchange cross-section data from that of the previous section in that there is a severe mismatch of binding energy on donor and receiver but nearer coincidence of donor electron speed and ion impact speed.

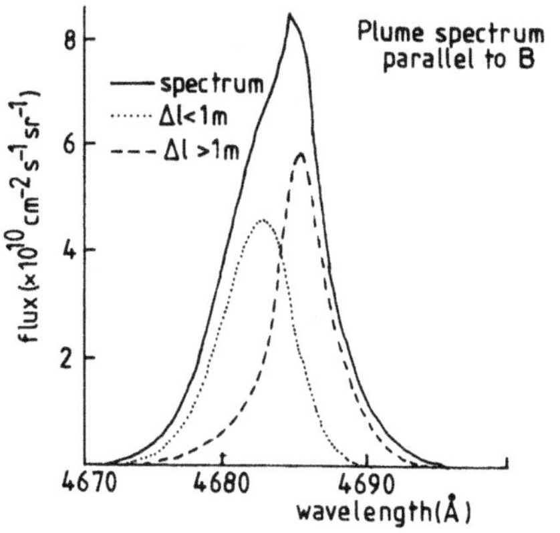

Figure 11. Model calculation of the slow He^{+1} plume emitting the HeII (n=4-3) (λ=4685.2Å) line. Note the assembly from near (slower) and distant (faster) ions and the co- / counter- rotation asymmetry.

THE ATOMIC DATA AND ANALYSIS STRUCTURE

At JET Joint Undertaking much attention has been given to atomic modelling. JET has promoted the production of key fundamental atomic collision data necessary for fusion and has made great efforts to ensure that comprehensive atomic models are in place to see these data into JET applications. The realisation of these efforts is the Atomic Data and Analysis Structure (ADAS). ADAS is a large interconnected set of computer codes and data collections for modelling the radiating properties of ions and atoms in plasmas and for assisting in the analysis and interpretation of spectral measurements. The two components of the package are firstly an interactive system, and secondly, a library of key subroutines. The interactive part provides immediate display of important fundamental and derived quantities used in analysis together with a substantial capability for exploring parameter dependencies and diagnostic predictions of atomic population and plasma models. The second part is non-interactive but provides a set of subroutines which can be accessed from the user's own codes to draw in necessary data from the derived ADAS database.

The various ADAS routines are essentially of two types, namely those which are interrogative on the ADAS database (both fundamental and derived parts) and those which execute atomic modelling calculations. The latter may relay substantial quantities of data for further processing or for addition to the derived database. In the interrogative codes, the principle objective has been to allow graphical display of any part of the database. All codes satisfy this quick look and check facility. However, there is a frequent need for some numerical values, of a rate coefficient or cross-section for example, at a particular set of temperatures or energies, possibly with a simple fit to the data. In ADAS interrogation codes, interpolation is performed on the source data from the database and these interpolated values are given as convenient printer tabulations.

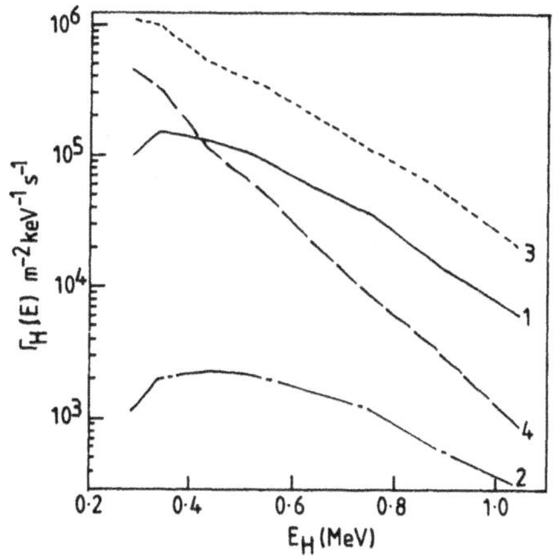

Figure 12. Neutral H fluxes detected by neutral particle analysers. 1) passive measurement; 2) radiative recombination contribution (calc.) ; 3) active measurement: 4) Estimate of direct CX contribution by electron exchange with beam atoms.

ADAS program series

The complete ADAS system is divided into code series according to task. ADAS series 1 is concerned with fundamental, that is essentially individual reaction data. This is to be distinguished from the composite effects of many processes which are the subject of the rest of ADAS. Ab initio calculations of individual reaction cross-sections or rates are not made available in ADAS. Rather ADAS series 1 allows preparation and assessment of such data from other sources for inclusion in the ADAS database. ADAS series 2 is concerned with evaluating excited populations of specific ions in a plasma environment and then their radiation emission. It relies on availability of a reaction rate data collection for the ion in the ADAS database although ADAS has some provision for generating an approximate collection of such data when this is not so. ADAS series 3 is concerned with those situations in a plasma when charge transfer from neutral hydrogen (or its isotopes) or neutral helium in beams is a primary mechanism. ADAS series 4 is concerned with the ground and metastable populations of ions in a plasma and particularly their preparation and calculation for dynamical plasma models. It therefore operates with generalised collisional dielectronic recombination and ionisation coefficients, associated power loss coefficients and metastable fractions. It also mediates a transfer from an isoelectronic sequence storage of data which is

Series	*Content*
ADAS1	Entry and validation of fundamental atomic data
ADAS2	Excited state populations of ions in a plasma
ADAS3	Charge exchange related emission
ADAS4	Recombination, ionisation and radiated power
ADAS5	Supplementary programs
ADAS6	Utilities

the preferred choice for atomic data preparation and manipulation to an isonuclear sequence storage preferred for applied studies in dynamical plasma models. ADAS series 5 provides a set of supplementary programs interrogating data collections in the derived database. The data classes addressed are those which have been found of particular importance for reduction of calibrated observed data and include 'ionisations per photon', photon emissivity coefficients and contribution functions. ADAS series 6 consists of extra utilities. Some further details of the ADAS codes applicable to the physics areas highlighted in the first sections of this paper are given in the following paragraphs.

Population processing relevant to the divertor.

ADAS series 2 programs are associated with two streams of population modelling. The first stream is concerned with modelling and prediction of detailed spectral line emission from plasmas and commences with a collection of 'complete' reaction data for an identified group of levels of an individual ion, called a *specific ion file* (type ADF04). Such a dataset can include state selective recombination, charge exchange recombination, electron impact ionisation and ion impact excitation coefficient data, but the primary content is electron impact excitation data. The electron impact excitation data in such files are individually prepared for the highest precision but can also be produced semi-automatically by direct calculation or, for members of prepared isoelectronic sequences, through extraction from *general z excitation files* (type ADF05) by the program ADAS203. The excited level populations are calculated at an array of temperatures and densities in the quasi-equilibrium approximation with respect to identified metastables of the ion. The basic program is ADAS205 and it produces plots and tables of populations and prepares files of metastable fractions. Also a temporary passing file of the complete population set is generated for use by program ADAS207. This program supports diagnostic line studies by plotting line emissivities and emissivity ratios as functions of temperature, density or as contours in the temperature/density plane. ADAS206 is functionally very similar to ADAS205 but specifically calculates and prepares output files of total line power and arbitrarily chosen specific line power.
The second stream is concerned with the modelling and prediction of effective recombination and ionisation coefficients in finite density thermal plasma. The main program is ADAS204 which carries out a *bundle-nS* population and generalised collisional-radiative calculations. It feeds directly to collisional-radiative recombination, ionisation, parent cross-coupling and recombination-cascade power coefficient files of type ADF10. Also it produces a low level *projection file* of type ADF17 for advanced studies. The most refined *photon emissivity*, *ionisation per photon* and *generalised collisional-radiative* data are produced along with a complete set of files of type ADF10 by the program ADAS208. This is an advanced version of ADAS205 which merges the key parts of the two streams. It draws necessary additional data on zero density ionisation coefficients (type ADF07), expansion (type ADF18) and condensation (type ADF17) data if required and as available.

Charge exchange processing for neutral beam.

The programs of series ADAS3 are associated with neutral beams of hydrogen or helium isotopes and there are two streams of modelling. The first stream is concerned with modelling and detailed spectral line emission from hydrogen-like impurity ions in a plasma following charge transfer from fast neutral beams. It commences with a collection of state selective charge transfer cross-section data at n, nl or nlm resolution (type ADF01) spanning an extended region of collision energies and n-shells and in some cases different sources. There are two processing options, namely for calculations in the ***bundle-nl*** (ADAS308) or ***bundle-nlj*** (ADAS306) approximations. The codes are specifically designed to assist in experimental data analysis and accept a number of experimental charge exchange line-of-sight integrated intensities from which they deduce an *emission measure* for the beam/plasma/viewing line intersection. They also allow some comparative assessment of theoretical charge exchange cross-sections with the experimentally implied ones, depending on the number of charge exchange lines presented. The programs also provide a detailed theoretical breakdown of the experimental charge exchange line intensities, together with plots of theoretical stick and Doppler broadened spectra. The programs include some capability for studying free-electron capture and electron impact driven line emission for these simple hydrogenic systems and also extend to lithium-like systems in fast beams, depending on the availability of fundamental data collections of type ADF01. Practical experimental data reduction requires effective emission coefficients for charge exchange lines as a function of relative collision energy for different plasma conditions. The programs ADAS308 and ADAS306 predict data for arbitrary lines at a fixed set of conditions. It is convenient to have available more automatic codes which generate look-up tables of these coefficients over ranges of plasma parameters without the intervening displays. This capability is provided by ADAS309 and ADAS307 for the bundle-nlj and bundle-nl pictures respectively. The latter are termed scanning versions of the codes and prepare directly data of type ADF18. The associated codes for interrogation of the fundamental state selective cross-section database and the effective charge exchange emission coefficient database are ADAS301 and ADAS303 respectively.

The second stream is concerned with modelling the attenuation of the beams and then the prediction of the emission spectrum of the beam atoms themselves. It commences with ion/atom collision cross-section data collections of type ADF02, stored separately for the hydrogen and helium beam reactants. Also electron collision data, type ADF04, first discussed in chapter 3 are utilised. There are separate processing programs for hydrogen and helium beams, ADAS310 and ADAS311 respectively. The primary calculations for the effective beam stopping coefficients are performed in a *bundle-nS* approximation and can generate extensive tabulations over plasma parameters (data type ADF21). ADAS304 is the interactive interrogation program for this data type.

The ADAS library and ADAS data formats.

An objective of ADAS is to provide a capability for embedding fundamental and derived atomic data from the database into a user's own measurement reduction and analysis codes. A library of subroutines is provided to draw data of the important ADAS types. Access is obtained by searching an ADAS computer module library. These independent extraction routines open appropriate files and are consistent with the related interactive interrogation programs. In most cases, they form primitive kernels at the heart of the interactive programs. Besides the extraction subroutines, there are additional subroutines providing equilibrium populations, equilibrium ionisation balance and so on.

There are 24 distinct data types in the ADAS databases. Each data set type has its layout and content precisely described. These prescriptions are called ADAS data formats or ADFs for short and must be rigidly adhered to in preparation of data for use by the ADAS package. These formats apply both to fundamental and derived data. A full description of all the ADAS data formats is not given here but are summarised as follows:

ADF01	bundle-n & bundle-nl charge exchange cross-sections
ADF02	ion impact cross-sections with named participant
ADF03	approximate form parameters for elements
ADF04	resolved specific ion data collections
ADF05	general z excitation data
ADF06	general z recombination/ionisation data
ADF07	direct resolved electron impact ionisation coefficients

ADF08	direct resolved radiative recombination coefficients
ADF09	direct resolved dielectronic recombination coefficients
ADF10	isoelectronic master files
ADF11	isonuclear master files
ADF12	charge exchange effective emission coefficients
ADF13	ionisation per photon coefficients
ADF14	thermal charge exchange coefficients
ADF15	photon emissivity coefficients
ADF16	generalised contribution functions
ADF17	condensed projection coefficients
ADF18	expansion definition data
ADF19	zero density radiative power coefficients
ADF20	G(Te) functions
ADF21	effective beam stopping coefficients
ADF22	effective beam emission coefficients
ADF23	state selective electron impact ionisation rate coefficients.
ADF24	state selective CX cross-sections for partially ionised ions.

CONCLUSIONS

There are very varied environments in tokamaks with quite different dominant atomic processes. In a number of these environments, ion/atom collisions are more important than electron/ion collisions. When modelling the radiating characteristics of these regions, comprehensive collisional-radiative calculations of populations and effective rate coefficients are usually necessary. Some situations such as HeII emission driven by beams are of long term importance with a steadily improving availability of atomic reaction data and sophistication of analysis. In other situations, such as divertors, exploration of novel materials means that interest in a particular species may be relatively short lived. Clearly, high quality calculations of atomic cross-sections which may take a year or more must be carefully targeted. Easy availability and utilisation of moderate precision atomic data for initial studies is helpful as well as more sophisticated data. At JET, the creation of a centrally supported and maintained atomic data and analysis structure, ADAS, orientated for direct experiment and model support has proved of great benefit in coping with the wide range of need. The translation of ADAS for operation on UNIX workstations is nearing completion. This is a jointly sponsored activity by the leading fusion laboratories in Europe.

REFERENCES

1. H. P. Summers , Adv. Atom. Mol. & Opt. Phys. 33: 275 (1994).
2. G.C. Vlases, Plasma Phys. Control. Fusion 35: B67 (1993).
3. L. D. Horton, JET Joint Undertaking Report JET-P(96)?? (1996).
4. P. Stangeby & J. D. Elder, J. Nucl. Mater. 196-198: 258 (1992).
5. H. P. Summers & W. J. Dickson, Nato Advanced Study Institute Series 296: 31 (1992) (W. G. Graham - ed., Plenum Press, New York)
6. C. F. Maggi, L. D. Horton, R. Konig et al.. JET Joint Undertaking Report JET-P(94)29: 187 (1994).
7. H. P. Summers, JET Joint Undertaking Report JET-IR(94)06 (1994).
8. K. H. Behringer, H. P. Summers, B. Denne et al., Plasma Phys. Control. Fusion 31: 2059 (1989).
9. W. Mandl, R. C. Wolf, M. G. von Hellermann et al., Plasma Phys. Control. Fusion 35: 1371(1993).
10. F. W. Bliek, H. O. Folkerts, R. Morgenstern et al., Nucl Instrum Meth B 98: 195 (1995).
11. U. Gerstel, L. Horton, H. P. Summers et al., JET Joint Undertaking Report JET-P(96)?? (1996).
12. A. Korotkov and A. Gondhalekar, Europhysics Conference Abstracts 18B part1: 266 (1994).

LASER-PLASMA INTERACTION IN THE FEMTOSECOND TIME REGIME

D. von der Linde

Institut für Laser- und Plasmaphysik, Universität Essen,
D-45117 Essen, Germany

INTRODUCTION

Progress in femtosecond laser technology[1] has made a strong impact on the field of laser-plasma interaction. With the development of chirped pulse amplification[2] and new broadband solid state laser materials, femtosecond laser pulses with peak power in the terawatt range and focused intensities greater than 10^{18} W/cm^2 can be generated with relatively small-scale, table-top laser systems. The electric field strengths corresponding to these intensities are comparable with the Coulomb field in atoms. During the interaction with such an intense laser pulse atomic, molecular or any condensed form of matter is very rapidly ionized and turned into a plasma. Thus the physics of laser-plasma interaction is of fundamental importance for a broad class of applications of intense femtosecond laser pulses.

Key issues of current studies of femtosecond laser-produced plasmas[3] include the physical mechanisms of the transformation of cold, solid material into a hot, dense plasma, the coupling of optical energy into the plasma, the dynamics of cooling and expansion, and nonlinear optical phenomena such as second and higher order harmonic generation. Last but not least, a hot, dense, short-lived microplasma represents an interesting point source of incoherent ultrashort X-ray pulses with photon energies ranging from the soft X-ray regime of a few hundred eV to hundreds of keV.

In this paper the following topics will be discussed. The first part deals with the generation of plasmas in dielectric solids during the interaction with femtosecond laser pulses and describes quantitative measurements of the threshold of plasma formation. Second harmonic generation from a developing surface plasma will be discussed in the second part. It will be shown that time-resolved measurements of the reflected second harmonic provide interesting insight into the transformation of matter from the solid state to a plasma. The final part describes the observation of optical harmonics of high order from solid targets by femtosecond laser pulses at intensities of approximately 10^{17} W/cm^2. High order harmonic generation from solid targets bears some potential for the generation of coherent femtosecond pulses in the vacuum UV and soft X-ray spectral region.

Photon and Electron Collisions with Atoms and Molecules
Edited by Burke and Joachain, Plenum Press, New York, 1997

PLASMA PRODUCTION ON SOLID TARGETS

When a plasma is generated by focusing an intense femtosecond laser pulse on the surface of a solid the time for plasma expansion during the interaction with the pulse is very short. An immediate consequence of this temporal constraint is that the electron density is very high, comparable with the electron density of the solid material. The plasma is confined to a thin surface layer, and the density drops from approximately solid density to vacuum in a distance much shorter than the wavelength of light. Thus high density and extremely sharp density gradients are distinctive features of plasmas produced by femtosecond laser pulses.

If an interaction of the peak of the laser pulse with a plasma of *solid density* is to be achieved the pulse must rise from an intensity level corresponding to the threshold of plasma formation to the maximum in a time much shorter than the time scale of plasma expansion. The necessary pulse rise time depends on the threshold of plasma formation of the target material. Optical breakdown and plasma formation in solid materials has been studied with nanosecond and picosecond laser pulses many years ago, and the physical mechanisms are well understood[4-7]. On the other hand, work on plasma formation in the femtosecond time regime has started only recently[8-11].

The thresholds of plasma formation can be obtained from temporally and spatially resolved measurements of the changes of the optical reflectivity[11]. The generation of a high density plasma in a dielectric solid leads to a sharp increase of the optical reflectivity when the free electron density reaches the critical density. This effect is an excellent probe of plasma formation and serves as an indicator of femtosecond laser-induced breakdown.

The experimental arrangement to measure the threshold of plasma formation on the surface of solid targets is shown in Fig. 1. A pump-probe scheme was used which provided both temporal and spatial resolution. An intense laser pump pulse was focused on the surface to generate the plasma (angle of incidence of approximately 50°). A much weaker probe pulse sent through a microscope objective illuminated a surface region overlapping the area of the pump pulse (approximately normal incidence). The microscope lens picked up the

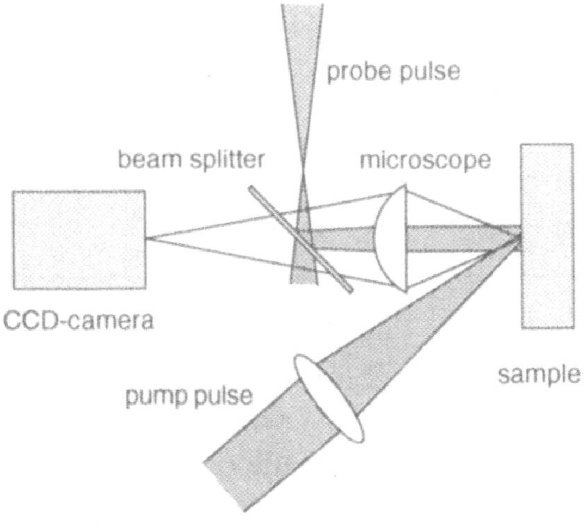

Figure 1. Experimental arrangement for femtosecond time-resolved optical microscopy

reflected probe light and produced an image of the surface on a digital CCD-camera. The polarizations of pump and probe were orthogonal, and a polarizer (not shown in Fig. 1) was used to block scattered light from the strong pump pulse. The experiments were performed in a vacuum chamber.

The following procedure was used to determine the threshold fluence for plasma formation. The method assumes that a dense plasma is formed when the threshold fluence is exceeded, causing a distinct jump of the optical reflectivity. The threshold fluence defines the characteristic diameter d of the surface area where the pump fluence is above threshold and the reflectivity is high. When d^2 is plotted as a function of the logarithm of the energy of the pump pulse one obtains a straight line*). The intersection with the horizontal axis gives the threshold energy E_{th}. The slope of the line is given by d_0^2, where d_0 is proportional to the Gaussian spot size of the beam. Note that the method determines both the threshold energy and the area of the pump pulse and thus the threshold fluence. With an independent measurement of the pulse duration the threshold intensity and electric field strength can be given. A very similar method has previously been used to measure the melting threshold in laser-induced solid-liquid transitions[12].

Threshold measurements were performed using laser pulses from an amplified CPM dye laser operating at 620 nm. The pulse duration (FWHM) was 120 fs. These pulses had a high peak-to-background contrast. The intensity of these laser pulses dropped from the peak down to 10^{-8} of the peak intensity in less than one picosecond. Pump and probe pulses were the same, differing only in total energy. For the threshold measurements the reflectivity was probed at a delay of 200 fs.

Results for optical glass, magnesium fluoride and sapphire are shown in Fig. 2, where the square of the measured diameter of the high reflectivity area is plotted as a function of the logarithm of the pump pulse intensity. The data can be fitted by a straight line. The intersections with the horizontal axis are the threshold intensities. Different slopes indicate some

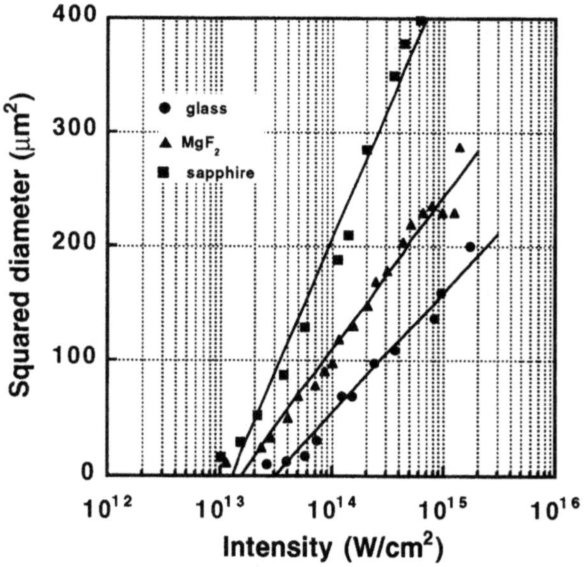

Figure 2. Squared diameter of the surface plasma plotted against peak laser intensity

---

* This behaviour follows from the Gaussian shape of the spatial beam profile

variation of the actual spot size of the pump beam. These variations are due to unavoidable uncertainties in the alignment of the pump focus. At low intensities the diameters of the high reflectivity area become comparable with the spatial resolution of the microscope, causing the observed deviations of the data from the straight line fit near threshold.

The results of the measurements of the surface breakdown threshold are summarized in Table I, where the threshold intensities and the corresponding electric field strengths are given for optical glass, magnesium fluoride, sapphire, and fused silica. As can be seen from the data, there is surprisingly little variation among the different materials. The threshold intensities and field strengths are approximately 10^{13} W/cm^2 and 10^8 V/cm, respectively.

The observation of relatively constant threshold values in the present experiments differs from previous measurements with nanosecond and picosecond pulses. For instance, the measured intensity thresholds varied by more than an order of magnitude over the family of alkali halide crystals[3]. However, these experiments had been designed to measure the thresholds of *intrinsic bulk breakdown.* The breakdown mechanism in these materials could attributed to avalanche ionization. For an *intrinsic* mechanism a systematic variation of the threshold in different materials is expected.

In present experiment *surface* breakdown was observed. No effort has been made so far to use surfaces with controlled physical and chemical composition. Thus the physical mechanism is most likely of extrinsic nature, and the measured breakdown thresholds are probably affected by surface imperfections such as scratches, cracks, grooves and chemical contamination[13].

In principle the same physical processes as in the bulk can also play a role in extrinsic surface processes. In fact, the measured breakdown field strengths in Table I are in reasonable agreement with extrapolations of nanosecond and picosecond avalanche ionization thresholds[5] to shorter times. Avalanche ionization probably plays an important role, but from the present data an unambiguous identification of the physical mechanisms of surface plasma formation is not possible.

Attempts were also made to produce plasmas in the bulk. The striking principal observation was that no indication of a fully developed bulk plasma and no material damage could be observed when ultrashort laser pulses (120 fs) were focused into the bulk. By contrast, pulses of the same energy stretched in time to 0.4 ns produced a bright spark and permanent volume damage. Close inspection revealed the well-known morphology of long-pulse laser damage[11]. From this result it must be concluded that with femtosecond laser pulses the concentration of optical energy necessary to produce breakdown in the bulk is never reached .

The failure of femtosecond laser pulses to produce a fully developed bulk plasma and damage can probably be attributed to self-defocusing of the laser beam in a very early stage of plasma formation. It can be readily shown that an electron density much less than the critical density is sufficient to cause strong enough beam defocusing to prevent breakdown and full ionization of the material.

Table 1. Threshold intensities and breakdown fields

Material	Intensity (W/cm^2)	Electric Field (V/cm)
Glass	$(2.8 \pm 0.4) \times 10^{13}$	1.00×10^8
Magnesium Fluoride	$(1.5 \pm 0.4) \times 10^{13}$	0.75×10^8
Sapphire	$(1.3 \pm 0.4) \times 10^{13}$	0.70×10^8
Fused Silica	$(1.0 \pm 0.4) \times 10^{13}$	0.61×10^8

SECOND HARMONIC GENERATION

As discussed in the preceding Section, a surface layer of solid material is ionized and transformed into a hot, dense plasma when an intense laser pulse with a solid target. Light incident on this plasma sheet is partially reflected, and upon reflection a small fraction of the light is frequency-doubled[14]. In this Section it will be shown that the reflected second harmonic reveals interesting details of the transformation of matter from its condensed phase to the plasma phase.

To see what kind of information may be obtained from second harmonic experiments, it is instructive to consider the plasma current density induced by the light. The current density can be expressed as follows[14]:

$$\mathbf{J} = -en\mathbf{v} = -en^{(0)}\mathbf{v}^{(1)} - e(n^{(1)}\mathbf{v}^{(1)} + n^{(0)}\mathbf{v}^{(2)}) + \dots \tag{1}$$

Here, n and **v** denote the electron density and velocity, respectively. The superscripts signify the order of the terms in a perturbation expansion.

The first term is responsible for the linear optical properties of the plasma, that is, reflection and absorption in the lowest order approximation. For a periodic perturbation at the frequency ω the second term gives rise to second harmonic generation. Note that the second order term depends on both velocity *and* density, whereas the first order current density is *independent* of the density perturbations. Thus information on induced *changes of the electron density* can be obtained from second harmonic experiments. This information is not available from the first order optical properties.

Before continuing the discussion of second harmonic generation, it may be useful to outline some basic features of the changes in the way the laser pulses interact with the material during the solid-to-plasma transition.

Solid-to-Plasma Transition

Light interacts primarily with the electrons of the material and changes the electron distribution. Here, we discuss the evolution of the electron density profile during the transformation of a solid into a plasma. In the second harmonic experiments described later the response of the material to p-polarized light (electric field vector in the plane of incidence) is of particular interest. The following discussion anticipates a pump-probe experiment in which a laser pump pulse generates a plasma, and a weak, p-polarized probe pulse interacts with the system after some delay time. It is convenient to represent the p-polarized electromagnetic field by its magnetic component. Referring to Fig. 3 the magnetic field of the probe light and the electron density profiles are represented by solid lines and dotted lines, respectively. Three different stages of the evolution can be distinguished.

(i) Suppose that the pump pulse is short enough that expansion of the material during ionization is negligible. Then, immediately after the ionization, one has a sharp drop of the electron density from solid density to vacuum. The detailed profile in the depth of the material is not important in this qualitative discussion, and the electron density can be represented by a step function (Fig. 3 a). Thus, as far as the optical properties and distribution of the electromagnetic fields are concerned, the situation is very similar to a metal. The incident light is strongly reflected at the boundary, and a standing wave pattern is formed outside due to the interference with the incident light. The field penetrates into the material over a distance given by the skin depth.

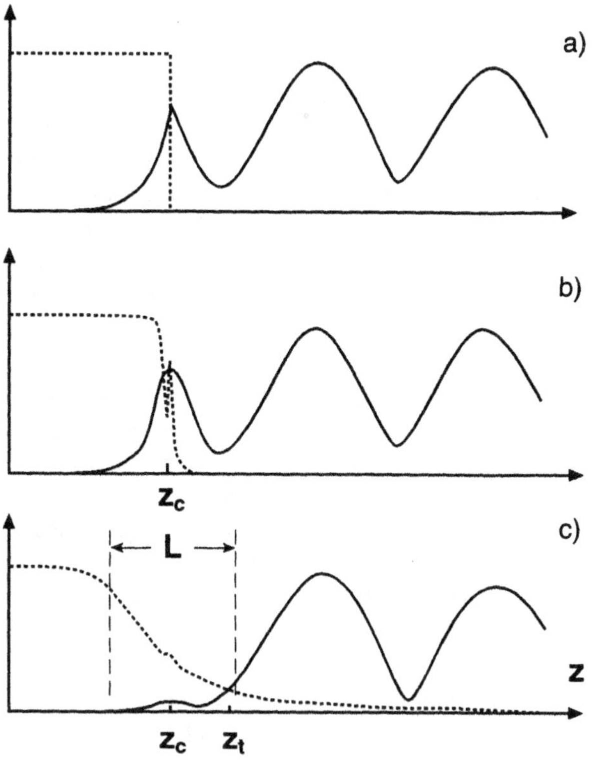

Figure 3. Qualitative picture of the optical magnetic field (solid line) and of the electron density profile (dotted line) for different stages of the evolution of the plasma. The z-direction is normal to the surface, and z_c indicates the position of the critical density surface. a) Initial stage with step-like plasma-vacuum boundary; b) Broadening of the boundary corresponding to $L << \lambda$; c) Broadening corresponding to $L > \lambda$.

(ii) During a short time of expansion the density step broadens to a finite thickness L. Fig. 3 b shows the case $L << \lambda$. The magnetic field is still quite similar to Fig. 3 a. However, a new feature appears in the density profile at z_c, the position of the critical density, where the normal component of the electric field couples resonantly with electron plasma oscillations[15] and drives strong electron density fluctuations. The resulting perturbations of the density profile are indicated in Fig. 3 b by a peak of the dotted line at z_c.

(iii) Finally, we consider the plasma sheet after expansion to a thickness greater than the wavelength. Because the electromagnetic field is evanescent for $z < z_t$[*)], the driving field for the excitation of electron plasma oscillations at the critical density is very weak. Thus, because the incident light is screened out when thickness of the plasma sheet increases, the resonant coupling of the light with electron plasma waves is disrupted.

It will be shown that the emergence of the plasma resonance during the formation of the plasma and its disappearance during expansion leads to striking effects in the time dependence of the second harmonic.

[*)] z_t is the so called classical turning point[15].

Time-Dependence of Second Harmonic Generation

For time-resolved measurements of the reflected second harmonic from femtosecond laser-produced plasmas a pump-probe scheme similar to Fig. 1 was used[16]. Because the reflected second harmonic is relatively weak, the CCD camera was replaced by a photomultiplier. Thus there was no spatial resolution in the second harmonic experiments.

Second harmonic generation was measured with laser pulses from an amplified CPM dye laser. The experimental parameters were as follows: Pulse duration 65 fs; spot diameter on the sample surface approximately 8 μm; angles of incidence 40 degrees and 50 degrees for the pump and the probe beam, respectively. The peak intensity on target of the s-polarized pump pulses was approximately 10^{16} W/cm^2, and the pump-probe energy ratio was 100:1.

The second harmonic generated by *p-polarized* probe pulses from a glass target is plotted against delay time in Fig. 4 a. It can be seen that there are two different features in the time dependence of the second harmonic: (i) a very sharp spike close to zero delay time, and (ii) a much broader structure with a maximum near 3 ps. The insert shows the first 500 fs on an expanded time scale. This representation of the data reveals that second harmonic generation is very weak near zero delay time and that the sharp maximum occurs *after* the pump pulse at a delay of 170 fs. Some resolution-limited substructure on a time scale of 50 fs can also be recognized. The measured ratio of the spike maximum to the maximum at 3 ps was about 4:1 (not resolved in the main part of Fig. 4 a).

The second harmonic produced by *s-polarized* laser pulses was much weaker. The measured ratio of the second harmonic for p- and s-polarized laser pulses, normalized to the same

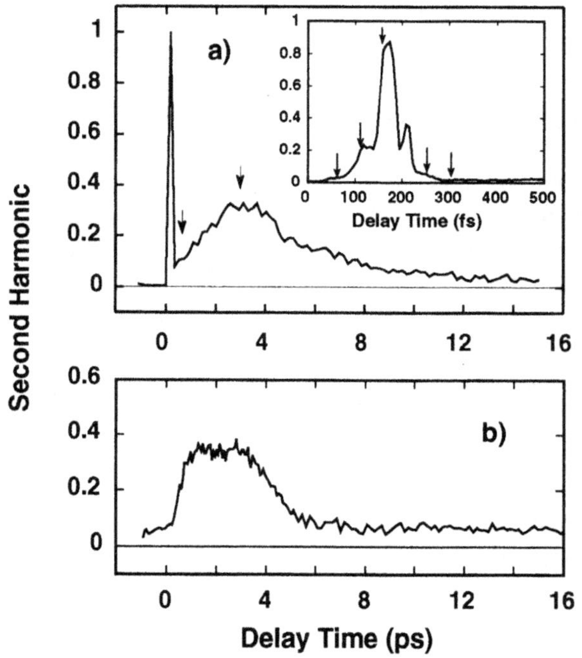

Figure 4. Second harmonic of the probe pulse plotted against delay time: a) p-polarization, b) s-polarization. The insert shows the second harmonic on an expanded time scale.

probe pulse energy, was 16:1. To increase the signal-to-background ratio, a greater probe pulse energy, corresponding to a ratio of 20:1, had to be used.

The result of pump-probe measurements with s-polarized probe pulses is shown in Fig. 4 b. The striking difference between the measurements with p- and s-polarization is the *absence of the sharp spike* in the case of s-polarization.

Interpretation of the Second Harmonic

The principal difference between the second harmonic generated by p- and by s-polarized laser pulses is the distinct sharp maximum of the second harmonic efficiency between 100 and 200 fs for p-polarization. This feature is absent in s-polarization.

A starting point for the explanation of the conspicuous second harmonic effects of p-polarized laser pulses is provided by the following circumstances:

(i) P-polarized light couples resonantly with electron plasma oscillations.
(ii) The evolution of the plasma outlined in Fig. 3 suggests that optimum conditions for resonant coupling exist for a certain thickness of the plasma layer $L \ll \lambda$.
(iii) Second harmonic generation is sensitive to density perturbations.

Thus it appears promising to look for an explanation in terms of the dependence of the second harmonic efficiency on the thickness of the plasma sheet L.

When the variation of the second harmonic with L is calculated from a perturbative model disagreement with the experimental results is obtained. Perturbation theory predicts a smooth variation of the second harmonic efficiency without a sharp maximum for short L. However, a perturbative treatment requires that the density fluctuations induced by the laser light be small compared with the unperturbed density. In typical experimental situations this condition is violated in the critical density region. At the critical density the electron plasma oscillations are driven so hard that the induced changes of the electron density become comparable with the unperturbed density. Under these conditions "breaking" of the electron plasma waves occurs[18]. An important consequence of wavebreaking is the opening up of a new channel of energy dissipation. The electrons of the breaking plasma wave are strongly accelerated in the direction perpendicular to the surface and gain substantial kinetic energy at the expense of wave energy[20]. This energy loss leads to a strong increase of the damping of the z-component of the collective electron motion and increases the absorption of the electromagnetic wave.

According to Rae and Burnett[21] the damping due to wavebreaking can be taken into account by a suitable modification of the dielectric function of the plasma. This is done by introducing anisotropic effective dielectric tensor[22]:

$$\varepsilon_{xx} = \varepsilon_{yy} = \varepsilon_1 = \varepsilon(\omega) = 1 - \frac{\omega_p^2}{\omega(\omega + i\nu)}; \quad \varepsilon_{zz} = 1 - \frac{\omega_p^2}{\omega(\omega + i\nu_{eff})} \tag{2}$$

Here, ω_p and ν are the local plasma frequency and the electron collision frequency, respectively. The matrix elements ε_{xx} and ε_{yy} of the diagonal tensor are given by the Drude dielectric function, while ε_{zz} is modified to take into account the increased damping associated with wavebreaking. This is done by replacing the collision frequency ν with a field-dependent effective damping constant,

$$\nu_{eff} = \max(\nu, \nu') \qquad \nu' = \sqrt{\frac{e|B_{1y}|\sin\theta}{mL}} \tag{3}$$

where B_{1y} is the optical magnetic field at the critical density at the onset of wavebreaking, and L and θ are the thickness of the plasma and the angle of incidence.

Physically, this procedure implies the following. For sufficiently weak light intensity the density perturbations are small and wavebreaking is absent. Under these condition one has ordinary collisional damping, $\nu_{eff} = \nu$. When the intensity is increased to the point where wavebreaking sets in, the additional collisionless damping is included by setting $\nu_{eff} = \nu'$.

To shed some light on how wavebreaking brings about an enhancement of the second harmonic, it is instructive to consider the calculated distribution of amplitude and phase of the second harmonic electromagnetic fields[20]. An example for $L/\lambda = 0.013$ and an exponential plasma profile, $n \propto \exp(-z/L)$, is shown in Fig. 5. The normalized amplitude and the phase of the second harmonic magnetic field are plotted against distance z. Figure 5 a is the low field case *without wavebreaking*. There is a sharp maximum of the second harmonic amplitude and a phase jump of approximately π at the critical density. The second harmonic emerging from the plasma ($z >> z_c$) is much smaller than the field at the critical density. The linear increase of the phase for $z >> z_c$ is, of cause, an indication of the free travelling second harmonic wave leaving the plasma.

The distribution of amplitude and phase are very different when the intensity is increased and *damping due to wavebreaking* is included (Fig. 5 b). For an incident intensity corresponding to 10^9 V/cm the damping is so large that the resonance is very strongly broadened. There is no longer a distinct maximum and a phase jump of the second harmonic at the critical density. The second harmonic amplitude has a high, nearly constant value, and the distribution of the phase in the plasma is uniform.

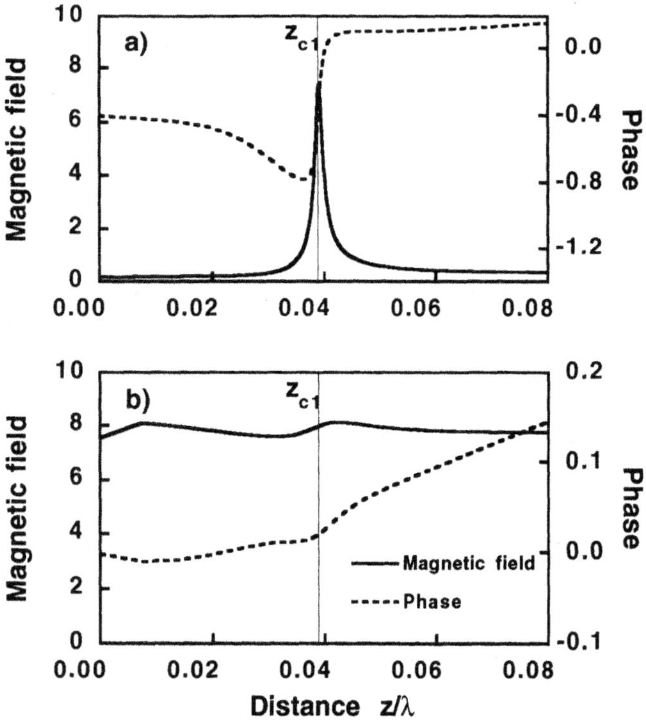

Figure 5. Amplitude and phase of the second harmonic for $L/\lambda = 0.013$ plotted against distance: a) low intensity limit without wavebreaking; b) incident laser intensity of 1.3×10^{13} W/cm^2, wavebreaking included.

The comparison of the field distributions suggests the following explanation of the effect of wavebreaking on second harmonic generation. At low intensity, in the absence of wavebreaking, there is a sharp resonance. The phase jump across the resonance causes a nearly perfect cancellation of the second harmonic generated in the vicinity of the critical density surface. The increase of damping associated with wavebreaking leads to a distinct broadening of the resonance and smoothing of the phase at higher intensity. As a result, the destructive interference is upset, and second harmonic radiation can leak out.

Thus one arrives at the following picture of the variation of second harmonic generation during the solid-to-plasma transition.

In the solid state regime corresponding to region (I) of Fig. 3 there is no electron plasma resonance and hence no effect of electron plasma oscillations on second harmonic generation. In the strongly expanded plasma corresponding to region (III) the light cannot penetrate to the resonant region near the critical density. The key point is that only in a narrow range of plasma thicknesses (region (II) in Fig. 3) electron plasma waves can be strongly excited and cause an enhancement of second harmonic generation by the mechanisms discussed here.

Comparison with Experimental Data

In this Section the measured dependence of the second harmonic on delay time is compared with the calculated second harmonic efficiency as a function of scale length *L*. The calculations are based on a perturbative model in which the effects of wavebreaking are simulated by an effective dielectric function as explained in the preceding paragraph. A basic assumption of the comparison is that the second harmonic efficiency of the plasma at a time *t* is the same as that of a static plasma with instantaneous scale length *L(t)*. The fit of the measured time dependence provides the expansion velocity of the plasma.

Examples of from two different experimental runs are shown in Fig. 6 a and b. In both cases the fundamental pulses were p-polarized. The angles of incidence were 45 and 60 degrees, respectively, and the probe pulse intensity corresponded to an electric field of 1.5×10^8 V/cm. The open circles and the solid line are the experimental data and the calculated second harmonic, respectively. The vertical scale is fitted by adjusting the height of the broad secondary maximum of the second harmonic. Good fits of the data are obtained for expansion velocities $v_{exp} = 2.5\times10^6$ cm/s and $v_{exp} = 4\times10^6$ cm/s.

There is good qualitative agreement of the experimental data and the model calculations. In particular, the conspicuous sharp spike and the broad secondary maximum of the measured second harmonic are well reproduced by the model. The measured and the calculated times of the sharp maximum are, respectively, 170 fs and 70 fs in Fig. 6 a, and 60 fs and 45 fs in Fig. 6 b. The measured and calculated ratios of the maxima are 4:2 and 2:4.

Keeping in mind the assumptions and approximations involved in the model, no effort was made to improve the agreement by other choices of the fitting parameters. For example, the height of the sharp spike is very sensitive to the value of the collision frequency, whereas the secondary maximum changes very little with damping. Thus the ratio could be adjusted by a suitable choice of the collision frequency.

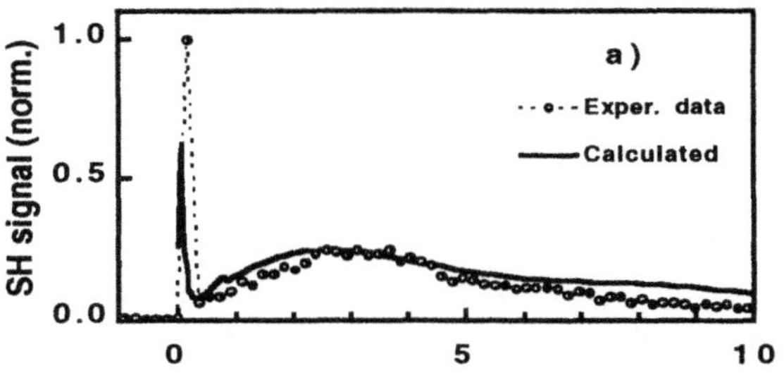

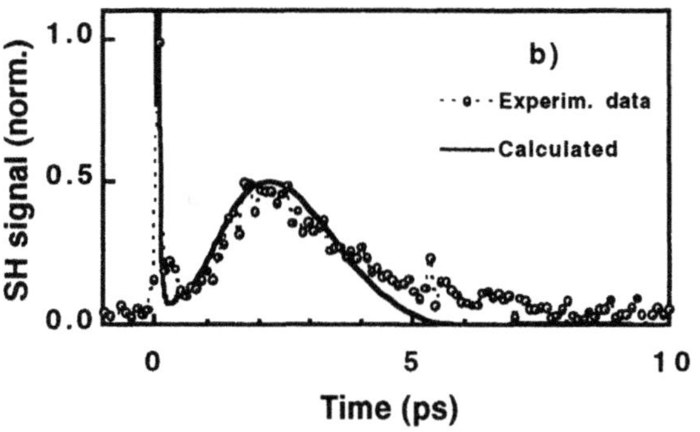

Figure 6. Comparison of the experimental data with the model calculations. Open circles: Measured second harmonic versus delay time for p-polarized laser pulses. Solid lines: Calculated second harmonic efficiency for an incident intensity of 3×10^{13} W/cm^2. Dotted line: guide to the eye. Angles of incidence: a) 45 °; b) 60°.

HIGH ORDER HARMONIC GENERATION FROM PLASMAS

The second harmonic generation from plasmas discussed in the preceding Section was obtained with laser pulses of relatively low intensity, $\approx 10^{14}$ W/cm^2. In this case laser-plasma interaction could still be treated by perturbation theory, except in the vicinity of the critical density surface. As pointed out in the introduction, much higher laser intensities are available today. Electric field strengths comparable with or greater than the Coulomb field in atoms can readily be reached in the focused femtosecond laser radiation. In this high field regime the interaction of light with matter can no longer be regarded as a perturbation, and new, high order nonlinear effects are expected to occur. One example is the generation of optical harmonics of high order. Extensive recent research has dealt with high order harmonic generation in rare gases[22]. The physical mechanism responsible for this harmonic generation is the nonlinear response of the *individual atoms*.

Optical harmonics of very high order were also observed more than a decade ago in *plasmas* produced by nanosecond laser pulses[23,24]. This type of harmonic generation was explained in terms of the anharmonicity of the laser-driven collective motion of the plasma electrons in a steep plasma density gradient[25,26]. Such step-like density profiles arise in an expanding plasma when the thermal pressure forces are balanced by the ponderomotive forces associated with intense laser light.

Femtosecond laser-surface interaction naturally provides plasmas with steep density gradients. This opens the possibility of generating coherent femtosecond ultraviolet or even soft X-ray pulses simply by irradiating the surface of solid targets with intense femtosecond laser pulses[27,28].

Measurement of High Order Harmonic Generation

This Section describes the experimental observation of high order harmonic generation from solid surfaces[28]. A schematic of the experimental arrangement is depicted in Fig. 7. Laser pulses from a titanium sapphire laser ($\lambda \approx 800$ nm) with pulse energies ranging from 10 mJ to 50 mJ were used. With different combinations of pulse energy, pulse duration, and focusing optics the peak intensity at the target surface could be varied between 10^{17} and 10^{18} W/cm^2. The incident laser beam was linearly polarized in the plane of incidence (p-polarization) and focused on the sample by means of an off-axis parabolic mirror or a magnesium fluoride lens at an angle of incidence of approximately 68 degrees. The use of reflective focusing optics avoided possible temporal and spatial distortions of the laser pulses resulting both from the linear optical dispersion and the optical nonlinearity of the lens material.

A small opaque disk was positioned near the surface of the focusing mirror or the lens to block the central portion of the incident beam (diameter approximately 2 cm). An aperture with an opening corresponding to the far field image of the disk blocked the reflected fundamental beam. The reflected harmonics generated on the surface passed through the opening of the aperture and struck a platinum coated, toroidal holographic grating (550 lines/mm). This arrangement protected the grating and the detector from being damaged by the strong reflected fundamental laser beam.

The harmonic spectra were recorded as follows. The grating produced a frequency-dispersed image of the target surface on the face of a phosphor screen. The magnification was approximately unity. The green phosphorescent light excited by the high energy harmonic photons was then imaged onto a photodiode array coupled to a multichannel analyser. A set of optical filters blocked the scattered fundamental laser light and the second harmonic.

The targets were optically polished bare glass substrates and glass substrates coated with 200 nm of aluminium. The target was raster-scanned during the experiment to provide a fresh sample surface for each laser pulse, and the experiments were performed under vacuum.

Two examples of typical harmonic spectra from glass are depicted in Fig. 8 and Fig. 9. Odd and even harmonics were observed up to a maximum order of 14 and 18. The peak at

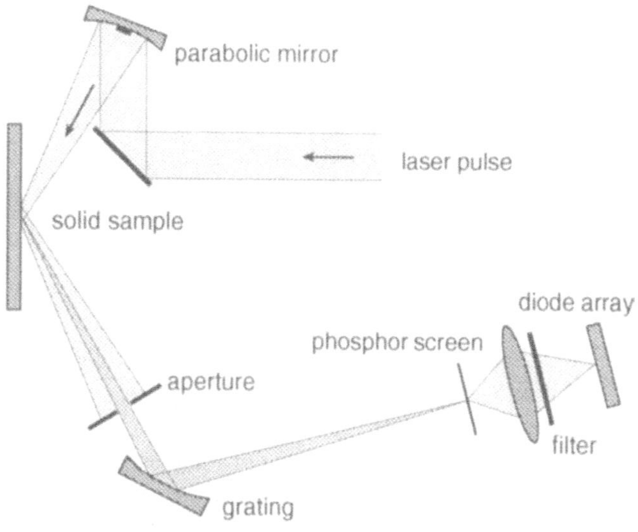

Figure 7. Experimental setup for measuring high order harmonic generation from solid targets.

38 nm in Fig. 9 is line emission from the plasma. The intensity on target was estimated to be approximately 10^{17} W/cm^2. Very similar spectra were observed with aluminium samples.

Figure 8 represents an example of very strong harmonic emission with a large harmonic-to-background ratio. Typically, the harmonics were superimposed on some background as shown in Fig. 9. The origin and detailed nature of the background effects are yet unknown, although optical emission of the plasma in the visible and ultraviolet undoubtedly contributed. The ratio of harmonics to background varied in different measurements. It was observed that the amount of background radiation and also the appearance of some plasma line emission depended on the temporal profile of the laser pulses. For instance, clean harmonic spectra with little background were observed with laser pulses of 120 fs duration and a temporal profile with an intensity drop of 10^5 in 1 ps. On the other hand, the spectra were dominated by background radiation and plasma line emission with only very weak harmonic lines when laser pulses of nominally 60 fs duration with an intensity drop of only about 3×10^{-3} in 1 ps were used. The intensity profiles for this comparison were estimated from autocorrelation measurements extended over a large intensity range.

The relative strength of the harmonic peaks in the spectra of Fig. 8 and 9 is mainly determined by the properties of the phosphor screen. When the spectral characteristic of the detection system is taken into account the variation of the harmonic signal in dependence of the harmonic order as shown in Fig.10 is obtained. Data from four different spectra are shown. Note that there is no plateau and no cutoff in the harmonic spectra. A plateau and a high frequency cutoff are characteristic features of harmonic generation in rare gases. As can be seen from Fig. 10 a relatively smooth roll-off corresponding an exponential decrease is observed in harmonic generation from solid targets.

The conversion efficiency of high order harmonic generation is relatively high. For the strongest harmonic spectra in these experiments the number of photons per pulse is estimated to be about 10^{11} and 5×10^9 at the tenth and the fifteenth harmonic, respectively, which corresponds to a photon conversion efficiency of approximately 10^{-6} and 5×10^{-8}. These numbers are subject to relatively large uncertainty, mainly because of the error in estimating the properties of the phosphor screen.

High frequency limit of harmonic generation

The maximum harmonic order for peak intensities of about 10^{17} W/cm^2 was 18, corresponding to a wavelength of $\approx$ 45 nm and a photon energy of 27.5 eV. Recent particle-in-cell (PIC) simulations by Gibbon[29] and by Lichters et al.[30] showed that a significant increase of the number of harmonics should occur when the intensity of the fundamental laser pulse exceeds 10^{18} W/cm^2. The increased efficiency of harmonic generation is due to relativistic effects which play an important role at such high intensities. Thus extension of harmonic generation to relativistic laser intensities should be very promising.

Experimentally, the following observations were made when the focused intensity on target was increased from 10^{17} W/cm^2 to about 10^{18} W/cm^2. At low intensities a clean, undisturbed reflected beam from the target surface was observed. The stable beam geometry readily permitted the separation of the harmonic light by an aperture as described above. The harmonic spectra shown in Fig. 8 and 9 were obtained under these conditions. An increase of the fundamental intensity lead to an increase of the plasma emission and of the background radiation, and, as a result, the ratio of harmonics to background decreased. At still higher intensities the reflected beam became unstable, and substantial beam distortions began to develop. For intensities around 10^{18} W/cm^2 the reflected light was no longer colli-

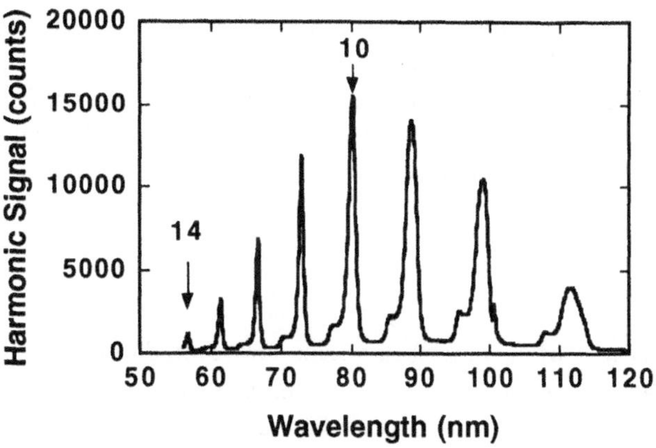

Figure 8. Example of an harmonic spectrum from glass. Laser energy 20 mJ.

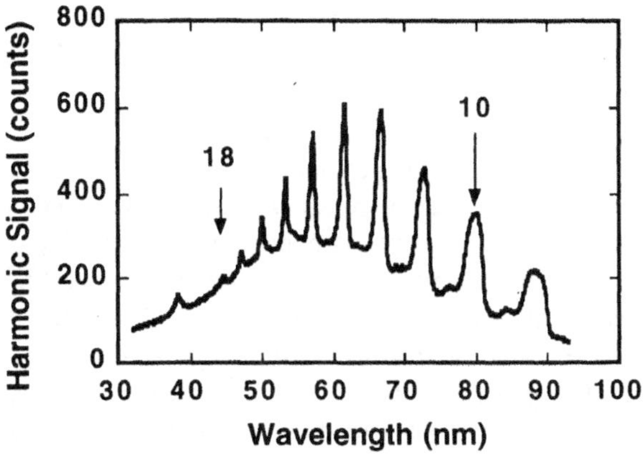

Figure 9. Example of an harmonic spectrum from glass showing background radiation and plasma line emission at 38 nm. Laser energy 20 mJ.

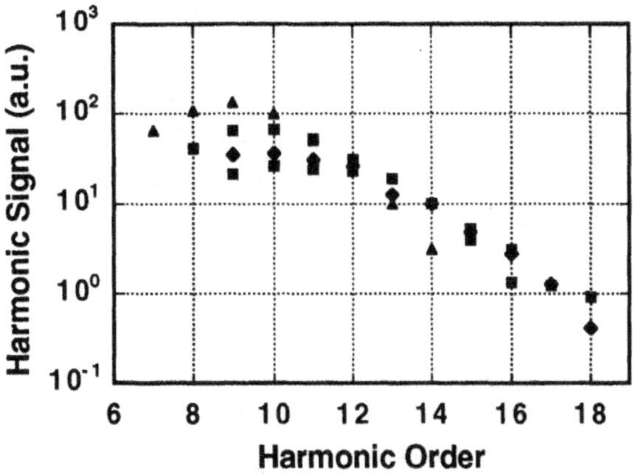

Figure 10. Harmonic signal of the various orders corrected for the spectral characteristic of the detection system. Data from four different spectra are shown.

mated, but spread out over a large solid angle around the specular direction. Under these conditions the generation of coherent high order harmonics disappeared. Because the measurements were restricted to a narrow solid angle around the specular direction, it cannot be excluded that some harmonic radiation was present in non-specular directions and larger solid angles, as observed in the experiments of Norreys et al.[31].

SUMMARY

The first part of this article described plasma formation during the interaction of femtosecond laser pulses with optically transparent solids. It was established that the plasmas were generated on the *surface* of the targets and that a well-defined intensity threshold of plasma formation exists. The threshold values in different dielectric materials were found to be very similar, close to 10^{13} W/cm^2. This result suggests that the threshold of plasma formation is determined by extrinsic surface properties.

It was not possible to generate a plasma by focusing laser pulses of 120 fs into the bulk. By contrast, plasma formation and bulk damage were observed when the pulse duration was increased to 0.4 ns. The absence of plasma formation and damage in the case of femtosecond pulses is tentatively explained by strong beam self-defocusing effects caused by negative contribution to the refractive index from photoexcited free carriers.

In the second part second harmonic generation during the formation and expansion of a femtosecond laser-produced plasma was considered. Second harmonic radiation produced by p-polarized laser pulses exhibits a very sharp maximum after an expansion time of about 150 to 200 fs. This effect is not observed with s-polarization, suggesting that excitation of electron plasma oscillations plays a role. The analysis indicated that for a thin sheet of plasma the excitation of electron plasma oscillations is strongly peaked for a narrow range of scale length of the order of a few percent of the wavelength. It was shown that breaking of electron plasma waves could lead to an enhancement of second harmonic generation. Model calculations based on these ideas showed good agreement with the experimental data. In particular, the model reproduces quite well the observed striking differences of the time dependence of the second harmonic for p-polarized and s-polarized laser pulses. These results suggest that the behavior of second harmonic generation provides direct evidence of the onset of wavebreaking during a specific stage of the plasma evolution at very short scale lengths.

The final part of the paper discussed the generation of harmonics of high order from solid surfaces at intensities of approximately 10^{17} W/cm^2. The characteristics of the observed harmonic generation can be summarized as follows. Generation of high order odd and even harmonics was observed only for p-polarized laser pulses at oblique incidence, while s-polarized light was ineffective. The harmonic radiation was confined to a collimated beam in the specular direction. A smooth, approximately exponential roll-off at high frequencies was measured. The maximum harmonic order was 18, corresponding to a wavelength of ≈ 45 nm and a photon energy of 27.5 eV. Insulating targets (glass) and metallic targets (aluminium films) gave very similar harmonic spectra. Evidence has been obtained that a sharp plasma-vacuum interface is crucial for harmonic generation.

Harmonic generation in the relativistic interaction regime promises higher efficiency and extension of harmonic radiation generation to much shorter wavelengths. Thus the generation of harmonics from solid targets could be a useful method for the production of coherent beams of ultrashort soft X-ray pulses.

ACKNOWLEDGEMENTS

The author is very much indebted to T. Engers for the calculations of second harmonic generation, and to H. Schüler and G. Jenke for their experimental contributions to this work. The hospitality and help of A. Mysyrowicz and A. Antonetti and their staff at the Laboratoire d'Optique Appliquée, Palaiseau, France, and the financial support of this work by the Deutsche Forschungsgemeinschaft and the European Commission are gratefully acknowledged.

REFERENCES

1. G. Mourou and D. Umstadter, Development and application of compact high intensity lasers, *Phys. Fluids* B 4: 2315 (1992).
2. P. Marine, D. Strickland, P. Bado, M. Pessot and G. Mourou, Generation of ultrahigh peak power pulses by chirped pulse amplification, *IEEE J. Quant. Electr.* QE-24:398 (1988).
3. See, e.g., High-Field Interaction and Short-Wavelength Generation, Feature Issue, *J. Opt. Soc. Am.* B 13: 51 (1996).
4. W. L. Smith, Laser-induced breakdown in optical materials, *Opt. Eng. 17*, 489 (1978).
5. N. Bloembergen, Laser-induced electrical breakdown in solids, *IEEE J. Quant. Electr.* QE-10, 375 (1974).
6. S. C. Jones, P. Bräunlich, R. Th. Casper, And X. A. Shen, Recent progress on laser-induced modifications and intrinsic bulk damage of wide-gap optical materials, *Opt. Eng. 28*, 1139 (1989).
7. M. J. Soileau, W. E. Williams, N. Mansour, and E. W. Van Stryland, Laser-induced damage and the role of self-focusing, *Opt. Eng. 28*, 1133 (1989).
8. D. Du, X. Liu, G. Korn, J. Squier, and G. Mourou, Laser-induced breakdown by impact ionization in SiO_2 with pulse widths from 7 ns to 150 fs, *Appl. Phys. Lett.* 64: 3071 (1994).
9. B. C. Stuart, M. D. Feit, S. Herman, A. M. Rubenchik, B. W. Shore, and M. D. Perry, Laser-induced damage in dielectrics with nanosecond to subpicosecond pulses, *Phys. Rev. Lett.* 74: 2248 (1995).
10. B. C. Stuart, M. D. Feit, S. Herman, A. M. Rubenchik, B. W. Shore, and M. D. Perry, Optical ablation by high-power short-pulse lasers, *J. Opt. Soc. Am.* B 13: 459 (1996).
11. D. von der Linde and H. Schüler, Breakdown threshold and plasma formation in femtosecond laser-solid interaction, *J. Opt. Soc. Am.* B 13: 216 (1996).
12. J. M. Liu, Simple technique for measurements of pulsed Gaussian-beam spot sizes, *Opt. Lett.* 7: 196 (1977).
13. N. Bloembergen, *Appl. Optics* 12: 661 (1973).
14. D. von der Linde, Second harmonic production from solid targets, in: *Laser Interactions with Atoms, Solids, and Plasmas*, R. M. More, ed., Plenum Publ. Corpor., New York, (1994) p. 207.
15. See, for example, W. L. Kruer, *The Physics of Laser Plasma Interaction*, Addison-Wesley, Redwood City (1988).
16. D. von der Linde, H. Schulz, T. Engers, and H. Schüler, Second harmonic generation in plasmas produced by intense femtosecond laser pulses, *IEEE J. Quant. Electr.* QE-28: 2388 (1992).
17. See, e.g., P. Mulser, Resonance absorption and ponderomotive action, in: *Physics of Laser Plasmas*, A. Rubenchik and S. Witkowski, eds., North-Holland, Amsterdam (1991), p. 435.
18. J. Albritton and P. Koch, Cold plasma wavebreaking: Production of energetic electrons, *Phys. Fluids* 18:1136 (1975).
19. S. C. Rae and K. Burnett, Reflectivity of steep gradient plasmas in intense subpicosecond laser pulses, *Phys. Rev.* A 44: 3835 (1991).
20. P. Audebert, F. Falliès, J. P. Geindre, J. Delettrez, A. Rousse, and J. C. Gauthier, Subpicosecond laser-produced plasma dynamics, *SPIE Proceedings* 2015: 205 (1993).
21. T. Engers, PhD thesis, University of Essen, Essen (1994).
22. See, e.g., *Atoms in Intense Fields*, M. Garvila, ed., Academic Press, Boston (1992).
23. R. L. Carman, D. W. Forslund, and J. M. Kindel, Visible harmonic generation as a way of measuring profile steepening, *Phys. Rev. Lett.* 46: 29 (1981).
24. R. L. Carman, C. K. Rhodes, and R. F. Benjamin, Observation of harmonics in the visible and ultraviolet created in CO_2-laser-produced plasmas, *Phys. Rev.* A 24: 2649 (1981).

25. B. Bezzerides, R. D. Jones, and D. W. Forslund, Plasma mechanism for ultraviolet harmonic generation due to intense CO_2 light, *Phys. Rev. Lett.* 49: 202 (1982).
26. C. Grebogi, V. K. Teripathi, and H. H. Chen, Harmonic generation of radiaton in a steep density profile, *Phys. Fluids* 26: 1904 (1983).
27. S. Kohlweyer, G. D. Tsakiris, C. G. Wahlström, C. Tillman, and I. Mercer, Harmonic generation from solid-vacuum interface irradiated at high laser intensities, *Optics Comm.* 117: 431 (1995).
28. D. von der Linde, T. Engers, G. Jenke, P. Agostini, G. Grillon, E. Nibbering, A. Mysyrowicz, and A. Antonetti, Generation of high order harmonics from solid surfaces by intense femtosecond laser pulses, *Phys. Rev.* A 52: 25 (1995).
29. P. Gibbon, "Harmonic generation from solid-vacuum interface irradiated at high laser intensities", *Phys. Rev. Lett.* 76: 50 (1981).
30. R. Lichters, J. Meyer-ter-Vehn, and A. Pukhov, Short pulse laser harmonics from oscillating plasma surfaces driven at relativistic intensity, *Phys. Plasmas*, in print.
31. P. A. Norreys, M. Zepf, S. Moustaizis, A. P. Fews, J. Zhang, P. Lee, M. Bakarezos, C. N. Danson, A. Dyson, P. Gibbon, P. Loukakos, D. Neely, F. N. Walsh, J. S. Wark, A. E. Angor, Efficient extreme UV harmonic generation from picosecond laser pulse interactions with solid targets, *Phys. Rev. Lett.* 76: 1832 (1996).

THE OPACITY AND IRON PROJECTS – AN OVERVIEW

Keith A. Berrington

Department of Applied Mathematics and Theoretical Physics
Queen's University, Belfast BT7 1NN, UK

INTRODUCTION

*Science profits from a two-objective aim.
One thing is to push back frontiers, understand
better the universe, its contents and its frame-
work of law. Objective two is to expand
quantitatively the base of science: to claim
new predictions, accurate data standards and
better models. These bring high rewards these days;
and in many applications, data pays.*

This theme – of the link between data needs and scientific progress – describes well the development of computational atomic physics, which is profiting from a systematic programme of research to provide the data required by applications as diverse as astrophysics and laboratory plasma technology. In particular, outstanding progress has been made by two international projects: the recently-completed Opacity Project (OP)[1–27] and the current IRON Project (IP)[28–48].

These projects will be described more fully in their respective sections, but as the names imply, the OP was to calculate radiative opacity data for atoms and atomic ions involved in stellar opacity; the IP is to calculate electron collisional data mainly for the iron group of elements. The two projects have much in common on the atomic physics front, using *ab initio* theoretical and state-of-the-art computational methods based on R-matrix techniques[49,50], as summarised in the next section, and both projects involve the calculation of large amounts of good atomic data.

But of equal importance, the execution of these projects has increased our understanding of collisional and radiative physics and led to many new developments. It is these aspects which will be discussed here; in particular, we shall examine the impact on the theory of photoionization.

Recent developments in atomic physics are straining the traditional divisions set by the collisional energy regime: low energy (R-matrix and other close-coupling methods), high energy (Born, distorted wave), negative energy (atomic structure). Two of these new developments concern us here. The first is of course the OP, which established that collisional (low energy) techniques could be used effectively to calculate the large amounts of bound-bound (negative energy) data required for stellar opacities. The second development arises from a concerted effort by a number of groups to solve the so-called 'intermediate energy' problem, around the ionization energy of an atom or ion where traditional low energy and high energy methods are inadequate at representing the infinity of open channels and the continuum. Major breakthroughs are now being made using converging basis set techniques, with methods such as convergent-close-coupling (CCC)[51] and R-matrix with pseudostates (RMPS)[52]. So far they have been applied to the calculation of electron impact cross sections for light atoms and ions. In the final section of this paper we look at the practical application of RMPS methods to resonances, and the incorporation of these methods into OP techniques for radiative processes. The aim is a new unified theory, which will further our understanding of atomic processes as well as enable the calculation of high precision atomic data.

THE R-MATRIX

Both the OP and the IP use similar *ab initio* techniques for calculating atomic data. These involve representing the electron–atom (or –ion) collision quantum-mechanically as a wavefunction Ψ, and solving Schrödinger's equation using the R-matrix method[49]. A configuration-interaction (CI) approach is combined with a close-coupling (CC) collisional model:

$$\text{CI:} \qquad \psi_i(x_1...x_N) = \sum_j \Phi_j(x_1...x_N)c_{ji} \tag{1}$$

$$\text{CC:} \qquad \Psi(x_1...x_{N+1}) = \mathcal{A}\sum_i \psi_i(x_i...x_N)F_i(x_{N+1}), \tag{2}$$

where ψ_i is a CI function for an N-electron "target" state, constructed using antisymmetrised configurations Φ_j with coefficients c_{ji} chosen to diagonalize $(\psi_i|H|\psi_{i'})$; F_i is a function for the added $(N+1)$th electron and $\mathcal{A}$ an anti-symmetrisation operator.

In the R-matrix method, Ψ is considered in two regions separated by a sphere of radius $r = a$, where r is the radial coordinate of the colliding electron and a effectively encloses the N-electron target.

- In the inner region $r < a$, the colliding electron must be considered indistinguishable from the atomic electrons, and a many-electron problem has to be solved. This is done in the most efficient way possible by expanding Ψ in terms of an energy-independent basis satisfying R-matrix boundary conditions. The equations are solved by a matrix diagonalization for all energies.

- In the outer region $r > a$, the colliding electron merely 'sees' a long range potential due to the atom, and Ψ can be written straightforwardly as a no-exchange close-coupling expansion. The resultant coupled differential equations are solved subject to open-channel (energetically allowed) or closed-channel boundary conditions as appropriate, for each value of the total energy E.

- Matching on the boundary $r = a$ yields the complete Ψ for all space.

The form of Ψ at infinity yields the observables, such as the cross section or collision strength $\Omega(E)$. The energy integral of $\Omega(E)$ over a Maxwellian distribution gives the collisional excitation rates normally used in astrophysics applications.

Ψ can also be used to obtain radiative data by calculating the appropriate dipole matrix elements. Bound states and their energies can be obtained from the discrete solutions Ψ given when all channels are closed. Thus oscillator strengths (f-values, bound-bound data), or photoionization cross sections (bound-free data) can be calculated from the same theoretical approach.

The method has some nice features. All of the important physics of a low energy collision are incorporated in an ab initio way. For example, the resonance structures which can so dominate a cross section at low energies arise naturally from such an approach. Note that the use of an energy-independent basis in the inner region makes the R-matrix method highly efficient for calculating data for large numbers of E values, an important feature for calculating large numbers of bound states for radiative calculations, or for calculating cross sections and $\Omega(E)$ over a fine E grid in order to represent complicated resonance structures.

A comprehensive and well-tested suite of computer programs[50] has been developed to calculate atomic data using the R-matrix method. These programs have a long history, and have involved many people in their development and use. The programs have been designed to deal with the general atom or ion, to calculate both collisional and radiative data. There are many options, for example to be able to incorporate relativistic effects via terms of the Breit-Pauli Hamiltonian. These programs are used worldwide, and are the standard tools used in the OP and IP projects.

THE OPACITY PROJECT (OP)

The opacity of a plasma is a measure of the absorption of radiation by the plasma and thereby controls the transport of radiation energy. Since most of the observable universe is in the form of plasma, and moreover the production and confinement of plasma under reactor conditions is the central aim of the fusion programme, it can be seen that accurate opacity data have wide applications in different areas of physics. For example in a star, where the radiation pressure gradient balances the gravitational force in hydrostatic equilibrium, the opacity of stellar material plays a fundamental role in the structure and evolution of stars and is involved in the instabilities which lead to stellar pulsation.

The *monochromatic opacity* κ_ν of a medium is the reciprocal of the mean free path of frequency ν. The *intensity of radiation* at position $\mathbf{r}$ and in a direction $\hat{\mathbf{s}}$ is described by a function $I_\nu(\mathbf{r}, \hat{\mathbf{s}})$. If d$s$ is an element of distance in the direction $\hat{\mathbf{s}}$, and j_ν is the *emissivity*, the equation of radiative transfer is

$$\frac{\mathrm{d}I_\nu}{\mathrm{d}s} = \kappa_\nu I_\nu + j_\nu \tag{3}$$

A medium of sufficiently low density may be considered as an assembly of 'particles': electrons, nuclei, atoms and molecules. For such a medium the opacity is

$$\kappa_\nu = \sum_n N(n)\sigma_\nu(n) \tag{4}$$

where $N(n)$ is the number of particles of type n per unit volume and $\sigma_\nu(n)$ is the cross section for absorption or scattering. In order to calculate κ_ν using (2) one needs to determine the populations $N(n)$ (the problem of the *equation of state*) and the cross sections $\sigma_\nu(n)$ (a problem of atomic or molecular physics).

Table 1: The Opacity and Iron Project teams. Contributors to the series of papers: 'Atomic Data for Opacity Calculations' (ADOC) in J.Phys. B; 'Equations Of State for stellar envelopes' (EOS) in Ap.J; 'Atomic data from the IRON Project' (IP) in A&A. **Affiliations:** Camb=Cambridge University IBMV=IBM Venezuela; IVIC=Instituto Venezolano de Investigaciones Cientificas; JILA=Joint Institute for Laboratory Astrophysics; MU=Munich University Institut fur A&A; OCA= Observatoire de la Côte d'Azur; OPM=Observatoire de Paris (Meudon); OSU=Ohio State University; Ox=Oxford; QUB=Queen's University Belfast; RHL=Royal Holloway London; UCL=University College London; UI=University of Illinois; UR=University of Rennes; VIL=Villafranca Satellite Tracking Station; V=Vilnius University.

Name	Affiliation	Opacity Project	Iron Project
M Bautista	OSU		IP XIII,XVI,XX
K A Berrington	QUB	ADOC II,XIV,XV	IP I,VIII,IX,XII,XV,XXI
P G Burke	QUB	ADOC II	
V M Burke	QUB	ADOC XVII	IP II
K Butler	MU	ADOC II,XIX	IP V
W Däppen	USCA	EOS II,III,IV	
W Eissner	QUB	ADOC	IP I
J A Fernley	UCL, VIL	ADOC VII	
M Galavis	IVIC		IP X
M Graziani	OSU		IP III
A Hibbert	QUB	ADOC VI,XXI	
D G Hummer	JILA, MU	EOS I-IV	IP I
R Kisielius	QUB, V, UCL		IP XV
M Le Dourneuf	UR		IP XIX
D J Lennon	QUB, MU		IP II
D Luo	JILA	ADOC X,XI	
A E Lynas-Gray	Ox	ADOC XXII	
H E Mason	Camb		IP XIV
C Mendoza	IBMV, IVIC	ADOC XIX	IP X
D Mihalas	UI	EOS I-IV	
B W Mihalas	UI	EOS III,IV	
S N Nahar	OSU	ADOC XVII,XX	IP XVI
P Norrington	QUB		IP XV
G Peach	UCL	ADOC IX	
J Pelan	QUB		IP IX,XII,XXI
A K Pradhan	OSU	ADOC X,XI,XVII,XX	IP I,III,VI,XIII,XVI
P Quinet	UR		IP XIX
H E Saraph	UCL	ADOC IX,X,XVI	IP I,IV,XI,XIV
P M J Sawey	QUB	ADOC XV	
M P Scott	QUB	ADOC XXI	
M J Seaton	UCL	ADOC I-V,VII-IX,XII-XIV,XXII	
P J Storey	UCL	ADOC II,X,XVI,XXII	IP I,XI,XIV
K T Taylor	RHL, QUB	ADOC II,III,VII,XVI	
J F Thornbury	QUB	ADOC VI	
J A Tully	OCA	ADOC XIV	IP I,IV
Yu Yan	UI NCSA	ADOC II,III,IV,X	
C J Zeippen	OPM	ADOC XIX	IP V,X,XIX
H Zhang	OSU		IP III,VI,XVIII

The total flux at temperature T is, integrated over frequency,

$$F(r) = -\frac{4\pi}{3}\frac{1}{\kappa_R}\frac{dB(T)}{dT}\frac{dT(r)}{dr} \tag{5}$$

where $B(T)$ is the intensity of black-body radiation. Here we have introduced the Rosseland mean opacity[53]

$$\frac{1}{\kappa_R} = \int_0^\infty \frac{1}{\kappa_\nu} f(u) du \qquad [u = \frac{h\nu}{kT}]. \tag{6}$$

Note that if κ_ν is zero over any finite range of ν, then $\kappa_R = 0$; therefore in calculating κ_R it is essential to 'plug all of the holes' in κ_ν.

The total monochromatic opacity κ_ν is obtained on summing the contributions from all of the radiative processes which could occur. The OP was a collaboration led by Professor Seaton involving ten groups from five countries (see Table 1 for details) to calculate using R-matrix techniques a vast amount of radiative data for atoms and ions.

Stellar envelopes are defined to be those parts of interiors for which the initial levels giving significant contributions to opacities are not markedly perturbed by the plasma environment, and for which equation (4) can be used. The OP work is concerned with opacities for envelopes. In equation (4) we use accurate cross sections for free atoms and we allow for plasma perturbations only in calculating the line profiles and the occupation probabilities. We do not include any molecular contributions. The main emphasis of the project was to obtain accurate atomic data which, incidentally, may be of value in a wide range of other problems. The methods used in the OP work may not give accurate results if employed in the deepest layers of stellar interiors.

The OP computed atomic data for opacity calculations for

- **H, He, Li, Be, B, C, N, O, F, Ne, Na, Mg, Al, Si, S, Ar, Ca** and **Fe**
- **energies** of terms having effective quantum numbers $\nu \leq 10$ and total angular momentum $L \leq 3$ or 4, all spin and parity combinations
- **gf-values** for all dipole transitions between these bound terms,

$$A_i + h\nu \rightarrow A_f \tag{7}$$

- **photoionization cross sections** from all calculated bound terms, tabulated on a grid of photon energies suitable to describe the resonance structure in sufficient detail to calculate reliable opacities,

$$A_i + h\nu \rightarrow A^+ + e^- \tag{8}$$

- **line broadening parameters**

The quality of the atomic data produced by the OP is discussed in a number of places[1–26]. It is therefore sufficient to give a summary of conclusions here.

The calculations are optimized for positive ions. For all ions up to the iso-electronic sequence of aluminium-like ions, the R-matrix calculations include all states belonging to ground complexes explicitly as target states in the expansion of the total wavefunction. For all positive ions in these systems the oscillator strengths and photoionization cross sections should be accurate to about 10 per cent, except for sensitive cases involving a lot of cancellation. The accuracy may be less good for some neutral atoms that have low-lying target states not belonging to ground complexes. The accuracy may also be less good for systems belonging to higher iso-electronic sequencies[23]. An independent approach (OPAL)[55] was taken at the Lawrence Livermore National Laboratory. The OP and OPAL opacities agree well within their range of validity. The new opacities are now being widely used in stellar structure, evolution and pulsation models.

Work still continues, particularly to improve the data for the first few ionization stages of iron and for the high energy behaviour of the photoionization cross section.

35 key research papers arising from the OP, including the J. Phys. B series *Atomic Data for Opacity Calculations*[1–22] and the ApJ series *Equations of State for Stellar Envelopes*, are reprinted in volumes 1 and 2 of *The Opacity Project.*[25,26] together with tables of calculated energies and oscillator strengths.

All numeric data are available from **TOPbase**[27], an on-line database at the Centre de Donnèes Astronomiques de Strasbourg (CDS). To use TOPbase through internet:

IP address: 130.79.128.5
account: topbase
password: seaton+

Anonymous ftp (same number) can be used to obtain the user guide.

THE IRON PROJECT (IP)

This new international project involves many of the OP participants (see Table 1). The aim is to systematically compute electron excitation cross sections for the iron group of elements, using R-matrix methods[28,50]. Particular attention is given to cross sections required for the interpretation of data from specific space observations, for example the Solar and Heliospheric Observatory (SoHO) and the Infrared Solar Observatory (ISO).

Atomic data for SoHO experiments were assessed at an international meeting hosted by the Rutherford Appleton Laboratory 26–27 March 1992[55]. The reviews revealed many deficiencies in electron impact excitation data currently available, including a lack of reliable data for n=2-4 transitions giving rise to UV lines, and the general inadequacy of Fe ion data. In some cases, an excitation rate is deduced from a cross section which has been calculated at just one impact energy – ie. ignoring the energy variation required to compute a thermally averaged rate – and that cross section may have a large error. For example, of the Fe ions to be observed in SoHO, Fe IX - XVI, only the excitation rates for Fe IX are known to sufficient accuracy for a sufficient number of transitions over a sufficient temperature range.

Predicted IR lines detectable by ISO include more than 100 lines from positive ions. Tully (1991, private communication) reviewed the atomic data requirements for these lines, and identified the need for much more accurate electron impact excitation rate coefficients for direct excitation, particularly of fine-structure levels, at low temperatures. Reliable excitation data hardly exist at all for low ionized stages of Ti, Cr, Mn, Co, Ni and Fe.

The IP is a significant organisational task. In order to manage a project of this size, and to provide a steady stream of results (a not insignificant consideration given research support constraints), the project is divided into a series of short-term and long-term goals. These goals can be summarised as follows:

Goal 1. Fine-structure transitions in ground state configurations.

Excitation cross sections are being calculated for fine-structure transitions in the ground configuration of all ions of astrophysical interest in the **B, C, O, F, Al, Si, S, Cl** iso-electronic sequencies, together with **Fe** I–VII (ie. ions with open p and d shells). These data are essential for the interpretation of IR lines to be observed by ISO, as well as for coronal spectra. The calculations are largely completed and are being published.

Goal 2. Fe ions.
Excitation cross sections are being calculated for all transitions involving states with principal quantum number up to and including $n = 4$. This is now underway, and will provide collisional rates for interpretation of observations from SoHO.

Further goals include: the calculation of collisional data for nearby elements, eg. Ti, V, Cr, Mn, Co, Ni; the extension of the Opacity Project calculations to radiative transitions beyond the electric dipole, in intermediate coupling, eg. **E2** and **M1** transitions; the calculation of radiative properties and energies for heavy ions of relevance to current beam-foil and other experimental investigations.

The IP is publishing its work in a series of papers in A&A under the generic title *Atomic data from the IRON Project*[28–48]. A&A makes the published tables available electronically via anonymous ftp **130.79.128.5** at the CDS.

TOWARDS HIGH PRECISION

Care is being taken to ensure that the new data calculated in the IP are accurate. This involves a number of checks, both of self-consistency within the calculation, and by comparison with other data where available. An important feature of the Project is the twice-yearly meeting, which acts as an important forum for exchanging ideas, criticising results, etc, as well as for managing the Project.

Papers already produced by the project give an indication of the accuracy obtainable with modern methods. Rather than giving specific details (these in any case can be found in the literature), it is worth making some general remarks on the expected sources of inaccuracy – and on techniques to minimise their effects.

1. Target state expansions.
   The collisional wavefunction involves the coupling of all the required initial and final target states, together with any other states that couple strongly to these. This could be a lot of states (!), and a compromise must be reached between what is necessary, and what is practicable. For ions, the omission of more highly excited states will lead to missing resonance structures in the calculation; such resonance structures could give important contributions to $\Omega(E)$ in the energy range of interest. A further consequence of omitting highly excited states is apparent at collisional energies above their threshold, where loss of flux into their channels would not be adequately accounted for. This is especially true in the so-called intermediate energy region, where ionization channels are also open. In practice the truncation of the target state expansion imposes an upper limit to the collision energy at which the calculation will be reliable, corresponding to the energy of the first omitted resonance or threshold.

2. Configuration-interaction (CI) effects in the target.
   A time-consuming (for the worker) part of a calculation on a given atomic system is the job of developing a good representation of the target wavefunction, before

the collision calculation can procede. Again compromises have to be made; each target state could be constructed from a very elaborate CI basis, but it is often not easy to carry such a complicated wavefunction through to a collision calculation. High quality experimental energy levels can be used in the calculation if available.

3. Relativistic effects.
   Fine-structure and other relativistic effects can be treated using the Breit-Pauli Hamiltonian or by a non-relativistic calculation in LS coupling followed by an algebraic transformation to more appropriate coupling schemes. The Breit-Pauli approach treats the relativistic effects more accurately, while the recoupling procedure allow more accurate wave functions for the target ion to be obtained. For most system the latter is sufficient, at least for the the iron group of elements.

4. Near-threshold resonances.
   This is related to the other points mentioned so far, in that to obtain highly accurate $\Omega(E)$ at near-threshold energies requires attention to both the target and the collisional calculations. A particular difficulty relates to near-threshold resonances; these are difficult to pin down accurately as small changes in the calculation method (the introduction of fine-structure effects for example), can change the position of any near-threshold resonance and hence the threshold behaviour of $\Omega(E)$. Hardly any of these resonance positions have been experimentally determined, so there is little scope for independent verification. In practice this places a lower limit on the temperature at which a calculation can be trusted. Resonance analysis is discussed further below.

5. High angular momenta and high-energy contributions.
   A more tractable problem. R-matrix calculations are normally, for practical reasons, carried out over a restricted range of collisional angular momentum and E. Facilities have been developed to 'top-up' $\Omega(E)$ for high angular momentum contributions, and for ensuring that electron excitation rates are calculated over a temperature range consistent with that of the calculation. These points may seem obvious, but have sometimes been overlooked in the past.

Autoionizing states can radically alter the low temperature behaviour of collision rates, and are a major contributor to opacity. In view of the importance here, and in other branches of atomic, molecular and optical science, a major new effort is being devoted to calculating data for such resonant states. A new procedure for analysing resonances in atomic and molecular collision theory is introduced, which exploits the analytic properties of R-matrix theory to obtain the energy derivative of the eigenphase, without assuming a pure Coulomb potential at large distances ('the QB method')[56]. Resonance properties can easily be extracted such as positions, widths, identifications, decay rates, etc. This new method, combined with convergent R-matrix techniques for the bound state problem, will enable a more unified and higher precision approach than hitherto for calculating atomic structure data for both bound and autoionizing states.

PHOTOIONIZATION AND RESONANCES

The aim of this section is to illustrate the some of the OP results and to discuss the methods used to produce large amounts of accurate photoionization cross sections. We first compare the OP data with a selection of experimental measurements, then we consider more closely the case of beryllium.

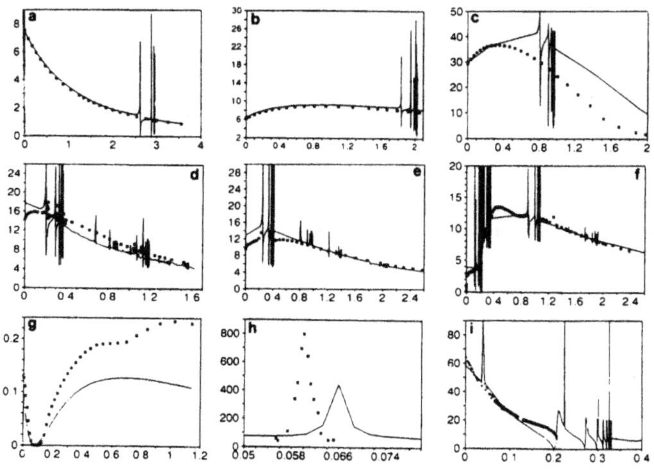

Figure 1: Photoionization cross section of some neutral atoms in units of $10^{-18}\mathrm{cm}^2$ as a function of final electron energy in Rydberg units. Comparison between OP calculations (—)[24] and experiments (dots): (a) He ground state by West and Marr (1976)[57]; (b) Ne ground state by West and Marr (1976)[57]; (c) Ar ground state by West and Marr (1976)[57]; (d) C ground state by Cantú *et al* (1981)[58]; (e) N ground state by Samson and Angel (1990)[59]; (f) O ground state by Angel and Samson (1988)[60]; (g) Na ground state by Hudson and Carter (1967)[61]; (h) Mg $3s3p\,^1P^o$ excited state by Bradley *et al* (1976)[62]; (i) Al ground state by Kohl and Parkinson (1973)[63] and Roig (1975)[64];

Comparisons between OP and experiments

A representative set of comparisons of photoionization cross sections is shown in figure 1, where the OP calculations are compared with absolute measurements. The agreement is found to be good except in the high-energy region of the ground state cross sections of Ar and Na; the latter discrepancy is believed to be due to experimental error.

Photoionization cross section measurements have been mainly for the ground state of neutrals. However, for Mg there is a measurement for photoionizing from an excited state ($3s3p\,^1P^o$). In this latter case, the cross section in the near-threshold region is dominated by a broad resonance ($3p^2\ ^1S^e$) as shown in figure 1(i), where a discrepancy is seen between the OP calculation[19] and the experimental measurements[62]. Although this experiment has been independently confirmed[65], no calculation reproduces the experimental resonance parameters with the desired accuracy. This presents a challenge for the new 'convergent' methods to be discussed below.

Beryllium

In the original OP work[14] Be was treated along with other members of the iso-electronic sequence, in a five-state approximation: that is, the Li-like target consisted of the $1s^22s, 2p, 3s, 3p, 3d$ states. The total photoionization cross section from the ground state of Be was calculated for photon energies from 0–3.5 Ryds, using R-matrix techniques. This energy range included the resonance structures associated with the target $n = 2$ and $n = 3$ states. As shown in figure 2, the $n = 2$ resonances have a major effect on the energy dependence of the cross section, the $n = 3$ resonances being somewhat minor features. Despite the truncation of the target state expansion

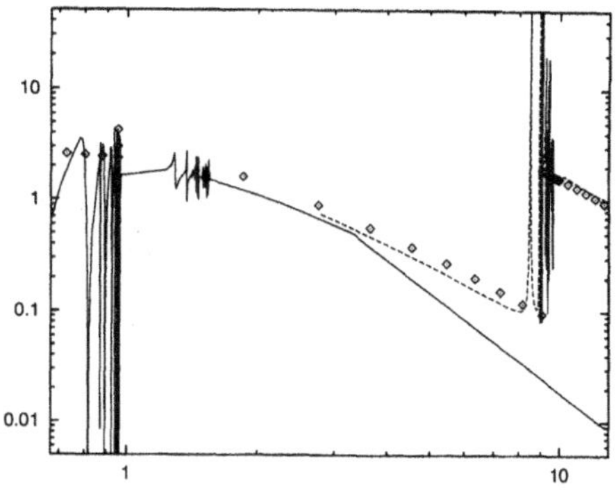

Figure 2: Photoionization cross section in units of $10^{-18}\mathrm{cm}^2$ for $\mathrm{Be}(1s^22s^2)$, as a function of photon energy in Rydberg units: — OP[14] (< 3.5 Ry.) and TOPBASE[27] (> 3.5 Ry.); - - - - present R-matrix calculation[68]; ◇ distorted-wave calculation[67].

at $n = 3$, we conclude from this that the OP calculation provided good accuracy in the energy range quoted.

Beyond 3.5 Ryds, an E^{-3} fit to the cross section was applied, to extend the energy range to arbitrarily large energies. This data was also stored in the on-line public data base TOPBASE[27], along with the R-matrix data, and this was probably a mistake as it seemed to have given the impression that the cross sections had the same accuracy throughout the energy range. In fact, for the purposes of calculating stellar envelope opacities, the high energy tail plays only a minor role; so a lesser accuracy here was acceptable to the OP, but may not be so for other applications.

As part of subsequent work to improve the OP database, we are doing further calculations to correct the high energy behaviour of the cross sections. In particular, we need to account for inner-shell processes that become significant at high energies. Two strategies are adopted. The first is to include inner-shell excited target states within the R-matrix expansion; the second is to use simpler distorted-wave (DW)[67] techniques to estimate the contribution of inner-shell excitation in the photoionization cross section. We show the results of both strategies in figure 2.

It is clear that there is substantial enhancement of the background cross section at the inner-shell thresholds, the first of which occurs at 9.1 Ryds ($1s2s^2$). This is due to transitions such as

$$h\nu + \mathrm{Be}(1s^22s^2) \rightarrow \mathrm{Be}^+(1s2s^2) + \mathrm{e}^-. \tag{9}$$

In addition, the R-matrix calculation shows significant resonance structure above 8.5 Ryds due to 'hollow atom' resonances such as $1s2s^2\epsilon l$.

Recent experiments agree well with theory in the qualitative description of these phenomena (Vo Ky *et al* 1992)[66]. However, in the R-matrix calculation reported by Vo Ky *et al* certain (pseudo) states were omitted, which is contrary to the philosophy of modern 'convergent' techniques, where it is known that the absolute value of the cross section can be affected by such omissions (the photoionization cross sections measurements were not absolute and so could not test this point). Specifically, the Be^+ target ionization threshold lies below the inner-shell thresholds, and account should be taken of the infinity of physical states of the form $1s^2nl$ omitted in the target expansion. This is analogous to the 'intermediate energy' problem in electron scattering, where pseudo-state expansions are being used with great success (eg. RMPS[52]). In our most recent work[68], represented by the dashed curve in figure 2, we converged the pseudo-state expansions to confirm the validity of the results.

For the Be-sequence, The TOPBASE data for the ground state photoionization cross sections at high energies are therefore being replaced with our R-matrix inner-shell calculations[69]. These are being used to benchmark DW calculations, the latter being the only computationally feasible method for more complex systems.

We now turn to the calculation of the energy of the lowest autoionizing state of Be ($1s^22p^2$), in order to illustrate how pseudo-state expansions can lead to high precision calculations. This autoionizing state lies just above the Be^+ ($1s^22s$) ionization threshold, and has been experimentally measured[70]. The objective then is to simultaneously improve the wavefunction of both the N-electron $1s^22s$ and the $(N+1)$-electron $1s^22p^2$ states systematically in order to converge the calculated value of the small energy difference between them.

To do this we start, as in the photoionization calculation described above, with $1s$, $2s$ and $2p$ orbitals optimised on the Be^+. We then introduce pseudo-orbitals $\bar{n}l$ which will improve the Be^+ state through greater CI, and which will also allow the formation of pseudo-states (eg. $1s^2\bar{n}l$) with which we will augment the R-matrix CC expansion. In the RMPS approach, the pseudo-orbitals are normally chosen as members of a complete set of functions confined within the R-matrix sphere. Although any complete set of functions could be chosen, the rate of convergence of the method is obviously a consideration. For our purposes, we found it was adequate to use a simple exponential times a power series:

$$P_{\bar{n}l}(r) = e^{-r} \sum_{m=1}^{\bar{n}-l} C_{ml} r^{m+l} \quad [\bar{n} > 2], \tag{10}$$

where the C_{ml} are uniquely defined by orthonormality conditions. Two R-matrix calculations were carried out, in which the following states were included:

1. $1s^22s$, $1s^22p$, $1s^2\bar{n}l$ ('frozen core' approximation);
2. $1s^22s$, $1s^22p$, $1s^2\bar{n}l$, $1s2s\bar{n}l$, $1s2p\bar{n}l$.

In each case, $0 \leq l \leq \bar{n}-1$. The $1s^22p^2$ resonance parameters were found by the 'QB' method[56], and the energy difference with the $1s^22s$ state is plotted in figure 3. Also shown on that figure are the experimental measurements[70] and the two most accurate of previous calculations[66,71] both of which use R-matrix techniques; the calculation of Miura *et al*[71] achieving the same high accuracy as ourselves. The convergence of our results with increasing $\bar{n}$ is striking, particularly with calculation 2. Note that the frozen core calculation converges, but to the wrong value. This points up the importance of including inner-shell correlation and continuum effects for high precision calculations of bound and continuum states[68].

FINALLY

This paper has described two collaborative projects on the interface of atomic physics and astrophysics, and has discussed some of the on-going developments. Both the OP and the IP are producing an immense amount of good atomic data of use in many applications. Both Projects are committed to release the data into the public domain, and, perhaps even more important, to explain the methods used and approximations made in obtaining the data, and the expected reliability of the results.

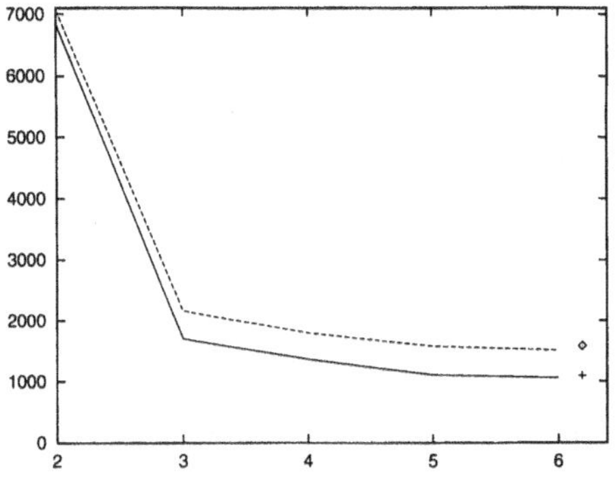

Figure 3: Position (cm^{-1}) of the Be resonance state $1s^2 2p^2\ {}^1S$ above the $1s^2 2s$ ionization threshold: — present RMPS[68] as a function of the highest $\bar{n}$ of the pseudostate expansion; - - - - as above, but maintaining a $1s^2$ frozen core; ◇ Vo Ky *et al* (1992)[66]; + Miura *et al* (1996)[71]. The experiment of Clark *et al* (1985)[70] is at $997 \pm 5 cm^{-1}$.

It is clear that the conception and execution of these projects was due to a synergism between astrophysicists and atomic physicists, harnessing expertise on both sides. It is hoped that this will continue in the future, and that other collaborative projects, or extensions to the present ones, will arise.

And here we note that our *fin de siècle* is witnessing the deployment of powerful new instruments for astronomical spectroscopy. For example ESA's programme (ISO and SOHO), and NASA's Great Observatories programme (HST, GRO, AXAF and SIRTF), covering the IR, optical, X-ray and γ-ray regions of the spectrum, will provide several significant new windows on the universe. The increasing sophistication, precision and range of observational techniques and of theoretical models create new demands for more and better data on atomic processes. In addition there are increasing applications for accurate atomic data in laboratory plasmas, whether these be in controlled fusion experiments or in technological and industrial plasmas.

After starting this talk with an *ottava rima*, this being a European conference, it would be fun to end with a romance. This concerns the story of Sleeping Beauty who is put to sleep for a hundred years – not just herself, but everyone in the castle. Here is how Tennyson, writing in 1830, viewed the situation:

> **The Day-Dream** (La Belle au Bois Dormant) – fragments
> *Well - were it not a pleasant thing*
> *To fall asleep with all one's friends,*
>
> . . . . . .
>
> *And wake on Science grown to more,*
> *On secrets of the brain, the stars;*
>
> . . . . . .
>
> *For we are Ancients of the earth,*
> *And in the morning of the times.*

Notice the two 'Grand Challenges' posed for science, still relevant today (the brain, the stars). Interestingly, it was almost one hundred years later when Rosseland(1924)[53] formulated his opacity, and it has taken a further 70 years to calculate stellar opacities *ab initio*, through the efforts of the OP. And of course the final couplet implies that what we do now, will all seem very primitive to future generations.

Acknowledgements

This research was supported by grants from EPSRC (GR/K24741), PPARC (GR/H93576) and EC (ERB CHRX CT920013).

REFERENCES

1. M.J.Seaton, Atomic data for opacity calculations: I. General description, *J. Phys. B: At. Mol. Phys.* 20: 6363 (1987).
2. K.A.Berrington, P.G.Burke, K.Butler, M.J.Seaton, P.J.Storey, K.T.Taylor and Yu Yan, Atomic data for opacity calculations: II. Computational methods, *J. Phys. B: At. Mol. Phys.* 20: 6379 (1987).
3. Yu Yan, K.T.Taylor and M.J.Seaton, Atomic data for opacity calculations: III. Oscillator strengths for C II, *J. Phys. B: At. Mol. Phys.* 20: 6399 (1987).
4. Yu Yan and M.J.Seaton, Atomic data for opacity calculations: IV. Photoionization cross sections for C II, *J. Phys. B: At. Mol. Phys.* 20: 6409 (1987).
5. M.J.Seaton, Atomic data for opacity calculations: V. Electron impact broadening of some C III lines, *J. Phys. B: At. Mol. Phys.* 20: 6431 (1987).
6. J.F.Thornbury and A.Hibbert, Atomic data for opacity calculations: VI. Static dipole polarisabilities of the ground state of the helium sequence, *J. Phys. B: At. Mol. Phys.* 20: 6447 (1987).
7. J.A.Fernley, K.T.Taylor and M.J.Seaton, Atomic data for opacity calculations: VII. Energy levels, f values and photoionization cross sections for He-like ions, *J. Phys. B: At. Mol. Phys.* 20: 6457 (1987).
8. M.J.Seaton, Atomic data for opacity calculations: VIII. Line-profile parameters for 42 transitions in Li-like and Be-like ions, *J. Phys. B* 21: 3033 (1988).
9. G.Peach, H.E.Saraph and M.J.Seaton, Atomic data for opacity calculations: IX. The lithium isoelectronic sequence, *J. Phys. B: At. Mol. Opt. Phys.* 21: 3669 (1988).
10. D.Luo, A.K.Pradhan, H.E.Saraph, P.J.Storey and Yu Yan, Atomic data for opacity calculations: X. Oscillator strengths and photoionization cross sections for O III, *J. Phys. B: At. Mol. Opt. Phys.* 22: 389 (1989).
11. D.Luo and A.K.Pradhan, Atomic data for opacity calculations: XI. The carbon isoelectronic sequence, *J. Phys. B: At. Mol. Opt. Phys.* 22: 3377 (1989).
12. M.J.Seaton, Atomic data for opacity calculations: XII. Line-profile parameters for neutral atoms of He, C, N and O, *J. Phys. B* 22: 3603 (1989).
13. M.J.Seaton, Atomic data for opacity calculations: XIII. Line profiles for transitions in hydrogenic ions, *J. Phys. B: At. Mol. Opt. Phys.* 23: 3255 (1990).
14. J.A.Tully, M.J.Seaton and K.A.Berrington, Atomic data for opacity calculations: XIV. The beryllium sequence, *J. Phys. B: At. Mol. Opt. Phys.* 23: 3811 (1990).
15. P.M.J.Sawey and K.A.Berrington, Atomic data for opacity calculations: XV. Fe I-IV *J. Phys. B: At. Mol. Opt. Phys.* 25: 1451 (1992).
16. H.E.Saraph, P.J.Storey and K.T.Taylor, Atomic data for opacity calculations: XVI. *Ab initio* calculations for Fe VIII and Fe VII, *J. Phys. B* 25: 4409 (1992).
17. V.M.Burke, Atomic data for opacity calculations: XVII. Calculation of line broadening parameters and collision strengths between n=2, 3 and 4 states in C IV, *J. Phys. B: At. Mol. Opt. Phys.* 25: 4917 (1992).
18. S.N.Nahar and A.K.Pradhan, Atomic data for opacity calculations: XVIII. Photoionization and oscillator strengths of Si-like ions Si^0, S^{2+}, Ar^{4+}, Ca^{6+}, *J. Phys. B: At. Mol. Opt. Phys.* 26: 1109 (1993).
19. K.Butler, C.Mendoza and C.J.Zeippen, Atomic data for opacity calculations: XIX. The magnesium isoelectronic sequence, *J. Phys. B: At. Mol. Opt. Phys.* 26: 4409 (1993).

20. S.N.Nahar and A.K.Pradhan, Atomic data for opacity calculations: XX. Photoionization cross sections and oscillator strengths for FeII, *J. Phys. B* 27: 429 (1994).
21. A.Hibbert and M.P.Scott, Atomic data for opacity calculations: XXI. The neon sequence, *J. Phys. B: At. Mol. Opt. Phys.* 27: 1315 (1994).
22. A.E.Lynas-Gray, M.J.Seaton and P.J.Storey, Atomic data for opacity calculations: XXII. Computations for 2 472 790 multiplet *gf* values in FeVIII to FeXIII, *J. Phys. B: At. Mol. Opt. Phys.* 28: 2817 (1995).
23. M.J.Seaton, Yu Yan, D.Mihalas and A.K.Pradhan, Opacities for stellar envelopes, *Mon. Not. R. Astron. Soc.* 266:805 (1994).
24. M.J.Seaton, C.J.Zeippen, J.A.Tully, A.K.Pradhan, C.Mendoza, A.Hibbert and K.A.Berrington, The Opacity Project – computation of atomic data, *Rev. Mex. Astron. Astrof.* 23:19 (1992). A.Lynas-Gray, C.Mendoza, and C.J.Zeippen (eds), Proc. Workshop on astrophysical opacities, IBM Venezuela Scientific Center, Caracas 15–19 July 1991, *Rev. Mex. Astron. Astrof.* Vol 23 (1992).
25. The Opacity Project Team, *The Opacity Project Vol 1*, Institute of Physics Publishing, Bristol and Philadelphia (1995).
26. The Opacity Project Team, *The Opacity Project Vol 2*, Institute of Physics Publishing, Bristol and Philadelphia (1996).
27. W.Cunto, C.Mendoza, F.Ochsenbein and C.J.Zeippen, TOPBASE, *A&A* 275: L5 (1993).
28. D.G.Hummer, K.A.Berrington, W.Eissner, Anil K. Pradhan, H.E.Saraph, J.A.Tully, Atomic data from the IRON Project: I. Goals and methods, *A&A* 279: 298 (1993).
29. D.J.Lennon, V.M.Burke, Atomic data from the IRON Project: II. Effective collision strengths for infrared transitions in carbon-like ions, *A&A Suppl. Ser.* 103: 273 (1994).
30. H.L.Zhang, M.Graziani and A.K.Pradhan, Atomic data from the IRON Project: III. Rate coefficients for electron impact excitation of Boron-like ions: Na VI, Mg VIII, Al IX, Si X, S X, Ar XIV, Ca XVI and Fe XXII, *A&A* 283: 319 (1994).
31. H.E.Saraph, J.A.Tully, Atomic data from the IRON Project: IV. Electron excitation of the $^2P(3/2) -^2 P(1/2)$ fine structure transition in fluorine-like ions, *A&A Suppl. Ser.* 107: 29 (1994).
32. K.Butler, C.J.Zeippen, Atomic data from the IRON Project: V. Effective collision strengths for transitions in the ground configuration of oxygen-like ions, *A&A Suppl. Ser.* 108: 1 (1994).
33. H.L.Zhang and A.K.Pradhan, Atomic data from the IRON Project: VI. Collision strengths and rate coefficients for FeII, *A&A* 293: 953 (1995).
34. Sultana N. Nahar, Atomic data from the IRON Project: VII. Radiative transition probabilities for FeII, *A&A* 293: 967 (1995).
35. K.A. Berrington, Atomic data from the IRON Project: VIII. Electron excitation of the $3d^4\ ^5D_J$ ground state fine structure transition in Ti-like ions VII, CrIII, MnIV, FeV, CoVI and NiVII, *A&A Suppl. Ser.* 109: 193 (1995).
36. J.C.Pelan and K.A.Berrington, Atomic data from the IRON Project: IX. Electron excitation of the $^2P_{3/2-1/2}$ fine structure transitions in chlorine-like ions from ArII to NiXII, *A&A Suppl. Ser.* 110: 209 (1995).
37. M.E.Galavis, C.Mendoza and C.J.Zeippen, Atomic data from the IRON Project: X. Effective collision strengths for infrared transitions in silicon- and sulphur-like ions, *A&A Suppl. Ser.* 111: 347 (1995).
38. H.E.Saraph and P.J.Storey, Atomic data from the IRON Project: XI. The $^2P_{1/2-3/2}$ fine-structure lines of ArVI, KVII and CaVIII, *A&A Suppl. Ser.* 115: 151 (1996).
39. K.A.Berrington and J.C. Pelan, Atomic data from the IRON Project: XII. Electron excitation of forbidden transitions in V-like ions MnIII, FeIV, CoV, and NiVI, *A&A Suppl. Ser.* 114: 367 (1995).

40. M.A.Bautista and A.K.Pradhan, Atomic data from the IRON Project: XIII. Electron excitation rates and emissivity ratios for forbidden transitions in NiII and FeII, *A&A Suppl. Ser.* 115: 551 (1996).
41. P.J.Storey, H.E.Mason, and H.E.Saraph, Atomic data from the IRON Project: XIV. Electron impact excitation for the FeXIV fine-structure transition $^2P^0_{1/2} - ^2P^0_{3/2}$, *A&A* 309: 667 (1996).
42. R. Kisielius, K.A.Berrington, and P.H.Norrington, Atomic data from the IRON Project: XV. Electron excitation of the fine-structure transitions in hydrogen-like ions HeII and FeXXVI, *A&A Suppl. Ser.* 118: 157 (1996).
43. Manuel A. Bautista, Atomic data from the IRON Project: XVI. Photoionization cross sections and oscillator strengths for FeV, *A&A Suppl. Ser.* 119: 105 (1996).
44. S.N.Nahar and A.K.Pradhan, Atomic data from the IRON Project: XVII. Radiative transition probabilities for dipole allowed and forbidden transitions in FeIII, *A&A Suppl. Ser.* 119: 507 (1996).
45. Hong Lin Zhang, Atomic data from the IRON Project: XVIII. Electron impact excitation collision strengths and rate coefficients for FeIII, *A&A Suppl. Ser.* 119: 523 (1996).
46. P.Quinet, M.Le Dourneuf, C.J.Zeippen, Atomic data from the IRON Project: XIX. Radiative transition probabilities for forbidden lines in FeII, *A&A Suppl. Ser.* (1996).
47. Manuel A. Bautista, Atomic data from the IRON Project: XX. Photoionization cross sections and oscillator strengths for FeI, *A&A Suppl. Ser.* (in press 1996).
48. J.Pelan and K.A.Berrington, Atomic data from the IRON Project: XXI. Electron excitation of fine-structure transitions involving the $3d^64s^2\ ^5D$ ground state and the $3d^74s\ ^5F$ metastable state of FeI, *A&A Suppl. Ser.* (in press 1996).
49. P.G. Burke and K.A. Berrington, *Atomic And Molecular Processes: An R-matrix Approach*, Institute of Physics Publishing, Bristol and Philadelphia (1993).
50. K.A.Berrington, W.B.Eissner and P.H.Norrington, RMATRX I, Belfast atomic R-matrix codes, *Comput. Phys. Commun.* 92: 290 (1995).
51. I.Bray and D.V.Fursa, Convergent close-coupling method: a "complete scattering theory"? *Phys. Rev. Lett.* 76:2674 (1996).
52. K. Bartschat, E.T. Hudson, M.P. Scott, P.G. Burke and V.M. Burke, Electron-atom scattering at low and intermediate energies using a pseudo-state/R-matrix basis, *J. Phys. B* 29:115 (1996). Convergent R-matrix with pseudo-states calculation for e–He collisions, *Phys. Rev. A* 54:R998 (1996).
53. S. Rosseland, *Roy. Astron. Soc. M. N.* 84:525 (1924).
54. C.Iglesias, F.J.Rogers and B.G.Wilson, Reexamination of the metal contribution to astrophysical opacity, *Astrophys. J.* 322:L45 (1987).
55. J.Lang (ed.), Proc. Atomic data workshop for SOHO, *Atom. Data Nucl. Data Tables*, 57 (1994).
56. L.Quigley and K.A.Berrington, The QB method: analysing resonances using R-matrix theory. Applications to C^+, He and Li, *J. Phys. B* 29:4529 (1996).
57. J.B.West and G.V.Marr, The absolute photoionization cross sections of helium, neon, argon and krypton in the extreme vacuum ultraviolet region of the spectrum, *Proc. R. Soc. Lond.* A 349:397.
58. A.M.Cantú, M.Mazzoni, M.Pettini and G.P.Tozzi, Photoionization spectrum of the ^3P ground state of neutral carbon, *Phys. Rev.* A 23:1223 (1981).
59. J.A.R.Samson and G.C.Angel, Single- and double-photoionization cross sections of atomic nitrogen from threshold to 31A, *Phys. Rev.* A 42:1307 (1990).
60. G.C.Angel and J.A.R.Samson, Total photoionization cross sections of atomic oxygen from threshold to 44.3A, *Phys. Rev.* A 38:5578 (1988).
61. R.D.Hudson and V.L.Carter, Atomic absorption cross sections of lithium and sodium between 600 and 1000A, *J.Opt.Soc.Am.* 57:651 (1967).

62. D.J.Bradley, C.H.Dugan, P.Ewart and A.F.Purdie, Absolute photoionization cross-section measurement of selectively excited magnesium, *Phys. Rev.* A 13:1416 (1976).
63. J.L.Kohl and W.H.Parkinson, Measurement of the neutral-aluminum photoionization cross-section and parameters of the $3p\ ^2P^o - 3s3p^2\ ^2S_{1/2}$ autoionization doublet, *Astrophys. J.* 184:641 (1973).
64. R.A.Roig, The photoionization spectrum of neutral aluminium, AlI, *J. Phys. B: At. Mol. Phys.* 8:2939 (1975).
65. R.E.Bonanno, C.W.Clark and T.B.Lucatorto, Multiphoton excitation of autoionizing states of Mg: line-shape studies of the $3p^2\ ^1S$ state, *Phys. Rev.* A 34:2082 (1986).
66. VoKy L, H.E.Saraph, W.Eissner, Z.W.Liu and H.P.Kelly, Inner-shell photoionization of beryllium, *Phys. Rev. A* 46:3945 (1992).
67. H.L.Zhang, Relativistic calculations for high energy photoionization cross sections, *in preparation* (1996).
68. K.A.Berrington, J.Pelan and L.Quigley, The calculation of atomic properties: resonances, f-values, photoionization and inner-shell processes, *in preparation* (1996).
69. K.A.Berrington, L.Quigley and H.L.Zhang, High energy photoionization cross sections for the Be-sequence, *in preparation* (1996).
70. C.W.Clark, J.D.Fassett, T.B.Lucatorto, L.J.Moore and W.W.Smith, Observation of autoionizing states of beryllium by resonance-ionization mass spectrometry, *J. Opt. Soc. Am. B* 2:891 (1985).
71. N.Miura, Y.Osanai, T.Noro and F.Sasaki, Theoretical determination of energies and widths of autoionizing states of the Be atom, *J. Phys. B* 29:2689 (1996).

POSITRON COLLISIONS WITH ATOMS

H.R.J. Walters[1], Ann A. Kernoghan[1], Mary T. McAlinden[2]
and C.P. Campbell[1]

[1] Department of Applied Mathematics and Theoretical Physics
The Queen's University of Belfast
Belfast BT7 1NN
United Kingdom

[2] Department of Applied Mathematics and Theoretical Physics
University of Cambridge
Silver Street
Cambridge CB3 9EW
United Kingdom
and
Gonville and Caius College
Cambridge CB2 1TA
United Kingdom

1. INTRODUCTION

The past few years have seen considerable advances in the experimental and theoretical study of positron - atom collisions. On the theoretical side these advances have been associated primarily with the development of coupled - state methods for positron collisions with one - electron atoms, i.e., with atomic hydrogen and, in a frozen core approximation, the alkali metals. For these systems there now exist computer programs capable of using large numbers of eigenstates and pseudostates. It is the achievements of this coupled - state approach to which we shall confine our attention here. However, in celebrating these recent successes, we would not wish to forget the earlier pioneering work of Wakid[1,2], Ghosh[3-20], McEachran and Stauffer[21-30], and their collaborators, using this form of approximation. Invaluable to the coupled - state work has been the existence of good variational calculations at low energies[31-54], especially those which go beyond the positronium formation threshold[34,41-45,47,48,50,52-54]. Last, but not least, the theory has profited enormously from the stimulation provided by experiment, particularly, in the present context, the experimental work of Charlton, Laricchia and co-workers at University College London, of Kauppila, Stein and co-workers at Detroit, and of Raith and co-workers at Bielefeld.

In what way does positron - atom scattering differ from electron - atom scattering? Consider positron scattering by ground state atomic hydrogen. The following processes are possible :

$$
\begin{array}{llll}
e^+ + H(1s) & \longrightarrow & e^+ + H(1s) & \text{Elastic scattering} \\
 & \longrightarrow & e^+ + H(nlm) & \text{Atom excitation} \\
 & \longrightarrow & e^+ + e^- + p & \text{Ionization} \\
 & \longrightarrow & p + \gamma \text{ rays} & \text{Annihilation} \\
 & \longrightarrow & Ps(nlm) + p & \text{Positronium formation}
\end{array} \qquad (1)
$$

The first three reactions are in common with electron scattering but the last two are particular to the positron. Of these two reactions, positronium formation is the most important, only at very low impact energies of the positron is annihilation significant. It is the existence of the positronium formation channels that really makes the difference. Positronium formation is a two - centre rearrangement process in which an electron is transferred from a bound orbital centred on the atomic nucleus to a bound orbital around the moving positron. The positronium atom thus created is in essence a very light hydrogen atom of mass¶ 2au and reduced mass 1/2au. In electron - atom scattering one has to deal with the rearrangement process of electron exchange between the incident electron and the target electrons, here, however, the exchange is associated only with a single centre, the atomic nucleus. Theoretically, this one - centre electron exchange is very much easier to treat than the two - centre positronium formation, that is the fundamental technical challenge posed by positron - atom scattering. In the coupled - state formalism the wave function expansion (see (9)) consists of terms corresponding to the positron - atom channels and other terms relating to the positronium formation channels. These two parts of the expansion are not orthogonal. This non - orthogonality results in properties that do not arise in a coupled - state calculation of electron - atom scattering where only one type of channel exists. An unanticipated consequence of this non - orthogonality was the appearance of strange new resonances above the ionization threshold in simple coupled - state calculations of positron - atom scattering. For a time it was thought that these might be real physical effects, it is now clear that they are a phenomenon of non - orthogonal coupled - state approximations. The history of this episode is traced in section 3. While we shall not be concerned with the annihilation process here, see (1), it is worthwhile to remark that it provides a very stringent test of the quality of the calculated collisional wave function. The annihilation cross section depends critically upon the correlation between the positron and the atomic electrons, requiring a knowledge of the values of the collisional wave function when the positron and electron positions coincide[41]. In a very recent paper, Van Reeth et al[55] have measured the annihilation γ - ray spectrum for positrons interacting with helium, obtaining excellent agreement over three orders of magnitude with results calculated from the elaborate variational wave functions of Van Reeth and Humberston[52,53] and establishing the non - Gaussian nature of this spectrum. For atoms containing two or more electrons, reactions more exotic than (1) are possible, for example, Ps^- formation, transfer ionization (in which the atom looses two electrons, one to positronium formation, the other electron being directly ionized)[56–58], multiple ionization, etc. Again, these are processes which will not be considered here. Experimentally, positron - atom collisions are considerably more difficult to study than electron - atom collisions because of the much lower intensity of

¶ Throughout this article we use atomic units (au) in which $\hbar = m_e = e = 1$. The symbol a_0 is used to denote the Bohr radius.

presently available positron beams (typically 10^{-5} that of an electron beam). Usually the positrons are obtained from a radioactive β^+ decay source such as Na^{22} or Co^{58}. The positrons emerge isotropically from these sources with energies in the region of a few hundred keV and so require considerable moderation, with consequent loss of flux, to obtain a well defined monoenergetic beam of several eV which can be used in an atomic collisions experiment. Another problem which complicates the experimental work is the detection of the neutral positronium atom.

We begin in section 2 with a description of the one - electron coupled - state model. Here we discuss the nature of the wave function expansion and describe how pseudostates are used to represent the continuum channels. Results for positron scattering by atomic hydrogen are presented in section 3 where we also catalogue the history of the "high - energy" resonance phenomenon. Positron scattering by the alkali metals is surveyed in section 4 and conclusions are presented in section 5.

2. COUPLED - STATE APPROXIMATION FOR POSITRON SCATTERING BY ATOMIC HYDROGEN AND THE ALKALI METALS

We model the alkali atom in a frozen core approximation. The state of the atom is now synonymous with the state of the valence electron. The interaction of the valence electron with the core is represented by the local central potential

$$\left(-\frac{1}{r_e} - V_e(r_e)\right) \tag{2}$$

where $\mathbf{r}_e$ is the position vector of the electron relative to the nucleus and where we have explicitly removed the long - range ionic tail $-1/r_e$ so that $V_e(r_e)$ is a non-Coulombic potential. Suitable model potentials of the form (2) for Li, Na, K, Rb and Cs may be found in the paper by Stein[59]. In the frozen core approximation (2) the atomic Hamiltonian becomes

$$H_A = -\frac{1}{2}\nabla_e^2 - \frac{1}{r_e} - V_e(r_e) \tag{3}$$

The Hamiltonian, H, describing the positron collision with the frozen core alkali is

$$H = -\frac{1}{2}\nabla_p^2 + H_A + \frac{1}{r_p} + V_p(r_p) - \frac{1}{|\mathbf{r}_p - \mathbf{r}_e|} \tag{4}$$

where $\mathbf{r}_p$ is the position vector of the positron relative to the nucleus and, in analogy with (2),

$$\left(+\frac{1}{r_p} + V_p(r_p)\right) \tag{5}$$

is the interaction of the positron with the atomic core. The coordinates $(\mathbf{r}_p, \mathbf{r}_e)$ used in writing (4) are those appropriate to positron - atom channels. When positronium is formed it is more appropriate to use the "positronium coordinates"

$$\mathbf{R} \equiv \frac{\mathbf{r}_p + \mathbf{r}_e}{2} \qquad \mathbf{t} \equiv \mathbf{r}_p - \mathbf{r}_e \tag{6}$$

Here $\mathbf{R}$ defines the centre of mass of the positronium relative to the nucleus and $\mathbf{t}$ is the positronium internal coordinate. In terms of (6), the Hamiltonian (4) may be equivalently written

$$H = -\frac{1}{4}\nabla_R^2 + H_{Ps} + \left(\frac{1}{|\mathbf{R}+\mathbf{t}/2|} + V_p(|\mathbf{R}+\mathbf{t}/2|)\right) - \left(\frac{1}{|\mathbf{R}-\mathbf{t}/2|} + V_e(|\mathbf{R}-\mathbf{t}/2|)\right) \tag{7}$$

In (7), $-\frac{1}{4}\nabla_R^2$ is the kinetic energy operator for the positronium centre of mass motion, H_{Ps} is the positronium Hamiltonian

$$H_{Ps} = -\nabla_t^2 - \frac{1}{t} \tag{8}$$

and the remaining terms give the interaction of the positronium with the ionic core of the atom. Positron collisions with atomic hydrogen are obtained as a special case of (2) to (8) by setting $V_e = V_p = 0$.

In the coupled - state approximation the collisional wave function Ψ for the system is expanded as

$$\Psi = \sum_n F_n(\mathbf{r}_p)\psi_n(\mathbf{r}_e) + \sum_m G_m(\mathbf{R})\phi_m(\mathbf{t}) \tag{9}$$

where the sum over n is over atom (valence electron) states ψ_n and the sum over m is over positronium states ϕ_m. These states may be either eigenstates or pseudostates. The first sum in (9) represents the atom channels, the second sum the positronium channels. It is interesting to note that if the sets of states ψ_n and ϕ_m were complete, then either

$$\sum_n F_n(\mathbf{r}_p)\psi_n(\mathbf{r}_e) \tag{10}$$

or

$$\sum_m G_m(\mathbf{R})\phi_m(\mathbf{t}) \tag{11}$$

alone would give an exact expansion of the system wave function Ψ. In this sense the expansion (9) is over complete, from which it is also clear that the positronium part of the expansion is not orthogonal to the atom part. In practical terms, however, we do not deal with complete sets of states, we are restricted to finite, and therefore incomplete, sets. What matters then is how rapidly the different forms of expansion (9), (10) and (11) converge as the number of states is increased. A priori, it is to be expected that a mixed expansion such as (9) will be more quickly convergent as both the atom channels and positronium channels are being directly represented, this is indeed found to be the case. However, some interesting and very informative calculations have been made using the single - centre expansion (10) and a large basis of atom pseudostates[60–62].

In principle, the expansions (9), (10) and (11) should include both bound and continuum states of the atom/positronium. In practice, it is at this moment not feasible to deal with continuum eigenstates. Instead we introduce pseudostates. These are constructed so that, together with the retained bound eigenstates, they diagonalize the atom/positronium Hamiltonian, i.e.,

$$\begin{aligned} \langle\psi_n|H_A|\psi_{n'}\rangle &= \epsilon_n\delta_{n,n'} & \langle\psi_n|\psi_{n'}\rangle &= \delta_{n,n'} \\ \langle\phi_m|H_A|\phi_{m'}\rangle &= E_m\delta_{m,m'} & \langle\phi_m|\phi_{m'}\rangle &= \delta_{m,m'} \end{aligned} \tag{12}$$

We do not distinguish between true eigenstates and pseudostates in our notation, using ψ_n and ϕ_m for both. A true eigenstate satisfies not only (12) but also

$$H_A\psi_n = \epsilon_n\psi_n \qquad H_{Ps}\phi_m = E_m\phi_m \tag{13}$$

While pseudostates obey (12), they do not satisfy (13). Figure 1 shows a typical eigenstate energy spectrum for the atom or positronium, consisting of a discrete part, containing an infinite number of states converging to the ionization threshold, followed by a continuous spectrum¶ The eigenvalues ϵ_n and E_m of (12) will in general be both

¶ Consistent with (3) and (8), the ionization threshold is taken to be at zero energy, bound eigenstates have negative energies, continuum eigenstates positive energies.

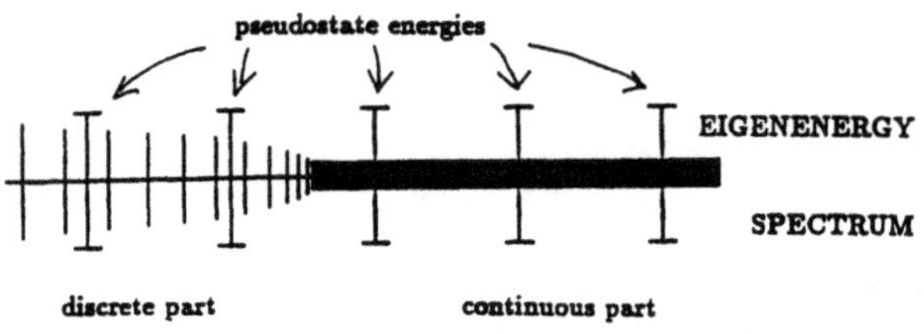

Figure 1.

positive and negative. When an eigenvalue corresponds to a pseudostate we shall call it the "pseudostate energy". Pseudostate energies will be distributed throughout both the discrete and continuous parts of the eigenstate energy spectrum as shown in figure 1. We can think of a pseudostate as being a "clump" or "distribution" over eigenstates with the average energy of the "clump" being the pseudostate energy[63]. Accordingly, we introduce an energy distribution function $f_n(\epsilon)$ for the pseudostate ψ_n by defining

$$f_n(\epsilon) = |\langle\psi_\epsilon|\psi_n\rangle|^2 \tag{14}$$

where by ψ_ϵ we mean an appropriately normalised¶ eigenstate, either bound or continuum, with energy ϵ and with the same angular momentum quantum numbers as the pseudostate ψ_n. Formula (14) is just the probability that ψ_n contains the eigenstate ψ_ϵ. In general the spectrum $f_n(\epsilon)$ will consist of discrete parts, corresponding to overlap of ψ_n with discrete eigenstates, and a continuous segment. The fraction of the continuum contained in the pseudostate ψ_n is

$$a_n = \int_0^\infty f_n(\epsilon)d\epsilon = 1 - \sum_m f_n(\epsilon_m) \tag{15}$$

where the sum on m in (15) is over all bound eigenstates with energy ϵ_m. The quality of a pseudostate set may be gauged by how well the pseudostate energies are distributed throughout the eigenstate spectrum, figure 1. As the number of pseudostates is increased the distribution of pseudostate energies can be made denser and so the division of the eigenstates into the "clumps" which we call pseudostates will become finer, with the result that we approach closer to the ideal of an eigenstate expansion. Surprisingly, in practical terms, this ideal is achieved more quickly than one might suspect, with a managable number of pseudostates giving a satisfactory representation for most purposes[64–66]. An inadequate density of pseudostates gives unphysical structures, called pseudostructures, in the calculated cross sections, this was a problem which dogged early pseudostate calculations of electron - atom scattering[67–71] which did not have the computing power to incorporate enough states. With an increasing density of pseudostates these pseudostructures get smaller, until eventually they become neglegible[64].

¶ (14) assumes that bound eigenstates ψ_ϵ are normalised to unity and continuum eigenstates to $\langle\psi_\epsilon|\psi_{\epsilon'}\rangle = \delta(\epsilon - \epsilon')$.

Pseudostates are usually constructed by diagonalizing the atom/positronium Hamiltonian in a basis of Slater type orbitals

$$r^n e^{-\lambda r/2} Y_{lm}(\hat{\mathbf{r}}) \tag{16}$$

where $\mathbf{r}$ stands for $\mathbf{r}_e$ or $\mathbf{t}$ as appropriate, l is the angular momentum of the state, n is an integer with $n \geq l$, and λ is a parameter which need not be the same for all basis functions (16). It is convenient if the basis can be expanded in a systematic way, for example, by using the same λ for all terms with the same l, taking $n = l, l+1, \ldots\ldots, N$, and letting N increase. However, as the powers of r^n go up, this leads to numerical linear dependence problems. This numerical difficulty can be overcome by adopting a Laguerre basis[64,65,72]

$$\left(\frac{\lambda^3 (n-l)!}{(n+l+2)!}\right)^{\frac{1}{2}} (\lambda r)^l L_{n-l}^{2l+2}(\lambda r) e^{-\lambda r/2} Y_{lm}(\hat{\mathbf{r}}) \qquad n = l, l+1, \ldots\ldots, N \tag{17}$$

where $L_\alpha^\beta(x)$ is a Laguerre polynomial as defined in Gradshteyn and Ryzhik[73]. Mathematically the basis (17) is identical to (16) for the same range of n, i.e., it includes exactly the same powers of r multiplying $e^{-\lambda r/2}$. Numerically, however, it is much more satisfactory in that the the functions (17) are mutually orthogonal for different n and the Laguerrre polynomials can be generated from a recurrence relation[73] instead of having to be evaluated as a sum of powers of r, which, numerically, would be equivalent to using (16). With the basis (17) there is no problem in diagonalizing the atom/positronium Hamiltonian, for a given l, using 100 or more states[74]. By contrast the basis (16) can fail if the number of states exceeds 15 or so.

Let us now return to the expansion (9). The coupled equations for the functions $F_n(\mathbf{r}_p)$ and $G_m(\mathbf{R})$ are obtained by substituting (9) into the Schrödinger equation with the Hamiltonian (4)/(7), projecting with $\psi_n(\mathbf{r}_e)$ and $\phi_m(\mathbf{t})$, and using (12). The resulting equations have the form

$$\left(\nabla_p^2 - 2V_p(r_p) + k_n^2\right) F_n(\mathbf{r}_p) = 2\sum_{n'} V_{nn'}(\mathbf{r}_p) F_{n'}(\mathbf{r}_p) + 2\sum_{m'} \int K_{nm'}(\mathbf{r}_p, \mathbf{R})\, G_{m'}(\mathbf{R})\, d\mathbf{R} \tag{18}$$

$$\left(\nabla_R^2 + p_m^2\right) G_m(\mathbf{R}) = 4\sum_{m'} U_{mm'}(\mathbf{R})\, G_{m'}(\mathbf{R}) + 4\sum_{n'} \int K_{n'm}^{\star}(\mathbf{r}_p, \mathbf{R})\, F_{n'}(\mathbf{r}_p)\, d\mathbf{r}_p \tag{19}$$

where $\star$ stands for complex conjugation and where we assume that the positron is incident with momentum $\mathbf{k}_0$ upon an atomic state of energy ϵ_0 so that

$$\frac{k_n^2}{2} + \epsilon_n = \frac{k_0^2}{2} + \epsilon_0 = \frac{p_m^2}{4} + E_m \tag{20}$$

The potentials $V_{nn'}(\mathbf{r}_p)$ and $U_{mm'}(\mathbf{R})$ are the direct potentials in the positron - atom and positronium channels respectively and are given by

$$V_{nn'}(\mathbf{r}_p) \equiv \left\langle \psi_n(\mathbf{r}_e) \left| \frac{1}{r_p} - \frac{1}{|\mathbf{r}_e - \mathbf{r}_p|} \right| \psi_{n'}(\mathbf{r}_e) \right\rangle \tag{21}$$

$$U_{mm'}(\mathbf{R}) \equiv \left\langle \phi_m(\mathbf{t}) \left| \left(\frac{1}{\left|\mathbf{R} + \frac{1}{2}\mathbf{t}\right|} - \frac{1}{\left|\mathbf{R} - \frac{1}{2}\mathbf{t}\right|} \right) \right| \phi_{m'}(\mathbf{t}) \right\rangle + \left\langle \phi_m(\mathbf{t}) \left| \left(V_p\left(\left|\mathbf{R} + \frac{1}{2}\mathbf{t}\right|\right) - V_e\left(\left|\mathbf{R} - \frac{1}{2}\mathbf{t}\right|\right) \right) \right| \phi_{m'}(\mathbf{t}) \right\rangle \tag{22}$$

However, the most interesting components of (18) and (19) are the non-local couplings $K_{nm}(\mathbf{r}_p, \mathbf{R})$, which, in the interests of conciseness, we shall not quote explicitly here[75]. It is these terms which give rise to positronium formation. The couplings $K_{nm}(\mathbf{r}_p, \mathbf{R})$ descibe positronium formation in the state ϕ_m by capture of an electron from state ψ_n of the atom, we call $K_{nm}(\mathbf{r}_p, \mathbf{R})$ a positronium formation kernel. It is the evaluation of these kernels and their incorporation into the coupled equations (18) and (19) which make the theoretical treatment of positron - atom collisions so much more difficult than the sister subject of electron - atom scattering. As stated in the "Introduction", the difficulty is the two - centre nature of $K_{nm}(\mathbf{r}_p, \mathbf{R})$. For an atom containing more than one electron[§] equation (19) is even more complicated. To the right hand side of (19) must then be added a term[¶]

$$\sum_{m'} \int L_{mm'}(\mathbf{R}, \mathbf{R}')G_{m'}(\mathbf{R}')d\mathbf{R}' \tag{23}$$

This term describes electron exchange between positronium in the state $\phi_{m'}$ and the atomic ion, resulting in positronium in the state ϕ_m. The positronium exchange kernels $L_{mm'}(\mathbf{R}, \mathbf{R}')$ are even more complicated to deal with[76] than the positronium formation kernels and little progress has been made in incorporating them into calculations[20,76]. The absence of these terms is one advantage of working with one - electron systems.

The coupled equations (18) and (19) are solved in partial wave form. Two main approaches are currently in use : (i) to solve the equations in coordinate space using the R - matrix technique[75–90] or the Harris - Nesbet variational method[91–98]; (ii) to transform the equations into momentum space and solve directly for the T - matrix[3–20,99–121]. The relative merits of these two approaches are discussed in reference [84].

3. POSITRON SCATTERING BY ATOMIC HYDROGEN

Positron scattering by atomic hydrogen is the simplest possible system and has therefore attracted most attention. The most recent coupled - state calculations on this system using the mixed expansion (9) have been made by Ghosh[11–14,16–19], Kernoghan, McAlinden and Walters[75,78,80,81,83,84,86], Higgins and Burke[89,90], Gien[91–98], Hewitt, Noble and Bransden[99,100], Mitroy[107–114,116,117,119,120], and their collaborators. The simplest form of the approximation (9) consists of retaining only the ground state of the atom and of positronium — this is known as the coupled - static approximation. For an atomic hydrogen target this corresponds to

$$\Psi = F(\mathbf{r}_p)\psi_{1s}(\mathbf{r}_e) + G(\mathbf{R})\phi_{1s}(\mathbf{t}) \tag{24}$$

and leads to the coupled equations

$$\begin{aligned} (\nabla_p^2 + k_0^2)F(\mathbf{r}_p) &= 2V_{1s,1s}(\mathbf{r}_p)F(\mathbf{r}_p) + 2\int K(\mathbf{r}_p, \mathbf{R})G(\mathbf{R})d\mathbf{R} \\ (\nabla_R^2 + p^2)G(\mathbf{R}) &= 4\int K(\mathbf{r}_p, \mathbf{R})F(\mathbf{r}_p)d\mathbf{r}_p \end{aligned} \tag{25}$$

In studying the S - wave cross section for Ps(1s) formation in this approximation, Higgins and Burke[89] made a totally unexpected discovery, they found a pronounced

§ Recall that in writing (18) and (19) we have modelled the alkalis as one - electron systems with a core potential.

¶ In writing (23) we have assumed that electron capture from the atom results in a unique ion state. More generally, G_m, $L_{mm'}$ and K_{nm} would have to carry indices i and i' as well as m and m' to indicate the state of the ion.

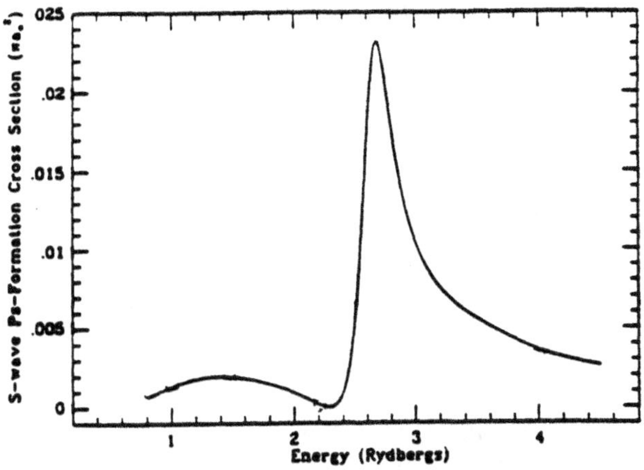

Figure 2. S - wave Ps(1s) formation cross section as calculated in the coupled - static approximation (24). Taken from Higgins and Burke[89].

resonance at an energy of 2.62 Ryd (35.6eV) and of width 0.31 Ryd (4.2eV), figure 2. Firstly, it was surprising to see a resonance at such a high energy, well above the ionization threshold, such a phenomenon is unknown in electron - hydrogen scattering. Secondly, the formation mechanism for this resonance was very unusual. Looking at equation (25) we see that there is no direct potential in the positronium channel, while the direct potential in the atom channel is the purely repulsive static potential $V_{1s,1s}$. The only way the resonance can therefore be trapped is through the action of the non - diagonal positronium formation coupling $K(\mathbf{r}_p, \mathbf{R})$; for this reason Higgins and Burke termed it a "coupled - channel shape resonance". Such a resonance formation mechanism was unknown in electron - atom scattering. At about the same time that Higgins and Burke discovered their resonance, McAlinden and Walters[76,77] were finding high energy resonances in coupled - static calculations of positron scattering by the noble gases. A little later, structures began to be seen in measurements of the elastic differential cross sections for Ar and Kr at fixed scattering angles[122–124] which, it was thought, might be signatures of high energy Higgins - Burke type resonances.

When the coupled - static approximation was enlarged to include the 2s and 2p states of atomic hydrogen and positronium[16,78,108,110,113], it was found that the Higgins - Burke resonance of figure 2 began to move to a substantially higher energy. Figure 3 shows the S - wave Ps(1s) formation cross section corresponding to figure 2 but now calculated in the Ps(1s, 2s, 2p) + H(1s, 2s, 2p) approximation[78] where 1s, 2s and 2p eigenstates of both positronium and atomic hydrogen are included in the wave function expansion (9). Here we again see the Higgins - Burke resonance, similar to figure 2, but now positioned at an energy of 51.05eV (3.75 Ryd), with a smaller width of 2.9eV (0.21 Ryd), and at a reduced height of 0.015 πa_0^2 compared with 0.023 πa_0^2 in figure 2. However, figure 3 now also shows a second resonance ! This resonance is much narrower than the original Higgins - Burke object but, like it, also appears above the ionization threshold, near 15eV. Altogether, in the Ps(1s, 2s, 2p) + H(1s, 2s, 2p) approximation, three resonances were found in the S - wave amplitudes, four resonances in the P - wave results, one resonance in the D - wave cross sections and one resonance in F - wave

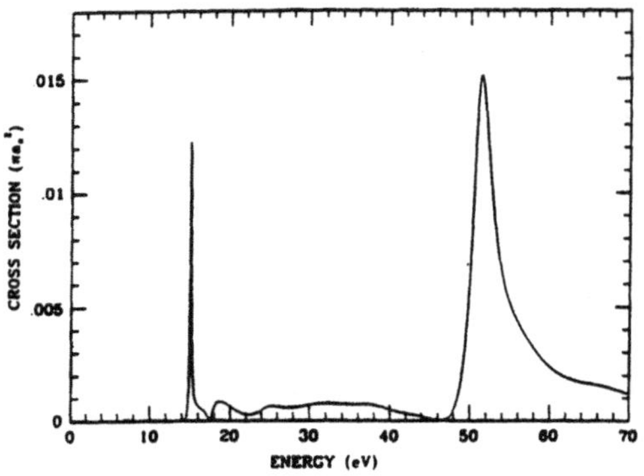

Figure 3. S - wave Ps(1s) formation cross section calculated in the 6 - state approximation[78] Ps(1s, 2s, 2p)+ H(1s, 2s, 2p).

scattering, all of these resonances being positioned above the ionization threshold[81,110,113]. This escalation in the number of resonances was rather curious and the fact that they always appeared above the ionization threshold lead Kernoghan et al[81] to wonder what would be the effect of ionization channels, which are not included in the coupled - static or Ps(1s, 2s, 2p) + H(1s, 2s, 2p) approximations, upon them. Ionization channels can be incorporated into the coupled - state approximation by using pseudostates. Kernoghan et al therefore set out to study the problem in an 18 - state Ps(1s, 2s, $\overline{3s}$, $\overline{4s}$, 2p, $\overline{3p}$, $\overline{4p}$, $\overline{3d}$, $\overline{4d}$) + H(1s, 2s, $\overline{3s}$, $\overline{4s}$, 2p, $\overline{3p}$, $\overline{4p}$, $\overline{3d}$, $\overline{4d}$) approximation in which six pseudostates, denoted by a bar, were added both to the positronium centre and to the atomic hydrogen centre. It was not clear at the start whether pseudostructure might defeat the purpose of the calculation, but it was the best that could be done at the time. Fortunately, the pseudostructure turned out to be small. The S - wave Ps(1s) formation cross section from this 18 - state calculation is compared with that from the earlier 6 - state Ps(1s, 2s, 2p) + H(1s, 2s, 2p) approximation, and with the accurate variational calculations of Brown and Humberston[44] in the Ore gap region (6.8 to 10.2eV), in figure 4. The 18 - state cross section shows some lumpy and spiky pseudostructure, but only to a small degree — there is no evidence of the two pronounced resonances near 15 and 50eV of the 6 - state approximation. Kernoghan et al also examined other transitions and higher partial waves, extending their search up to 100eV, they concluded that the coupling to the ionization channels had made the high energy resonances disappear (or, more tentatively, that they had been very considerably reduced). They suggested that the resonances were an attempt by simple coupled - state approximations to represent ionization channels missing from the system wave function expansion (9). Thus, for example, the 1s - 2s - 2p positronium component of the 6 - state Ps(1s, 2s, 2p) + H(1s, 2s,2p) approximation would have a non - zero projection on to the atomic hydrogen continuum and would therefore try to represent this continuum as best as possible within the restrictions of the 6 - state formalism, resulting in the above ionization threshold resonances that had been observed. This compensatory affect would seem to be a feature of non - orthogonal expansions[81]. Later hyperspherical calculations of Igarashi and Toshima[125] and Zhou and Lin[126] also failed

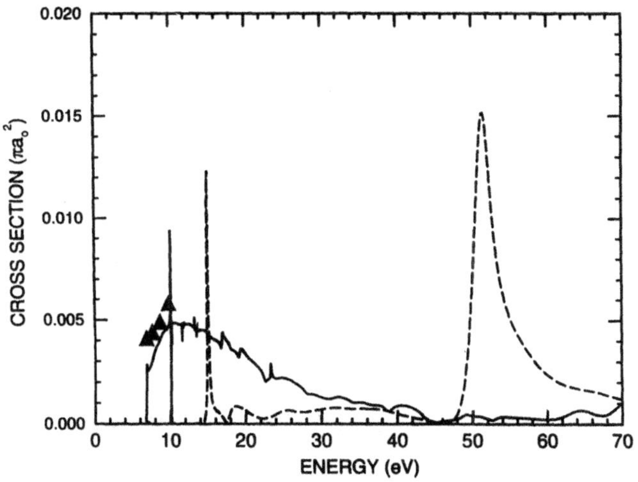

Figure 4. S - wave Ps(1s) formation cross section : solid curve, 18 - state approximation; dashed curve, 6 - state approximation; triangles, accurate variational results of Brown and Humberston[44].

to see any high energy resonances. Although these hyperspherical approximations involve coupled states, the expansion is in terms of orthogonal states and so the non - orthogonal compensation mechanism does not apply. Finally, even the tenuous experimental support for these resonances was removed when re-measurement of the elastic differential cross section for e^+ - Ar scattering[127,128] failed to find the structures seen earlier[122,123]. It now seems clear that the resonances above the ionization threshold are features of small basis coupled - state calculations.

These results are consistent with a re-discovered theorem due to Simon[129,130] which states that resonances in any many - particle system experiencing only Coulomb forces cannot exist above the threshold for complete disintegration of the system, i.e., there can be no resonances above the ionization threshold in e^+ + H scattering ! However, there appeared to be a contradiction to Simon's theorem in the form of an experimental result for e^- + H^- scattering which showed two resonances above the $3e^-$ + p threshold[131–133]. A more recent experiment by Andersen et al[134] has failed to detect these resonances, whose absence is also consistent with a recent theoretical analysis of Robicheaux et al[135]. Real resonances do, however, exist in the e^+ + H system. Thus, as predicted by Mittleman[136], real Feshbach resonances, similar to those in electron scattering, exist near excitation thresholds[75,78,81,83,95,97,98,113,116]. Such a resonance is clearly visible near the H(n = 2) threshold at 10.2eV in the 18 - state calculation of figure 4. This calculation also shows a much smaller, but real, resonance spike near the Ps(n = 2) threshold at 11.9eV, figure 4. A lot of work on resonances in the e^+ + H system has been done by Y.K. Ho[137,138].

Encouraged by the low degree of pseudostructure in the elastic and Ps(1s) partial wave cross sections, Kernoghan et al[83] went on to explore the 18 - state Ps(1s, 2s, $\overline{3s}$, $\overline{4s}$, 2p, $\overline{3p}$, $\overline{4p}$, $\overline{3d}$, $\overline{4d}$) + H(1s, 2s, $\overline{3s}$, $\overline{4s}$, 2p, $\overline{3p}$, $\overline{4p}$, $\overline{3d}$, $\overline{4d}$) approximation in more detail. Their results for positronium formation and total scattering, for which experimental data were available, are shown in figure 5. The calculated cross sections exhibited in these figures have been smoothed to remove small amplitude pseudostructure. From figure 5(a) it is seen that positronium formation is almost completely into the 1s ground

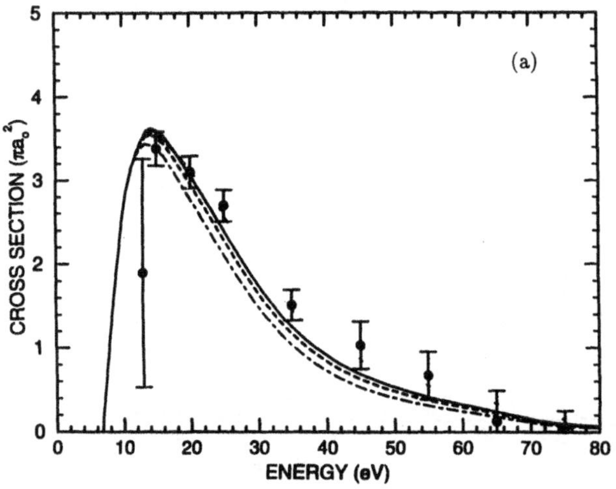

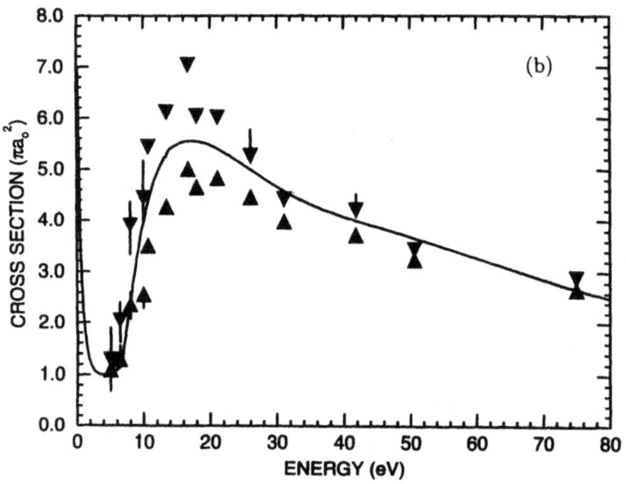

Figure 5. Cross sections for (a) positronium formation and (b) total scattering as calculated in the 18 - state approximation of Kernoghan et al[83]. In (a) : dash - dot curve, Ps(1s) cross section; dashed curve, Ps(1s) + Ps(n=2) cross section; solid curve, total positronium formation cross section (employing n^3 scaling for Ps($n \geq 3$)); solid circles, experimental data of Weber et al[139]. In (b) : solid curve, calculated total cross section; solid triangles up (down), lower (upper) bound measurements of Zhou et al[141].

state. Figure 5(a) also shows the experimental data of Weber et al[139,140] for total positronium formation. The agreement between theory and experiment is reasonable, but could be better. For total scattering, figure 5(b), there is agreement with the upper and lower bound measurements of Zhou et al[141] except near 30 and 50eV. Most intriguing, however, was a tentative estimate of the ionization cross section, this is

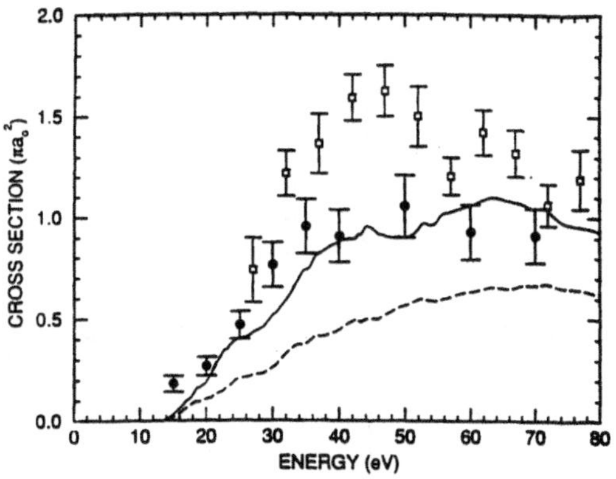

Figure 6. Ionization cross section as calculated in the 18 - state approximation of Kernoghan et al[83] : solid curve, full cross section (26); dashed curve, contribution (27) from atomic pseudostates alone; solid circles, experimental data of Jones et al[142]; open squares, experimental data of Weber et al[139].

shown in figure 6. Kernoghan et al[83] calculated this ionization cross section using the ansatz

$$\sigma_{ion} = \sum_i a_i \sigma_H(i) + \sum_j a_j \sigma_{Ps}(j) \tag{26}$$

where the sum on i,(j) runs over the $\overline{3s}$, $\overline{4s}$, $\overline{3p}$, $\overline{4p}$, $\overline{3d}$ and $\overline{4d}$ pseudostates of atomic hydrogen (positronium), $\sigma_H(i)$ ($\sigma_{Ps}(j)$) is the cross section for exciting the ith (jth) pseudostate of atomic hydrogen (positronium) and a_i (a_j) is the probability (15) that the ith (jth) pseudostate overlaps the eigenstate continuum¶. It is seen from figure 6 that the calculated cross section is in quite good agreement with the measurements of Jones et al[142] but in disagreement with the data of Weber et al[139,140] which lie somewhat higher. Pseudostructure has not been smoothed out of figure 6, hence the lumpiness. Also shown in figure 6 is the contribution to σ_{ion} coming from the atomic hydrogen pseudostates alone, ie, see (26),

$$\sum_i a_i \sigma_H(i) \tag{27}$$

This amounts to little more than 50% of σ_{ion}, demonstrating that, in the energy range shown, the ionization is carried almost equally by the positronium and atomic hydrogen pseudostates.

In view of the relative success of the 18 - state approximation, Kernoghan et al[86] decided to embark upon a much larger pseudostate calculation in the hope of reducing the pseudostructure to a neglegible level and also with the purpose of getting a better estimate of the ionization cross section. Because the atomic hydrogen and positronium components of the system wave function expansion (9) are not orthogonal, it is possible

¶ For the states used by Kernoghan et al[83] in their 18 - state approximation, a_i has the same value for corresponding hydrogen and positronium pseudostates.

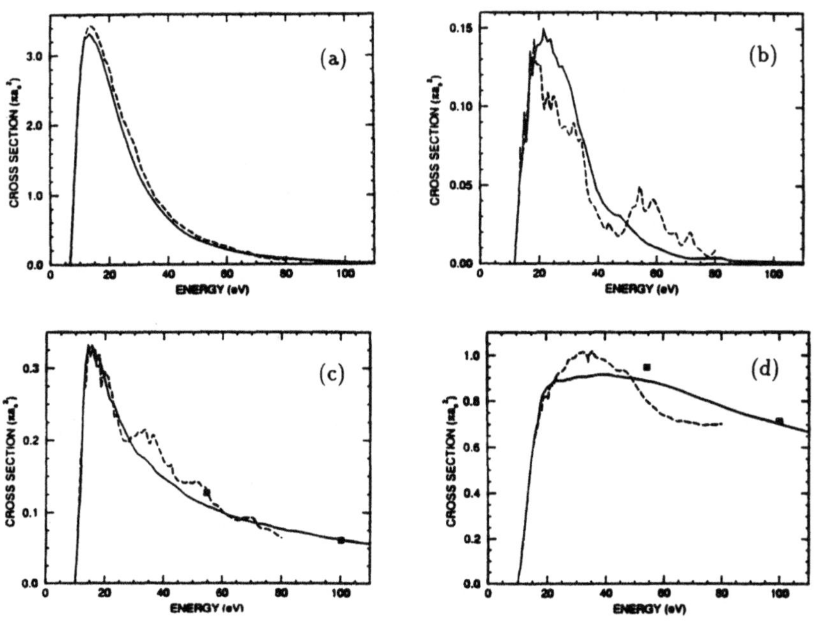

Figure 7. Comparison of 33 - state[84,86] and 18 - state[83] results for (a) Ps(1s) formation, (b) Ps(2p) formation, (c) H(2s) excitation and (d) H(2p) excitation : solid curve, 33 - state calculation; dashed curve, 18 - state calculation; solid squares, multipseudostate close - coupling approximation of Walters[143].

that the ansatz (26) may involve double counting between the positronium and the atomic hydrogen contributions. To avoid this ambiguity Kernoghan et al[86] decided to use pseudostates only on the atom centre, in this case the ansatz (26) reduces to

$$\sigma_{ion} = \sum_i a_i \sigma_H(i) \tag{28}$$

and there can be no double counting. Experience with the 18 - state approximation had shown that positronium formation was dominated by the 1s state, figure 5(a), it was therefore felt that inclusion only of the 1s, 2s and 2p eigenstates of positronium in the expansion (9) might be sufficient. Kernoghan et al[86] therefore decided upon the 33 - state approximation Ps(1s, 2s, 2p) + H(1s, 2s, $\overline{3s}$ to $\overline{9s}$, 2p, $\overline{3p}$ to $\overline{9p}$, $\overline{3d}$ to $\overline{9d}$, $\overline{4f}$ to $\overline{9f}$), the hydrogen eigenstates and pseudostates being constructed by diagonalizing the atomic hydrogen Hamiltonian in the basis¶

$$r^n e^{-r} Y_{lm}(\hat{\mathbf{r}}) \qquad n = l, l+1, \ldots\ldots, 9 \tag{29}$$

This corresponds to taking $\lambda = 2$ in (16) for each orbital angular momentum l.

¶ This diagonalization actually generates states up to $\overline{10s}$, $\overline{10p}$, $\overline{10d}$ and $\overline{10f}$. However, the $\overline{10}$ states have rather high energies and so were omitted from the coupled - state calculation on the basis that their contribution should not be important.

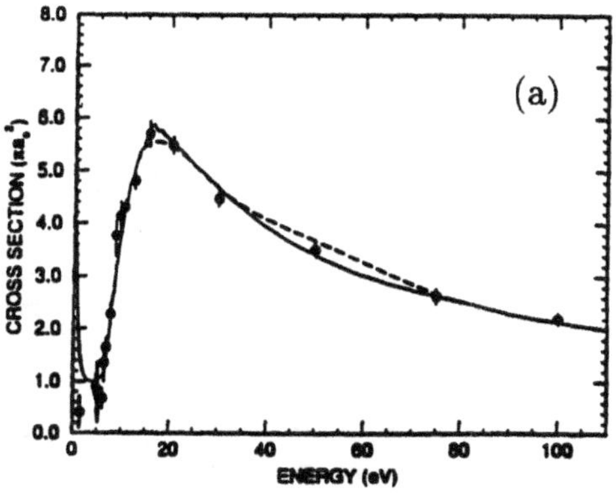

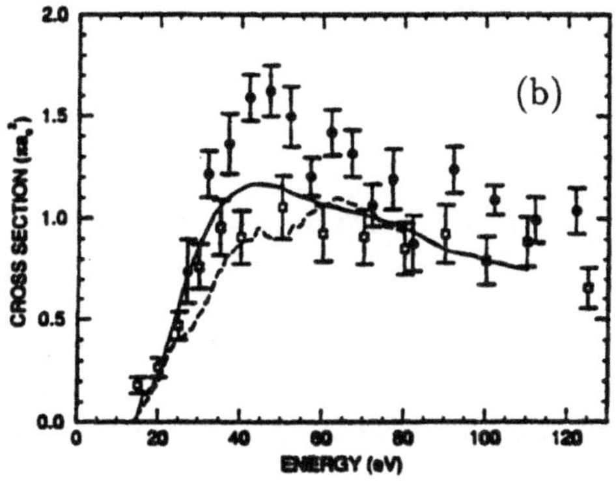

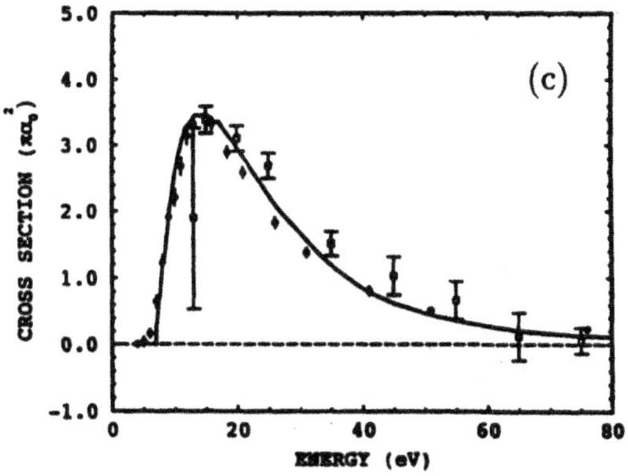

Figure 8. 33 - state results[84,86] for (a) the total cross section, (b) ionization and (c) total positronium formation : solid curve, 33 - state approximation; dashed curve, 18 - state approximation of Kernoghan et al[83]. In (a) solid circles are experimental data of Stein et al[144]. In (b) solid circles are experimental data of Weber et al[139] while open squares are measurements of Jones et al[142]. In (c) solid circles are data of Zhou et al[145] and open squares are results of Weber et al[139].

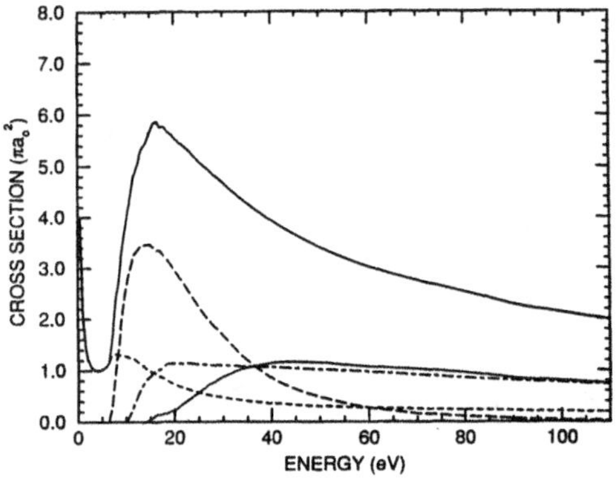

Figure 9. Cross sections in the 33 - state approximation[84,86] : solid curve, total cross section; long - dashed curve, total positronium formation; short - dashed curve, elastic scattering; dash - dot curve, H(2p) excitation; dotted curve, ionization.

In figure 7 we compare the 18 - state and 33 - state results for Ps(1s) and Ps(2p) formation and for excitation of the hydrogen atom to the 2s and 2p states. In this figure there has been **no** smoothing of pseudostructure in the calculated cross sections. The eveness of the 33 - state results contrasts with the "lumpy" pseudostructure in the 18 - state cross sections for Ps(2s) formation and H(2s) and H(2p) excitation. Clearly increasing the number of pseudostates, from 12 in the 18 - state approximation to 27 in the 33 - state approximation, has significantly reduced the amount of pseudostructure. Interestingly, however, we see that a suitable average through the pseudostructure in the 18 - state cross sections would give an answer close to the 33 - state results.

Figure 8 compares the 18 - state cross sections of figures 5(b) and 6 with the new 33 - state calculations for total scattering and for ionization. For total scattering there is not much difference between the two sets of cross sections, but where there are differences, the 33 - state results have moved into agreement with the very recent measurements of Stein et al[144§], indeed the 33 - state calculations are in almost perfect accord with these new measurements. Figure 8(b) also shows that the, now unambiguous, 33 - state ionization cross section is not greatly different from the original tentative 18 - state estimate of figure 6, in particular the 33 - state numbers again prefer the data of Jones et al[142] over those of Weber et al[139,140]. Finally, in figure 8(c) we show the 33 - state cross section for total positronium formation. This differs only marginally from the 18 - state result of figure 5(a) and consequently the agreement with the data of Weber et al[139,140¶] is the same as in that figure. However, also shown in figure 8(c)

§ These measurements have been confirmed by the even more recent experiment of Zhou et al[145].

¶ We take this opportunity to correct an error in reference [86]. The data of Weber et al quoted in figure 3 of this paper are not those of reference [139] as stated there, but rather earlier preliminary measurements. The correct comparison of the results of Weber et al[139] with the 33 - state approximation is now seen in figure 8(c) of the present work.

are some very recent data of Zhou et al[145]. These new measurements are in impressive agreement with the 33 - state calculation.

Figure 9 shows the total cross section and its main components as calculated in the 33 - state approximation. We see that between 8.5 and 35eV positronium formation is the main contributor. Above 35eV ionization and H(2p) excitation are largest and of comparable size. Because of the very good agreement between the 33 - state numbers and experiment described above, and because of the agreement on average between the 33 - state results and those of the 18 - state approximation, we believe that the new 33 - state calculations shown in figure 9 now give the main cross sections for positron scattering off ground state atomic hydrogen to a high degree of accuracy.

4. POSITRON SCATTERING BY THE ALKALI METALS

As the ionization potential of an alkali atom is typically around 5eV, and as the binding energy of Ps(1s) is 6.8eV, unlike atomic hydrogen, Ps(1s) can be formed for all impact energies of the positron on the alkali atom, i.e., the process is exothermic. A consequence of this is that the Ps(1s) formation cross section becomes infinite as $1/k_0$ at zero impact energy[47,78]. As in electron scattering, and for collisions with the ground state, we expect the dominant atom states in a coupled - state approximation to be the ground state and the first excited p - state, we also anticipate that ionization will be comparatively small. So far, these predictions have been largely borne out by calculation. Figure 10 shows some very recent calculations[88] of positron scattering by ground state Li in the energy range 0.5 to 60eV. We see that at low energies elastic scattering is dominant¶, with this dominance passing directly to the Li(2s) $\rightarrow$ Li(2p) excitation with increasing energy; ionization is almost neglegible on the scale of the other cross sections. Comparing figures 9 and 10, we note that the Li cross sections are more than an order of magnitude larger than those for atomic hydrogen, also, unlike atomic hydrogen, the total positronium formation cross section is never dominant¶.

A complete set of calculations now exists for all of the alkali metals Li, Na, K, Rb and Cs[75,78,84,85,87,88,103,104,115]. In figures 11 and 12 we show results for positronium formation and total scattering calculated in the approximations[75,78,84,85,87,88] :

(i) Ps(1s, 2s, 2p) + Li(2s, 2p, 3s, 3p, 3d)
(ii) Ps(1s, 2s, 2p) + Na(3s, 3p, 3d, 4s, 4p)
(iii) Ps(1s, 2s, 2p, 3s, 3p, 3d) + K(4s, 4p, 5s, 5p, 3d)
(iv) Ps(1s, 2s, 2p, 3s, 3p, 3d) + Rb(5s, 5p, 6s, 6p, 4d)
(v) Ps(1s, 2s, 2p, 3s, 3p, 3d) + Cs(6s, 6p, 7s, 7p, 5d)

Figure 11(a) shows a dramatic collapse in the ground state Ps(1s) formation cross section on ascending the series from Li to Cs, with a corresponding inflation in the cross section for positronium formation in excited states, figure 11(b). The overall effect is to give a total positronium formation cross section which is peaked at an energy of about 1eV for Li and Na, but which displays a broad maximum near 6eV for K, Rb and Cs, this is illustrated for Li and Cs in figure 11(c). Figure 12 also indicates a similar division between the total cross sections for Li and Na, on the one hand, and those for K, Rb and Cs on the other hand. The Li and Na total cross sections are peaked near 1eV while the K, Rb and Cs cross sections, again, peak near 6eV. The broad maximum in these last three cases derives primarily, although not entirely[75,85,87], from the peak in the excited state positronium formation cross sections of figure 11(b).

¶ Except, of course, at very low energies where the Ps(1s) formation cross section becomes infinite as $1/k_0$, see above.

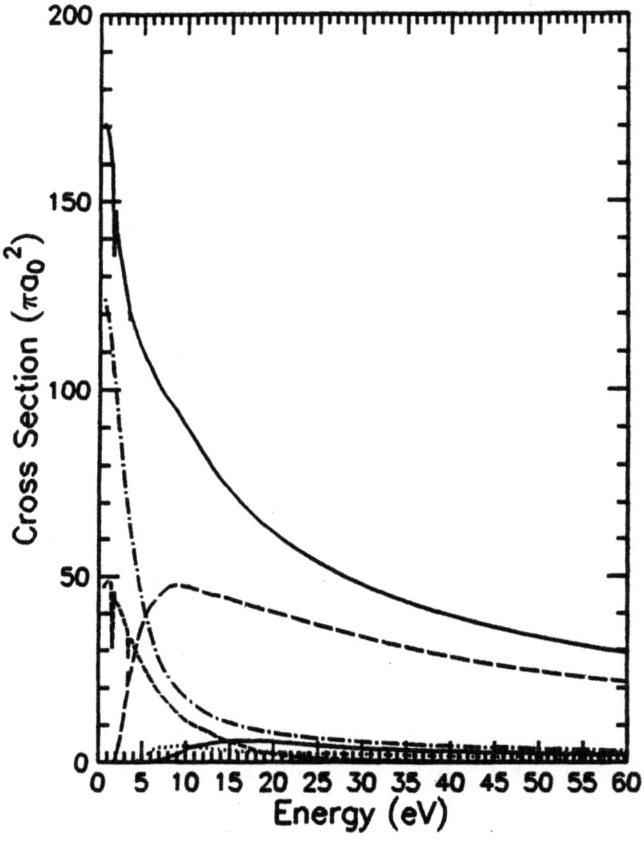

Figure 10. Cross sections for positron scattering by Li in a Ps(1s, 2s, 2p) + Li(2s, 3s, $\overline{4s}$ to $\overline{9s}$, 2p, 3p, $\overline{4p}$ to $\overline{9p}$, 3d $\overline{4d}$ to $\overline{9d}$, $\overline{4f}$ to $\overline{9f}$) approximation[88] : solid curve, total cross section; dash - dot curve, elastic scattering; long - dashed curve, Li(2s) → Li(2p); short - dashed curve, total positronium formation; dotted curve, Li(2s) → Li(n=3); (lower) solid curve, ionization.

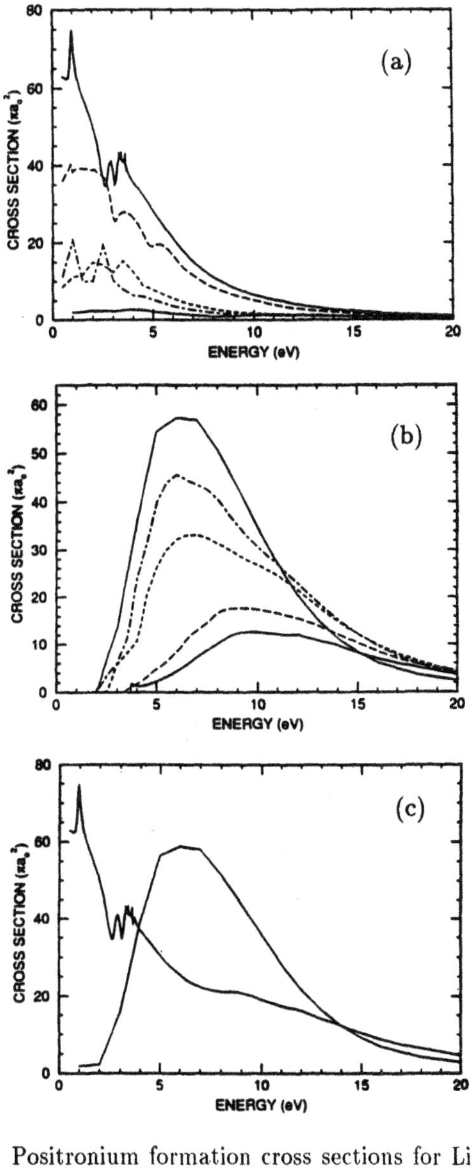

Figure 11. Positronium formation cross sections for Li, Na, K, Rb and Cs : (a) Ps(1s) formation; (b) positronium formation in excited states; (c) total positronium formation. Curves : solid, Li; long - dashed, Na; short - dashed, K; dash - dot, Rb; dotted, Cs.

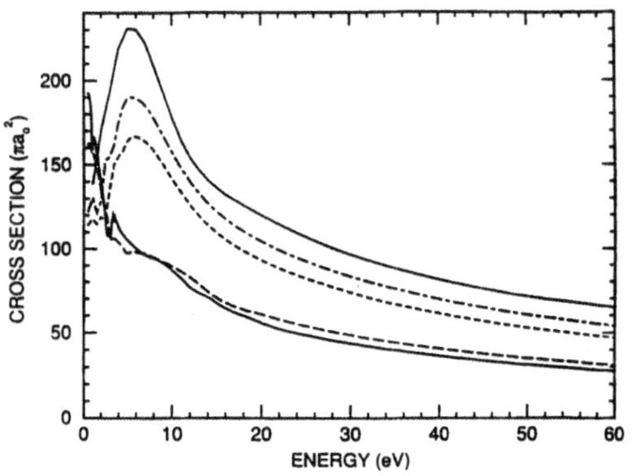

Figure 12. Total cross sections for positron scattering by Li, Na, K, Rb and Cs. Curves : solid, Li; long - dashed, Na; short - dashed, K, dash - dot, Rb; dotted, Cs.

Measurements of total positronium formation and total scattering have been made for Na, K and Rb targets by the Detroit group[144,146–151]. Generally the agreement between theory and experiment is good[75,78,84,85,87]. We illustrate the situation for K in figure 13. Here we see good agreement between the calculated cross section, in approximation (iii), and the lower bound measurements for positronium formation[150], figure 13(a), as well as with the total cross section[147]¶, figure 13(b). Figure 13 also shows how the total cross section is composed out of its components. There is, however, one serious exception to this pattern of agreement, it occurs for positronium formation in Na[75]. Figure 14 compares the calculated total positronium formation cross section in approximation (ii) for Na with the upper and lower bound measurements of Zhou et al[150]. Below 10eV the calculations are in clear disagreement with the experimental data, lying significantly below the experimental lower bound. Surprisingly, theoretical results from Hewitt et al[103] in the slightly simpler Ps(1s, 2s, 2p) + Na(3s,3p,4s,4p) approximation are in agreement with the measurements. We shall return to this case in figure 18.

Until very recently, and with the exception of a few low partial waves for e^+ - Li scattering[82], there were no substantial coupled - pseudostate calculations for positron scattering by the alkali metals. This situation has now been remedied for positron scattering by Li[88], Na[152] and K[153]. A sample of these results is shown in figures 15 to 19.

We start with the earlier partial wave results of Kernoghan et al[82] for e^+ - Li scattering at energies less than 2.75eV ($k_0 \leq 0.45$au). Figure 15 shows the S - wave and P - wave cross sections for elastic scattering and Ps(1s) formation respectively. This

¶ The experimental total cross section shown in figure 13(b) has been corrected for missing forward elastic scattering and normalised upwards by a factor of 1.10 in order to give the best visual fit to the theory (see reference [85]). The change in normalisation lies well within the quoted 21% normalisation error on the experimental data[147]

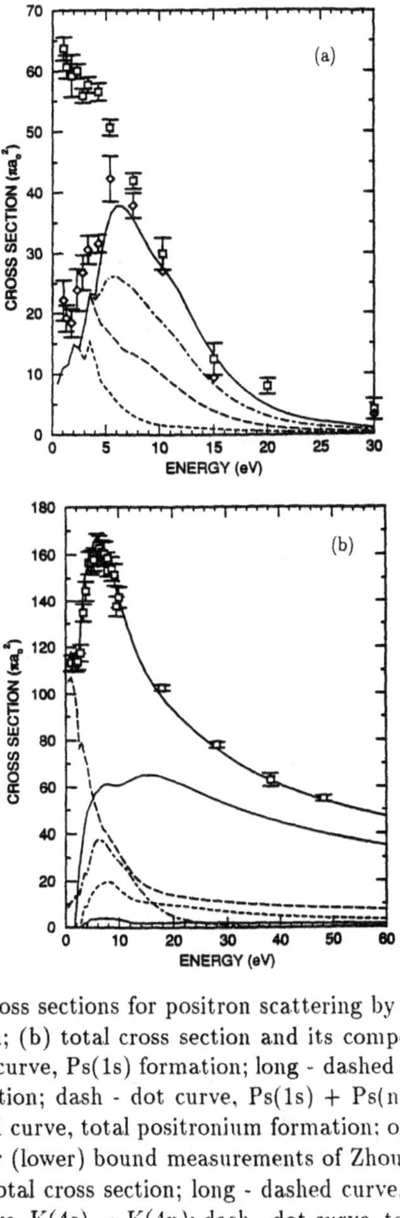

Figure 13. Cross sections for positron scattering by K : (a) positronium formation; (b) total cross section and its components. In (a) : short - dashed curve, Ps(1s) formation; long - dashed curve, Ps(1s) + Ps(n=2) formation; dash - dot curve, Ps(1s) + Ps(n=2) + Ps(n=3) formation; solid curve, total positronium formation; open squares (diamonds), upper (lower) bound measurements of Zhou et al[150]. In (b) : solid curve, total cross section; long - dashed curve, elastic scattering; dotted curve, K(4s) → K(4p); dash - dot curve, total positronium formation; short - dashed curve, K(4s) → K(3d); (lower) solid curve, K(4s) → K(n=5); open squares, experimental data of Parikh et al[147] corrected for forward elastic scattering and renormalised upwards by a factor of 1.10 as described in reference [85].

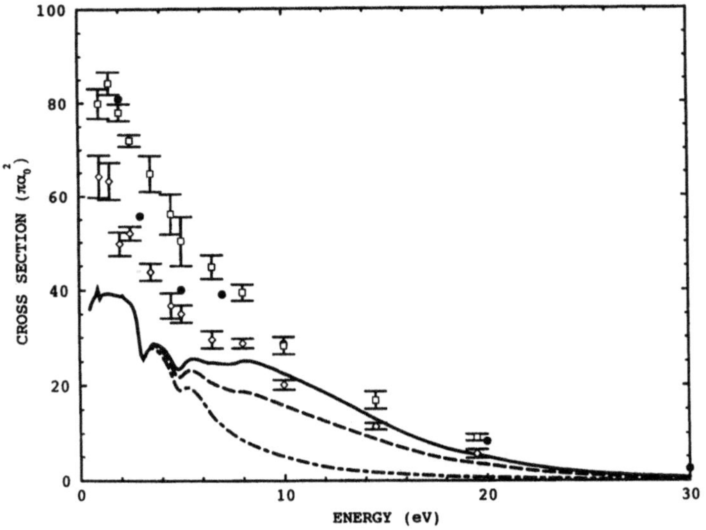

Figure 14. Positronium formation in Na. Curves are results in the eigenstate approximation (ii) : dash - dot curve, Ps(1s) formation; dashed curve, Ps(1s) + Ps(n=2) formation; solid curve, total positronium formation (using n^3 scaling for Ps(n $\geq$ 3)). Open squares (diamonds) are upper (lower) bound measurements of Zhou et al[150]. Solid circles are results of Hewitt et al[103] for total positronium formation in the Ps(1s, 2s, 2p) + Na(3s, 3p, 4s, 4p) approximation (including n^3 scaling for Ps(n $\geq$ 3)).

figure compares calculations in Ps(1s, 2s, 2p) + Li(2s, 2p) and Ps(1s, 2s, 2p) + Li(2s, 2p, 3s, 3p, 3d) eigenstate approximations with results in the corresponding Ps(1s, 2s, $\overline{3s}$, $\overline{4s}$, 2p, $\overline{3p}$, $\overline{4p}$, $\overline{3d}$, $\overline{4d}$) + Li(2s, 2p) and Ps(1s, 2s, $\overline{3s}$, $\overline{4s}$, 2p, $\overline{3p}$, $\overline{4p}$, $\overline{3d}$, $\overline{4d}$) + Li(2s, 2p, 3s, 3p, 3d) pseudostate approximations, where $\overline{3s}$, $\overline{4s}$, $\overline{3p}$, $\overline{4p}$, $\overline{3d}$ and $\overline{4d}$ pseudostates have been added to the positronium centre. For future reference we shall designate these four approximations as 3PS2ATOM¶, 3PS5ATOM, 9PS2ATOM and 9PS5ATOM respectively. Also shown in figure 15 are the accurate variational numbers of Watts and Humberston[47,50]. The convergence to the variational answers produced by the inclusion of the pseudostates is clear. The eigenstate approximations do not deviate greatly from the pseudostate and variational results for elastic scattering, figure 15(a), but give cross sections which are too large for Ps(1s) formation, figure 15(b).

The 9PS5ATOM pseudostate approximation of Kernoghan et al[82] has been continued up to higher impact energies by McAlinden et al[88] and evaluated for the full range of partial waves. In addition, McAlinden et al have made calculations in a Ps(1s, 2s, 2p) + Li(2s, 3s, $\overline{4s}$ to $\overline{9s}$, 2p,3p, $\overline{4p}$ to $\overline{9p}$, 3d, $\overline{4d}$ to $\overline{9d}$, $\overline{4f}$ to $\overline{9f}$) approximation (3PS29ATOM) which is an analogue of the highly successful 33 - state approximation of section 3 for atomic hydrogen. The 3PS29ATOM approximation employs pseudostates only on the atom centre and so is "complementary" to the 9PS5ATOM approximation which uses pseudostates only on the positronium centre. Like the 18 - state and 33 - state expansions of section 3, the 9PS5ATOM approximation shows some degree of

¶ Meaning 3 positronium states and two atom states etc.

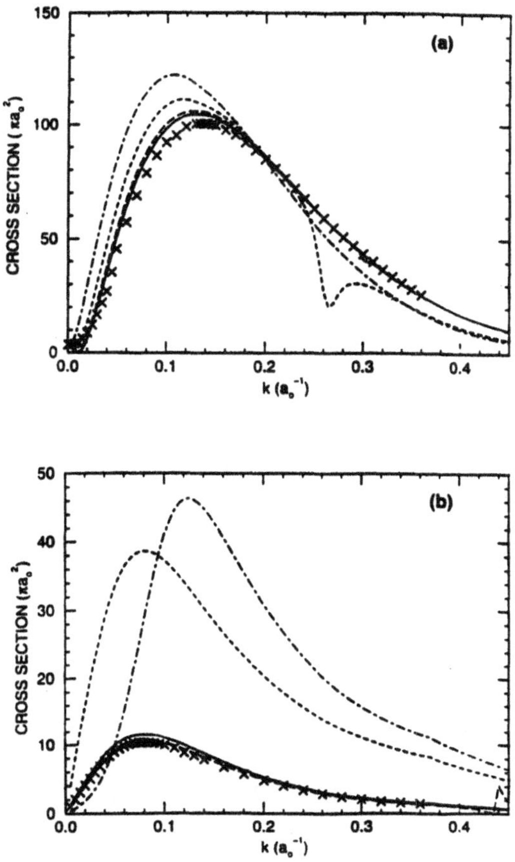

Figure 15. Partial wave cross sections for positron scattering by Li, plotted against the momentum k of the incident positron : (a) S - wave elastic scattering; (b) P - wave Ps(1s) formation. Approximations : dash - dot curve, 3PS2ATOM; short - dashed curve, 3PS5ATOM; long - dashed curve, 9PS2ATOM; solid curve, 9PS5ATOM. Crosses are variational results of Watts and Humberston[47,50].

pseudostructure while the larger 3PS29ATOM approximation gives results which are noticeably smoother. Similar to the case of atomic hydrogen, averaging through the pseudostructure in the 9PS5ATOM cross sections generally leads to very good agreement with the 3PS29ATOM numbers. This situation is illustrated in figure 16(a) for Li(2s) $\rightarrow$ Li(2p) excitation, the pattern is similar to that seen in figure 7(d) for atomic hydrogen. For elastic scattering and Ps(1s) formation the pseudostructure in the 9PS5ATOM cross sections is quite small and the 9PS5ATOM curve essentially sits on top of its 3PS29ATOM counterpart, this is shown for Ps(1s) formation in figure 16(b). Such agreement between two complementary pseudostate approximations creates considerable confidence in the computed results, so much so, that we believe that

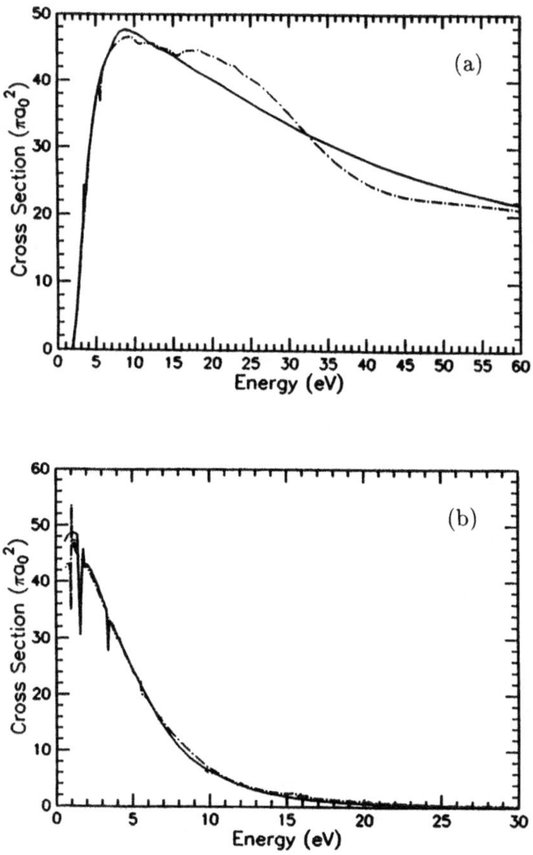

Figure 16. Positron scattering by Li[88] : (a) Li(2s) → Li(2p); (b) Ps(1s) formation. Solid curve, 3PS29ATOM approximation; dash - dot curve, 9PS5ATOM approximation.

the 3PS29ATOM approximation now gives the main cross sections for positron scattering off ground state Li to a high degree of accuracy, at least within the frozen core model of section 2. Figure 10 summarises the principal results of the 3PS29ATOM approximation. We note, as anticipated earlier, that ionization is only a small component of the total scattering. Taking the 3PS29ATOM approximation as the standard, the calculations of McAlinden et al also demonstrate that the 3PS2ATOM and 3PS5ATOM eigenstate approximations can give the elastic, Li(2s) → Li(2p) excitation, and total scattering cross sections correct to better than 20%. However, the eigenstate predictions for positronium formation are not quite so impressive. Figure 17 compares total positronium formation cross sections calculated in the 3PS2ATOM, 3PS5ATOM and 3PS29ATOM approximations. It is clear that the eigenstate approximations give a

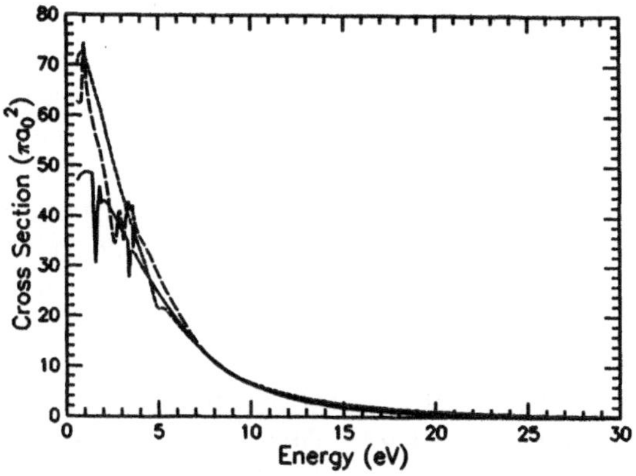

Figure 17. Total positronium formation cross section for positron scattering by Li. Approximations : solid curve, 3PS29ATOM; long - dashed curve, 3PS5ATOM; short - dashed curve, 3PS2ATOM.

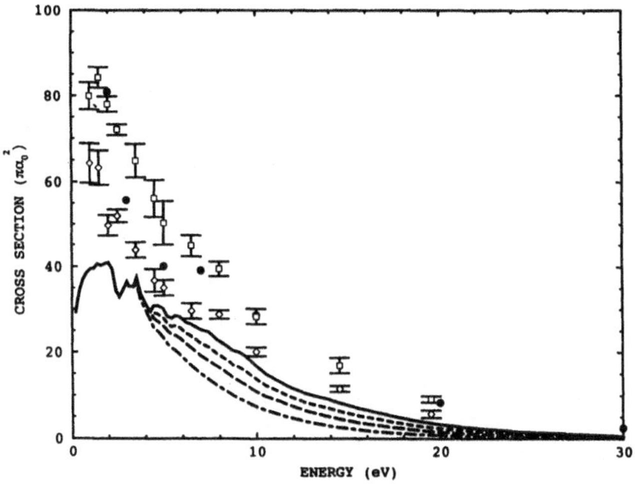

Figure 18. Positronium formation in Na. Curves are results in the 6PS27ATOM pseudostate approximation : dash - dot curve, Ps(1s) formation; long - dashed curve, Ps(1s) + Ps(n=2) formation; short - dashed curve, Ps(1s) + Ps(n=2) + Ps(n=3) formation; solid curve, total positronium formation (including n^3 scaling for Ps($n \geq 4$)). Open squares (diamonds) are upper (lower) bound measurements of Zhou et al[150]. Solid circles are results of Hewitt et al[103] for total positronium formation in the Ps(1s, 2s, 2p) + Na(3s, 3p, 4s, 4p) approximation (including n^3 scaling for Ps($n \geq 3$)).

generally larger result. That the eigenstate cross sections are too large at low energies, less than 2.75eV, where only Ps(1s) can be formed, was to be anticipated from the partial wave results of figure 15.

Analogous calculations in Ps(1s, 2s, 2p) + Na(3s, 3p) (3PS2ATOM) and Ps(1s, 2s, 2p) + Na(3s, 3p, 3d, 4s, 4p) (3PS5ATOM) eigenstate approximations and in a Ps(1s, 2s, 2p, 3s, 3p, 3d) + Na(3s, $\overline{4s}$ to $\overline{9s}$, 3p, $\overline{4p}$ to $\overline{9p}$, 3d, $\overline{4d}$ to $\overline{9d}$, $\overline{4f}$ to $\overline{9f}$) (6PS27ATOM) pseudostate approximation have been made by Campbell et al[152]. The results are similar in pattern to those just described for Li. One exception is the low energy Ps(1s) formation cross section which, unlike figure 17, is of comparable size in all three approximations. The most interesting outcome of these calculations is the 6PS27ATOM prediction for the total positronium formation cross section, this is shown in figure 18. Here we see that inclusion of pseudostates in the approximation has not resolved the discrepancy between theory and experiment which we first saw in figure 14. In view of the success of the analogous 33 - state approximation for atomic hydrogen (section 3) and of the analogous 3PS29ATOM approximation for Li, this difference between theory and experiment must be treated as serious, despite the support given to the experiment by the simple Ps(1s, 2s, 2p) + Na(3s, 3p, 4s, 4p) approximation of Hewitt et al[103].

Finally, in figure 19 we show a pseudostate calculation of positron scattering by K. The approximation, Ps(1s, 2s, $\overline{3s}$, $\overline{4s}$, 2p, $\overline{3p}$, $\overline{4p}$, $\overline{3d}$, $\overline{4d}$) + K(4s, 4p, 5s, 5p, 3d), is the analogue of the 9PS5ATOM pseudostate approximation for Li discussed above. As described there, and as seen in figure 16(a), this approximation is not free from pseudostructure but its average should give a good estimate of the true cross sections.

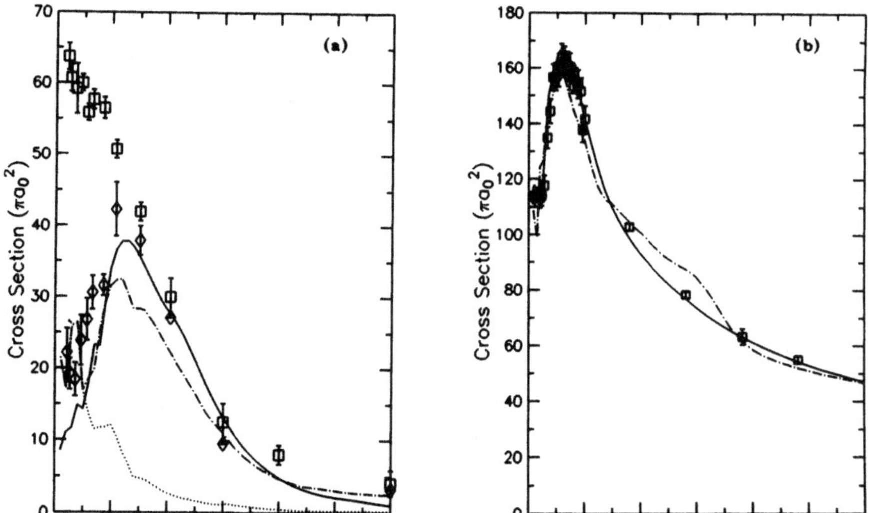

Figure 19. Positron scattering by K : (a) positronium formation; (b) total cross section. In (a) : solid curve, total positronium formation cross section in Ps(1s, 2s, 2p, 3s, 3p, 3d) + K(4s, 4p, 5s, 5p, 3d) eigenstate approximation; dash - dot curve and dotted curve, total positronium formation and Ps(1s) formation cross sections respectively in the Ps(1s, 2s, $\overline{3s}$, $\overline{4s}$, 2p, $\overline{3p}$, $\overline{4p}$, $\overline{3d}$, $\overline{4d}$) + K(4s, 4p, 5s, 5p, 3d) pseudostate approximation. In (b) : solid curve, eigenstate approximation; dash - dot curve, pseudostate approximation. Experimental data as in figure 13.

The interesting question is whether the pseudostate results will support the dramatic growth in excited state positronium formation which the eigenstate approximations (i) to (v) predicted as starting at K, figure 11. Figure 19 compares the pseudostate results for positronium formation and total scattering with the Ps(1s, 2s, 2p, 3s, 3p, 3p) + K(4s, 4p, 5s, 5p, 3d) eigenstate approximation and with the experimental data of Zhou et al[150] and Parikh et al[147] shown earlier in figure 13. From figure 19(a) we see that the pseudostate calculation¶ supports the eigenstate prediction of a large growth in excited state positronium formation, although the pseudostate numbers are not in quite such good accord with the experimental data of Zhou et al[15] for total positronium formation. Figure 19(b) shows the total cross section. Here we see the pseudostate curve entwined around the eigenstate cross section and, on average, in fairly good agreement with it. The two important results of the eigenstate approximation shown in figures 13(a) and 13(b) are therefore more or less confirmed by the pseudostate calculation.

5. CONCLUSIONS

We hope that in this short review we have been able to convince the reader of the power of coupled - state methods in treating positron - atom collisions. We believe that the 33 - state approximation described in section 3 now gives the main cross sections for positron scattering off ground state atomic hydrogen to a high degree of accuracy, the convergence between theory and experiment in this case is particularly convincing. Likewise, the 3PS29ATOM and 3PS27ATOM results for positron - Li and positron - Na scattering are thought to be highly reliable within the context of the frozen core model. The simpler eigenstate approximations for the alkalis seem to be roughly correct and, in particular, the eigenstate predictions of substantial growth in excited positronium formation at K appear to be confirmed by the tentative pseudostate calculation of reference [153]. It remains, however, to extend the large pseudostate calculations to the heavier alkalis K, Rb and Cs, this is in progress. Beyond this, the next step is obviously to deal with many - electron targets, attacking the very difficult problem of the positronium exchange kernels, see (23).

ACKNOWLEDGEMENTS

We are indebted to G. Laricchia for advice on experimental aspects of positron - atom collisions, to S. Zhou, H. Li, W.E. Kauppila, C.K. Kwan and T.S. Stein for allowing us to use their measurements in advance of publication, and to K. Higgins, P.G. Burke and Journal of Physics B : Atomic, Molecular and Optical Physics for permission to reproduce figure 2.

REFERENCES

1. S.E.A. Wakid and R.W. LaBahn, Positronium formation in positron - hydrogen collisions, *Phys. Rev. A* 6:2039 (1972).
2. S.E.A. Wakid, Resonances in low - energy positron - hydrogen collisions, *Phys. Lett. A* 54:103 (1975).

¶ Formula (6) of reference [83] has been used to calculate the total positronium formation cross section.

3. P. Mandal, A.S. Ghosh and N.C. Sil, e^+ - He scattering using an integral form of the close - coupling approximation, *J. Phys. B* 8:2377 (1974).
4. P. Mandal, D. Basu and A.S. Ghosh, Positron - helium scattering including polarization effects, *J. Phys. B* 9:2633 (1976).
5. D. Basu, G. Banerji and A.S. Ghosh, Positron - hydrogen scattering using an integral approach, *Phys. Rev. A* 13:1381 (1976).
6. S. Guha and A.S. Ghosh, Positron - lithium - atom collisions using the two - state approximation, *Phys. Rev. A* 23:743 (1981).
7. A.S. Ghosh, N.C. Sil and P. Mandal, Positron - atom and positron - molecule collisions, *Phys. Repts.* 87:313 (1982).
8. P. Khan, S. Dutta and A.S. Ghosh, Positron - lithium scattering using the eigenstate expansion method, *J. Phys. B* 20:2927 (1987).
9. K.P. Sarkar, M. Basu and A.S. Ghosh, Positron - sodium and positron - lithium scattering using CCA, *J. Phys. B* 21:1649 (1988).
10. K.P. Sarkar and A.S. Ghosh, Elastic differential cross sections for positron - lithium and positron - sodium scattering, *J. Phys. B* 22:105 (1989).
11. M. Basu, M. Mukherjee and A.S. Ghosh, Positron - hydrogen scattering including short - and long -range correlations, *J. Phys. B* 22:2195 (1989).
12. M. Mukherjee, M. Basu and A.S. Ghosh, Positron - hydrogen scattering at intermediate energies using the close - coupling approximation, *J. Phys. B* 23:757 (1990).
13. M. Basu, M. Mukherjee and A.S. Ghosh, Positron - hydrogen scattering below the first excitation threshold, *J. Phys. B* 23:2641 (1990).
14. M. Mukherjee and M. Basu, Positon - hydrogen scattering at intermediate energies using CCA, *Z. Phys. D* 21:19 (1991).
15. M. Basu, A.S. Ghosh, Positron - lithium scattering at low and intermediate energies, *Phys. Rev. A* 43:4746 (1991).
16. N.K. Sarkar, M. Mukherjee, M. Basu and A.S. Ghosh, Resonance in positron - hydrogen scattering at medium energy, *J. Phys. B* 26:L427 (1993).
17. N.K. Sarkar, M. Basu and A.S. Ghosh, Resonance just above the Ps - formation threshold in positron - hydrogen scattering, *J. Phys. B* 26:L799 (1993).
18. N.K. Sarkar, L.K. Jha and A.S. Ghosh, Positron - hydrogen scattering below Ps - formation threshold using CCA, *Ind. J. Phys. B* 67:189 (1993).
19. N.K. Sarkar and A.S. Ghosh, Positron - hydrogen scattering at low and medium energies using CCA, *J. Phys. B* 27:759 (1994).
20. M. Kahali, S.K. Adhikari, D. Basu and A.S. Ghosh, Positron - helium scattering in the close coupling approach, *Chem. Phys. Lett.* 239:344 (1995).
21. M.A. Abdel - Raouf, J.W. Darewych, R.P. McEachran and A.D. Stauffer, Application of the coupled - static approximation to positron - hydrogen inelastic scattering, *Phys. Lett. A* 100:353 (1984).
22. S.J. Ward, M. Horbatsch, R.P. McEachran and A.D. Stauffer, Theoretical studies of low - energy positron - alkali atom scattering, in : *Atomic Physics with Positrons*, J.W. Humberston and E.A.G. Armour, eds, Plenum, New York (1988).
23. S.J. Ward, M. Horbatsch, R.P. McEachran and A.D. Stauffer, Close - coupling approach to positron scattering from potassium, *J. Phys. B* 21:L611 (1988).
24. S.J. Ward, M. Horbatsch, R.P. McEachran and A.D. Stauffer, Close - coupling approach to positron scattering for lithium, sodium and potassium, *J. Phys. B* 22:1845 (1989).
25. S.J. Ward, M. Horbatsch, R.P. McEachran and A.D. Stauffer, Resonances in low - energy positron scattering from Li, Na and K, *J. Phys. B* 22:3763 (1989).
26. S.J. Ward, M.Horbatsch, R.P. McEachran and A.D. Stauffer, Close - coupling calculations of low - energy positron - alkali collisions, *Nucl. Instrum. Methods B* 42:472 (1989).

27. R.P. McEachran, M. Horbatsch and A.D. Stauffer, Positron scattering from rubidium, *J. Phys. B* 24:1107 (1991).
28. K.M. De Vries, K. Bartschat, R.P. McEachran and A.D. Stauffer, Polarization of light emitted after positron impact excitation of alkali atoms, *J. Phys. B* 25:L653 (1992).
29. K. Bartschat, R.P. McEachran and A.D. Stauffer, Positron scattering from alkalis and noble gases – suggestions for testing theoretical predictions, *Hyperfine Interactions* 73:99 (1992).
30. K. Bartschat, K.M. De Vries, R.P. McEachran and A.D. Stauffer, Differential cross sections for elastic and inelastic positron scattering from alkali atoms, *Hyperfine Interactions* 89:57 (1994).
31. C. Schwartz, Electron scattering from hydrogen, *Phys. Rev.* 124:1468 (1961).
32. R.L. Armstead, Electron - hydrogen scattering, *Phys. Rev.* 171:91 (1968).
33. A.K. Bhatia, A. Temkin, R.J. Drachman and H. Eisereke, Generalized Hylleraas calculation of positron - hydrogen scattering, *Phys. Rev. A* 3:1328 (1971).
34. J. Stein and R. Sternlicht, Inelastic S - wave positron - hydrogen scattering, *Phys. Rev. A* 6:2165 (1972).
35. J.W. Humberston and J.B. Wallace, The elastic scattering of positrons by atomic hydrogen, *J. Phys. B* 5:1138 (1972).
36. J.W. Humberston, The scattering of low energy positrons by helium, *J. Phys. B* 6:L305 (1973).
37. A.K. Bhatia, A. Temkin and H. Eisereke, Rigorous precision p - wave positron - hydrogen scattering calculations, *Phys. Rev. A* 9:219 (1974).
38. D. Register and R. Poe, Algebraic variational method – a quantitative assessment in e^+ - H scattering, *Phys. Lett. A* 51:431 (1975).
39. R.I. Campeanu and J.W. Humberston, The scattering of p - wave positrons by helium, *J. Phys. B* 8:L244 (1975).
40. R.I. Campeanu and J.W. Humberston, The scattering of s - wave positrons by helium, *J. Phys. B* 10:L153 (1977).
41. J.W. Humberston, Theoretical aspects of positron collisions in gases, *Adv. At. Mol. Phys.* 15:101 (1979).
42. J.W. Humberston, Positronium formation in s - wave positron - hydrogen scattering, *Can. J. Phys.* 60:591 (1982).
43. J.W. Humberston, Positronium formation in s - wave positron - hydrogen scattering, *J. Phys. B* 17:2353 (1984).
44. C.J. Brown and J.W. Humberston, Positronium formation in positron - hydrogen scattering, *J. Phys.B* 18:L401 (1985).
45. J.W. Humberston, Positronium – its formation and interaction with simple systems, *Adv. At. Mol. Phys.* 22:1 (1986).
46. U. Roy and P. Mandal, Elastic P - wave phaseshifts in positron - hydrogen - atom collisions using the Schwinger principle, *J. Phys. B* 23:L55 (1990).
47. M.S.T. Watts and J.W. Humberston, Low energy s - wave positron - lithium scattering, *J. Phys. B* 25:L491 (1992).
48. U. Roy and P. Mandal, Positronium formation using the multichannel Schwinger variational principle, *Phys. Rev. A* 48:233 (1993).
49. U. Roy and P. Mandal, Elastic D - and higher - partial - wave shifts in positron - hydrogen - atom collisions using Schwinger's principle, *Phys. Rev. A* 48:2952 (1993).
50. J.W. Humberston and M.S.T. Watts, Positron - lithium scattering with the inclusion of positronium formation, *Hyperfine Interactions* 89:47 (1994).
51. P. Van Reeth and J.W. Humberston, The use of inexact helium wave functions in positron - helium scattering, *J. Phys. B* 28:L23 (1995).
52. P. Van Reeth and J.W. Humberston, Positronium formation in low energy s - wave positron - helium scattering, *J. Phys. B* 28:L511 (1995).

53. J.W. Humberston and P. Van Reeth, Positronium formation in low - energy positron - helium scattering, *Can. J. Phys.* 74:335 (1996).
54. P. Van Reeth and J.W. Humberston, A partial wave analysis of positronium formation in positron - helium scattering, *J. Phys. B*, to be published.
55. P. Van Reeth, J.W. Humberston, K. Iwata, R.G. Greaves and C.M. Surko, Annihilation in low - energy positron - helium scattering, *J. Phys. B* 29:L465 (1996).
56. T. Falke, W. Raith and M. Weber, Obsevation of structure in differential transfer ionization of argon atoms by positrons, *Phys. Rev. Lett.* 75:3418 (1995).
57. M. Weber, T. Falke, W. Raith and U. Wesskamp, Differential ionization in e^+ - Ar scattering, *Can. J. Phys.* 47:468 (1996).
58. T, Falke, W.Raith and M. Weber, Structure in differential transfer ionization : a Thomas resonance ?, *Can. J. Phys.* 74:540 (1996).
59. M. Stein, Pseudo - potential approach to the relativistic treatment of alkali atoms, *J. Phys. B* 26:2087 (1993).
60. K. Higgins, P.G. Burke and H.R.J. Walters, Positron scattering by atomic hydrogen at intermediate energies, *J. Phys. B* 23:1345 (1990).
61. I. Bray and A.T. Stelbovics, Convergent close - coupling calculations of low - energy positron - atomic - hydrogen scattering, *Phys. Rev. A* 48:4787 (1993).
62. I. Bray and A.T. Stelbovics, Calculation of the total and total ionization cross sections for positron scattering on atomic hydrogen, *Phys. Rev. A* 49:R2224 (1994).
63. H.R.J. Walters, Excitation and ionization of atoms by electrons of medium to high energies, in : *Electronic and Atomic Collisions*, H.B. Gilbody, W.R. Newell, F.H. Read and A.C. Smith, eds, North - Holland, Amsterdam. (1988).
64. I. Bray and A.T. Stelbovics, Explicit demonstration of the convergence of the close - coupling method for a Coulomb three - body problem, *Phys. Rev. Lett.* 69:53 (1992).
65. I. Bray and A.T. Stelbovics, Convergent close - coupling calculations for electron - hydrogen scattering, *Phys. Rev. A* 46:6995 (1992).
66. I. Bray and A.T. Stelbovics, Calculation of the total ionization cross section and spin asymmetry in electron - hydrogen scattering from threshold to 500eV, *Phys. Rev. Lett.* 70:746 (1993).
67. P.G. Burke and T.G. Webb, Electron scattering by atomic hydrogen using a pseudo - state expansion III. Excitation of 2s and 2p states at intermediate energies, *J. Phys.B* 3:L131 (1970).
68. P.G. Burke and J.F.B. Mitchell, Electron scattering by atomic hydrogen using a pseudo - state expansion IV. The convergence of the S - state expansion at intermediate energies, *J. Phys. B* 6:320 (1973).
69. D.H. Oza and J. Callaway, Pseudostate expansions in a simplified model of electron - hydrogen scattering, *Phys. Rev. A* 27:2840 (1983).
70. D.H. Oza, Convergence of pseudostate expansions in electron - hydrogen scattering, *Phys. Rev. A* 30:1101 (1984).
71. J. Callaway, Scattering of electrons by atomic hydrogen at intermediate energies : Elastic scattering and n = 2 excitation from 12 to 54eV, *Phys. Rev. A* 32:775 (1985).
72. A.T. Stelbovics, L^2 discretisation of the Coulomb problem in an orthonormal Laguerre function basis, *J. Phys. B* 22:L159 (1989).
73. I.S. Gradshteyn and I.M. Ryzhik, *Table of Integrals, Series, and Products*, Academic Press, New York (1980).
74. I. Bray, D.A. Konovalov and I.E. McCarthy, Convergence of an L^2 approach in the coupled - channel optical - potential method for e - H scattering. *Phys. Rev. A* 43:1301 (1991).
75. A.A. Kernoghan, *Positron Scattering by Atomic Hydrogen and the Alkali Metals*, PhD Thesis, The Queen's University of Belfast, (1996).

76. M.T. McAlinden, *Atomic Collisions Involving Positrons*, PhD Thesis, The Queen's University of Belfast, (1992).
77. M.T. McAlinden and H.R.J. Walters, Positron scattering by the noble gases, *Hyperfine Interactions* 73:65 (1992).
78. M.T. McAlinden, A.A. Kernoghan and H.R.J. Walters, Cross - channel coupling in positron - atom scattering, *Hyperfine Interactions* 89:161 (1994).
79. M.T. McAlinden and H.R.J. Walters, Differential cross sections for elastic scattering and positronium formation for positron collisions with Ne, Ar, Kr and Xe, *Hyperfine Interactions* 89:407 (1994).
80. A.A. Kernoghan, M.T. McAlinden and H.R.J. Walters, On structures seen in the Ps(1s) + H(1s, $\overline{2p}$) approximation to positron scattering by atomic hydrogen, *J. Phys. B* 27:L211 (1994).
81. A.A. Kernoghan, M.T. McAlinden and H.R.J. Walters, Resonances above the ionization threshold in positron - hydrogen scattering, *J. Phys. B* 27:L543 (1994).
82. A.A. Kernoghan, M.T. McAlinden and H.R.J. Walters, Scattering of low energy positrons by lithium, J. Phys. B 27:L625 (1994).
83. A.A. Kernoghan, M.T. McAlinden and H.R.J. Walters, An 18 - state calculation of positron - hydrogen scattering, *J. Phys. B* 28:1079 (1995).
84. H.R.J. Walters, A.A. Kernoghan and M.T. McAlinden, Overview of the present theoretical status of positron - atom collisions, *AIP Conf. Proc.* 360:397 (1995).
85. M.T. McAlinden, A.A. Kernoghan and H.R.J. Walters, Positron scattering by potassium, *J. Phys.B* 29:555 (1996).
86. A.A. Kernoghan, D.J.R. Robinson, M.T. McAlinden and H.R.J. Walters, Positron scattering by atomic hydrogen, *J. Phys.B* 29:2089 (1996).
87. A.A. Kernoghan, M.T. McAlinden and H.R.J. Walters, Positron scattering by rubidium and caesium, *J. Phys.B* 29:3971 (1996).
88. M.T. McAlinden, A.A. Kernoghan and H.R.J. Walters, Positron scattering by lithium, *J. Phys. B*, to be published.
89. K. Higgins and P.G. Burke, Positron scattering by atomic hydrogen including positronium formation, *J. Phys. B* 24:L343 (1991).
90. K. Higgins and P.G. Burke, Positron scattering by atomic hydrogen including positronium formation, *J. Phys. B* 26:4269 (1993).
91. G. Liu and T.T. Gien, Variational close - coupling calculation for positron - hydrogen scattering at low energies, *Phys. Rev. A* 46:3918 (1992).
92. T.T. Gien and G. Liu, Variational close - coupling calculation for D - wave positron - hydrogen scattering, *Phys. Rev. A* 48:3386 (1993).
93. T.T. Gien, Structures at intermediate energies in positron - hydrogen scattering, *J. Phys. B* 27:L25 (1994).
94. T.T. Gien and G. Liu, On S - wave resonances at energies just above the Ps formation threshold, *J. Phys. B* 27:L179 (1994).
95. T.T. Gien, S - wave resonances below the n = 2 threshold of hydrogen in positron - hydrogen scattering, *J. Phys.B* 28:L313 (1995).
96. T.T. Gien, Six - state algebraic close - coupling calculation of positron - hydrogen scattering at low energies, *Phys. Rev. A* 52:1186 (1995).
97. T.T. Gien, Sequences of threshold resonances in e^+ - H scattering, *J. Phys. B* 29:2127 (1996).
98. T.T. Gien, Feshbach resonances in positron - hydrogen collisions, *Can. J. Phys.* 74:343 (1996).
99. R.N. Hewitt, C.J. Noble and B.H. Bransden, Positronium formation in positron - hydrogen scattering, *J. Phys. B* 23:4185 (1990).
100. R.N. Hewitt, C.J. Noble and B.H. Bransden, Intermediate energy resonances in positron - hydrogen scattering, *J. Phys. B* 24:L635 (1991).

101. R.N. Hewitt, C.J. Noble and B.H. Bransden, He(2^1S, 2^1P) excitation and positronium formation in positron - helium collisions at intermediate energies, *J. Phys. B* 25:557 (1992).
102. B.H. Bransden, C.J. Noble and R.N. Hewitt, On the reduction of momentum space scattering equations to Fredholm form, *J. Phys. B* 26:2487 (1993).
103. R.N. Hewitt, C.J. Noble and B.H. Bransden, Positron collisions with alkali atoms at low and intermediate energies, *J. Phys. B* 26:3661 (1993).
104. R.N. Hewitt, C.J. Noble and B.H. Bransden, Coupled - channel calculations of positron - atom scattering at intermediate energies, *Hyperfine Interactions* 89:195 (1994).
105. R.N. Hewitt, C.J. Noble and B.H. Bransden, The scattering of positrons and positronium by atoms in the close - coupling approximation, *Acta Phys. Pol. A* 88:139 (1995).
106. R.N. Hewitt, C.J. Noble and B.H. Bransden, Coupled - channel calculations of e^+ - Mg scattering, *Can. J. Phys.* 74:559 (1996).
107. J. Mitroy, Close coupling theory of positron - hydrogen scattering, *Aust. J. Phys.* 46:751 (1993).
108. J. Mitroy, Another s - wave resonance in positron - hydrogen scattering, *J. Phys. B* 26:L625 (1993).
109. J. Mitroy, Close coupling calculations of positron - hydrogen scattering at low energies, *J. Phys.B* 26:4861 (1993).
110. J. Mitroy and A.T. Stelbovics, Resonance structure in the positron - hydrogen system above the ionization threshold, *J. Phys. B* 27:L55 (1994).
111. J. Mitroy and H.V. von Geramb, The Ps - p system in the coupled static model, *J. Phys. B* 27:L427 (1994).
112. J. Mitroy and A.T. Stelbovics, Positronium and Antihydrogen formation in the e^+ - H and Ps - $\bar{p}$ systems, *Phys. Rev. Lett.* 72:3495 (1994).
113. J. Mitroy and A.T. Stelbovics, The positron - hydrogen system in a six state model, *J. Phys.B* 27:3257 (1994).
114. J. Mitroy, L. Berge and A.T. Stelbovics, Positron - hydrogen scattering at low energies, *Phys. Rev. Lett.* 73:2966 (1994).
115. J. Mitroy and K. Ratnavelu, Close coupling theory of positron scattering from alkali atoms, *Aust. J. Phys.* 47:721 (1994).
116. J. Mitroy and K. Ratnavelu, The positron - hydrogen system at low energies, *J. Phys. B* 28:287 (1995).
117. J. Mitroy, Large basis calculation of positron - hydrogen scattering at low energies, *Aust. J. Phys.* 48:645 (1995).
118. J. Mitroy, Positronium - proton scattering at low energies, *Aust. J. Phys.* 48:893 (1995).
119. K. Ratnavelu, J. Mitroy and A.T. Stelbovics, Positron - hydrogen and positronium - proton scattering at intermediate and high energies, *J. Phys. B* 29:2775 (1996).
120. J. Mitroy, An L^2 calculation of positron - hydrogen scattering at intermediate energies, *J. Phys. B* 29:L263 (1996).
121. J. Mitroy, K. Ratnavelu and A.T. Stelbovics, Unusual nature of positronium - proton elastic scattering, *Phys. Rev. A* 54:1720 (1996).
122. W.E. Kauppila and T.S. Stein, Experimental investigations of positron differential elastic scattering, *Hyperfine Interactions* 73:87 (1992).
123. L. Dou, W.E. Kauppila, C.K. Kwan and T.S. Stein, Observation of structure in intermediate - energy positron - argon differential elastic scattering, *Phys. Rev. Lett.* 68:2913 (1992).
124. L. Dou, W.E. Kauppila, C.K. Kwan, D. Przybyla, S.J. Smith and T.S. Stein, Evidence for resonances and absorption effects in positron - krypton differential - elastic - scattering measurements, *Phys. Rev. A* 46:R5327 (1992).

125. A. Igarashi and N. Toshima, Hyperspherical coupled - channel study of positronium formation, *Phys. Rev. A* 50:232 (1994).
126. Y. Zhou and C.D. Lin, On the non-existence of resonances above the ionization threshold in positron - hydrogen scattering, *J. Phys. B* 28:L519 (1995).
127. R.M. Finch, Á. Kövér, M. Charlton and G. Laricchia, Positron - argon elastic, positronium formation and ionizing collisions at 60^0, *J. Phys. B* 29:L667 (1996).
128. W. E. Kauppila, C.K. Kwan, D. Przybyla, S.J. Smith and T.S. Stein, Positron - inert - gas - atom elastic DCS measurements *Can. J. Phys.* 74:474 (1996).
129. B. Simon, Absence of positive eigenvalues in a class of multiparticle quantum systems, *Math. Ann.* 207:133 (1974).
130. B. Simon, Resonances and complex scaling : A rigorous overview, *Int. J. Quantum Chem.* 14:529 (1978).
131. D.S. Walton, B. Peart and K. Dolder, Structure observed during detailed measurements of detachment from H^- by electron impact, *J. Phys. B* 3:L148 (1970).
132. D.S. Walton, B. Peart and K. Dolder, A measurement of cross sections for detachment from H^- by a method employing inclined ion and electron beams, *J. Phys. B* 4:1343 (1971).
133. B. Peart and K. Dolder, Measurements which indicate the existence of a second short - lived state of H^{2-}, *J. Phys. B* 6:1497 (1973).
134. L.H. Andersen, D. Mathur, H.T. Schmidt and L. Vejby - Christensen, Electron impact detachment of D^- : Near - threshold behaviour and the nonexistence of D^{2-} resonances, *Phys. Rev. Lett.* 74:892 (1995).
135. F. Robicheaux, R.P. Wood and C.H. Greene, Simplest doubly charged negative ion : Nonexistence of H^{2-} resonances, *Phys. Rev. A* 49:1866 (1994).
136. M.H. Mittleman, Resonances in proton - hydrogen and positron - hydrogen scattering, *Phys. Rev.* 152:76 (1966).
137. Y.K. Ho, The method of complex coordinate rotation and its application to atomic collision processes, *Phys. Repts.* 99:1 (1983).
138. Y.K. Ho, Resonances in positron - atom scattering, *Hyperfine Interactions* 73:109 (1992).
139. M. Weber, A. Hofmann, W. Raith and W. Sperber, Results of the Bielefeld - Brookhaven e^+ - H experiment, *Hyperfine Interactions* 89:221 (1994).
140. W. Raith, A. Hofmann, K.G. Lynn and M. Weber, The e^+ - H experiment of the Bielefeld - Brookhaven collaboration, *Can. J. Phys.* 74:361 (1996).
141. S. Zhou, W.E. Kauppila, C.K. Kwan and T.S. Stein, Measurements of total cross sections for positrons and electrons colliding with atomic hydrogen, *Phys. Rev. Lett.* 72:1443 (1994).
142. G.O. Jones, M. Charlton, J. Slevin, G. Laricchia, Á. Kövér, M.R. Poulsen and S. Nic Chormaic, Positron impact ionization of atomic hydrogen, *J. Phys. B* 26:L483 (1993).
143. H.R.J. Walters, Positron scattering by atomic hydrogen at intermediate energies : 1s → 1s, 1s → 2s and 1s → 2p transitions, *J. Phys. B* 21:1893 (1988).
144. T.S. Stein, J. Jiang, W.E. Kauppila, C.K. Kwan, H. Li, A. Surdutovich and S. Zhou, Measurements of total and (or) positronium - formation cross sections for positrons scattered by alkali, magnesium, and hydrogen atoms, *Can. J. Phys.* 74:313 (1996).
145. S. Zhou, H. Li, W.E. Kauppila, C.K. Kwan and T.S. Stein. Measurements of total and positronium formation cross sections for positrons and electrons scattered by hydrogen atoms and molecules, to be published.
146. C.K. Kwan, W.E. Kauppila, R.A. Lukaszew, S.P. Parikh, T.S. Stein, Y.J. Wan and M.S. Dababneh, Total cross - section measurements for positrons and electrons scattered by sodium and potassium atoms, *Phys. Rev. A* 44:1620 (1991).
147. S.P. Parikh, W.E. Kauppila, C.K. Kwan, R.A. Lukaszew, D. Przybyla, T.S. Stein and S. Zhou, Towards measurements of total cross sections for positrons and electrons scattered by potassium and rubidium atoms, *Phys. Rev. A* 47:1535 (1993).

148. C.K. Kwan, W.E. Kauppila, S.P. Parikh, T.S. Stein and S. Zhou, Measurements of total and positronium - formation cross sections for the scattering of positrons by alkali atoms, *Hyperfine Interactions* 89:33 (1994).
149. W.E. Kauppila, C.K. Kwan, T.S. Stein and S. Zhou, Evidence for channel - coupling effects in positron scattering by sodium and potassium atoms, *J. Phys. B* 27:L551 (1994).
150. S. Zhou, S.P. Parikh, W.E. Kauppila, C.K. Kwan, D. Lin, A. Surdutovich and T.S. Stein, Measurements of positronium formation cross sections for positron scattering by K, Na and Ar atoms, *Phys. Rev. Lett.* 73:236 (1994).
151. A. Surdutovich, J.Jiang, W.E. Kauppila, C.K. Kwan, T.S. Stein and S. Zhou, Measurements of positronium - formation cross sections for positrons scattered by Rb atoms, *Phys. Rev. A* 53:2861 (1996).
152. C.P. Campbell, A.A. Kernoghan, M.T. McAlinden and H.R.J. Walters, Positron scattering by sodium, to be submitted for publication.
153. M.T. McAlinden, A.A. Kernoghan and H.R.J. Walters, in preparation.

SCATTERING OF POLARIZED ELECTRONS FROM ATOMS AND MOLECULES

G. F. Hanne

Physikalisches Institut
Universität Münster
Wilhelm-Klemm-Str. 10
D-48149 Münster

INTRODUCTION: HISTORICAL REMARKS AND BASIC PRINCIPLES

Let me start with some historical remarks before I will concentrate on recent experimental advances on scattering experiments involving polarized electrons. Collisions of electrons with atoms and molecules have been studied for about one hundred years. The general behaviour of the cross section for those collisions is plotted on Figure 1 where some of the pioneering work is also mentioned. Lenard (since 1893) studied total cross sections and found a strong decrease of the cross section with increasing energy. Franck and Hertz (1913) discovered discrete excitation levels, whereas Ramsauer and Townsend (1920) found that some targets exhibit a deep cross section minimum at very low energies, which must be explained by the wave nature of the electron beam. Schulz (1963) discovered sharp resonance structures which are temporary negative ion states.

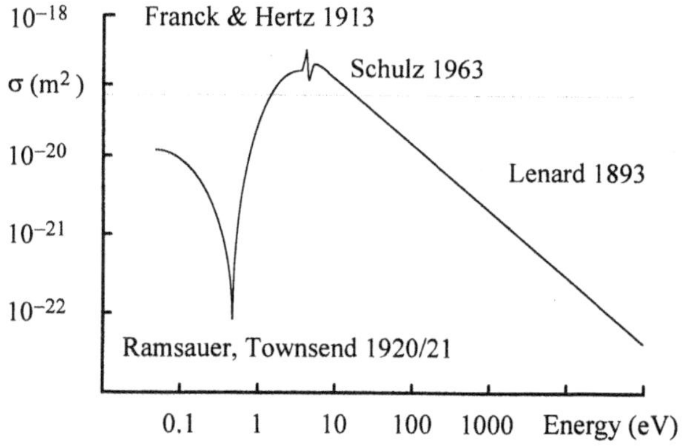

Figure 1. General behavior of electron-atom/molecule scattering cross sections

Photon and Electron Collisions with Atoms and Molecules
Edited by Burke and Joachain, Plenum Press, New York, 1997

With respect to the topic of my talk it is very interesting that the excitation of mercury in a Franck-Hertz experiment does strongly depend on spin effects,[1] but that was not known to Franck and Hertz or anyone else at that time.

The particular aspect that I want to present in this review talk came into play about 60 years ago when Mott - based on Diracs theory of the electron - predicted that fast electrons that are scattered twice from heavy nuclei should show a left-right asymmetry after the second scattering process. A number of experimental investigations failed to show this asymmetry, and it took almost 15 years until Shull, Chase and Myers[2] confirmed Motts prediction.

The physical origin of this asymmetry is the spin-orbit interaction. The experimental scheme of such an experiment is illustrated in Figure 2 (see Kessler's textbook[3]). In the first scattering process the scattered electrons become polarized. What does that mean? We call an ensemble of electrons polarized, if there is a preferential direction for which the number of electron with spin parallel to this direction ($N_\uparrow$) is different from that with spin antiparallel ($N_\downarrow$). In our case the preferential direction is normal to the first scattering plane and the polarization function S_P that that describes the polarization of the first scattering process is called Sherman function:

$$P = \frac{N_\uparrow - N_\downarrow}{N_\uparrow + N_\downarrow} = S_P(\vartheta_1) \tag{1}$$

With such a polarized beam the second scattering shows a left-right asymmetry

$$\frac{N_l - N_r}{N_l + N_r} = S_A(\vartheta_2)\cdot P\ . \tag{2}$$

For those who are not familiar with this effect I will explain it below.

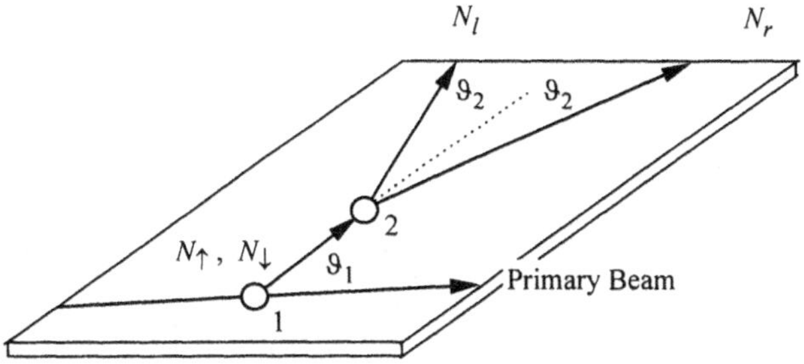

Figure 2. Scheme of a double scattering experiment (Mott scattering)

It is obvious that the left-right asymmetry observed in the second scattering process can be used to measure an unknown polarization, provided the asymmetry function is well known. This is what we call a Mott detector.

Nowadays we do not perform double-scattering experiments, but use photocathodes to produce polarized electrons with which we study various types of collisions. A scheme of a modern experiment is shown in Figure 3. With two detectors we measure for example the number of electrons scattered to the left and to the right through the same angle. The electrons are released from a GaAs photocathode and after an electrostatic deflection by 90° their polarization is measured with a Mott detector.

Let me state some experimental details. Modern sources of polarized electrons based on photoemission from stressed GaAs crystals that are irradiated with circularly polarized laser light may produce beam currents of several μamps with up to 80% polarization. The actual current at the targets depends, of course, on the required beam properties such as angular and energy spread. The currents of such beams are, however, comparable with those from conventional thermionic cathodes.

Only little progress, however, has been achieved for efficient polarization detectors. The somewhat bulky Mott detector - operating at about 100 keV - is still the most efficient detector where e. g. only 0.4% of the electrons can be detected with a asymmetry function of about 25%. We are still waiting for new ideas!

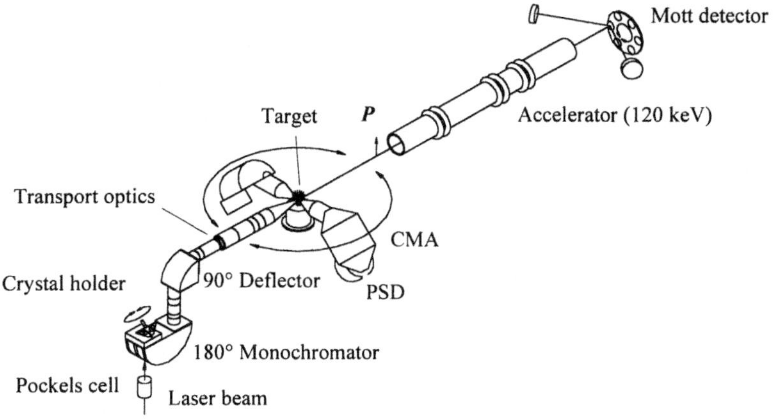

Figure 3. Scheme of a modern 'double' scattering experiment (Münster apparatus)

A typical result that we obtained recently[4] with this apparatus is shown in Figure 4. We measured the asymmetry function S_A for elastic collisions of slow polarized electrons from mercury, thallium and lead atoms as a function of the scattering angle. These three atoms have similar nuclear charges ranging from $Z = 80$ to $Z = 82$ but quite different configurations of the valence electrons.

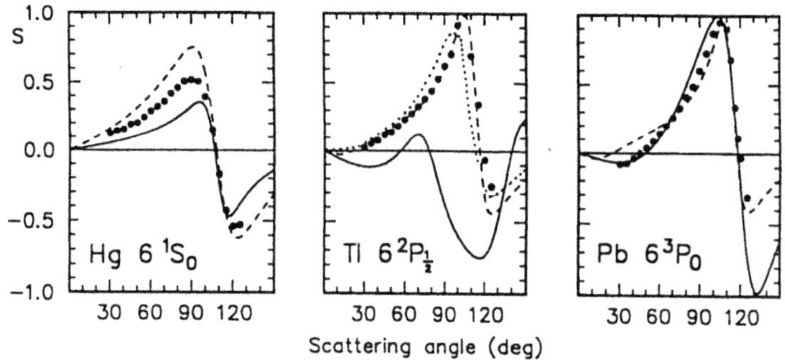

Figure 4. Asymmetry function S_A for elastic scattering of polarized electrons (E = 1.5 eV) from Hg, Tl and Pb atoms. • Experimental results,[4] Theory: —— Bartschat,[5] ----- Fritsche et al,[6] ····· Bartschat[7]

On these results there are three things that I want to point out. The first observation is that the asymmetry function shows large values up to unity. This is in strong contrast to Motts anticipation that such asymmetries - which are basically relativistic effects - should be strong only for fast relativistic electrons and not for such slow electrons.

The second point is that mercury is a closed-shell atom for which potential scattering approaches - including standard spin-orbit terms - may work well. But what is the influence of the open shells of thallium and lead on this type of experiment? We shall discuss this later.

And the third thing that I want to point out here is that the calculations[5-7] are in qualitative agreement with the experimental data, however, the quantitative agreement is less satisfactory. In particular, the theoretical procedure does not converge. That means: You can try to improve your calculation, and you may get quite different results, even worse than before! There are, however, systems where the situation is much better nowadays.

Before I show further examples I will make a few general remarks about the experimental methods and spin-dependent observables.

SPIN-DEPENDENT OBSERVABLES

Our goal is to disentangle the various scattering channels and interactions that govern the collisions. How can we probe collisions of electrons with atoms and molecules on the most fundamental level, i. e. as detailed as possible? The answer is illustrated in Figure 5: We must perform a state selection before and after scattering! Thus we may not only use a polarized incident electron beam, but also state-selected targets before the collision. The system forms a collision complex, and a scattered electron beam with a polarization different of that of the incident beam may be obtained. And also the target will be left in quite different states including excited or ionized states with some angular momentum orientation and alignment. Photon or electrons that are emitted from the target could be observed in coincidence with the scatterd electrons which allows one to probe the final channels in great detail. However, it is not feasible to date to measure the electron polarization after scattering in a coincidence experiment, because the efficiency of such detectors is too small to give reasonable high coincidence signals.

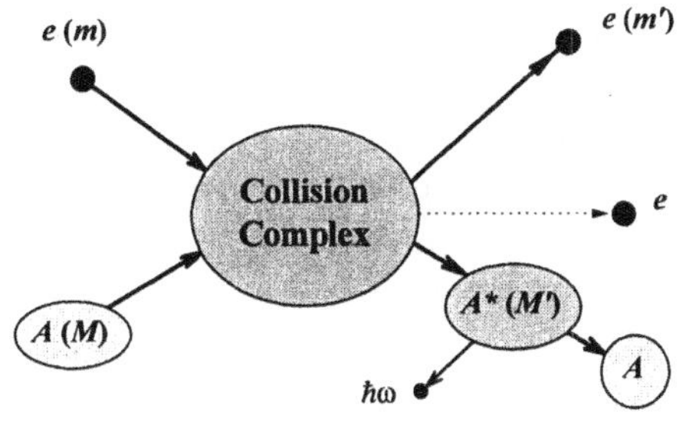

Figure 5. Scheme of state selected electron-atom/molecule collisions

How can we describe all this in formal terms? All observables can be expressed by complex scattering amplitudes

$$\langle J' M' \boldsymbol{k}' m' | \boldsymbol{T} | J M \boldsymbol{k} m \rangle, \tag{3}$$

that are characterized by the quantum numbers of the initial and final states involved. Here J and M mean total angular momentum and magnetic quantum number of the target before scattering, and $\boldsymbol{k}$ and m mean wave vector and magnetic spin quantum number of the incident electron, respectively. The primes denote the corresponding quantities after scattering. The description of a simple transition from a target state with angular momentum $J = 0$ to a final state with angular momentum $J = 1$ may require up to six different amplitudes, i. e. up to eleven independent parameters! It is obvious that a complete experimental determination of all these parameters is a painful work, and there are only few examples for which that goal has been achieved.

NATURE OF SPIN EFFECTS

The spin of the colliding electrons is involved in two ways: Exchange and the Pauli principle makes the cross sections dependent on the relative orientation of the spins of the colliding electrons whereas the spin-orbit interaction is an explicit spin-dependent force that results in a fine-structure interaction within the target and Mott scattering effects for the continuum electron.

We can distinguish three different situations:

(i) The spin-orbit interaction of the entire system (continuum electrons plus target electrons) is so small that it does not influence the collision process. In that case spin effects are only caused by exchange, and *LS* coupling holds for the entire system. If the collisions lead to transitions between certain target states that belong to a fine-structure multiplet the (small) spin-orbit splitting is accounted for by an angular-momentum coupling procedure.

(ii) We assume that the spin-orbit interaction of the continuum electron is still negligible. A weak violation of *LS* coupling in the target can be accounted for by a description in the intermediate-coupling scheme, where the different fine-structure states have approximately the same radial wave functions. Spin effects are still caused only by exchange, but *LS* coupling is violated for the entire system. An example of such a situation are the $2p^6$–$2p^5 3s$ transitions in Ne ($Z = 10$) where an intermediate-coupling scheme applies. In certain situations, e.g. at high energies and small scattering angles, exchange is negligible and, thus, the scattering is not spin dependent. In that case no effect from the violation of *LS* coupling is observed. But this, of course, does not mean that *LS* coupling holds in this case; the target states must always be described in the intermediate-coupling scheme.

(iii) The *LS* coupling in the target is strongly violated, so that in the description of the intermediate-coupling scheme different radial wave functions must be used. In that case the spin-orbit interaction of the continuum electrons is likely to be important as well. This could be an adequate description for the heavy targets xenon and mercury.

Case (i) results in certain relationships for spin-dependent observables that are only valid in *LS* coupling. Thus a violation of *LS* coupling results in a violation of such relationships which can be observed experimentally. However, it is, in general, not possible to distinguish case (ii) from case (iii).

There are a number of investigations in which polarized electrons have been used to probe collisions in some detail. I don't have the time to review all these activities, so I shall concentrate on only a few examples to highlight recent advances and the potential of these methods.

OBSERVATION OF EXCHANGE

Let me start with an example that shows the relative importance of exchange. The polarization P' of a scattered electron beam is different from the initial polarisation P if, for example, spin selected primary electrons are exchanged with target electrons of opposite spin direction. Of course, such processes can be observed only for paramagnetic targets with unsaturated spin in the initial and/or the final state.

We did such measurements and the experimental scheme is shown in Figure 6 with the source of polarized electrons, the scattering region and the Mott detector, where the initial and final polarizations can be determined. Polarized electrons are scattered from a target where an electron optical system with rotabable deflectors allows one to change the scattering angle. The polarizations P and P' are determined for different scattering angles.

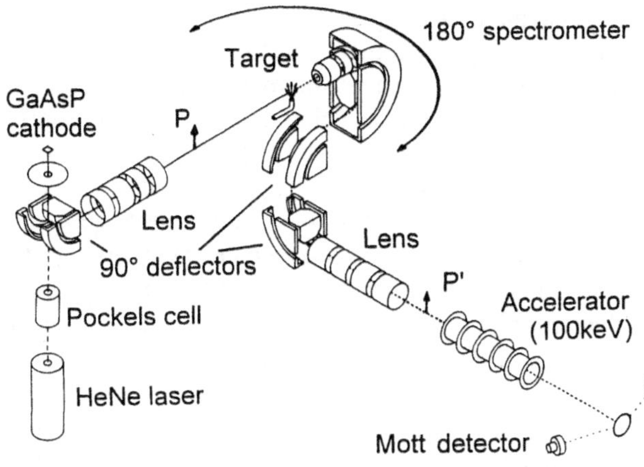

Figure 6. Scheme for direct observation of exchange processes

A ratio *P'/P* below unity is a direct observation of exchange processes, provided explicit spin-dependent forces can be neglected. This is true for light open-shell targets such as sodium atoms, nitric oxide or oxygen molecules.

Experimental results obtained by Hegemann et al[8] from Münster and theoretical data[9-11] for elastic collisions of polarized electrons with these targets are shown in Figure 7.

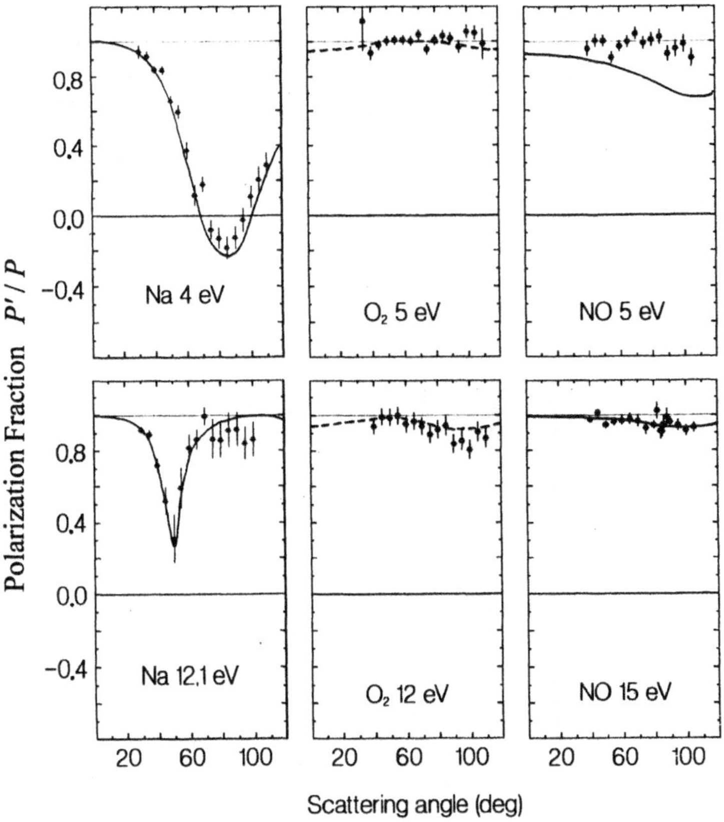

Figure 7. Polarization fraction *P'/P* for different open-shell targets. • Experiment[8]; theory: — Na (Bray and McCarthy[9]), — NO (da Paixão et al[10]), ---- O_2 (Fullerton et al[11])

Let me make two statements about these results.

The first statement is that for sodium atoms strong exchange effects are observed and predicted for large scattering angles whereas the overall effect for the molecular targets at comparable energies is quite small. The calculations of da Paixão et al[10] and Fullerton et al[11] - this is a cooperation of the Belfast school and a theory group in Münster - predict, however, that there exist certain orientations of the molecular axis and rotational states where much larger effects should be observed. In our experiment we could not prepare such situations, so future work is left to ambitious experimentalists. At the moment we concentrate on experiments where different paramagnetic atomic targets are compared with each other to get a clear cut picture of the relative importance of exchange collisions with atomic targets. It may well be, that alkalis are exceptional targets for such large exchange effects.

The second statement is that there is very good agreement between the theory of Bray and McCarthy[9] from Australia and our data for the sodium target. There are also a number of beautiful experiments of the NIST group at Gaithersburg[12] - which I cannot show here - that are perfectly well described by these calculations. Sodium is an example where there has been a breakthrough during the last five years or so with convergent close-coupling calculations that work also at intermediate energies where proper calculations were a difficult task in the past.

SPIN-ORBIT-EFFECTS

Now I shall demonstrate that such a perfect agreement is the exception rather than the rule if we look at more complex targets. For elastic collisions I have already shown an example at the beginning. The situation is even worse if we go to inelastic collisions.

In Figure 8 are shown results of the left-right asymmetry function S_A for excitation of various levels of xenon atoms. The experiment was carried out by Dümmler et al[13] at Münster. The energy of 40 eV is well above the excitation threshold, so distorted-wave approaches should apply.

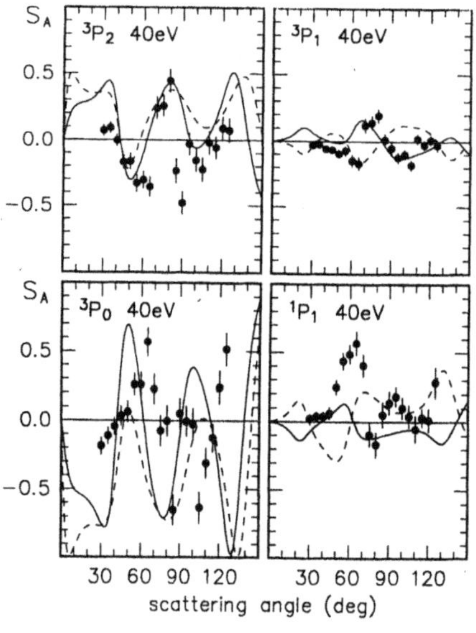

Figure 8. Asymmetry function S_A for various levels of xenon at 40 eV. • Experiment,[13] Theory: —— DWBA calculation by Bartschat and Madison,[14] ----- DWBA calculation by Zuo et al[15]

What we see is, that large asymmetries are observed and predicted; however, the agreement between experiment and calculations is not satisfactory. It is a little bit surprising that the two distorted-wave calculations by Bartschat and Madison[14] from Des Moines and by Zuo et al[15] from Toronto differ so much from each other, although these approaches are quite similar: Both calculations include exchange as well as the spin-orbit orbit interaction of the continuum electron and the fine-structure coupling within the target. The relativistic distorted-wave calculation of Zuo et al[15] makes also use of relativistic target orbitals.

This is a further example for a system where theories do not converge and are far away from being satisfactory.

At this point it may be necessary to discuss the physical mechanisms that produce such asymmetries as shown in Figure 8.

The way the spin-orbit interaction of the continuum electron works is well known (Figure 9a): The magnetic moment $\langle \mu_s \rangle$ of the electron moving in the field of the nucleus experiences a magnetic interaction $\langle \mu_s \cdot \boldsymbol{B}_\ell \rangle$ that depends on its relative orientation to the magnetic field vector $\boldsymbol{B}_\ell$ which is perpendicular to the scattering plane. This results e.g. in a spin polarization of scatterd electrons that are initially unpolarized or in up-down or left-right scattering asymmetries of electrons that are initially polarized perpendicular to the scattering plane.

A quite different effect is observed if the orbital angular momentum of an target with unsaturated spin is involved in the collision in the initial and/or the final state. The scattering cross section may be different for left- and right-oriented target orbitals, i. e. there is a preferred orientation $\langle \boldsymbol{L} \rangle$ of the orbital angular momentum vector of the target electrons perpendicular to the scattering plane during the scattering. If we select a certain fine-structure level, the spin of the target is coupled to that preferential orbital orientation, and thus the spin $\langle \boldsymbol{S} \rangle$ of the target is also oriented as indicated in Figure 9b. Now: If excitation by exchange shall work well we must offer continuum electrons of that preferred spin direction. With other words, there is also a spin-up down asymmetry or a left-right asymmetry if that mechanism is involved.

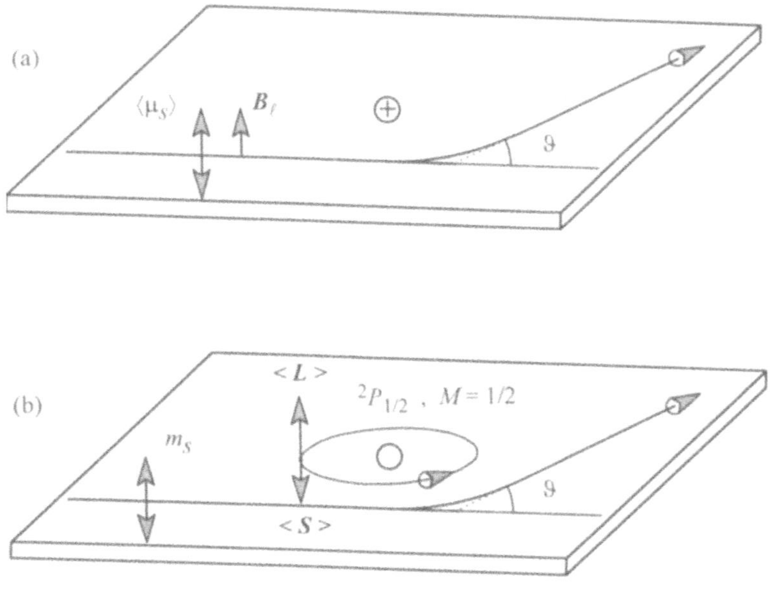

Figure 9. Simple picture of spin-orbit effects. (a) Spin-orbit interaction, (b) target orientation effect

The two mechanisms illustrated in Figure 9 have quite different origins: The spin-orbit interaction probes the field near the nucleus whereas the orientation effect is dominated by the interaction of the continuum electron with the valence electrons of the target. Nevertheless they act in the same manner. And if they could act simultaneously, e. g. in heavy targets, they will interfere and it is in general not possible to distinguish these two mechanisms that are certainly present in the results for Tl and Pb from Figure 4.

For inelastic collisions one can, however, compare different scattering channels. For example, the spin up-down asymmetry for the excitation from a $^1S_0$ state to a $^3P_{J'}$ state with $J' = 0$ should be of opposite sign and twice as large as that with $J' = 2$ if only the collisionally induced orientation effect is important:

$$\text{LS-coupling:} \qquad S_A(J'=0) = -S_A(J'=2) \tag{4}$$

A critical check of that relation will indicate how the two mechanisms contribute to the observed spin effects. I will not go into detail here, but as a general feature it can be stated that in inelastic scattering at low energies spin effects are in general dominated by the interaction with the valence electrons and not by the continuum spin-orbit interaction.

There are a variety of further investigations in which the interference between exchange and spin-orbit interaction has been studied for complex targets. I want to draw your attention on two of them. The goal of a series of experiments of Klose and Kessler[16] from Münster in which the change of the polarization vector $\boldsymbol{P} = (P_x, P_y, P_z) \rightarrow \boldsymbol{P}'$

$$\boldsymbol{P}' = \frac{1}{1+P_y S_A}\begin{bmatrix} T_x P_x + U_{xz} P_z \\ T_y P_y + S_P \\ T_z P_z - U_{zx} P_x \end{bmatrix} \tag{5}$$

has been measured is to get a more detailed picture about the relative importance of the two scattering mechanisms that I have mentioned just before.

Another very interesting experiment has been performed by Leuer *et al*[17] from Bielefeld. In that investigation polarized electrons (polarization P_e) are scattered from polarized cesium atoms (P_A), and three independent asymmetries can be calculated from four different cross sections with different relative orientations of the spins, $\sigma(\uparrow\uparrow)$, $\sigma(\downarrow\downarrow)$, $\sigma(\uparrow\downarrow)$ and $\sigma(\downarrow\uparrow)$:

$$\frac{\sigma(\uparrow\downarrow)+\sigma(\uparrow\uparrow)-\sigma(\downarrow\uparrow)-\sigma(\downarrow\downarrow)}{\sigma(\uparrow\downarrow)+\sigma(\downarrow\uparrow)+\sigma(\uparrow\uparrow)+\sigma(\downarrow\downarrow)} = S_A \cdot P_e \tag{6a}$$

$$\frac{\sigma(\uparrow\downarrow)+\sigma(\downarrow\uparrow)-\sigma(\uparrow\uparrow)-\sigma(\downarrow\downarrow)}{\sigma(\uparrow\downarrow)+\sigma(\downarrow\uparrow)+\sigma(\uparrow\uparrow)+\sigma(\downarrow\downarrow)} = A_{ex} \cdot P_e \cdot P_A \tag{6b}$$

$$\frac{\sigma(\uparrow\uparrow)+\sigma(\downarrow\uparrow)-\sigma(\uparrow\downarrow)-\sigma(\downarrow\downarrow)}{\sigma(\uparrow\downarrow)+\sigma(\downarrow\uparrow)+\sigma(\uparrow\uparrow)+\sigma(\downarrow\downarrow)} = A_{SO} \cdot P_A \tag{6c}$$

Here the first arrow refers to the direction of the electron polarization with respect to the scattering plane, whereas the second arrow is that for the atoms. The spin up-down asymmetry S_A is sensitive to the spin-orbit interaction alone whereas the asymmetry A_{ex} is caused by exchange. Note that S_A is the asymmetry function that also describes the left-right asymmetry (2). The third observable is particularly interesting: The interference between exchange and spin-orbit interaction produces an *atomic* spin up-down asymmetry A_{SO} that were absent otherwise.

The results of all these investigations are not shown here, but it can be stated that the theoretical description of such systems needs still to be improved.

SPIN-DEPENDENT TRIPLE DIFFERENTIAL CROSS SECTIONS

After some examples for elastic and inelastic collisions let us now discuss ionization processes. Ionization processes can be studied by socalled (e,2e) experiments, where scattered and ejected electron are detected in coincidence (Figure 10). This yields triple differential cross sections that may depend on the spin orientations of the colliding electrons.

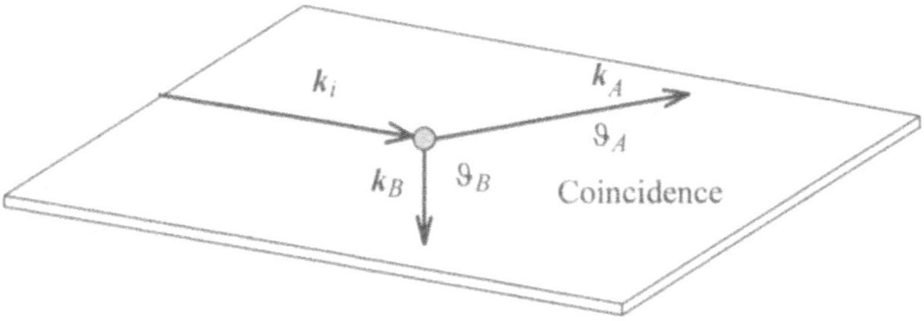

Figure 10. Scheme of an (e,2e) experiment

Two types of spin asymmetries can be evaluated from a measurement of the coincidence cross sections. If polarized electrons are scattered by polarized atoms a spin antiparallel-parallel asymmetry

$$\frac{\sigma(\uparrow\downarrow)-\sigma(\uparrow\uparrow)}{\sigma(\uparrow\downarrow)+\sigma(\uparrow\uparrow)} = A_{ex}P_eP_A \qquad (7)$$

of the coincidence signal will be found: The cross section for electron and target spin parallel will, in general, differ from that for antiparallel spins due to exchange effects or - in other words - due to the fact that the Pauli principle requires spatial wave functions of different symmetry for the two spin orientations. If spin-orbit effects play a significant role one may also obtain a spin up-down asymmetry

$$\frac{\sigma(\uparrow)-\sigma(\downarrow)}{\sigma(\uparrow)+\sigma(\downarrow)} = S_A \cdot P_e \qquad (8)$$

in the coincidence signal for scattering of polarized electrons from unpolarized atoms. In that case, for polarized atoms, the asymmetry A_{ex} may also depend on the orientation of the initial polarization vectors with respect to the scattering plane. In (8) the directions up and down refer to the normal on the scattering plane.

Results for an experiment of the first type (asymmetry A_{ex}) for lithium at 54.4 eV and a symmetric energy sharing (24.5 eV) have been obtained recently by the Bielefeld group[18]. Large spin antiparallel-parallel asymmetries for ionization of lithium atoms are observed and are predicted by distorted wave calculations.[19] A convolution of the theoretical results from Zhang *et al*[19] from Belfast that includes the experimental resolution shows fair agreement with the experimental data.

A quite different effect can be studied in (e,2e) collisions with heavy rare gas targets such as xenon. Such targets are, of course, initially unpolarized, but in the ionization process a hole in the *p* shell is created, and, if we resolve one of the fine-structure states, that hole may be orbital and spin oriented by the mechanism that I explained before. That means, that also the spin of the ejected electron can be oriented, because the spins compensated each other initially. Thus it makes a difference in the triple differential ionization cross section whether the spin of the incident electron is initially up or down, i. e.

parallel or antiparallel to the spin of the ejected electron! From an experimental point of view, (e,2e) studies with xenon and krypton targets are of particular interest, since their fine-structure splittings of 1.31 eV and 0.67 eV, respectively, can be resolved without major difficulties.

First experiments have been performed by a group in Australia[20] at 147 eV incident energy and by our group[21] at 40 eV and 200 eV. Results that have been obtained in our group with the apparatus shown in Figure 3 are presented in Figure 11. Spin effects are clearly seen, in agreement with predictions,[22,23] but there is a clear discrepancy between experiment and theory at 40 eV. At this energy and angle range ($E_1 = E_2$, $\vartheta_1 = 40°$) there is a significant momentum transfer to the ion, and this is obviously a situation that is not well described by the distorted wave approaches, whereas at 200 eV and the energies and angles studied there ($E_1 = 138$ eV, $\vartheta_1 = 30°$) the momentum transfer is small and the theory works well. Similar findings have been reported by the Australian group[24]. Further experimental studies are desirable. Relativistic effects (continuum spin-orbit interaction and/or relativistic orbitals) are obviously not essential for the observed asymmetries.

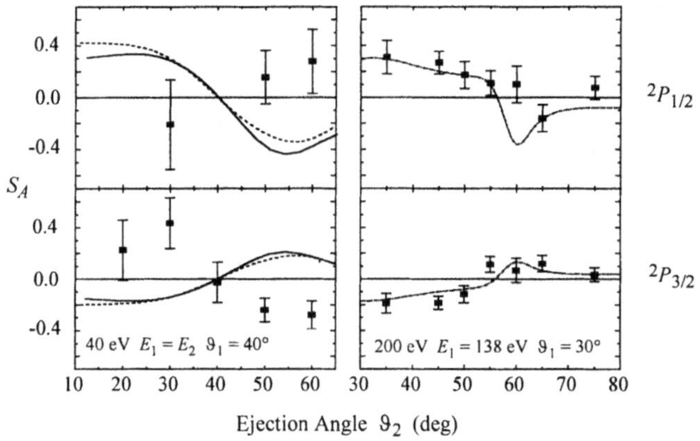

Figure 11. Spin asymmetry function S_A for electron-impact ionization of xenon (J = 1/2 or 3/2 core) at 40 and 200 eV. ■ Experiments[21]; Theory 40 eV: — relativistic DWBA calculation[22], ---- nonrelativistic DWBA calculation[23a], 200 eV: relativistic DWBA calculation (convoluted with the angular resolution)[23b].

A further experiment with Ag has been performed in Tübingen.[25] Spin up-down asymmetries for excitation of inner-shell electrons are clearly observed and predicted[26,27] for electrons that are emitted into large angles (recoil peak). This is caused by the spin-orbit interaction that these electrons experience in the field of the nuclei of Ag atoms.

It is interesting to note that the orbital orientation effect may contribute as well for ionization of electrons from subshells with $\ell > 0$. This has been predicted recently[28] and is the subject of a new experimental investigation.[29]

ELECTRON-PHOTON COINCIDENCES

A quite different method to probe spin-dependent collisions are electron-photon coincidences. In inelastic collisions, the magnetic sublevels of the target may show an anisotropic population. This will, in general, result in an anisotropic shape and in an orientation of the charge-cloud distribution of the excited target electrons. If we detect scattered electrons and emitted photons in coincidence, we will select excited atoms that have scattered the electrons into a particular direction.

Such an ensemble of excited atoms will emit polarized light, where the linear polarization components will give us a picture of the shape of the charge cloud, i. e. the alignment, whereas the circular polarization components will tell us something about the target orientation after scattering. We are particularly interested in how the incident electron's spin affects the resulting charge cloud distribution.

In Figure 12 is shown a scheme of our apparatus. We can map the excited target by means of a photon detector that looks perpendicular to the scattering plane and another detector - not shown here - that looks parallel to the scattering plane. The coincidence rates $\dot{N}(P_e, \alpha)$ thus obtained depend on the electron polarization P_e and the angle α of the light polarizer. From these coincidence rates a variety of observables (generalized Stokes parameters) can be determined.

I should mention that you can, of course, measure the polarization of the emitted light without detecting the scattered electrons. Such experiments with polarized electrons have been proposed about 25 years ago by Farago and Wykes,[30] and we use them frequently for calibration purposes.[31]

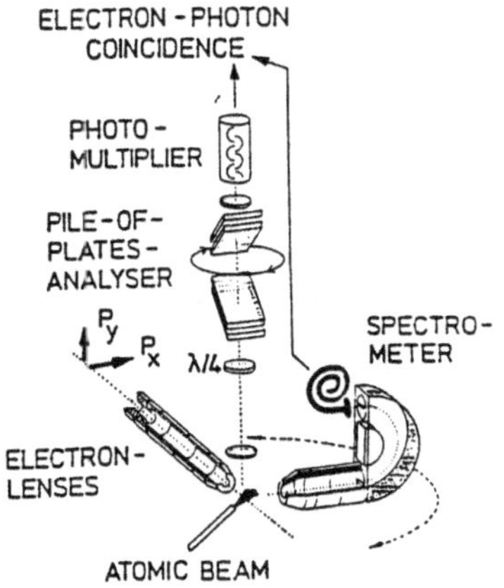

Figure 12. Scheme of an electron-photon coincidence experiment with polarized electrons. The source of polarized electrons is similar to that shown in Figure 3.

Here I want to review electron-photon coincidence studies. We did such experiments with mercury and xenon atoms and as an example I show some results of experimentally determined charge cloud distributions for mercury (Figure 13).[32] We studied the excitation of the $6\ ^3P_1$ state of mercury at 8 eV. The excitation energy of this famous Franck-Hertz transition is about 4.9 eV, thus the electrons are scattered with a final energy of about 3.1 eV. The scattering angle is 20°.

It is clearly seen that the shape of the resulting charge cloud of the excited target electron depends strongly on the spin orientation of the incident electron. Also the alignment angle γ through which the charge cloud is tilted from the incident electron beam direction is strongly dependent on the initial spin orientation. The size of the charge cloud distribution is a measure of the excitation cross section. So these distributions indicate also the magnitude of the cross sections for electrons with spin up and spin down. The average of the two charge cloud distributions gives the distribution for unpolarized electrons.

The agreement between experimental data and data from an *R*-matrix calculation by Bartschat[33] is quite good, as can be seen from another parameter of the same experiment. From measurements of the circular polarization we can extract the orientation of the target electrons, and this orientation is plotted in Figure 14 as a function of scattering angle for different spin orientations of the incident electrons.

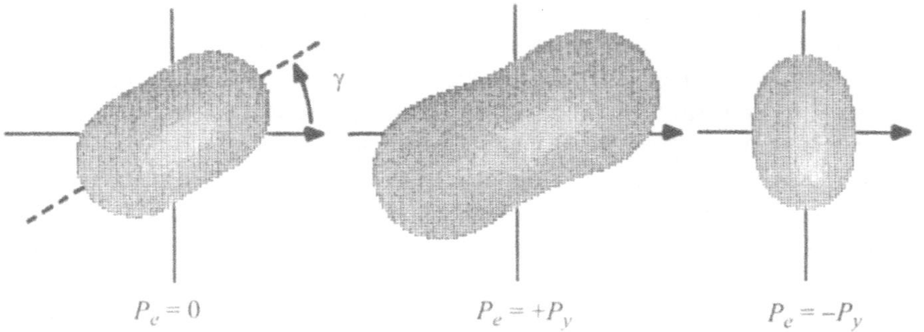

Figure 13. Charge cloud distributions for Hg atoms (after electron-impact excitation of the 6 $^3P_1$ state with polarized electrons at $E = 8$ eV and $\vartheta = 20°$). View perpendicular to the scattering plane

For s - p transitions there exists an orientation propensity rule that states that the orientation of a charge cloud is positive for scattering of unpolarized electrons at small angles. This is obviously not true for spin-resolved collisions where the electron spin is initially down.

A careful analysis of our data by Andersen *et al*[34] shows that in the case of the 6 $^3P_1$ excitation of mercury spin flips are very likely for spin-down electrons, but those spin-down electrons whose spin is not flipped tend to transfer a negative angular momentum to the atom! This is indicated in Figure 14. Unfortunately, only a few data points are available for the spin-resolved orientation parameter. Thus we plan to revive that investigation.

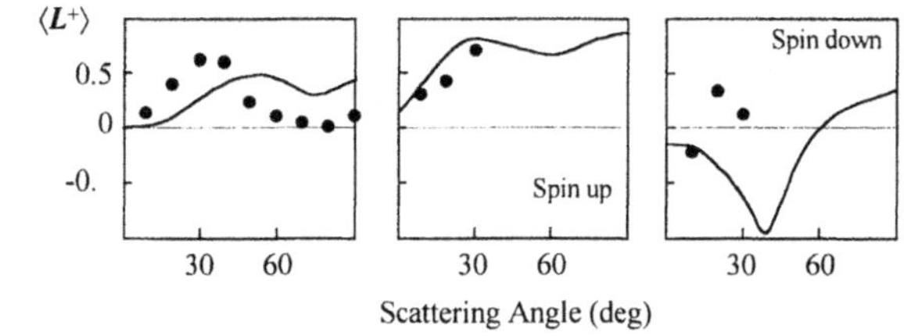

Figure 14. Spin-resolved orientation for excitation of Hg (6 $^3P_1$ state)[34]

First results for similar experiments for xenon have been obtained recently.[35]

SCATTERING OF POLARIZED ELECTRONS FROM CHIRAL MOLECULES

Let me finally present another interesting experiment from the Münster group. In that experiment the transmission of longitudinally polarized electrons through a target of unoriented chiral molecules has been measured. That transmission may depend on the helicity of the electrons, as has been suggested by Farago,[36] an effect that has been called electron optic dichroism. From the intensities of the transmitted currents for longitudinal polarizations $+P$ and $-P$ an asymmetry

$$A = \frac{I(P) - I(-P)}{I(P) + I(-P)} \tag{9}$$

can be determind.

Recent investigations by Tim Gay's group in Nebraska[37] and by the group at Münster[38] show that such helicity effects are well below a limit of 10^{-4} for unoriented camphor vapor in agreement with theoretical estimates of Blum and Thompson.[39]

If, however, chiral molecules containing high-Z atoms are studied, much larger effects could be expected.[39] This is because the chiral structure of the regular (non parity violating) electrostatic force in conjunction with the spin-orbit force can produce electron optic dichroism. Resonances, i. e. temporary negative ion states, are expected to enhance the polarization effect.

The experiment was performed with L- and D-handed $Yb(hfc)_3$ molecules that contain ytterbium ($Z = 70$) surrounded by three camphor-like ligands.

The transmission asymmetry (9) was determined by the transmitted intensities measured for electrons with positive and negative helicity.

A result of this experiment is shown in Figure 15. Two averaged energy scans ranging from 0.5 to 10 eV are plotted for the two enantiomers. The measured asymmetries differ significantly from zero and have opposite signs for L- and D-enantiomers, as required by symmetry. Pronounced maxima and minima of the asymmetry A can be seen at 1.7 and 2.6 eV where also sharp resonances are detected.

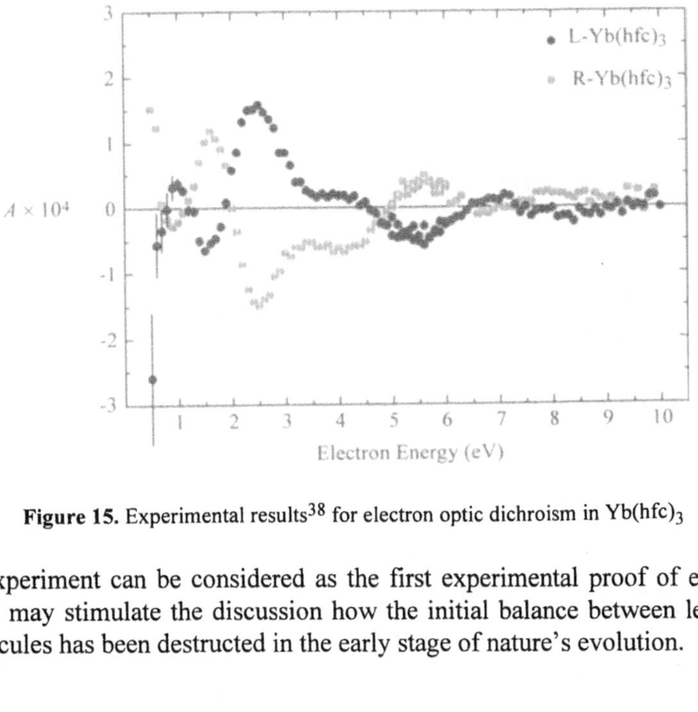

Figure 15. Experimental results[38] for electron optic dichroism in $Yb(hfc)_3$

This experiment can be considered as the first experimental proof of electron optic dichroism. It may stimulate the discussion how the initial balance between left and right-handed molecules has been destructed in the early stage of nature's evolution.

CONCLUDING REMARKS

The use of polarized electrons and atoms allows one to probe collisions of electrons with atoms and molecules in great detail, in particular by measurement of various spin asymmetries - the socalled generalized S-T-U parameters - (e,2e) and electron-photon coincidences. In particular, these methods can help

- to study separately reaction channels over which the average is usually performed, and, in processes not too complicated, to obtain the maximum possible information,
- to study the influence of the weak spin-orbit interaction, and to disentangle it from exchange and the Coulomb interaction by which it is usually masked.

Whereas for light one-electron atoms, such as sodium, convergent close-coupling calculations give excellent agreement with experimental data, the description of complex targets is, in general, still unsatisfactory.

There is still work left to be done in the future. Of course the experimental methods should be developed further, and a variety of complex targets should be studied with improved experimental and theoretical methods. Of particular interest are high-spin targets such as Eu ($^8S_{7/2}$)[40], Cr ($^7S_3$)[41] or Mn ($^6S_{5/2}$)[42] which are the subject of various investigations in Münster.

Very little experimental and theoretical work on spin effects in electron-molecule scattering is available. First investigations show that features obtained with atomic targets are not necessarily obtained with molecular targets. For example, the $C\ ^3\Pi_u$ - $B\ ^3\Pi_g$ transition in N_2 is not circularly polarized if excited by polarized electrons from the singlet ground state.[43] A nonzero circular polarization of this line should be anticipated, however, from the findings for similar singlet-triplet excitations in helium atoms.[31] More work on molecular targets seems highly desirable.

Acknowledgements. The experimental work performed in Münster was supported by the Deutsche Forschungsgemeinschaft in Sonderforschungsbereich 216 *Polarization and Correlation in Atomic Collision Complexes*.

REFERENCES

1. G. F. Hanne, *Am. J. Phys.* 56: 696 (1988)
2. C. G. Shull, C. T. Chase, and F. E. Myers, *Phys. Rev.* 63: 29 (1943)
3. J. Kessler, *Polarized Electrons,* 2nd edn, Springer, Berlin (1985)
4. M. Dümmler, M. Bartsch, H. Geesmann, G. F. Hanne, and J. Kessler, *J. Phys. B* 25: 4281 (1992)
5. K. Bartschat, *J. Phys. B* 23: 2341S (1990)
6. L. Fritsche, C. Kroner, and Th. Reinert, *J. Phys. B* 25, 4287 (1992)
7. K. Bartschat, *private communication,* (1993)
8. T. Hegemann, S. Schroll, and G. F. Hanne, *J. Phys. B* 26: 4607 (1993)
9. I. Bray and I. E. McCarthy, *Phys. Rev. A* 47: 317 (1993)
10. F. da Paixão, M. A. P. Lima, and V. McCoy, *Phys. Rev. Lett.* 68: 1698 (1992)
    F. da Paixão and M. A. P. Lima, *Phys. Rev. A* 53: 1400 (1996)
11. C. M. Fullerton, G. Woeste, D. G. Thompson, K. Blum, and C. J. Noble, *J. Phys. B* 27: 185 (1994)
12. S. R. Lorentz, R. E. Scholten, J. J. McClelland, M. H. Kelley, and R. J. Celotta, *Phys. Rev.* A 47: 3000 (1993) and further references therein.
13. M. Dümmler, G. F. Hanne, and J. Kessler, *J. Phys. B.* 28: 2985 (1995)
14. K. Bartschat and D. H. Madison, *J. Phys. B* 21: 2621 (1988)
15. T. Zuo, R. P. McEachran, A. D. Stauffer, *J. Phys. B* 25: 3393 (1992)
16. M. Klose, *Thesis, University of Münster* (1995)
17. B. Leuer, G. Baum, L. Grau, R. Niemeyer, W. Raith, and M. Tondera, *Z. Phys. D* 33: 39 (1995)
18. G. Baum, W. Blask, P. Freienstein, L. Frost, S. Hesse, W. Raith, P. Rappolt, and M. Streun, *Phys. Rev. Lett.* 69: 3037 (1992)
19. X. Zhang, T. Colm, T. Whelan, and H. R. J. Walters, *J. Phys. B* 25: L457 (1992)
20. X. Guo, J. M. Hurn, J. Lower, S. Mazevet, Y. Shen, E. Weigold, B. Granitza, and I. E. McCarthy, *Phys. Rev. Lett.* 76: 1228 (1996)
21. Th. Simon, Ch. Mette, and G. F. Hanne, ICPEAC XIX, Abstracts of Contributed Papers, 224 (1995)
22. Mazevet, I. E. McCarthy, and E. Weigold ICPEAC XIX, Abstracts of Contributed Papers, 220 (1995)

23a. S. Jones, D. H. Madison, and G. F. Hanne, *Phys. Rev. Lett.* 72: 2554 (1994)

23b. D. H. Madison, V. D. Kravtsov, and S. Jones, private communication (1995)

24. J. Lower, private communication (1995)
25. H.-Th. Prinz, K.-H. Besch, and W. Nakel, *Phys. Rev. Lett.* 74: 243 (1995)
26. R. Tenzer and N. Grün, *Physics Letters* A194: 300 (1994)
27. D. H. Jakubaßa-Amundsen, *J. Phys. B* 28: 259 (1995)

28. S. Keller, private communication (1995)
29. W. Nakel, private communication (1995)
30. P. S. Farago and J. Wykes, *J. Phys. B* 2: 747 (1969)
31. T. J. Gay, *J. Phys. B* 16: 553 (1983)
32. M. Sohn and G. F. Hanne, *J. Phys. B* 25: 4627 (1992)
33. K. Bartschat, private communication (1991)
34. N. Andersen, K. Bartschat, J. T. Broad, G. F. Hanne, and M. Uhrig, *Phys. Rev. Lett.* 76: 208 (1996)
35. M. Uhrig, G. F. Hanne, and J. Kessler, *J. Phys. B* 27: 4009 (1994)
36. P. S. Farago, *J. Phys. B* 14: L743 (1981)
37. K. W. Trantham, M. E. Johnston, and T. Gay, *J. Phys. B* 28: L543 (1995)
38. S. Mayer and J. Kessler, *Phys. Rev. Lett.* 74: 4803 (1995)
39. K. Blum and D. Thompson, *J. Phys. B* 22: 1823 (1989)
40. B. Santjer and H. Merz, *private communication*
41. G. F. Hanne, J. J. McClelland, R. E. Scholten, and R. J. Celotta, *J. Phys. B* 26: L753 (1993)
42. R. Meintrup, *Thesis*, *University of Münster*
43. C. Mette, S. Fischer, and G. F. Hanne, *unpublished* (1994)

CHIRAL EFFECTS IN ELECTRON SCATTERING BY MOLECULES

D.G.Thompson[1] and K.Blum[2]

[1]Department of Applied Mathematics and Theoretical Physics
Queen's University of Belfast
Belfast BT7 1NN
N.Ireland

[2]Institut für Theoretische Physik I
Universität Münster
Wilhelm-Klemm-Straβe 9
48149 Münster
Germany

1 Introduction

The study of chiral or "handed" effects holds a particular fascination, as well as being of fundamental importance. Nature abounds with chiral objects but unfortunately, from a computational point of view, they are quite complicated, as can be seen immediately from their definition. A chiral molecule is distinguishable from its mirror image, or equivalently, distinguishable from its image under spatial inversion; i.e., we have "left" handed and "right" handed distinct versions of the same molecule. This means it must have at least four non-coplanar atoms, and there must be no planes of reflection symmetry, or centres of inversion.

In some of our work we have introduced model molecules of the form AH_3; A is a heavy atom at the centre of a coordinate system and the H are along the three coordinate axes. The molecule is chiral if the three A - H bond lengths are different and we note that such a molecule has no symmetry (it belongs to the C_1 symmetry group). However chiral molecules can have C_n or D_n axes of rotation. As an example consider the molecule H_2S_2 for which we have also done calculations. If we set up the molecule with (x,y,z) coordinates for each of the nuclei as follows:

$$S_1 : (0,0,z_1),\ \ S_2 : (0,0,-z_1),\ \ H_1 : (x,y,z),\ \ H_2 : (x,-y,-z)$$

then we can see that a rotation of π about the x axis (x $\rightarrow$ x, y $\rightarrow$ -y, z $\rightarrow$ -z) leaves the molecule unchanged; the x axis is a C_2 axis of symmetry. However there are no reflection planes of symmetry or centres of inversion and the molecule is chiral.

The initial interest in electron scattering by chiral molecules arose from biological considerations. Many of the molecules in living organisms are chiral and purely from chemical considerations we would expect that for any given molecule there would be equal numbers of left and right handed forms present. This is not so - amino acids and sugars are present in only one form. The reason for this "homochirality" is still a matter of debate though the present concensus appears to be that it was just a matter of chance.

However several possible causal mechanisms have been put forward (c.f review by Keszthelyi[1]). One is that the interaction between longitudinally polarised electrons and a medium of left and right handed molecules is asymmetric; the suggestion is that if we pass longitudinally polarised electrons through a medium which initially contains equal numbers of left and right handed molecules there will be asymmetric destruction of the two molecules and eventually one form will dominate. A rough theoretical estimate by Hegstrom[2] indicated that the effect should be very small and indeed the experimental evidence for it is rather ambiguous.

There have been experiments involving positronium formation in the scattering of polarised positrons by chiral molecules[1]. We would expect different formation cross-sections for left and right handed molecules. Again the experimental evidence is not conclusive but the size of the effect seems to agree fairly well with a rough theoretical analysis[2].

There are other rather exotic effects. Bisi et al[3] allowed positrons to slow down in a chiral medium in the presence of a magnetic field. The polarisation of the positrons, just before positronium formation depends significantly on the handedness of the molecule. Garay et al[4] find that Cerenkov radiation emitted when longitudinally polarised electrons pass through a chiral medium has a similar dependence.

However in this paper we will concentrate on a simpler effect for which there is now definite experimental evidence. This is the electron scattering analogue of the light scattering effect known as circular dichroism. In "electron dichroism" the attenuation of a beam of longitudinally polarised electrons depends on whether the polarisation is parallel or anti-parallel to the beam direction and on whether the molecules are right or left handed. Let $I^{+(-)}_{L(R)}$ be the intensity of the attenuated beam, initial polarisation +P(-P), after passing through a medium of L[eft](R[ight]) handed molecules. One measure of the asymmetry is the attenuation fraction A:

$$
\begin{aligned}
A &= (I_L^+ - I_L^-) \,/\, (I_L^+ + I_L^- &\quad (1)\\
&= -(I_R^+ - I_R^-) \,/\, (I_R^+ + I_R^-) &\quad (2)\\
&= (I_L^+ - I_R^+) \,/ (I_L^+ + I_R^+) &\quad (3)
\end{aligned}
$$

We also find that an initially unpolarised beam passing through a chiral medium becomes **longitudinally** polarised.

In light scattering we also have an effect known as optical activity. For electron scattering we have an analagous effect, "electron activity", in which transversely polarised electrons are passed through a chiral medium; the polarisation vector rotates about the beam direction, the direction of rotation depending on the handedness of the molecules.

These effects were first discussed and analysed by Farago[5,6], invoking symmetry arguments. The analysis was extended by Blum and Thompson[7] and by Thompson and Kinnin[8]. It is instructive to look briefly at the symmetry arguments as they throw light on the physics of the problem.

Consider the elastic scattering of polarised electrons by closed shell molecules. Let us denote by m_0 and m_1 the components of the initial and final electron spins, and by

$f(m_1, m_0)$ the corresponding scattering amplitude. The scattering is described by the matrix M

$$M = \begin{bmatrix} f(1/2, 1/2) & f(1/2, -1/2) \\ f(-1/2, 1/2) & f(-1/2, -1/2) \end{bmatrix} \tag{4}$$

This 2x2 matrix can be expressed in terms of the Pauli spin matrices σ. For forward scattering Farago expressed M in the form

$$M = f\,1 + h\,\sigma.\hat{\mathbf{k}} \tag{5}$$

where 1 is the 2x2 unit matrix and $\mathbf{k}$ is the electron wave vector. Let us take the quantisation axis along $\mathbf{k}$ so that

$$M = f1 + h\sigma_3 = \begin{bmatrix} f+h & 0 \\ 0 & f-h \end{bmatrix} \tag{6}$$

We can see from equation 5 that M is invariant under time-reversal ($\sigma \to -\sigma, \mathbf{k} \to -\mathbf{k}$), but not under spatial inversion ($\sigma \to \sigma, \mathbf{k} \to -\mathbf{k}$). We also note that the structure of the scatterer has been ignored; indeed the above M is only appropriate for scattering by spherically symmetric targets with an interaction which is invariant under time-reversal but not spatial inversion. Farago obtained expressions for the chiral observables in terms of h; if $h = 0$ there is no chiral effect.

Blum and Thompson[7] corrected this in two ways, first by taking the structure of the molecule into account and second by considering the interaction to be invariant under spatial inversion as well as time reversal, since the parity violating terms of the Hamiltonian must be negligible (we will be considering electron energies of only a few eV). Blum and Thompson introduced a molecular frame $(\mathbf{e_1}, \mathbf{e_2}, \mathbf{e_3})$ and obtained

$$M = g_0 1 + \sum_{i=1}^{3} g_i\,(\mathbf{k}\mathbf{x}\mathbf{e_i}).\sigma + g_4\,(\mathbf{e_1}\mathbf{x}\mathbf{e_2}).\mathbf{e_3}\,(\hat{\mathbf{k}}.\sigma) \tag{7}$$

There are five g_i functions but only four of them are independent; they are proper scalars, invariant under rotations, time reversal and spatial inversion; we note in particular that $(\mathbf{e_1}\mathbf{x}\mathbf{e_2}).\mathbf{e_3}$ is a pseudo scalar, changing sign under spatial inversion.

The spin-flip amplitudes $f(m_1, m_0), m_1 \neq m_0$, coming from the middle term of equation 7, are no longer zero, and this has an effect on the analysis of the attenuation experiment for oriented molecules; however the spin-flip amplitudes, averaged over all molecular orientations, are zero. Thus for randomly oriented ensembles of chiral molecules we can effectively take M to be the same as equation 6 with h replaced by $g_4(\mathbf{e_1}\mathbf{x}\mathbf{e_2}.\mathbf{e_3})$.

Farago carried out two experiments in an attempt to confirm electron dichroism. In the first (Beerlaage et al[9]) an unpolarised beam was scattered by a gas of camphor molecules ($C_{10}H_{16}O$) at 25 eV. The search for an in-plane component of polarisation of the scattered beam was unsuccessful. At 25 eV the effect would certainly be less than the precision available to experimentalists at that time. In the second experiment Campbell and Farago[10,11] a longitudinally polarised beam was passed through camphor at 5 eV. An attenuation fraction (c.f equations 1-3) of the order of 10^{-3} was obtained.

In the meantime there had been a theoretical discussion of the effect (Rich et al[12]) which gave an order of magnitude estimate rather smaller than this experimental result. It was suggested that the asymmetry was caused by a spin-orbit perturbation of the molecular orbitals. For example, for a closed shell system there will be subshells containing two electrons, usually with the same spatial wave-function but opposite spins

($m_s = \pm 1/2$); the spin-orbit interaction will mix the spins, producing an asymmetry. Detailed calculations were not made. (For further details see section 5.)

Fandreyer et al[13] introduced a different model, including a spin-orbit interaction between the incident electron and the nuclei. Their results for a model molecule BiH_3 showed that the effect could be very significant if the molecule contained a nucleus of high Z and the scattering contained narrow resonance regions. The model has been extended very successfully to real chiral molecules (Smith et al[14,15] - again see section 5 for more details).

The suggestion has also been made that the effect is due to a spin-other-orbit interaction and a calculation within a resonance model has been carried out by Gallup[16].

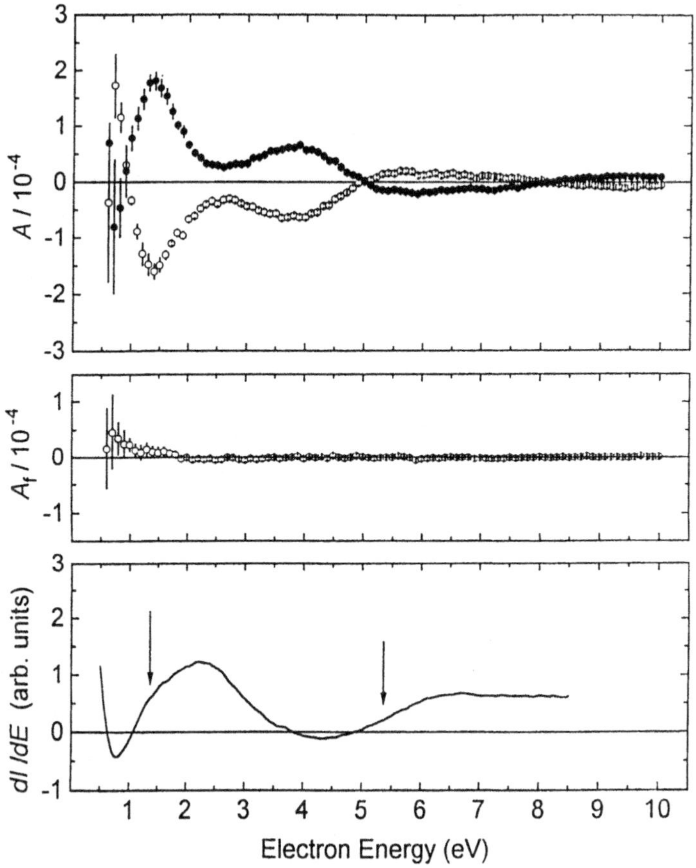

Figure 1. Results of Meyer et al[18] for bromocamphor. Upper diagram: attenuation asymmetry A, solid and open circles for the two forms of the molecule. Middle: instrumental asymmetry. Bottom: derivative dI/dE of the transmitted current. The resonance energies are indicated by arrows.

Very recently the Campbell-Farago experiment has been repeated. Although they were not able to see an effect for camphor Meyer and Kessler[17] and Meyer et al[18] have obtained positive results for molecules containing a heavy nucleus with order of magnitude 10^{-4}. (c.f also Trantham et al[19]). We show their results for bromocamphor, $C_{10}H_{14}OBr$, in figure 1. Note the variation with energy, with A taking both positive and negative values, implying that this type of experiment cannot identify the handedness of the target molecule.

We now proceed to discuss the attenuation experiment in theoretical detail. In section 2 we analyse the experiment in terms of forward scattering amplitudes. In section 3 we derive expressions for these amplitudes in terms of a scattering matrix, continuing the discussion of their relevance to the attenuation experiment in section 4. In section 5 we discuss the physical cause of the effect and present some results, concluding with a brief summary in section 6.

2 Analysis of the attenuation experiment

Recall that we are considering the elastic scattering of polarised electrons by closed shell molecules, described by the matrix M of equation 4. We will use the more compact form

$$M = \begin{bmatrix} f_{11} & f_{12} \\ f_{21} & f_{22} \end{bmatrix} \tag{8}$$

Usually the observables in a spin polarised experiment can be analysed by an application of density matrix theory to the matrix M. The position is complicated here becaused the scattering is in the forward direction and we need to consider interference between scattered and unscattered beams.

Let us consider a polarised electron beam incident in the z direction,wave number k on a slab of gas, width Δz, of infinite extent perpendicular to the beam direction, and containing n molecules per unit volume. We introduce the matrix

$$N = 1 + i\,\gamma\,\Delta z\,M \tag{9}$$

where $\gamma = 2\,\pi\,n\,/\,k$. N is the factor by which the incident electron state is changed by the slab of gas. (c.f. Thompson and Kinnin[8]) Let I^0 and $\mathbf{P^0}$ be the initial intensity and polarisation of the electron beam. The intensity I^f and polarisation $\mathbf{P}^f$ after passing through the slab can be obtained from

$$I^f = I^0\,tr[N(1+\mathbf{P}^0.\sigma)N^\dagger]/2 \tag{10}$$
$$I^f\,\mathbf{P}^f = I^0\,tr[N(1+\mathbf{P}^0.\sigma)N^\dagger\sigma]/2 \tag{11}$$

where the elements of σ are the Pauli spin matrices. (c.f Farago[6] and Kessler[20].) In general these expressions lead to four coupled first order linear differential equations for I, IP_x, IP_y, IP_z. The analysis is appropriate for dealing with scattering by oriented molecules, but in this paper we will deal only with ensembles of randomly oriented molecules. It has been shown from symmetry arguments (Blum and Thompson[7]), and we will show directly in section 4, that the spin-flip amplitudes averaged over all orientations are zero. We can thus effectively take N to be

$$\begin{bmatrix} 1+i\gamma f_{11}\Delta z & 0 \\ 0 & 1+i\gamma f_{22}\Delta z \end{bmatrix} \tag{12}$$

where f_{11} and f_{22} are now to be taken as averages over all molecular orientations. Instead of four coupled equations we obtain two sets of two coupled equations, the first for I and IP_z giving a description of electron dichroism, the second for IP_x and IP_y giving a description of electron activity.

It is now straightforward to show that $N(1+\mathbf{P}.\sigma)N^\dagger$ is

$$\begin{bmatrix} (1+P_z)(1-i\gamma f_{11}^*\Delta z)(1+i\gamma f_{11}\Delta z) & (P_x-iP_y)(1-i\gamma f_{22}^*)(1+i\gamma f_{11}\Delta z) \\ (P_x+iP_y)(1-i\gamma f_{11}^*\Delta z)(1+i\gamma f_{22}\Delta z) & (1-P_z)(1-i\gamma f_{22}^*\Delta z)(1+i\gamma f_{22}\Delta z) \end{bmatrix} \tag{13}$$

Neglecting terms second order in f we obtain from equations 10 and 11

$$I^f = I^0 - \gamma\,\Delta z\, I^0[Im(f_{11}+f_{22}) + P_z^0 Im(f_{11}-f_{22})] \tag{14}$$
$$I^f P_z^f = I^0 P_z^0 - \gamma\,\Delta z\, I^0[Im(f_{11}-f_{22}) + P_z^0 Im(f_{11}+f_{22})] \tag{15}$$

Taking the limit Δz tending to zero we obtain differential equations for I and IP_z.

$$dI/dz = -\gamma I\, Im(f_{11}+f_{22}) - \gamma(IP_z)\, Im(f_{11}-f_{22}) \tag{16}$$
$$d(IP_z)/dz = -\gamma I\, Im(f_{11}-f_{22}) - \gamma(IP_z)\, Im(f_{11}+f_{22}) \tag{17}$$

Adding we obtain

$$d(I+IP_z)/dz = -2\gamma(I+IP_z)Imf_{11}$$

and subtracting,

$$d(I-IP_z)/dz = -2\gamma(I-IP_z)Imf_{22}$$

which are easily integrated to give

$$I + IP_z = I^0(1+P_z^0)e^{-z_1 Imf_{11}} \tag{18}$$
$$I - IP_z = I^0(1-P_z^0)e^{-z_1 Imf_{22}} \tag{19}$$

where $z_1 = 2\gamma z = 4\pi nz/k$. Thus the final intensity I becomes

$$I = I^0[e^{-z_1 Imf_{11}} + e^{-z_1 Imf_{22}}] + I^0 P_z^0[e^{-z_1 Imf_{11}} - e^{-z_1 Imf_{22}}] \tag{20}$$

From equation 1 the attenuation fraction becomes

$$A = P_z^0\,[e^{-z_1 Imf_{11}} - e^{-z_1 Imf_{22}}]/[e^{-z_1 Imf_{11}} + e^{-z_1 Imf_{22}}] \tag{21}$$
$$= -P_z^0 tanh[z_1 Im(f_{11}-f_{22})/2] \tag{22}$$
$$\approx -P_z^0\, z_1\, Im(f_{11}-f_{22})/2 \tag{23}$$

provided $\mid f_{11} - f_{22} \mid \ll 1$, which is what we would expect.

There are several alternative, but equivalent, expressions in the literature. From the optical theorem we can express the total cross section, Q, as

$$Q = (4\pi/k)Imf$$

so that

$$2I = I^0(1+P_z^0)e^{-nzQ_{11}} + I^0(1-P_z^0)e^{-nzQ_{22}} \tag{24}$$

and

$$A = -P_z^0 tanh[nz(Q_{11}-Q_{22})/2] \tag{25}$$
$$\approx -P_z^0\, n\, z\, (Q_{11}-Q_{22})/2 \tag{26}$$

Defining $Q = (Q_{11}+Q_{22})/2$ we have

$$A \approx -P_z^0\, Q\, n\, z\, X \tag{27}$$

where $X = (Q_{11}-Q_{22})/(Q_{11}+Q_{22})$. Again using the Optical theorem we have

$$X = Im(f_{11}-f_{22})/Im(f_{11}+f_{22}) \tag{28}$$

If I(0) is the final intensity for an unpolarised beam we note that

$$A \approx P_z^0\, ln[I(0)/I_0]\, X \tag{29}$$

In experiments it is usual to have

$$| P_z^0 \, ln[I(0)/I_0] | \approx 1$$

so that $A \approx X$.

We have noted that there is possible second experiment in which the final longitudinal polarisation is measured for an initially unpolarised beam. We note from equations 18 and 19 that the induced polarisation can be obtained from

$$2IP_z = I^0[(1 + P_z^0)e^{-z_1 Imf_{11}} - (1 - P_z^0)e^{-z_1 f_{22}}] \tag{30}$$

If $P_z^0 = 0$ then

$$\begin{aligned} P_z &= [e^{-z_1 Imf_{11}} - e^{-z_1 Imf_{22}}]/[e^{-z_1 Imf_{11}} + e^{-z_1 Imf_{22}}] && (31) \\ &= tanh[z_1 Im(f_{11} - f_{22})/2] && (32) \end{aligned}$$

The two electron dichroism experiments are related by

$$A = -P_z^0 \; P_z \tag{33}$$

We can also obtain expressions for the electron activity effect. Using equation 11 we obtain two coupled equations for IP_x and IP_y.

$$\begin{aligned} d(IP_x)/dz &= -\gamma(IP_x)Im(f_{11} + f_{22}) + \gamma(IP_y)Re(f_{11} - f_{22}) && (34) \\ d(IPy)/dz &= -\gamma(IP_x)Re(f_{11} - f_{22}) - \gamma(IP_y)Im(f_{11} + f_{22}) && (35) \end{aligned}$$

from which we easily obtain

$$I(P_x + iP_y) = I^0(P_x^0 + iP_y^0)e^{-z_1 Im(f_{11}+f_{22})/2}e^{iRe(f_{11}-f_{22})/2} \tag{36}$$

Choosing initial transverse polarisation $P_x^0 = 0$ and $P_y^0 = 1$ we obtain the rotation angle θ from $tan\theta = P_x/P_y$ giving

$$\theta = z_1 Re(f_{11} - f_{22})/2 \tag{37}$$

3 The scattering amplitude

We will consider that the interaction between the electron and the molecule is given by a potential $V(\mathbf{r}, \sigma)$ which may be non-local and spin-dependent. The wave function $F(\mathbf{r}, \sigma)$ of the incident electron will satisfy an equation of the form

$$[-(1/2)\nabla^2 - (1/2)k^2 + V(\mathbf{r}, \sigma)]F(\mathbf{r}, \sigma) = 0 \tag{38}$$

Let us choose to expand $F(\mathbf{r}, \sigma)$ in an uncoupled basis:

$$F(\mathbf{r}, \sigma) = \sum_{lmm_s} f_{lmm_s}(r) r^{-1} Y_l^m(\hat{\mathbf{r}}) \, \chi_{m_s}(\sigma) \tag{39}$$

We can derive a set of coupled (integro) differential equations for the f_{lmm_s}:

$$(d^2/dr^2 + k^2 - l(l+1)/r^2) f_{lmm_s} = \sum_{\lambda\mu\mu_s} V_{lmm_s,\lambda\mu\mu_s} f_{\lambda\mu\mu_s}(r) \tag{40}$$

The form of the V matrix elements is discussed further in section 5.

The equations for f_α, ($\alpha \equiv lmm_s$) provide a set of linearly independent solutions which we will label with $\beta(\equiv \lambda\mu\mu_s)$. The S matrix is obtained in the usual way from the asymptotic form of the $f_{\alpha\beta}$:

$$e^{-i(kr-l\pi/2)}\delta_{\alpha\beta} - S_{\alpha\beta}e^{i(kr-l\pi/2)} \tag{41}$$

The general solution can be written

$$F_\alpha(r) = \sum_\beta a_\beta f_{\alpha\beta} \tag{42}$$

and the total solution

$$F(\mathbf{r},\sigma) = \sum_\alpha F_\alpha(r)r^{-1}Y_l^m(\hat{\mathbf{r}})\chi_{m_s}(\sigma) \tag{43}$$

The coefficients a_β are chosen so that the asymptotic form of the $F(\mathbf{r},\sigma)$ is

$$e^{i\mathbf{k}.\mathbf{r}}\chi_{M_i} + \sum_{M_f} f_{sc}(M_f, M_i)e^{ikr}r^{-1}\chi_{M_f} \tag{44}$$

where M_i and M_f specify the initial and final electron spin states repectively and $f_{sc}(M_f, M_i)$ is the scattering amplitude.

The quantisation axis for m and m_s is the z axis of a body-fixed coordinate system (xyz) while the quantisation axis for M_i and M_f is the Z axis of a space-fixed coordinate system (XYZ); (xyz) is obtained from (XYZ) by a rotation through Euler angles $\Omega \equiv (\alpha\beta\gamma)$. Using

$$\chi_M = \sum_m \chi_m\, D^{1/2\,*}_{Mm}(\Omega) \tag{45}$$

and

$$Y_l^m(\hat{\mathbf{r}}) = \sum_q Y_l^q(\hat{\mathbf{s}})\, D^l_{qm}(\Omega) \tag{46}$$

where the D are rotation matrices (c.f. Brink and Satchler[21]), $\hat{\mathbf{s}}$ is with respect to the space-fixed axes and $\hat{\mathbf{r}}$ with respect to the body fixed axes, the scattering amplitude for an electron incident along Z is

$$\begin{aligned} f_{sc}(M_f, M_i; \hat{\mathbf{s}}, \Omega) &= \sum_{\alpha\beta q} (1/ik)\,[\pi(2l+1)]^{1/2}\, i^{l-\lambda}\, Y_\lambda^q(\hat{\mathbf{s}}) \\ &\quad D^{l\,*}_{0m}\, D^\lambda_{q\mu}\, D^{1/2\,*}_{M_i m_s}\, D^{1/2}_{M_f\mu_s}\, T_{\beta\alpha} \end{aligned} \tag{47}$$

where $T = S - 1$.(c.f Greer and Thompson[22].)

We can note immediately that if the electron molecule interaction is spin independent then

$$\begin{aligned} T_{lm1/2,\lambda\mu1/2} &= T_{lm-1/2,\lambda\mu-1/2} \\ T_{lm1/2,\lambda\mu-1/2} &= 0 \\ T_{lm-1/2,\lambda\mu1/2} &= 0 \end{aligned}$$

from which it follows that

$$\begin{aligned} f_{sc}(1/2,1/2) &= f_{sc}(-1/2,-1/2) \\ f_{sc}(1/2,-1/2) &= 0 \\ f_{sc}(-1/2,1/2) &= 0 \end{aligned}$$

and there are no chiral effects.

4 Application to the attenuation experiment

We have seen that we need the forward scattering amplitudes averaged over all molecular orientations. Noting that

$$Y_l^q(\hat{\mathbf{0}}) = \delta_{q0}\,[(2l+1)/4\pi]^{1/2}$$

and using

$$\begin{aligned}\int D^{l\,*}_{om}\, D^{\lambda}_{0\mu}\, D^{1/2\,*}_{M_i m_s} D_{M_f \mu_s} d\Omega = \\ 8\pi^2(-1)^{m+\mu_s-M_f}\sum_S (2S+1)\begin{pmatrix} l & \lambda & S \\ 0 & 0 & 0\end{pmatrix}\begin{pmatrix} l & \lambda & S \\ -m & \mu & M_S\end{pmatrix} \\ \begin{pmatrix} 1/2 & 1/2 & S \\ -M_i & M_f & 0\end{pmatrix}\begin{pmatrix} 1/2 & 1/2 & S \\ -m_s & \lambda_s & -M_S\end{pmatrix}\end{aligned}$$

we can immediately infer that

$$\begin{aligned}\langle f_{sc}(M_f, M_i)\rangle &= (1/8\pi^2)\int f_{sc}(M_f, M_i)\, d\Omega \\ &= 0 \end{aligned} \tag{48}$$

if $M_f \neq M_i$; i.e. the spin-flip amplitudes averaged over all molecular orientations vanish.

The averaged non-flip amplitude becomes

$$\begin{aligned}\langle f(M,M)\rangle &= \sum_{\alpha\beta S}(1/2ik)[(2l+1)(2\lambda+1)]^{1/2}\, i^{l-\lambda}\,(-1)^{m+\mu_s-M}\, T_{\beta\alpha}\,(2S+1) \\ &\begin{pmatrix} l & \lambda & S \\ 0 & 0 & 0\end{pmatrix}\begin{pmatrix} l & \lambda & S \\ -m & \mu & M_S\end{pmatrix}\begin{pmatrix} 1/2 & 1/2 & S \\ -M & M & 0\end{pmatrix}\begin{pmatrix} 1/2 & 1/2 & S \\ -m_s & \lambda_s & -M_S\end{pmatrix}\end{aligned}$$

It is easy to see that the $S = 0$ contribution is the same for both values of M, so that the difference between the two non-flip amplitudes becomes

$$f_{11} - f_{22} = \sum_{\alpha\beta} P_{\alpha\beta}\, T_{\beta\alpha} \tag{49}$$

where

$$\begin{aligned}P_{\alpha\beta} &= (1/2ik)[(2l+1)(2\lambda+1)]^{1/2}\, i^{l-\lambda}\,(-1)^{-1/2+m+\mu_s}\, 6^{1/2}\begin{pmatrix} l & \lambda & 1 \\ 0 & 0 & 0\end{pmatrix} \\ &\begin{pmatrix} l & \lambda & 1 \\ -m & \mu & M_S\end{pmatrix}\begin{pmatrix} 1/2 & 1/2 & 1 \\ -m_s & \mu_s & -M_S\end{pmatrix}\end{aligned} \tag{50}$$

This expression is of course zero when spin dependent interactions are neglected. We can also show it is zero for interactions leading to a symmetric T matrix. This can be seen by interchanging the dummy indices α and β in equation 49; i.e.

$$f_{11} - f_{22} = \sum_{\beta\alpha} P_{\beta\alpha}\, T_{\alpha\beta} \tag{51}$$

But

$$P_{\beta\alpha} = (-1)^{l-\lambda}\, P_{\alpha\beta} \tag{52}$$

i.e since $\mid l - \lambda \mid = 1$

$$f_{11} - f_{22} = -\sum_{\alpha\beta} P_{\alpha\beta}\, T_{\alpha\beta} \tag{53}$$

In section 4.1 we show that $T_{\alpha\beta} = T_{\beta\alpha}$ for molecules with a reflection plane of symmetry, which implies that $f_{11} - f_{22} = 0$, i.e. no chiral effect. We also show that $T_{\alpha\beta} = 0$ for molecules with a centre of inversion when $\mid l - \lambda \mid = 1$; again $f_{11} - f_{22} = 0$. For a molecule like H_2S_2 with the x axis a C_2 axis of symmetry

$$T_{\beta\alpha} = (-1)^{(l+m+m_s)-(\lambda+\mu+\mu_s)} T_{\alpha\beta} \tag{54}$$

Since in our expression for $f_{11} - f_{22}$ we have $m + m_s = \mu + \mu_s$ we see clearly that $T_{\beta\alpha} = -T_{\alpha\beta}$ for the H_2S_2 matrix elements of interest.

4.1 Properties of the T matrix

First let us consider an operator M_{op} which is invariant under time-reversal; i.e.

$$\theta = \theta^\dagger \, M_{op}^\dagger \, \theta \tag{55}$$

where θ is the anti-unitary time-reversal operator. Consider matrix elements in a $Y_l^m \, \chi_{m_s}$ basis. We have (e.g. Taylor[23])

$$\begin{aligned} \langle \lambda\mu\mu_s \mid M_{op} \mid lmm_s \rangle &= \langle \lambda\mu\mu_s \mid \theta^\dagger M_{op}^\dagger \theta \mid lmm_s \rangle \\ &= (-1)^{l+1/2-m-m_s+\lambda+1/2-\mu-\mu_s} \langle \lambda - \mu - \mu_s \mid M_{op}^\dagger \mid l - m - m_s \rangle^* \\ &= (-1)^{\lambda+\mu+\mu_s-(l+m+m_s)} \langle l - m - m_s \mid M_{op} \mid \lambda - \mu - \mu_s \rangle \end{aligned} \tag{56}$$

Now return to the scattering amplitudes $f(M_f, M_i)$ which can be considered to be the matrix elements of an operator M_{op} in a basis of spin states $\mid M \rangle$ in the space-fixed frame. Let us transform to the uncoupled basis $\mid lmm_s \rangle$ in the body-fixed frame.

$$\begin{aligned} f(M_f, M_i) &= \langle M_f \mid M_{op} \mid M_i \rangle \\ &= \sum_{\lambda\mu\mu_s} \sum_{lmm_s} \langle M_f \mid \lambda\mu\mu_s \rangle \, \langle \lambda\mu\mu_s \mid M_{op} \mid lmm_s \rangle \, \langle lmm_s \mid M_i \rangle \end{aligned}$$

Now $\mid M \rangle = \sum_m D_{Mm}^{1/2\,*}(\Omega) \;\mid m \rangle$ and thus $\langle m \mid M \rangle = D_{Mm}^{1/2\,*}$ and

$$f(M_f, M_i) = \sum_{\lambda\mu\mu_s} \sum_{lmm_s} D_{M_i m_s}^{1/2\,*} \, D_{M_f \mu_s}^{1/2} \, Y_l^{m*}(\hat{\mathbf{k}}) \, Y_\lambda^\mu(\hat{\mathbf{r}}) \, \langle \lambda\mu\mu_s \mid M_{op} \mid lmm_s \rangle \tag{57}$$

Comparing this expression with equation 47, noting that the spherical harmonics here are with respect to the body-fixed frame, we see that

$$\langle \lambda\mu\mu_s \mid M_{op} \mid lmm_s \rangle = (2\pi/ik) \, i^{l-\lambda} \, T_{\lambda\mu\mu_s, lmm_s} \tag{58}$$

It follows immediately that

$$T_{\lambda\mu\mu_s, lmm_s} = T_{l-m-m_s, \lambda-\mu-\mu_s} \, (-1)^{\mu+\mu_s-m-m_s} \tag{59}$$

We will now consider the effect of other symmetry operations which can be represented by, say, a unitary operator R; then $M_{op} = R^\dagger M_{op} R$. We will consider

$$\langle \lambda\mu\mu_s \mid M_{op} \mid lmm_s \rangle = \langle \lambda\mu\mu_s \mid R^\dagger M_{op} R \mid lmm_s \rangle \tag{60}$$

for a number of different R, and also use the result from time-reversal, equation 59.

(a) *Molecules with a reflection plane of symmetry.* For simplicity choose this plane to be the x-z plane. With angles (θ, ϕ) becoming $(\theta, -\phi)$, the spherical harmonics $Y_l^m(\theta, \phi)$ transform into $(-1)^m Y_l^{-m}$ under R. For the spin functions we note that the reflection is equivalent to an inversion plus a rotation of π about y (c.f Messiah[24]). We find that $| m_s\rangle$ transforms into $(-1)^{-1/2-m_s} \mid -m_s\rangle$ and $\langle m_s \mid$ into $(-1)^{1/2+m_s}\langle -m_s \mid$, so that

$$T_{\lambda\mu\mu_s,lmm_s} = (-1)^{\mu+\mu_s-m-m_s} T_{\lambda-\mu-\mu_s,l-m-m_s} \tag{61}$$

Using equation 59 we see that

$$T_{\lambda\mu\mu_s,lmm_s} = T_{lmm_s,\lambda\mu\mu_s} \tag{62}$$

(b) *Molecules with inversion symmetry.* With (θ,ϕ) becoming $(\pi-\theta,\phi+\pi)$ the Y_l^m transform into $(-1)^l Y_l^m$, and since the spin is unchanged we have

$$T_{\lambda\mu\mu_s,lmm_s} = (-1)^{l+\lambda} T_{\lambda\mu\mu_s,lmm_s} \tag{63}$$

which of course implies that $\mid l-\lambda \mid$ is even for non zero T matrix element. However we note that in expression 50 $\mid l-\lambda \mid = 1$.

(c) *Chiral molecules with a C_2 axis of symmetry.* Assume that the x axis is the C_2 rotational axis of symmetry, which is consistent with the calculations of Smith et al[15]. We note that (θ,ϕ) becomes $(\pi-\theta,-\phi)$, so that Y_l^m transforms into $(-1)^l Y_l^{-m}$. The spin function $\mid m_s\rangle$ transforms into $i \mid -m_s\rangle$, $\langle m_s \mid$ into $-i\langle -m_s \mid$. We obtain

$$T_{\lambda\mu\mu_s,lmm_s} = (-1)^{\lambda+l} T_{\lambda-\mu-\mu_s,l-m-m_s} \tag{64}$$

and using equation 59 we have

$$T_{\lambda\mu\mu_s,lmm_s} = (-1)^{\lambda+\mu+\mu_s-l-m-m_s} T_{lmm_s,\lambda\mu\mu_s} \tag{65}$$

Thus when $\mu+\mu_s = m+m_s$ and $\mid \lambda - l \mid = 1$, as in equation 50, we have

$$T_{\lambda\mu\mu_s,lmm_s} = -T_{lmm_s,\lambda\mu\mu_s} \tag{66}$$

In scattering theory it is usual to introduce a K matrix related to S by $S = (1-iK)^{-1}(1+iK)$. Since S is unitary it is easy to show that K is Hermitian. In most scattering calculations K is also real, from which it can be shown that S and T are symmetric. However this is not the case for electron scattering by chiral molecules using the uncoupled (lmm_s) basis of this paper. Greer and Thompson[22] have discussed the potentials within an exchange model and shown that the potential matrix elements in this basis are hermitian, but in general not real.

5 The physical cause of the effect

We continue our discussion of the scattering of electrons by rigid fixed molecules which have a closed shell configuration in their ground state. We consider the molecule to have N electrons, spatial and spin coordinates $\mathbf{x_i} \equiv \mathbf{r_i}, \sigma_i$ and M nuclei, coordinates $\mathbf{R_j}$ and charges Z_j. The total wavefunction for the system of molecule plus electron ,$\Psi(\mathbf{x}_1, ...\mathbf{x}_{N+1})$, can be expanded in terms of molecular eigenstates, Φ_j, in the usual way.

As a first approximation let us retain only the ground state, Φ_0 in this expansion, i.e. we take (dropping the suffix)

$$\Psi(\mathbf{x}_1, ...\mathbf{x}_{N+1}) = \sum_{i=1}^{N+1} (-1)^{i+1}\, \Phi(i^{-1})\, F(\mathbf{x}_i) \tag{67}$$

where (i^{-1}) represents all coordinates $\mathbf{x}_1, ...\mathbf{x}_{N+1}$ except $\mathbf{x}_i$. Assuming Φ is antisymmetric for interchange of electrons the form of equation 67 ensures that Ψ is also antisymmetric. Equations for F, the scattered electron function, can be obtained from

$$\int \Phi^*(\mathbf{x}_1...\mathbf{x}_N)\,[H(1,...,N+1) - \mathcal{E}]\,\Psi(\mathbf{x}_1...\mathbf{x}_{N+1})\,d\mathbf{x}_1...d\mathbf{x}_N = 0 \qquad (68)$$

Since the Hamiltonian H is symmetric for interchange of electrons this reduces to

$$\begin{aligned} 0 &= \int \Phi([N+1]^{-1})^*\,[H(1,...,N+1] - \mathcal{E}] \\ &\quad [\Phi([N+1]^{-1})\,F(\mathbf{x}_{N+1}) - N\,\Phi(N^{-1})\,F(\mathbf{x}_N)]d\mathbf{x}_1...d\mathbf{x}_N \end{aligned} \qquad (69)$$

Write the Hamiltonian as

$$H(1....N+1) = H_{mol}(1...N) + V_{int}(1,...,N+1) \; - 1/2\,\nabla^2_{N+1} \qquad (70)$$

where H_{mol} is the molecular Hamiltonian and V_{int} is the interaction potential between the electron and the molecule. V_{int} is the sum of a spin- dependent term V_{sd}, discussed below, and a spin independent term V_{el}:

$$V_{el} = -\sum_{i=1}^{N} \mid \mathbf{r}_i - \mathbf{r}_{N+1} \mid^{-1} + \sum_{j=1}^{M} Z_j / \mid \mathbf{R}_j - \mathbf{r}_{N+1} \mid \qquad (71)$$

Let us assume that Φ is given by a single determinant of single-electron functions $u_\alpha(\mathbf{r},\sigma)$.

$$\Phi(\mathbf{x}_1, ...\mathbf{x}_N) = N^{-1/2}\,\epsilon_{\alpha\beta...\pi}\,u_\alpha(1)\,u_\beta(2)...u_\pi(N) \qquad (72)$$

where $\epsilon_{\alpha...\pi}$ equals 1 if $\alpha...\pi$ is an even permutation of 1 . . . N, equals -1 for odd permutations, and equals zero otherwise. Assuming

$$\begin{aligned} H_{mol}(1,...,N)\,\Phi(\mathbf{x}_1...\mathbf{x}_N) &= E_1\,\Phi(\mathbf{x}_1...\mathbf{x}_N) &(73)\\ \mathcal{E} &= E_1 + k^2/2 &(74) \end{aligned}$$

and neglecting V_{sd} and any spin-dependent terms in H_{mol} equation 69 reduces to the usual Static-Exchange or Hartree-Fock approximation.

$$[(-1/2)\nabla^2 - (1/2)k^2 + V_{st}(\mathbf{r})F(\mathbf{r}) = \int K(\mathbf{r},\mathbf{s},\sigma)\,F(\mathbf{s},\sigma)\,ds\,d\sigma \qquad (75)$$

The static potential is

$$V_{st}(\mathbf{r}) = \sum_\alpha \int u^*_\alpha(\mathbf{s},\sigma)\; \mid \mathbf{r} - \mathbf{s} \mid^{-1}\; u_\alpha(\mathbf{s},\sigma)\,ds\,d\sigma + \sum_j Z_j / \mid \mathbf{R_j} - \mathbf{r} \mid \qquad (76)$$

and the exchange kernel is

$$K(\mathbf{r},\mathbf{s},\sigma) = \sum_\alpha u^*_\alpha(\mathbf{s},\sigma)\; \mid \mathbf{r} - \mathbf{s} \mid^{-1}\; u_\alpha(\mathbf{r},\sigma) \qquad (77)$$

As we have seen this equation alone will not give any chiral effect. There are several ways we can include spin-dependent potentials in order to obtain a positive effect.

(a) Spin-orbit interaction involving the incident electron.

Consider the spin-orbit interaction between the incident electron and the nuclei. For simplicity let the interaction be with just one nucleus, assumed to have a large value of Z; if this nucleus is at the centre of our coordinate system

$$v_{sd} = \xi(r)\,\mathbf{l}_{N+1}.\mathbf{s}_{N+1} \qquad (78)$$

with $\xi(r) = Z/(2\,c^2\,r_{N+1}^3)$

With the chiral fields generated by the static and exchange terms above this model will produce positive chiral effects. The intego-differential equation has so far only been solved for small achiral molecules (Greer and Thompson[22]); chiral effects are of course only present for oriented systems.

In the application to the model chiral molecule BiH_3 Fandreyer et al[13] neglected exchange and solved the equation in a distorted wave approximation. Their results for the X of equation 28 are shown in Table 1. We recall that

$$X = Im(f_{11} - f_{22})/Im(f_{11} + f_{22})$$

and that it is usually approximately equal to the attenuation fraction (c.f. equation 1).

Table 1. Results for the X of equation 28 for the model molecule BiH_3

Energy in eV	.1	.5	1.0	2.0	3.0	4.0	6.0	8.0	10.0
$X\,.\,10^4$	-76.3	-44.3	-10.3	2.13	1.64	.13	-.58	-.44	-.47

The model is interesting in that there is resonance structure at very low electron energies which causes considerable enhancement of the chiral effect. For energies greater than 2 eV the size of the effect is comparable to the experimental results for molecules containing a heavy atom. We note again that the attenuation fraction oscillates as a function of energy, i.e the scattering experiment can not be used to obtain the handedness of the molecule.

In more recent work Smith et al[14,15] the Continuum Multiple Scattering Method has been extended to include the spin orbit interaction between the incident electron and each nucleus. Each nucleus is surrounded by a sphere; the scattering problem is solved within each sphere and between the spheres, solutions being joined on the sphere boundaries by contraints of continuity. The spin-orbit interaction can be easily included on each nucleus but exchange is neglected. Preliminary results for H_2S_2 are shown in Table 2 for the parameter X.

Table 2. Results for the X of equation 28 for H_2S_2.

Energy in eV	1.0	2.0	4.0	6.0	8.0	10.0
$X\,.10^5$	2.0	4.5	-1.6	-1.4	.17	-.54

The results are smaller than the most recent experimental measurements for molecules with high Z atoms, by a factor of 4. Calculations with molecules containing heavier atoms may lead to a greater chiral effect. The variation with energy is very similar to the results for the model molecule. We note that Hayashi[25] has used an approximate version of this technique but does not discuss forward scattering.

(b) Spin-orbit involving the molecular electrons

To be consistent we should also include the spin-orbit interaction between the molecular electrons and the nuclei, i.e. there should be an additional spin-orbit term in the molecular Hamiltonian H_{mol}. The effect of such a term is to change the form of the molecular orbitals u_α. We can expand the u_α in the same way as the continuum functions in equation 39, i.e.,

$$u_\alpha(\mathbf{r},\sigma) = \sum_{lmm_s} u^{(\alpha)}_{lmm_s}(r)\,r^{-1}\,Y_l^m(\hat{\mathbf{r}})\,\chi_{m_s}(\sigma) \tag{79}$$

If spin-dependent forces are neglected the function will have a definite spin quantum number, $m_s = \pm 1/2$; inclusion of the spin-orbit interaction mixes the spins.

The electronic part of the static potential (c.f. equation 76) can be written

$$\sum_{\alpha} \int u_\alpha^*(\mathbf{s},\sigma) \mid \mathbf{r}-\mathbf{s} \mid^{-1} u_\alpha(\mathbf{s},\sigma)\, d\mathbf{s}\, d\sigma = \sum_{LM} v_{LM}(r)\, Y_L^M(\hat{\mathbf{r}}) \tag{80}$$

where

$$\begin{aligned} v_{LM}(r) &= \sum_{\alpha} \sum_{lmm_s} \sum_{\lambda\mu} (-1)^{\mu} \left[\frac{(2l+1)(2\lambda+1)4\pi}{(2L+1)}\right]^{1/2} \begin{pmatrix} L & \lambda & l \\ 0 & 0 & 0 \end{pmatrix} \begin{pmatrix} L & \lambda & l \\ -M & \mu & -m \end{pmatrix} \\ &\int u_{lmm_s}^{(\alpha)*}(s)\, \gamma_L(r,s)\, u_{\lambda\mu m_s}^{(\alpha)}(s)\, ds \end{aligned} \tag{81}$$

and where $\gamma_L(r,s) = r^L/s^{L+1}$ if $r < s$, and equals s^L/r^{L+1} if $s < r$. The contribution to the potential matrix element in the radial equations (c.f. equation 40) is

$$\begin{aligned} U_{lmm_s,\lambda\mu\mu_s} &= \sum_{LM} \int Y_l^{m*}(\hat{\mathbf{r}})\, \chi_{m_s}^{\dagger}(\sigma)\, v_{LM}(r)\, Y_L^M(\hat{\mathbf{r}})\, Y_\lambda^\mu(\hat{\mathbf{r}})\, \chi_{\mu_s}(\sigma)\, d\hat{\mathbf{r}}\, d\sigma \\ &= \delta_{m_s\mu_s} \sum_{L} [(2L+1)(2l+1)(2\lambda+1)]^{1/2}(1/4\pi)^{1/2}\, (-1)^m\, v_{LM}(r) \\ &\quad \begin{pmatrix} L & \lambda & l \\ 0 & 0 & 0 \end{pmatrix} \begin{pmatrix} L & \lambda & l \\ M & \mu & -m \end{pmatrix} \end{aligned} \tag{82}$$

We note that this form for the matrix elements does not depend on whether spin-orbit is included in the calculation of the molecular orbitals; since the potential is zero if $m_s \neq \mu_s$, and $U_{1/2,1/2} = U_{-1/2,-1/2}$, this amendment to the static interaction does not lead to a positive chiral effect.

The exchange contribution to the potential in the radial equations comes from

$$\int Y_l^m(\hat{\mathbf{r}})\, \chi_{m_s}^{\dagger}(\sigma) \sum_{\alpha} \int u_\alpha^*(\mathbf{s},\sigma) \mid \mathbf{r}-\mathbf{s} \mid F(\mathbf{s},\sigma)\, d\mathbf{s}\, u_\alpha(\mathbf{r},\sigma)\, d\hat{\mathbf{r}}\, d\sigma \tag{83}$$

This reduces to $\sum_{\lambda\mu\mu_s} \int K_{lmm_s,\lambda\mu\mu_s}(r,s)\, f_{\lambda\mu\mu_s}(s)\, ds$ where

$$\begin{aligned} K_{lmm_s,\lambda\mu\mu_s}(r,s) &= \sum_{\alpha} \sum_{l_1 m_1} \sum_{l_2 m_2} \sum_{pq} (-1)^{m_1+m_2} [(2\lambda+1)(2l_2+1)(2l_1+1)(2l+1)]^{1/2} \\ &\quad u_{l_1 m_1 \mu_s}^{(\alpha)*}(s)\, \gamma_p(r,s)\, u_{l_2 m_2 m_s}^{(\alpha)}(r) \begin{pmatrix} p & l_2 & l \\ q & m_2 & -m \end{pmatrix} \begin{pmatrix} p & l_2 & l \\ 0 & 0 & 0 \end{pmatrix} \\ &\quad \begin{pmatrix} p & \lambda & l_1 \\ -q & \mu & -m_1 \end{pmatrix} \begin{pmatrix} p & \lambda & l_1 \\ 0 & 0 & 0 \end{pmatrix} \end{aligned} \tag{84}$$

The effect of the spin-orbit included in the molecular orbitals is to have radial functions $u^{(\alpha)}$ and thus the K matrix elements which depend on the spin quantum number. If the spin-orbit interaction is neglected $m_s = \mu_s$, the molecular radial functions do not depend on the spin quantum number and

$$K_{lmm_s,\lambda\mu m_s} = K_{lm-m_s,\lambda\mu-m_s} \tag{85}$$

$$K_{lmm_s,\lambda\mu\mu_s} = 0 \quad m_s \neq \mu_s \tag{86}$$

Thus inclusion of spin-orbit in the molecular Hamiltonian will lead to chiral effects but only through the exchange term. Rich et al[12] (c.f also Hegstrom[2]) have discussed this but except for order of magnitude estimates there have not been any calculations to date.

(c) The spin-other-orbit interaction.
Gallup[16] has pointed out that the incident electron causes considerable molecular distortion and the electron spin can interact with these induced currents through a spin-other-orbit potential. We can include this suggestion in our scattering model by considering more terms in our eigen function expansion of the total wavefunction.

Inclusion of other states (either physical or pseudo-states) allows for the effect of molecular distortion and leads to the solution of coupled integro- differential equations for the scattered functions $F_i(\mathbf{r}, \sigma)$. Formally the procedure is equivalent to adding an "optical potential" to V_{st} in equation 75 which may be non-local and complex. For incident energies below the first excitation threshold this extra term, which can be viewed as producing distortion through "virtual excitations", gives a potential which varies as r^{-4} as r, the incident electron coordinate, tends to infinity, and has a size proportional to the polarisability of the molecule.

A simpler way of dealing with this term is to use perturbation theory to obtain the distortion of the molecule due to the incident electron and then to calculate the effect of this distortion back on the incident electron. We replace Φ in the second line of equation 69 by $\Phi + \Phi^{(1)}$ where $\Phi^{(1)}$ is the perturbation correction to the ground state. $\Phi^{(1)}$ will have the form

$$\Phi^{(1)} = \sum_{i=1} \langle \Phi_i \mid V_{int} \mid \Phi_0 \rangle / (E_0 - E_i) \tag{87}$$

where the sum is over all excited molecular states and V_{int} is the spin independent part of the electron interaction with the molecule. There will be an extra direct "polarisation" potential, $\langle \Phi_0 \mid V_{int} \mid \Phi^{(1)} \rangle$ and a new spin-dependent term $\langle \Phi_0 \mid V_{soo} \mid \Phi^{(1)} \rangle$ where V_{soo} is the spin-other-orbit correction to the Hamiltonian. (There will also be extra exchange terms). The spin-other-orbit term has the form

$$\sum_{i=1} \langle \Phi_i \mid V_{int} \mid \Phi_0 \rangle \langle \Phi_0 \mid V_{soo} \mid \Phi_i \rangle \, / \, (E_0 - E_i) \tag{88}$$

Gallup has pointed out that this expression has similarities to the form obtained from optical activity theories. The first integral contains matrix elements involving the electric dipole operator and the second matrix elements involving the magnetic dipole operator. Gallup approximated the above ideas in a resonance model and carried out a calculation for propylene epoxide (C_3H_6O), obtaining a chiral effect of the order of 10^{-4}.

6 Conclusions

We have carried out a full theoretical analysis of the electron dichroism effect which has been recently confirmed experimentally. This leads to an expression for the attenuation of the electron beam in terms of the **forward** scattering amplitudes for the non-flip processes. Calculations for H_2S_2 using a scattering model which includes the spin-orbit interaction between the incident electron and all the nuclei have produced encouraging results which compare favourably with experiment in size and energy dependence.

We hope to extend the H_2S_2 calculations to general angle scattering for which we have carried out a full theoretical analysis (Johnston et al[26]), and to investigate other chiral molecles and also oriented achiral molecules (c.f. Greer and Thompson[22]).

Acknowledgements

We would like to thank Professor J.Kessler for interesting discussions and for permission to use his experimental results before publication. We acknowledge financial assistance from EC-HCM Network Contract Number ERB CHRX CT 920013, Nato Collaborative Research Grant CRG930056, and Sonderforschungsbereich 216 Bielefeld-Münster.

References

1. L.Keszthelyi *Quarterly Review of Biophysics.* 28:473(1995)
2. R.A.Hegstrom, *Nature* 297:643(1982)
3. A.Bisi. N.Gambari and L.Zappa, *Il Nuovo Cimento D.* 14:617(1992)
4. A.S.Garay, J.A.Ahlgren-Beckendorf and B.Gazdy, *Phys.Rev.A* 48:3008(1994)
5. P.S.Farago, *J.Phys.B:At.Mol.Phys.* 13:L567(1980)
6. P.S.Farago, *J.Phys.B.At.Mol.Phys.* 14:L743(1981)
7. K.Blum and D.Thompson, *J.Phys.B:At.Mol.Opt.Phys.* 22:1823(1989)
8. D.Thompson and M.Kinnin, *J.Phys.B.At.Mol.Opt.Phys.* 28:2473(1995)
9. M.J.M.Beerlage, P.S.Farago and M.J.Van der Wiel, *J.Phys.B:At.Mol.Phys.* 14:3245(1981)
10. D.M.Campbell and P.S.Farago, *Nature* 318:52(1985)
11. D.M.Campbell and P.S.Farago, *J.Phys.B.At.Mol.Phys.* 20:5133(1987)
12. A.Rich, J.v.House and R.A.Hegstrom, *Phys.Rev.Lett.* 48:1341(1982)
13. R.Fandreyer, D.Thompson and K.Blum, *J.Phys.B:At.Mol.Opt.Phys.* 23:3031(1990)
14. I.M.Smith, D.G.Thompson and K.Blum, in: *Abstracts of contributed papers, XIX ICPEAC* (1995)
15. I.M.Smith, D.G.Thompson and K.Blum, *to be published*
16. G.A.Gallup, in: *Electron Collisions With Molecules,Clusters and Surfaces.* H.Ehrhardt and L.A.Morgan eds., Plenum,New York(1994)
17. S.Meyer and J.Kessler, *Phys.Rev.Lett.* 74:4803(1995)
18. S.Meyer, C.Nolting and J.Kessler, *J.Phys.B:At.Mol.Opt.Phys.* 29:3497(1996)
19. K.W.Trantham, M.E.Johnston and T.J.Gay, *J.Phys.B:At.Mol.Opt.Phys.* 28:L543(1995)
20. J.Kessler *Polarised Electrons.* Springer,Berlin(1985)
21. D.M.Brink and G.R.Satchler, *Angular Momentum, 3rd edn.* Clarendon, Oxford(1993)
22. R.Greer and D.Thompson, *J.Phys.B:At.Mol.Opt.Phys.* 28:4667(1995)
23. J.Taylor *Scattering Theory.* Wiley,New York(1972)
24. A.Messiah, *Quantum Mechanics.* North Holland,Amsterdam(1961)
25. S.Hayashi *J.Phys.B:At.Mol.Opt.Phys.* 21:1037(1988)
26. C.Johnston, K.Blum and D.Thompson, *J.Phys.B:At.Mol.Opt.Phys.* 26:965(1993)

INDEX

Series Publications

Below is a chronological listing of all the published volumes in the *Physics of Atoms and Molecules* series.

ELECTRON AND PHOTON INTERACTIONS WITH ATOMS
Edited by H. Kleinpoppen and M. R. C. McDowell

ATOM–MOLECULE COLLISION THEORY: A Guide for the Experimentalist
Edited by Richard B. Bernstein

COHERENCE AND CORRELATION IN ATOMIC COLLISIONS
Edited by H. Kleinpoppen and J. F. Williams

VARIATIONAL METHODS IN ELECTRON–ATOM SCATTERING THEORY
R. K. Nesbet

DENSITY MATRIX THEORY AND APPLICATIONS
Karl Blum

INNER-SHELL AND X-RAY PHYSICS OF ATOMS AND SOLIDS
Edited by Derek J. Fabian, Hans Kleinpoppen, and Lewis M. Watson

INTRODUCTION TO THE THEORY OF LASER–ATOM INTERACTIONS
Marvin H. Mittleman

ATOMS IN ASTROPHYSICS
Edited by P. G. Burke, W. B. Eissner, D. G. Hummer, and I. C. Percival

ELECTRON–ATOM AND ELECTRON–MOLECULE COLLISIONS
Edited by Juergen Hinze

ELECTRON–MOLECULE COLLISIONS
Edited by Isao Shimamura and Kazuo Takayanagi

ISOTOPE SHIFTS IN ATOMIC SPECTRA
W. H. King

AUTOIONIZATION: Recent Developments and Applications
Edited by Aaron Temkin

ATOMIC INNER-SHELL PHYSICS
Edited by Bernd Crasemann

COLLISIONS OF ELECTRONS WITH ATOMS AND MOLECULES
G. F. Drukarev

THEORY OF MULTIPHOTON PROCESSES
Farhad H. M. Faisal

PROGRESS IN ATOMIC SPECTROSCOPY, Parts A, B, C, and D
Edited by W. Hanle, H. Kleinpoppen, and H. J. Beyer

RECENT STUDIES IN ATOMIC AND MOLECULAR PROCESSES
Edited by Arthur E. Kingston

QUANTUM MECHANICS VERSUS LOCAL REALISM: The Einstein-Podolsky-Rosen Paradox
Edited by Franco Selleri

ZERO-RANGE POTENTIALS AND THEIR APPLICATIONS IN ATOMIC PHYSICS
Yu. N. Demkov and V. N. Ostrovskii

COHERENCE IN ATOMIC COLLISION PHYSICS
Edited by H. J. Beyer, K. Blum, and R. Hippler

ELECTRON–MOLECULE SCATTERING AND PHOTOIONIZATION
Edited by P. G. Burke and J. B. West

ATOMIC SPECTRA AND COLLISIONS IN EXTERNAL FIELDS
Edited by K. T. Taylor, M. H. Nayfeh, and C. W. Clark

ATOMIC PHOTOEFFECT
M. Ya. Amusia

MOLECULAR PROCESSES IN SPACE
Edited by Tsutomu Watanabe, Isao Shimamura, Mikio Shimizu, and Yukikazu Itikawa

THE HANLE EFFECT AND LEVEL CROSSING SPECTROSCOPY
Edited by Giovanni Moruzzi and Franco Strumia

ATOMS AND LIGHT: INTERACTIONS
John N. Dodd

POLARIZATION BREMSSTRAHLUNG
Edited by V. N. Tsytovich and I. M. Ojringel

INTRODUCTION TO THE THEORY OF LASER–ATOM INTERACTIONS (Second Edition)
Marvin H. Mittleman

ELECTRON COLLISIONS WITH MOLECULES, CLUSTERS, AND SURFACES
Edited by H. Ehrhardt and L. A. Morgan

THEORY OF ELECTRON–ATOM COLLISIONS, Part 1: Potential Scattering
Philip G. Burke and Charles J. Joachain

POLARIZED ELECTRON/POLARIZED PHOTON PHYSICS
Edited by H. Kleinpoppen and W. R. Newell

INTRODUCTION TO THE THEORY OF X-RAY AND ELECTRONIC SPECTRA OF FREE ATOMS
Romas Karazija

VUV AND SOFT X-RAY PHOTOIONIZATION
Edited by Uwe Becker and David A. Shirley

DENSITY MATRIX THEORY AND APPLICATIONS (Second Edition)
Karl Blum

SELECTED TOPICS ON ELECTRON PHYSICS
Edited by D. Murray Campbell and Hans Kleinpoppen

PHOTON AND ELECTRON COLLISIONS WITH ATOMS AND MOLECULES
Edited by Philip G. Burke and Charles J. Joachain

COINCIDENCE STUDIES OF ELECTRON AND PHOTON IMPACT IONIZATION
Edited by Colm T. Whelan and H. R. J. Walters

GPSR Compliance

The European Union's (EU) General Product Safety Regulation (GPSR) is a set of rules that requires consumer products to be safe and our obligations to ensure this.

If you have any concerns about our products, you can contact us on ProductSafety@springernature.com

In case Publisher is established outside the EU, the EU authorized representative is:

Springer Nature Customer Service Center GmbH
Europaplatz 3
69115 Heidelberg, Germany

Batch number: 10755805

Printed by Printforce, the Netherlands